T0332003

Oar Feet and Opal Teeth

Oar Feet and Opal Teeth

About Copepods and Copepodologists

CHARLES B. MILLER

OXFORD
UNIVERSITY PRESS

OXFORD
UNIVERSITY PRESS

Oxford University Press is a department of the University of Oxford. It furthers the University's objective of excellence in research, scholarship, and education by publishing worldwide. Oxford is a registered trade mark of Oxford University Press in the UK and certain other countries.

Published in the United States of America by Oxford University Press
198 Madison Avenue, New York, NY 10016, United States of America.

Library of Congress Cataloging-in-Publication Data
Names: Miller, Charles B., 1940- editor.
Title: Oar feet and opal teeth : about copepods and copepodologists / [edited by] Charles B. Miller.
Description: New York, NY : Oxford University Press, [2023] | Includes bibliographical references.
Identifiers: LCCN 2022039027 (print) | LCCN 2022039028 (ebook) | ISBN 9780197637326 (hardback) | ISBN 9780197637340 (epub)
Subjects: LCSH: Copepoda.
Classification: LCC QL444.C7 O27 2023 (print) | LCC QL444.C7 (ebook) | DDC 595.3/4—dc23/eng/20221027
LC record available at https://lccn.loc.gov/2022039027
LC ebook record available at https://lccn.loc.gov/2022039028

DOI: 10.1093/oso/9780197637326.001.0001

Printed by Integrated Books International, United States of America

Contents

Preface

In the early 1960s I started sampling zooplankton from oceans and estuaries. Almost consistently the most numerous and most readily named animals in the catches have been copepods, members of a large and diverse class of Crustacea. I learned about some of them, those living their whole lives swimming in the sea. My interests came to focus on their life histories: spawning, hatching, molting through distinctive phases, forming their teeth, migrating, maturing, mating, and again, spawning. By 1970, I had a university job teaching biological oceanography and generating research projects. Teaching about zooplankton required learning quite generally about their biology, about things I did not work on myself, like modes of feeding. Copepods dominate the zooplankton, so that learning was mostly about them. Most of my research also connected me to them.

A recurring experience for people who study copepods—copepodologists—is that most people we meet have never heard of copepods: our neighbors, our dentist, even our children's biology teacher. They have no idea copepods exist, what they are like, or that they are key components of all wet habitats (except perhaps clouds). Unlike insects, no images spring to mind. So in 2017, I decided to write a book to introduce them more widely. This is it. Many chapters include some biological background to make their topics accessible. There are quasi-literary asides in places to leaven the loaf.

In 2017 I attended the 13th International Conference on Copepoda, held at a hotel in Los Angeles. Those meetings are organized by the World Association of Copepodologists. Yes, that exists, and such meetings are held somewhere about every three years (delayed lately by the COVID-19 epidemic). There, I met with and interviewed some of the leading copepod biologists. For other interviews I traveled to meet the subjects, or used online "facetime." Those interviews provided biographical materials for characterizing the life histories of people doing this work at a professional level. The selection is much narrower than I would like. Most, but not all, of the stories are about people I have worked with or followed closely because of my professional connections (e.g., people trained by Bruce Frost). A majority are from the United States or Canada, but specimens of about ten other nationalities are considered. There are several thousand important copepodologists active now or recently (thinking in decades), and the sampling here is quite parochial. If you are a copepodologist who feels left out of the book, I'm sorry. You almost certainly deserved to be in it.

The subjects covered are listed in the table of contents. Far from everything known about copepods is covered. As stated in the first chapter, the subject is the free-living copepods of lakes and oceans. Parasitic and meiofaunal groups are only mentioned, and somebody else should write semipopular books about them. The focus is on the animals themselves, on aspects of their individual lives, much more so than on their place in the ecology of their habitats. There are good books about those ecosystem dynamics; just not this one. Later in the book the focus shifts to the methods and results of molecular genetics applied to copepods. I have reached for an understandable level of explanation. Knowing DNA sequences in detail has provided a new lens for discerning the long evolutionary history of copepods and their phylogeny. The tentative genetic conclusions have also been guided by recent and very refined morphological comparisons. Those back chapters tell how that cross fertilization has been accomplished. Extraordinary insights have already emerged from the combined efforts, and more are already in lab notebooks. We start to see how the orders and families of copepods have diverged through time.

I hope you enjoy reading *Oar Feet and Opal Teeth,* and I hope you feel like recommending it to others you think would enjoy it as well.

Acknowledgments

I thank all the copepodologists whom I interviewed for the personal insights they offered, for their photographs, and for checking what I wrote about them and their scientific contributions. Reading on, you will meet them. Among those copepodologists I especially thank Russ Hopcroft, for sharing so many of his beautiful copepod photos with me and the anticipated readers. Thanks also to all the other photographers credited in figure captions. All were generous with permission. Thank you Jeremy Lewis and Michelle Kelley at Oxford University Press for taking on this project and seeing it through. Thank you Martha Clemons, my wife, and for some years a valued coauthor contributing to life history studies of *Neocalanus*. Thanks to her for staying the course with me through the multiyear adventure that writing *Oar Feet and Opal Teeth* has been. Thanks to everyone who has taught me important things, and special thanks to John McGowan, my graduate school advisor. Thanks for the honest arguments and good company to the crews of my scientific colleagues at Oregon State University, Scripps, Woods Hole, Plymouth Marine Lab, National Institute of Water and Atmospheric Research in New Zealand, and the universities of Washington, Tokyo, Hiroshima, and Tromsø. Indeed, I am grateful for my many scientific and personal connections with marine and copepod scientists around the world. Thanks everyone.

Finally, warm thanks to copy editor Anne Sanow for hundreds of "serial commas," for cooling my outbursts of enthusiasm (and cynicism), for slogging through every detail of the references.

Charlie Miller // Biddeford, Maine // 29 November 2022

1

Planktonic Copepods Have Those Oar Feet

Defining Plankton

Many of the organisms living in water are grouped as *plankton*. The word was taken from the ancient Greek word πλαγκτοσ (planktos). It is a beautiful word meaning "that which must wander or drift," things like clouds. Oversimplifying current classifications considerably, the plankton includes bacteria, algae (the phytoplankton), and a wide range of animal types, the zooplankton. Scientists who call themselves planktologists sample these organisms in several ways. The larger zooplankton (0.1 mm to 10 cm) are most commonly sampled with fine-mesh nets (example in Figure 1.1) that are hauled through the water like large, conical kitchen strainers, usually with mouth openings from about 15 cm to 2 meters. Pumps and water-sampling bottles are also used. When nets with mesh holes a millimeter or less across are pulled through ocean water, the catch most often will be dominated by small, torpedo-shaped crustaceans of the sub-class Copepoda, familiarly called *copepods*.

Because copepods make up half or more of such catches, especially in numbers, many limnologists, oceanographers, and zoologists who take an interest in plankton find themselves intensely engaged with the life and times of copepods. Results of their studies are the subject of this book, along with some notes on the lives and times of these copepodologists. In some places I will emphasize my own experiences of discoveries, few as those seem from the vantage point of age. However, I can speak to those from the distorted memory of having been there. I have also interviewed a number of the significant students of copepods, and I will represent some of their experiences as they told them to me.

Again, copepods are among the most abundant and most diverse multicellular (metazoan) animals in the ocean. They are also abundant in lakes and are found in underground streams; in the water around tropical, epiphytic bromeliads; even as very small versions living in the interstitial water among the grains of sand in beaches, seafloors, and lakebeds. Some, like salmon lice, are parasites, and it is likely that the commensal and parasitic forms exceed all the others in species diversity. They will need a book of their own (by somebody else). The free-living forms are beautiful, so study of them is an aesthetic as well as scientific pleasure.

Oar Feet and Opal Teeth. Charles B. Miller, Oxford University Press. © Oxford University Press 2023.
DOI: 10.1093/oso/9780197637326.003.0001

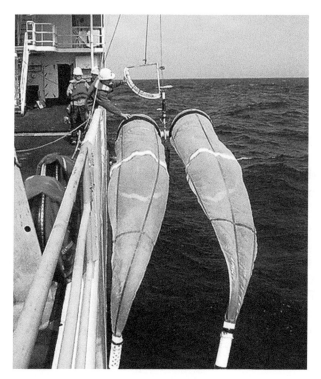

Figure 1.1 A paired "Bongo" net system ready to tow with the ship underway, obliquely down and back up. Suspension from the central axle avoids having towing bridles in front of the net mouths. Two mesh sizes can be used. Catches are taken by unlatching the white "cod ends" from the tails. The initial bongo designs were drawn in the mid-1960s by John McGowan of Scripps Institution of Oceanography. Photo taken by Andrew King on a cruise for a California Current Ecosystem Research project.

Copepods have an evolutionary history dating way back, likely to the Cambrian period (541 to 485 million years ago), the time when metazoans generally began to leave a fossil record. According to paleontologists, some tiny Cambrian fossils look like the toothed edges of copepod jaws,[1] though not all copepodologists think those bits could only be from copepods. Unlike the much later emerging insects, which molecular genetic data[2] suggest derived from crustacean ancestors, copepods are known all but exclusively to those copepodologists and ecologists who study them. However, they are fascinating and well worth learning about, hence this book. Copepodologists? Really? Yes, there are some, and there is even a World Association of Copepodologists with more than 500 members.

Like the term "plankton," the name copepod derives from Greek words: "cope" (κουπι) for oar and "pod" (ποδι) for foot: oar-foot. According to David Damkaer,[3] the premier historian of copepod studies, the name was first applied by Henri Milne Edwards in 1830. He was a French carcinologist at a time when reasonably accurate relationships among crustacean groups were first emerging into view. In those days essentially every well-educated person knew something of both Greek and Latin, and they combined Greek or Latin roots to name almost everything they felt needed naming. We still do that today, though very few of us actually know either ancient language.

Copepods are surprisingly unfamiliar to most people, surprising because of their abundance and nearly universal presence in natural waters. Well, they are small. They go unnoticed by human swimmers, though small fish can spot them and then eat them. There is virtually no mention of them in nonscientific literature. Long ago I asked a friend, J. Dennis Evans, then an English Instructor at Oregon State University, whether there were any poetic or other references to copepods in English literature. As it happened, he was writing a thesis[4] on the works of the American poet A. R. Ammons. In "Hymn IV," Ammons[5] is cursing God (or some omnipotent entity, addressed as "you") for all the evils from disease to "life's all-clustered grief." Here (with permission) are those three stanzas:

> I hold you responsible for
> every womb's neck
> clogged with
> killing growth
>
> . . .
>
> I keep you existent at least as
> a ghost crab
> moon-extinguished his crisp
> walk silenced on broken shells
>
> answering at least as
> the squiggling copepod
> for the birthing and aging of
> life's all-clustered grief

Ammons at the time (1957 or 1958) seems to have run out of full stops and commas. Why "the squiggling copepod" should answer for any of life's grief he did not explain, though apparently he was only keeping some God existent because he/she/it was as trivial as a copepod for explaining the problem of evil (that is, why does evil exist, if there is a big-G God?). At least (to borrow Ammons's recurring limit to the minimal), he clearly knew copepods exist, he mentions them elsewhere. He majored in biology at college and worked for decades at Cornell University, where a limnologist maybe mentioned them during lunch at the Faculty Soup Club. Unlike insects, lobsters, crabs, shrimp, and crayfish, all of which have wider poetic representation, a squiggling specimen seems to be *it* for copepods, at least (again), in literary English. Copepodologists constantly meet people not familiar with copepods. Reading on will remove you from that crowd.

Meet One of the Larger Copepod Species

Figure 1.2 is a photo of the last larval stage, the fifth "copepodite," of *Neocalanus plumchrus*, a freely swimming copepod that lives in the subarctic reaches of the Pacific, waters north of 45°. It has none of the primary or secondary sexual features of an adult copepod, so it shows well the more basic features of this crustacean class. Copepods have the general features of that class, so like all arthropods they have plastic-like, chitin sheaths as outer skeletons. Chitin serves as armor and as other interfaces with the outside world, even in some species as eye lenses. Implications of this crop up in every aspect of their lives. Chitin is a starch-like polymer of glucose molecules, but each of those includes an amine (NH_3) group, which allows crosslinking into pliable sheets as well as chains.

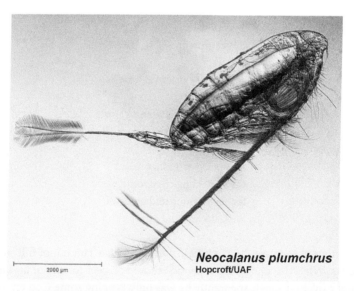

Neocalanus plumchrus
Hopcroft/UAF

2000 μm

Figure 1.2 The fifth copepodite (last larval stage) of a relatively large calanoid copepod: not counting the long tail setae, the total length is around 5 mm (2,000 μm is 2 mm). Photo by Russell Hopcroft (University of Alaska, Fairbanks), one of the outstanding photographers of planktonic organisms. With his kind permission.

As depicted (Figure 1.2), the head end is at the right, the tail at the left. The main body, called the fore-body or *prosome*, has the drag-reducing shape and smooth surface of a somewhat bulky torpedo. The carapace or shell over the head segments is fused into a continuous skin of chitin. Each prosome segment bears a pair of limbs on the ventral side. Five pairs of limbs at the back are the oar feet attached below the thorax, so they are termed "thoracic legs." Chitin over the first thoracic segment (T1) is continuous with the fused carapace of the head (also known as the *cephalosome*, "head body"). The last four thoracic segments are articulated with each other and the cephalosome, so the thorax can bend a little.

Limbs beneath the head are specialized in different ways: for sensing odor and motion in the water, for slow swimming, and for feeding. Much of the specialization between genera and species is located in the leading antennules (the long tubes turned below the body in Figure 1.2) and among the next five pairs of appendages, the mouth parts, each adapted for a specific aspect of feeding. At the left is the urosome, a tubular tail section terminated by two short, pad-like appendages, the caudal furcae. Those are tipped by the two sides of a fan of setae, which in a few genera includes a very long medial pair of setae. In *N. plumchrus* those are orange, with ribbon-like setules or side whiskers. Once biologists have written out a genus name, as in *Neocalanus plumchrus*, they allow themselves to abbreviate it to one or two letters.

A First Example of a Copepodologist, Russell Hopcroft

The meaning of "copepodologist" is obvious: somebody who studies copepods. Copepods vary greatly in where and how they live, though only in wet places. They also vary in their sizes and shapes, especially the shapes of parasitic forms. Thanks to all that variation, copepodologists come in many versions: basic taxonomists, evolution specialists (phylogeneticists), biomechanicians, limnologists, oceanographers, parasitologists (including fisheries experts), and recently molecular biologists. I will introduce some of those. Russell Hopcroft (Figure 1.3) is an ocean ecologist and zooplankton photographer, who lives in Fairbanks, Alaska, teaching at the branch there of the state university. He studies zooplankton all around that huge peninsula from Prince William Sound and along the southern coast, up through the Bering and Chukchi Seas, even spreading out around the entire Arctic Ocean. He did not start with such high latitude interests. Russ was born in Exeter, Ontario, a town of about 500, and moved at age five to live near his grandparents in a town of about 6,000. He went through most of his early schooling there, and then the family followed his father as he shifted among several jobs. They moved to Toronto and later west to near Hamilton. He recalls early interest in the bugs and tadpoles of nearby freshwater, often catching

Figure 1.3 Russell Hopcroft at his microphotography station, with a Leica stereoscope with megapixel digital cameras. Both Russ Hopcroft and the scope-camera were aboard the University of Alaska ship *RV Sikuliaq*, preparing for a cruise from Seward across the Gulf of Alaska shelf. His t-shirt indicates a majority interest in respect to his planktonic friends. Photo by the author.

and watching them in jars and wading pools. He thinks an influence may have come from Jacques Cousteau's television programs about oceans and ocean life. He knew entering high school that marine biology would be his life interest and he prepared for college accordingly, as well as learning dive skills. Included was a gap year to think about life's possibilities. College, starting in 1978, was at the University of Guelph, a public college offering specialization in the subject.

During his senior year of college Russ started to work with Professor John Roff. Their first joint publication in 1985 reported the respiration of a rock-clinging sea cucumber[6] measured during a study trip to the Huntsman Centre on Passamaquoddy Bay in New Brunswick. Roff also gave him access to an early desktop computer and some video-microscopy gear; the challenge was to develop a length-measuring system for tiny animals like copepod nauplii. Russ had a major part in that, and the system is still in use.[7] At the time Roff was engaged in an extended series of plankton studies in and near Kingston Harbor, Jamaica, a partnership between University of Guelph and the marine labs of the University of the West Indies. He took on Russ as grad student, who took him on as an advisor, and they worked together on tropical plankton until about 1997, though more papers came out later. It wasn't all copepods; there were also studies of phytoplankton and sedimenting particles that were reported in his master's thesis.

It is a badly kept secret that marine labs powerfully foster romance among their human occupants, as well as allowing study of exotic aspects of mating among snails, whales, and, of course, copepods. Another of Roff's students was Cheryl Clarke, who worked mostly on components of the plankton other than copepods.[8] She and Russ had met at the Huntsman Centre. Interest continued in Guelph and Jamaica. They completed their Jamaica-based MS degrees in 1990 and married in 1991. Cheryl pushed more than Russ (according to him), that he continue for his doctorate. He did, turning to the population dynamics and production rates of copepods. "Production" in ecology refers to how much organic matter accumulates in an organism or population, a central interest for Roff and so for his students. Tropical nearshore plankton are predominantly small, including copepods like *Oithona*, *Euterpina*, and *Corycaeus* that mature at about 1 mm. So Russ became a refined microscopist in order to study aspects of growth and production among these small crustaceans, emphasizing the effects of body size on growth rates. He finished in 1997.[9] Some of his results[10] concerned growth rates and fecundity of small copepod varieties that carry their eggs to hatching in sacs attached to their urosomes.

As Dr. Hopcroft, Russ moved with Cheryl to Moss Landing, California, where he had been offered a postdoctoral position, limited by contract to two years, at the Monterey Bay Aquarium Research Institute (MBARI). Their first child, a daughter, was born there, and their twin boys were born in Alaska in 2000. Cheryl's training in zooplankton research was not set aside, and she now works with Russ as his lead technician at the University of Alaska in Fairbanks (UAF), appearing in bylines as C. Clarke. At MBARI, Russ worked with Bruce Robison, who largely led the institute's submersible and remotely operated vehicle (ROV) studies of deep-sea animals. Using an ROV, they photographed, collected, and later described four previously unknown but relatively large (0.5 to 1 cm) deep-living larvaceans, gelatinous zooplankters that feed by spreading elaborate filtering structures made of mucous around themselves.[11] (You could learn about those by looking up the paper; they are not copepods, so we're not going into more detail here.)

Back on the job market, Russ applied for an opening at UAF that came about because the lead zooplankton worker there retired. As he puts it, most of the smaller oceanographic institutes in the United States and Canada have one or only a couple of positions in each subject matter "slot." He got an offer to fill the UAF zooplankton slot, and he took it. There are not many zooplankton slots in the United States or Canada, and Russ is still in the one at Fairbanks. Siting an oceanography department over 600 kilometers from the sea in a city with brutally cold winters (including rare days below -50°C) surely once had a political logic. No other explanation serves. The university, with about 10,000 students, is one of the city's principal industries. Russ says that apart from the cold and the mosquito swarms in spring, it is a good place to live and raise a family. He and Cheryl have done that; the twins are now in college.

The cover of a recent novel, *pH*, by Nancy Lord, offers a picture by Russ of a planktonic snail, *Limacina*. That was chosen because ocean acidification (lower pH) threatens erosion of their shells. The novel has two principal characters. Jackson Oakley, an ocean-acidification chemist, works against his discipline in return for oil company payoffs, and Ray Berringer, who is rather like Russ in several respects, leads numerous ocean cruises, trains graduate students, and works with talented postdocs (many of whom are women), is recognized for excellent plankton photography, and has a supportive family. It's a good yarn, though the fictional Ray Berringer probably had more struggles with university administration than the real-life Russ does. In the novel the struggles came about because of the payoffs and other sick deals connected to Oakley. Possibly Dr. Hopcroft likes beer as much as did Dr. Berringer.

When we met at Seward aboard UAF's *R/V Sikuliaq* for an interview in early July of 2019, Russ was just back from one of the regular transect cruises for a Long-Term Ecosystem Research (LTER) program, sampling at stations from the mouth of Resurrection Bay across the shelf to deep water over the Aleutian Trench. The goal is a long record of changes in the ecosystems of the Alaskan coastal currents and the northern Gulf of Alaska. In two days he would be sailing again, leading a cruise involving an ROV study of the fauna on and above a sea mount, helping a student sample large jellyfish with a specially designed net, and sampling with Petra Lenz and Vittoria Roncalli (of the University of Hawaii) to obtain some *Neocalanus flemingeri* from depths below 1,000 m for study of gene activity related to their summer resting stage (see Chapter 16). In the years since 2000, when he started at UAF, Russ has spent long weeks and months in northern waters. Many good papers have come from those cruises, many of them about copepods, most coauthored with eminent copepodologists (e.g., Ksenia Kosobokova of Shirshov Institute of Oceanology, Moscow) or his students.

Drawings of Copepods

Color photographs are lovely and informative, especially those by Russ. However, the traditions for depicting animals developed before cameras, and drawings remain the medium of exchange among copepod taxonomists. Like the photographs, they strongly flatten rounded shapes into a two-dimensional projection, but in the case of taxonomic work they are usually just outlines with the segment edges drawn in. For the whole animal, such a drawing (Figure 1.4) is termed a *habitus* drawing, often a side view. Copepods in a preserved plankton sample mostly lie on their sides, so the *habitus* is what you generally see through a stereomicroscope before your probes toss an animal into other postures. This drawing, by Gayle Heron from a paper by Bruce Frost,[12] is unusually complete. Often the aim is just to show the shape of the body, the prosome and urosome, so the legs and setae are left out. Getting the body shape about right is easy, especially by using a camera lucida (or by tracing a photograph). Adding the realism of legs, and particularly to include setae, takes a real artist. We have some among us yet, though Heron has retired. Rony Huys of the British National Museum and several others keep great drawings coming.

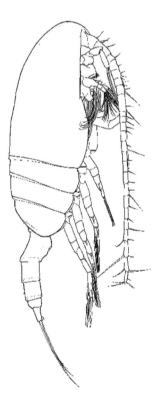

Figure 1.4 *Pseudocalanus moultoni*, female, from Frost,[12] a *habitus* drawing by Gayle Heron. The total length of this specimen, not including the tail setae (at left), was 1.65 mm. Permission, *Canadian Journal of Zoology.*

Oar Feet

Let's look at more construction details of a copepod starting with the thoracic legs, those oars. The paired left and right thoracic legs are also known as pereiopods: walking legs, from shrimp terminology, often abbreviated as P1 to P5. A photograph of the P4, or fourth from the front, from a specimen of *N. plumchrus*, is labeled in Figure 1.5. The paired left and right legs are mirror images in shape and are attached to each other below the body by a coupler plate (formally, an "intercoxal sclerite"). It forces them to move together. These legs are used for amazingly rapid escape or prey-chase swimming. When a predator is sensed moving nearby, certainly by tiny water motions sensed by the long antennules, the legs swing backward in sequence from back (P5) to front (P1). Each leg, acting as an oar, accelerates the copepod forward, which with some luck is away from an approaching predator or toward potential prey. Nothing is foolproof for protecting prey from predators. For example, a fish can actually

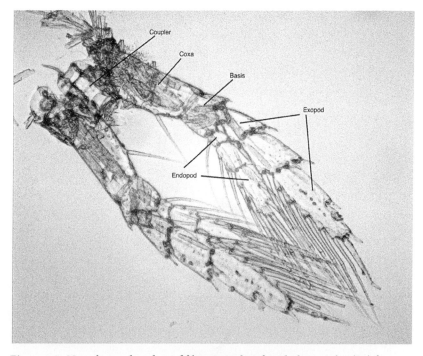

Figure 1.5 *Neocalanus plumchrus*, fifth copepodite: fourth thoracic leg (P4) from a preserved specimen. Segment names are given. Dissection and micrograph graciously provided by Atsushi Yamaguchi (University of Hokkaido). Some setae are broken, which is common for net-caught copepods.

take advantage by approaching the head, sucking in a little, and the copepod "escapes" right into its mouth.

When all of the legs have pushed back against the water and point aft (as in Figure 1.4), they swing forward (as in Figure 1.2) for another series of thrusts. The elaborate jointing of the legs facilitates that return. The muscles for the return stroke pull the coupler forward, and the leg is feathered back to reduce water resistance, or drag. Human rowers rotate their paddles on the back stroke, "feathering" them, to slip them forward through waves. On copepod legs, all the segments bend backward at their joints to present less surface to the water ahead. The joints can bend because the hard casings of the segments are joined around their edges by thin and flexible connectors (*arthrodial membranes*) with folds inside the articulations. The joints do not bend forward on the power strokes, because there are hard stops on their forward edges to prevent it. Tiny muscles in the segments closer to the body (termed *proximal*), visible in the micrograph as pale tan stripes, attach to the inner surface and reach through the joint to just inside next (more *distal*) segment, straightening the leg at the end of the return stroke. The main pressure against the water of a whole leg in the power stroke comes from large, fan-shaped muscles converging from the upper wall of its thoracic segment to attachments in the most proximal joint in the leg (the *coxa*). There are, of course, retractor muscles to help the leg back forward, but probably some of that return force comes from springiness of the exoskeleton.

Indeed, to the uninitiated there is a sort of mumbo-jumbo tone to the terminology (cephalosome, coxa, arthrodial, proximal, distal). All science can be like that, offering many odd terms. You might think the vocabulary is a sort of code designed to turn away all but the cognoscenti. And you might be right! In legal practice (*habeas corpus*) that is a major purpose. I introduce terms here so you can start to sense what the chatter is like at a conference of copepodologists.

A few more things about these oar feet; they have two branches (that is, they are *biramous*), a feature of some limbs in other crustacean groups. In copepods the outer branch (*exopod*) of each leg is longer than the inner branch (*endopod*). In many copepod species both branches have three segments, but in others some segments are fused; the legs have lost one or more articulations. The branches of some thoracic legs in different species have evolved various specialized shapes or tools. The P4 shown in the photo is close to the primitive original, or at least we (all those copepodologists) think so. The outer tips of the exopod segments have articulated spines. How those function is uncertain, but possibly they enable shedding of tiny eddies during the power strokes.

The medial sides of the exopods and both sides of the endopods bear many setae, long bristles that are part of the exoskeleton, and those bristles have short, paired side branches called setules. The arrangement is a little like the barbs and barbules on bird feathers. As the fan of setae is spread by drag from the water during the power stroke, the array becomes a sort of webbed extension of the oar. An aspect of fluid flow, surprising to most people, is that flow along a solid surface is restrained by viscosity such that layers right at that surface do not move. There is no "slip." That was initially proved for flowing water by the German fluid dynamicist Ludwig Prandtl,[13] who published his results in 1904. Relative velocity accelerates progressively from zero at the surface out into the main flow as a boundary layer.

Mimi Koehl and Rudi Strickler[14] suggested much later (in 1981) that the boundary layers around setae, and more so around setules, are close enough together to overlap and let very little water through, even as a limb is moving rapidly. Thus, the setules effectively create an almost solid surface, like oar blades, and their flexibility lets them bend to return forward by sliding almost lengthwise through the water. A little (only that) more explanation of the different sources of fluid drag is offered in a box at the end of the chapter.

The Tail Fan

In the side view of the *N. plumchrus* copepodite (Figure 1.2) above, the tail fan is just a line. A dorsal view (Figure 1.6) shows that the urosome carries a fan of setae spreading from two blobs, the *caudal furcae*, that extend from the last segment of the urosome (location of the anus, and thus the "anal segment"—What, not Greek or Latin? Well, it is from Latin). The long muscles in the prosome and urosome can flip this fan like a fish swishing its tail. If the copepod were still this would have little effect, the viscosity of the water would hold the body almost in place, but once the thoracic legs have the whole body moving, the final swish of the fan can accelerate it further. Again, at the smaller scale of the setae and setules, the boundary layers would make it an almost solid sculling blade.

In the early 1990s, two biological oceanographers who did not usually work directly with copepods, Victor Smetacek from Germany and the late Peter Verity from the United States, were nevertheless talking about them.[15] They must have asked each other why so many variations have evolved in the anterior feeding limbs and in the life ways of planktonic copepods, species varying in adult size by about a hundred-fold, when there are only minor variations in the pattern of smoothly tapered prosomes, oar-like thoracic legs cocked forward when resting and a tail fan at the rear. Their answer was that all of them have the need in common for extremely accelerated escape jumps. Surely they were right.

Calanus hyperboreus
Hopcroft/UAF/NOAA/CoML

Figure 1.6 *Calanus hyperboreus*, just the anal segment, two pad-like extensions (the caudal furcae) and five stiff setae on each side. Two smaller midline setae bend together and touch along the midline. The lateral five setae have dense lines of long setules on each side that fill the plane as a tail fan. The scale in red is hard to read here; it is 500 μm (0.5 mm). Photo by Russell Hopcroft.

Escape: How Far, How Fast?

So, how fast exactly (or approximately) do copepods move when they feel something approach? Ed Busky and colleagues[16] used high frame-frequency TV recordings to study jumps of several species of *Acartia* (about 1.0 mm long) and found speeds up to 500 body-lengths per second, or around 50 cm per second. They did not jump very far, about 6–8 mm, using 1 to 9 thrust sequences, but probably that is far enough and fast enough to provide substantially improved survival (otherwise they would evolve so as not to waste the necessary energy). They also mentioned what happens in all species: the initial acceleration swings the antennules back into the position shown in Figure 1.2.

I am fearlessly using metric units here. All scientists use them, and it becomes very hard for us to think, at least for scientific subjects, in what are now almost exclusively "American" units: inches, feet, cubic feet, and so on. Metric units are better because of the ten-fold ratios between units adjacent with differing names (1 cm = 10 mm; 1 liter = 1,000 milliliters), and because they give Americans a language for measurements in common with the rest of the world. American readers, stick with me in this; it isn't difficult.

Many workers beside Ed Busky have taken a turn at estimating the speed and power of copepod escape jumps, with an early entry from William Vlymen,[17] who examined the drag encountered by *Labidocera trispinosa* as it decelerated to rest after a leap. In succeeding years and decades movies and video cameras with better and better slow-motion capability have been used to follow copepod movement in aquaria that are thin from front to back (to keep the copepods almost in focus). Leonid Svetlichny, working in Crimea, provided one of the best studies.[18] He enlarged vertical and horizontal dimensions of his thin aquarium enough to see how many thrust cycles his copepods applied in rushing away from alarms. Thirty years ago he had a movie camera that could record up to 3,500 frames per second. You can buy one of those SKS-1M cameras on eBay, maybe even film for it.

Later Svetlichny worked with Thomas Kiørboe[19] of Denmark (see Chapter 11) using smaller aquaria and sufficiently motion-slowing video (1,200 frames per second) to characterize the cycles of thoracic leg power strokes in several species. A graph of their results (Figure 1.7) for *Calanus helgolandicus* shows its speed pulses, velocity rising as the legs swing back, falling off as they are feathered forward for the next cycle. Notice the very short durations of each cycle, rising up to about 80 cm/second, then falling to 20 or 40 in about 10 milliseconds (10

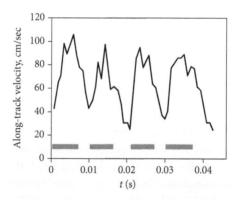

Figure 1.7 *Calanus helgolandicus.* Initial velocity cycles of an escape jump. The x-axis is time in seconds. The black bars are the intervals in which the thoracic legs thrust back in sequence (P5, P4, ..., P1). During the gaps between bars the legs feather into cocked position forward. From Svetlichny et al.[19] Permission, *Journal of Experimental Biology.*

ms = 0.01 seconds). You cannot blink that fast (~100 ms at ~20 cm/second). The peak speeds of successive cycles slow slightly. I added up the distances covered as the speeds varied: 2.7 cm in 40 milliseconds. In his earlier work Svetlichny[18] showed that C. helgolandicus (~3.2 mm) can sustain that rush for a full second, though the peak velocities of successive thrust cycles fall off, as they started to do in the graphed data, decreasing to approximately 40 cm/second before thrusts ceased. The final distance when again at rest can be nearly a meter, maybe 300 body lengths.

Petra Lenz and colleagues,[20] also using video, measured top speeds of Calanus finmarchicus (~3 mm prosome) in the same range, around 80 cm per second. So an adult Calanus arrives almost a meter away from an attacking fish, when viscosity drags it again to a complete stop. Visual paths are short in water, and shorter for nearly transparent (Figure 1.2) and tiny objects, so the jump provides a substantial gain in safety. Just as important, Lenz showed that an escape response to a new vibrational stimulus initiates in about two milliseconds for Calanus, faster than the proverbial greased lightning. That quickness comes from very rapid transmission of motor impulses from the antennules along "giant" neurons in the ventral nerve cord. All of the invertebrates with their main along-body nerve below the gut (annelids, molluscs, arthropods) have these giant neurons carrying the news of danger to the muscles in order to initiate immediate movement. In copepods those nerves are also short. Repeating the cycling requires extremely fast recharge of muscles with fuel, with the adenosine-triphospate (ATP) molecules that are the energy currency of all animals. Lots of mental exercise has come from calculating the power output capacity of copepod muscles, showing that it is probably the greatest (per gram!) in the animal kingdom.

There are refinements. Petra Lenz and colleagues discovered in the late 1990s[21] that some copepods (e.g., Calanus, Bestiolina, and Parvocalanus) have myelinated axons, whereas some other copepods do not (e.g., Acartia, Eurytemora, and Centropages). Axons with myelin are likely, based on data from animals generally, to have much faster impulse conduction than bare nerve fibers. More recently, Lenz, with Ed Busky and other colleages,[22] conducted experimental observations with high-speed cameras by applying pulsed and directional hydrodynamic disturbances to copepods from both groups. Copepodites of "amyelinate" species escaped in the direction they were already headed when the pulse hit them. The danger of that, noted above, is that a signal from in front can cause them to swim toward an attacker. Copepodites of two myelinate species, Bestiolina and Parvocalanus, swam away from the stimulus source regardless of where it hit them. The myelin evidently provides more time for processing the signals from setal sensors around the body in order to leap away.

Resting

Finally, moving all the way back, the center of the tail fan of *N. plumchrus* has those two amazing setae, termed the II bristles, with orange, ribbon-like setules (Figure 1.2). Most of the time copepods are not escaping; they are hanging nearly motionless in the water. Mark Benfield[23] measured the orientations of 152 individual older copepodites and adults of *Calanus finmarchicus* (a family-level relative of *N. plumchrus*) from pictures of them captured by high-speed video cameras towed through the ocean while recording. It is possible for such cameras to record from far enough ahead that the pressure wave from their approach does not obviously disturb animals until after their images are recorded. A strong majority of specimens were oriented head up, tail down, with the antennules extending to either side. That is also how live specimens orient in aquaria when eating (or, for all we know, just thinking). If a preserved specimen of *N. plumchrus* with intact II bristles is released tail first into water, the ribbons spread out into a sort of three-dimensional feather (much as in Figure 1.2), a sort of sea anchor resisting sinking. If you pull that specimen forward with forceps through the water, the ribbons fold back along the bristle, reducing their drag. That is surely to accommodate the need for high escape velocity. Similar vertical sea anchors are found in other families, notably the Euchaetidae (*Euchaeta* and *Paraeuchaeta*). You will meet them further on.

Enough about the back end; let's move now to the front.

Box 1   About Drag

Motion of an animal through water is resisted by drag of two kinds. Both require work to overcome the resistance. *One*, the water has mass, so it must be accelerated out from in front of the animal, and then accelerated in the opposite direction to fill the space the animal has just left. Basically, the latter is done by accelerating the body forward, pushing against the inertia of the water. Work must be done to impart new momentum to the water in the way, force motion where there was none before. The rule is Newton's Second Law: *Force needed = Mass × Acceleration.*

Applying that to fluid motion is more complicated than calculating the speed of a ball falling through a vertical vacuum, but it can be done. The amount of work needed to overcome this *inertial drag* depends on how far the water must move, farther if a beast is bigger and on how fast it is moving. For a given shape, inertial drag increases as the *square* of the velocity. It also depends on the density of the water. Heavier water that is colder or saltier takes more work than does warmer, fresher and thus lighter water.

Two, water is sticky: the molecules are attracted to each other and to the animal's surfaces. So, moving through them requires work against *viscous drag,* work to stretch and rearrange those attractive forces. The amount of that work is greater if the intermolecular forces are greater: stronger in cold water than warm. Intermolecular forces are always present, and the faster an animal swims the greater the rate of rearrangements, the resulting drag increasing *proportionally* with velocity.

The difference in velocity's effect on inertial and viscous drag means that viscous drag becomes *relatively* less significant as a copepod accelerates. But as it slows, not expending propulsive power, viscous drag quite rapidly hauls it to a stop. Copepods usually are a little denser than water and tend to sink, but slowly, and viscous drag dominates. So, when resting, viscosity tends to hold them almost in place. They are in exactly the right size range for this to work well. They are also the right size for inertial drag to make their thoracic legs effective oars, their tail fan effective for sculling.

When swimming, people experience drag dominated by inertial effects, including long glide paths. A viscosity-dominated regime has seemingly counterintuitive effects, such as miniscule glide paths, essentially none. Another key effect in respect to swimming is that the viscous drag along the length of a moving seta (or any cylinder) is half that of flow around it. So, small crustaceans push setae side-on through water during power strokes and slide them through lengthwise on return strokes. That is also why protozoan flagellae, at sizes smaller than setae, can act as propellers. More effects of viscosity will come up in respect to feeding mechanics (in Chapter 7).

References

1. Harvey, T. H. P., & B. E. Pedder (2014) Copepod mandible palynomorphs from the Nolichucky shale (Cambrian, Tennessee): implications for the taphonomy and recovery of small carbonaceous fossils. *Palaios* 28(5): 278–284.
2. Regier, J. C., J. W. Shultz, A. Zwick, A. Hussey, B. Ball, R. Wetzer, J. W. Martin, & J. W. Cunningham (2010) Arthropod relationships revealed by phylogenomic analysis of nuclear protein-coding sequences. *Nature* **463**: 1079–1083.
3. Damkaer, D. (2002) *The Copepodologist's Cabinet, A Biographical and Bibliographical History.* Two volumes to date. American Philosophical Society, Philadelphia.
4. Evans, J. Dennis (1978) *The Book of Laws Founded Against Itself: The Poetry of A.R. Ammons, 1951–1976.* PhD dissertation, University of California, Berkeley.
5. West, Robert M., ed. (2017) "Hymn IV." In *The Complete Poems of A. R. Ammons, Vol. 1, 1955–1977.* W.W. Norton & Co., New York.

6. Hopcroft, R. R., D. B. Ward, & R. C. Roff (1985) The relative significance of body surface and cloacal respiration in *Polus fabricii* (Holothuroidea: Dendrochirotida). *Canadian Journal of Zoology* 64: 2878–2881.

7. Roff, J. C., & R. R. Hopcroft (1986) High precision microcomputer based measuring system for ecological research. *Canadian Journal of Fisheries and Aquatic Sciences* 43: 2044–2048.

8. Clarke, C., & R. C. Roff (1990) Abundance and biomass of herbivorous zooplankton off Kingston, Jamaica, with estimates of their annual production. *Estuarine, Coastal and Shelf Science* 31: 423–437.

9. Hopcroft, R. R. (1997) Size-Related Patterns in Growth Rate and Production of Tropical Marine Planktonic Communities Along a Trophic Gradient. PhD dissertation, University of Guelph.

10. Hopcroft, R. R., & J. C. Roff (1996) Zooplankton growth rates: diel egg production in the copepods *Oithona, Euterpina* and *Corycaeus* from tropical waters. *Journal of Plankton Research* 18: 789–803.

11. Hopcroft, R. R., & B. H. Robison (2005) New mesopelagic larvaceans in the genus *Fritillaria* from Monterey Bay, California. *Journal of the Marine Biological Association, U.K.* 85: 665–678.

12. Frost, B.W. (1989) A taxonomy of the marine calanoid copepod genus *Pseudocalanus*. *Canadian Journal of Zoology* 67: 525–551.

13. Anderson, John D. Jr. (2005) Ludwig Prandtl's boundary layer. *Physics Today* 58: 42–48.

14. Koehl, M., & J. R. Strickler (1981) Copepod feeding currents: food capture at low Reynolds number. *Limnology and Oceanography* 26: 1062–1073.

15. Verity, P., & V. Smetacek (1996) Organism life cycles, predation and the structure of marine ecosystems. *Marine Ecology Progress Series* 130: 277–293.

16. Buskey, E. J., P. H. Lenz, & D. K. Hartline (2002) Escape behavior of planktonic copepods in response to hydrodynamic disturbances: high speed video analysis. *Marine Ecology Progress Series* 235: 135–146.

17. Vlymen, W. (1970) Energy expenditure of swimming copepods. *Limnology and Oceanography* 15(3): 348–356.

18. Svetlichny, L. (1987) Speed, force and energy expenditure in the movement of copepods. *Oceanology* 27: 497–502.

19. Svetlichny, L., P. S. Larson, & T. Kiørboe (2018) Swim and fly: escape strategy in neustonic and planktonic copepods. *Journal of Experimental Biology* 221: doi:10.1242/jeb.167262.

20. Lenz, P. H., A. E. Hower, & D. K. Hartline (2004) Force production during pereiopod power strokes in *Calanus finmarchicus*. *Journal of Marine Systems* 49: 133–144.

21. Davis, A. D., T. M. Weatherby, D. K. Hartline, & P. H. Lenz. (1999) Myelin-like sheaths in copepod axons. *Nature* 398: 571.

22. Buskey, E. J., J. R. Strickler, C. J. Bradley, D. K. Hartline, & P. H. Lenz (2017) Escapes in copepods: comparison between myelinate and amyelinate species. *Journal of Experimental Biology* 220: 754–758.

23. Benfield, M., C. S. Davis, & S. M. Gallager (2000) Estimating the in-situ orientation of *Calanus finmarchicus* on Georges Bank using the Video Plankton Recorder. *Plankton Biology and Ecology* 47: 69–72.

2

The Front End

Olfaction, Motion Sensors, Feeding Limbs Teeth

The head end, or cephalosome ("head body") of a generalized copepod, one much like *Calanus helgolandicus* (~2.5 mm) that is common in British waters and south to Spain, was drawn side view (Figure 2.1) and published in 1928 by H. Graham Cannon.[1] His picture shows the general layout of the complex tangle of limbs for dealing with aspects of copepod life other than escape and reproduction. Things could get technical here, and I will go there later. Instead, let's skip to the mandible (labeled in Figure 2.1) to finish explaining the book title.

Barbara Sullivan Finds Opals in Copepod Mouths

The most usual format of the mandible (Figure 2.2[2]) includes the coxa, articulating with the copepod's ventral surface, and a palp. The mandibular palp is two basis segments and two segmented branches: exopod and endopod. It extends below and alongside the body (Figure 1.2) and can swing forward and back, moving water toward the mouth or away from it. That is part of feeding-current generation. The coxae bear relatively long side branches (*edites*), known as the mandibular *gnathobases*, ending in chewing surfaces that are the copepod's jaws. Left and right gnathobases extend under the labrum (a fat front lip) and meet inside the mouth. They are tilted enough (Figure 2.3) that the little spine, shown at the bottom in Figure 2.2, is termed the dorsal seta.

Barbara K. Sullivan (Figure 2.4), a graduate student working with me at Oregon State University (OSU) in the early 1970s, discovered in the course of studying the gut contents of arrow worms (chaetognaths) that teeth of many copepods have opal crowns,[3] which led to the title of this book. Discoveries are not always planned; often enough they just happen along the way. Arriving at OSU after my undistinguished postdoctoral year in New Zealand (neither my advisor there nor I had any clear idea what I should do as a postdoc), I needed to find research projects. That went quickly enough, but I was expected also to begin advising graduate students. There is no training whatever for that, except having been one. It was also expected that I would generate stipends for students

Oar Feet and Opal Teeth. Charles B. Miller, Oxford University Press. © Oxford University Press 2023.
DOI: 10.1093/oso/9780197637326.003.0002

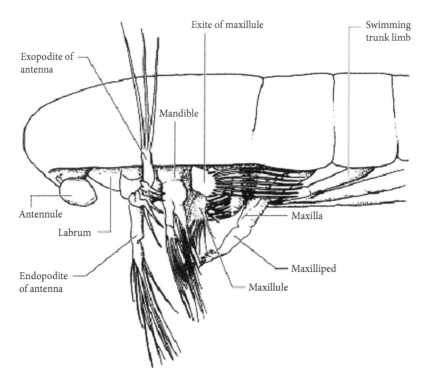

Figure 2.1 Lateral view of the head of *Calanus* late larva (*copepodite*) showing the positions of the limbs located behind the antennules (just one segment of that shown; see Figure 1.2 for them) and ahead of the thoracic feet (here, "swimming trunk limb"). The labels show the standard terminology. Drawing by H. G. Cannon.[1] Permission, *Journal of Experimental Biology.*

from grants. In a year or two I did get some modest research grants, and I helped the oceanography department admit a few students.

Among those was Barbara Sullivan, who had already completed an MS degree in environmental engineering from Johns Hopkins University (JHU). She had an excellent academic record, though she told me in an interview decades later that I was inappropriately contemptuous of her JHU thesis. I admit that easy contempt was part of my Scripps education; I'm still recovering. The very idea of environmental engineering was foreign to me; wasn't everything from farming to forestry (and mining and road paving) environmental engineering? No. About selecting that program she recently wrote me this:

The year 1970, when I began graduate school at Johns Hopkins University, was a critical turning point in awareness of human impact on the environment.

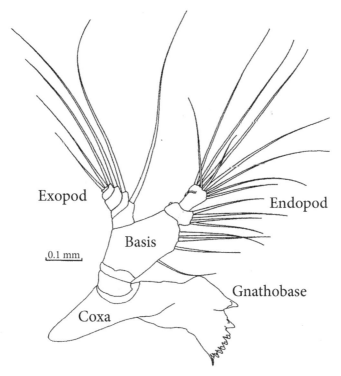

Figure 2.2 A generalized layout of a calanoid copepod mandible. From an Australian species, *Boeckella major*. From Green and Shiel.[2] Permission, *New Zealand Journal of Marine and Freshwater Research*.

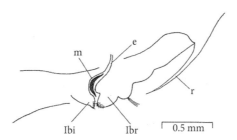

Figure 2.3 Sagittal section, from the rostrum (r) at the right past the mouth (m), showing the orientation of the gnathobase (in black) of *Neocalanus cristatus*. The mouth is closed anteriorly by the labrum (lbr) and posteriorly by the left and right labial palps (lbi). Those palps have a mid-line groove between them through which the maxillules (see below) push food items. The esophagus (e), directed dorsally, includes a toothed "gastric mill," and opens into the stomach. From Sullivan et al.[3] Permission, *Marine Biology*.

Figure 2.4 Barbara Sullivan-Watts out by the seashore on a summer day. With Permission from Dr. Sullivan.

Anyone who was an ecologist at heart, as I was, was tuned into the serious environmental pollution issues that caused the general populace, especially the socially active young, to explode in frustration, culminating in the First Earth Day. So evident had air and water pollution become, so clear was our ignorance of toxic substances (we had absorbed that from Rachel Carson) that protest was our only hope to force a better future. In addition to marching and holding teach-ins, I wanted to be part of the scientific solution.

At JHU Barbara studied impacts of thermal pollution from releases of power-plant cooling water. She says that the application of science to such practical problems was and remains a key value to her. Her interest in oceans was partly based on a lot of sailing experience with friends and a Girl Scout group while growing up on the north shore of Long Island. Also, her mother was a fan of Rachel Carson, loving and talking about *The Sea Around Us*. She says her interest in science started with her father. He was a self-trained expert who ran an electronics repair business. He had wide interest in how all sorts of things work, interest he shared often and in detail with Barbara. She attended a Catholic high school, with her interest in biology starting there, and Catholic University in Washington, DC, where they filled in physics and chemistry. She credits both for excellent basic science preparation, particularly from her genetics professor. She married just before finishing college and followed her husband to Baltimore. Once there she became the first woman student in that Johns Hopkins Environmental Engineering program. Eventually they moved to Oregon, where her husband got a job in a location nearly matching her new graduate appointment. Barbara was not the first woman to study oceanography at OSU, but she was the only one in her class and probably because of that, was left out of at least some of the early collaborations.

In 1971–1972 and now, most of an oceanography graduate student's first year is taken up with core courses: basic physics of currents and waves, chemistry of seawater, seafloor geology, and ocean ecology. The candidates come in with bachelors and sometimes master's degrees in one of those basic disciplines, but they all take those core courses, helping each other with the underlying basic sciences. There is a great deal that is useful for any aspiring oceanographer to know about all of these interacting aspects of marine science. During that beginning and on through her whole degree program she had the added time drain of needing to commute to school from 30 miles away. But she got the grades and did well enough to pass an oral preliminary examination. Those were then an ordeal of several hours, facing five professors, each with different ideas of what a student should demonstrate during the exam. For women students in the 1970s the challenge usually involved some version of facing down a row of older men, uniformly with professor-sized egos. The mind can go blank when such a panel asks you to explain the basic principle of the Coriolis effect, or the bioamplification of toxins along a plankton-to-fish food chain. With some practice, Barb passed and in the next season proceeded to some practical experience. She was part of several zooplankton sampling cruises, some of them long runs out into the Pacific, others along the Oregon coast. Looking at plankton alive and dead led to an interest in the soft-bodied predators like ctenophores, jellyfish, and arrow worms (more formally chaetognaths).

Samples available to Barbara from the Gulf of Alaska (50°N, 145°W, a spot in the middle of watery nowhere known as Station P) had come from a summer cruise I had made there with helpers. We towed bongo nets (see Figure 1.1) at eight levels down to 500 m. The vertical distributions of most of the zooplankton caught in the nets had been worked out by Christopher Marlowe,[4] a fellow OSU grad student, and Barbara decided to examine the chaetognaths closely. Almost all of them were either *Sagitta elegans* (Figure 2.5) or *Eukrohnia hamata*. Her questions became: What do they eat? Do they eat what passes by them in the proportions passing, or do they pick and choose? How does feeding shift with the worms' size (age), depth, and time of day? So she sorted them out, re-estimated their density, and checked their gut contents.

Copepods are the main prey of these small fanged predators (Figure 2.4). Copepod bodies are ingested whole and passed along the worm's straight gut to where most of the digestion occurs, that is, just ahead of the anus. Barbara cut across them ahead of this digestive area and squeezed out the contents. Frequently copepod gnathobases were all that she found left from their meals

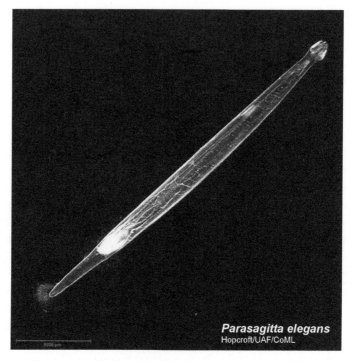

Figure 2.5 Photograph of *Parasagitta elegans*, an arrow worm common in the North Pacific and North Atlantic oceans. Head at right, tail fan at left. The white mass just ahead of the tail section is a copepod under digestion. The worm's chitinous fangs are articulated beside the jawless mouth and here are folded against the head. There are long, transparent fins along the side, not showing in the picture. This specimen is immature; testes in the tail section and ovaries alongside the gut were not yet developed. Photo by Russell Hopcroft.

before capture. Scanning electron microscopy (SEM) had just become available on campus, so Barbara cut jaws from all of the abundant copepod species in the same samples, stuck them on aluminum disks, and took them across the street to the mini-SEM, which was all OSU could afford at the time. Al Soeldner, a very helpful electron microscopist, coated them with spray-on gold and took their pictures (similar to Figures 2.6 and 2.15).

Indeed, different species had distinctive jaws and particularly tooth shapes. The teeth were clasped at the base by grooves like jewelry bezels in the chitin. The crowns often cracked in the vacuum of the SEM, and the broken surfaces appeared to be minerals. Several tests exposing teeth to fumes from hydrofluoric acid removed the mineral, while other acids did not, suggesting they were silica, and we called them "glass teeth." Later work with energy x-ray spectroscopy

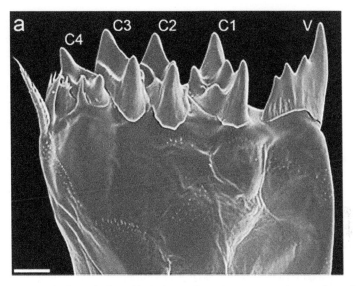

Figure 2.6 Toothed gnathobase blade from *Rhincalanus gigas*, a large calanoid copepod from the Southern Ocean. The dorsal seta is at the left, the ventral, most blade-like tooth (V) at the right. Seating of the opal crowns in the chitin is clearly outlined. The white scale bar is 25 μm. From Jan Michels et al.[5] Permission, *Zoology* (Elsevier).

agreed, and the teeth appeared to be "doped" with zinc. Some electron scattering studies of ultra-thin sections suggested the teeth were amorphous, that is, *opal* (hydrated crystals of silicic acid). Barbara published the story,[3] acknowledging that the great Russian biological oceanographer Vladimir Beklemishev[6] had determined several decades earlier that copepod teeth are siliceous, also by dissolving them in hydrofluoric acid.

SEM results in hand, Barbara could identify with a light microscope the co-pepod species and development stage of a jaw's original owner, and thus she could examine the preferences (only vaguely the same as preferring chocolate to vanilla) of chaetognaths among potential prey species. To aid in that, the two bongo nets producing her samples had different mesh sizes: 183 μm and 333 μm. She took her prey abundance estimates from the finer mesh samples (dominated by a small species, *Oithona similis*) and took her predator arrow worms from those retained by coarser mesh. That reduced effects on her data from the worms feeding on smaller animals in the dense slurry of the samples. To the same end, she only examined prey that were at least a bit digested. Some of the results mostly concern chaetognaths, and those interested can find them in the paper Barbara wrote later from her dissertation.[7]

However, insights about the prey were important, too. The smaller and numerically dominant copepods, those *Oithona similis*, were about the proportion of prey in the guts predicted by their relative abundance in the water column, particularly for *Sagitta* near the surface at night. An arrow worm, at least when hungry, must dart at most potential prey whose swimming eddies move the hair-like vibration sensors on the worm's skin, catching some of them. A disproportionate lot of the jaws came from *Metridia pacifica* (Figure 2.7), suggesting that it is a preferred prey.

It is clear why. If a gently collected net sample containing *Metridia* is diluted and examined alive in a small aquarium (no lenses needed), the older copepodites will be in constant, swirling motion, doing rapidly repeated loop-the-loops. *Metridia* are alert, quick and hard to catch with a pipette and suction bulb, but there they are in the meals of chaetognaths, fish, shrimp, and more. They not only have strong, directed escape swimming, but when escaping they also shoot brilliant blue bioluminescence into their wake from pairs of cells on their dorsal thorax and urosome. A blinding flash is left behind as a decoy, much like a squid leaving an ink blob as it jet propels away from a predator. More than any other copepod family, the metridiids (*Metridia, Pleuromamma*) consistently spend the daylight hours at considerable depth in very dim light. They surely are avoiding visual predators like fish. At night, they migrate 300 or 400 m (200,000 body lengths) to near the surface to feed and then back down at dawn.

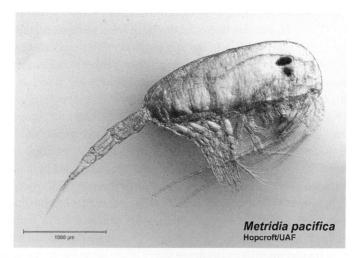

Metridia pacifica
Hopcroft/UAF

Figure 2.7 Lateral view of a female *Metridia pacifica*. The genus is entirely unpigmented and as close to transparent as tissue can be; the photo depends on refraction effects. Food in the gut does add one or a few opaque spots. Length from head to the ends of the caudal furcae is around 3 mm. Photo by Russ Hopcroft.

They are almost completely transparent: that is, they have every known predator avoidance scheme in play, *except* that they keep moving all the time, resembling inebriated stunt pilots. Apparently, that can be fatal for zooplankters. So why do they do it? We do not know, but since it is so often fatal, it must be important.

Let's give Barbara and teeth a break in order to look at other parts of the head.

Information Gathering

The cephalosomes of most pelagic (free-swimming) copepods have a fused, smoothly rounded carapace over six or sometimes seven segments, exactly like that of *N. plumchrus* (see Figure 1.2). There are also pointy-headed species. There is a sort of suture around the lower edge of this dome-like head-body, attaching it to a ventral sequence of sternite plates. The anterior "legs" articulate with those plates, moved by muscles extending through openings in the plates to the carapace walls. When a copepodite molts to the next and larger stage, that carapace to sternite suture breaks, opening a deep V-shaped slot pointing toward the back of the prosome. The animal then shimmies through the slot, forward out of the old exoskeleton, pulling its legs from their own chitin cases (more on molting eventually).

Depending from the midline at the bottom edge of the forehead are a pair of spines, sometimes curving back, termed the *rostrum*, which is not a stand for tiny orators. However, lexicographers tell us that the Romans used the prow ornaments of captured ships to decorate their oratory pulpits. Various leading bits of some crustaceans and insects may bear some resemblance to a prow ornament, though that is not very obvious for copepods. For copepods the term "rostrum" carries over from features on other arthropods.

Rolf Eloffsson of Sweden diagrammed[8] the midline nerves emerging forward from the brain (Figure 2.8), which sits on the floor of the head in front of the esophagus. They branch for many species into a tiny whisker above the rostrum, termed the *frontal organ*. In many species the nerve continues around (Figure 2.8B) into the larger, curved tines of the rostrum. Elofsson sliced ultrafine sections of the two projections from heads embedded in plastic and examined them with transmission electron microscopy. Do not underestimate the difficulty of passing a microtome knife through such small features buried in miniscule bits of almost opaque plastic. He did it, finding structures inside the nerves termed *ciliary dendrites* (Figure 2.9).

At least for organs on an animal's surface, these ultrastructural patterns are a good indication of odor sensing capability, which probably is the role of both the frontal organ and rostrum. Similar nerves are found at the front tips of many "primitive" crustacean groups, and the sensory ends are often in a so-called

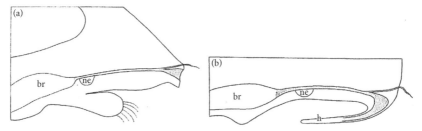

Figure 2.8 Schematic drawings of sagittal (midline) sections of (**a**) *Euchaeta norwegica* and (**b**) *Calanus finmarchicus*, showing a medial nerve passing from the brain (**br**) into the frontal organ (the hair-like whisp at the right) and the ventral "horn" **h** (from Eloffson,[8] who called the rostrum a "horn"). The nerve passes between the so-called naupliar eyes (**ne**), which are photoreceptors present in copepods from the earliest larvae (nauplii) to adulthood. Permission, from *Acta Zoologica* (Wiley).

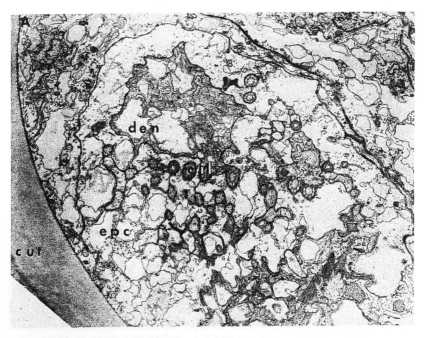

Figure 2.9 Transmission electron micrograph of a section through the rostrum of *Calanus finmarchicus*, showing one of its nerve fibers, a dendrite (den), a modified epidermal cell (epc) surrounding it and adjacent to the cuticle (cut). The cilia (cil) of the ciliary dendrite are the small, darkly stained circles with the microtubules of the axonemes appearing as inner rings. From Elofsson.[8] Permission, *Acta Zoologica* (Wiley).

cavity organ, because they end against an anterior pit in the forehead, as they do in the predatory copepod *Euchaeta norwegica* (Figure 2.8A). Leading with an olfactory "nose" is a general design feature of animals, certainly an instance of recurringly convergent evolution. Fish, for example, have anterior head pits (nares) that are lined with olfactory nerve endings. Dogs follow very attentively behind their leading noses.

Different animals have different olfactory repertoires, the range of molecules that can be scented. The number of scents that crustacea can detect is smaller than the olfactory repertoire of land animals, including insects and vertebrates. All of the molecules stimulating lobster and copepod neural discharges must be water soluble. For large crustaceans, like spiny lobsters (species of *Panulirus*), with forests of olfactory setae on their antennules, that range can be tested[9] and proves to be amino acids, some small peptides (short chains of linked amino acids), a few other amines, nucleotides ("building block" molecules of DNA), adenosines (AMP), and some sugars. For copepods this repertoire is more difficult to study.

Nevertheless, experiments by Greg Teegarden[10] and Jiarong Hong et al.[11] show that toxins from red-tide dinoflagellates can be detected by copepods using these anterior "noses," and possibly other toxins can be detected as well. Can the ammonia in fish urine or some breakdown product of fish mucus generate copepod nerve impulses? Jonathan Cohen and Richard Forward[12] have shown that *Calanopia americana* (an ~2 mm, nearshore copepod) will migrate upward when downward directed illumination dims. Such movement is related to daily ("diel") vertical migration (we will get to that) upward to feed at night. They will *not* rise in the presence of extracts from the mucus of fish or ctenophores (predatory jellies). However, such responses to so-called kairomones (scents conveying information but not necessarily benefitting the "sender") have not been found consistently in behavioral observations of copepods. There also are olfactory sensors on some of the feeding limbs.

Elofsson's[8] drawings (Figure 2.8) show small blebs labeled *ne* for *naupliar eyes* (i.e., larval eyes). He has also examined those eyes in ultra-fine detail, and he[13] now terms them "frontal eyes," since in copepods and other very early branches of the crustacean family tree, they persist through all the stages from the initial nauplius larva to the adult. They are resorbed in juveniles of some, but not all, of the shrimps and crabs. There are actually five separate eyes in the frontal eye complex: paired lateral eyes, two cells termed Glickhorn's organs, and a central ventral eye (Figure 2.10).

The components of the lateral and ventral eyes are layered in a cup shape: a reflective tapetum around a light absorbing pigment layer and nerve cell bodies that fire when the very dark pigment (a rhodopsin) absorbs light. The cup is

(a)   (b)

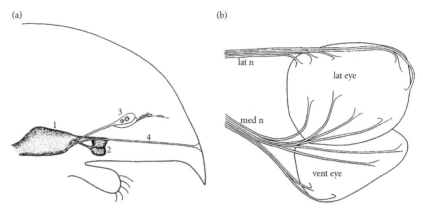

Figure 2.10 Diagram of the frontal eye layout in *Pareuchaeta norvegica* from Elofsson.[13] (a) Locations of visual organs (2 and 3) ahead of the brain (1) and adjacent to the olfactory nerve (4). The uppermost cells (3) are also termed Glickhorn's organ, which has been shown to be a light sensor. (b) The right pair of frontal eyes. The more dorsal pair are also farther toward the side than the ventral pair. The lateral (lat n) and medial (med n) nerves carry signals to a visual ganglion in the central front of the brain. Permission, *Arthropod Structure & Development* (Elsevier).

filled by several large lensing cells. The lateral eyes look dorsally; the ventral eye ventrally. Since copepods can be oriented variously, this is not consistently the same as up versus down. The Glickhorn's organs (Figure 2.10A, no. 3) are also eyes, constituted of a few cells with two nuclei, whatever that implies. They have a pigmented end, or rhabdom. In at least one species, *Calanus marshallae*,[14] this "auxiliary eye" is much larger and spreads against the smoothly curved frontal dome of its head. All taken together, this array of eyes must provide fairly elaborate information about the field of light surrounding a copepod. Asked directly what they see, however, copepods say nothing. They do respond to variations in surrounding light, migrating into or away from beams (sometimes swarming like beetles and moths around a door light), migrating up into dim light above and down away from brightening light above. Possibly they also respond to nearby flashes of bioluminescence. Studies of these responses go back many decades, but the character of the actual cognition is more difficult to study than that in people, or even dogs and cats.

In some copepods the frontal eyes have become much more elaborated. In juvenile and adult *Epilabidocera* species, for example, the lateral eyes are displaced upward to the top of the head (Figure 2.11), where the exoskeleton has warped down into the spaces next to them and wrapped into a spherical lens. Spheres are the nearly standard shape of image-forming eye lenses in fish, squids, and even a

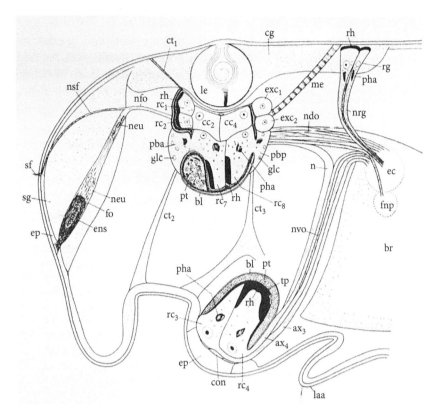

Figure 2.11 Diagram showing the eye structures of *Epilabidocera longipedata* from an anatomical study by Tai Soo Park[15] (a copepodologists considered further in Chapter 3). The lateral eye is adjacent to the surface of the head, which has formed a spherical lens for it. The ventral eye is directed ahead and down. A few of Park's codes: **br** = brain; **cc** = crystalline cell; **ec** = eye center of the brain; le = lens; me = eye muscle; **n, nvo, ndo, nrg** = nerve; **pt** = pigmented cell; **rc** = receptor neuron; **rg** = Glickhorn's organ; **rh** = rhabdom (black representing visual pigment); **tp** = tapetum. Copyright at *La Cellule* has been allowed to lapse, most recently Éditions Nauwelaerts.

few planktonic polychaete worms (e.g., Alciopidae). Notice the eye muscle (me) that can shift the eye around its lens.

Antennules

Just behind and to each side of the rostrum, the *antennules* are attached, long, unbranched tubes with many articulated segments. Bare setae, with no setules, are set at several angles on all or most of those canister-like divisions. A few setae at the tips do have setules (Figure 1.2), they are probably drag enhancers to slow sinking of animals in the "Kalanus position." Recall Mark Benfield's towed-video-camera result that most resting or feeding individuals of *Calanus finmarchicus* orient head-up, tail-down with the antennules extending straight to the sides. During escape jumps, the antennules are swept back and down into the position shown for *Neocalanus* in Figure 1.2.

This "arms-out posture" may have been noticed early on, perhaps as early as 1819 by William Leach, who established the genus name.[16] Marshall and Orr[17] recalled the story or myth that Alexander the Great, riding victorious into Persia or India, came upon a man, a yogi, standing in the sun bare-chested, head back and arms out to the side. Supposedly, Alexander asked the man (had he already found translators?) what he was up to. The man, whose name the Macedonian's tame secretary somehow learned was Kalanus, told Alex to stop blocking the sun. Leach could have named the genus for him based on the shared posture, but he did not state that in his paper. David Damkaer[18] found little basis for this story, suggesting the name may have been a typesetting error, given Leach's difficult handwriting. Moreover, there were some other and similar oddities in Leach's naming practices. For example, Damkaer also points out that *Anilocra*, *Canolina*, *Cirolana*, *Conilana*, *Nelocira*, *Nerocila*, *Olencira*, and *Rocinela* for isopod genera are all near anagams of Carolina, though no woman of that name is recorded as associated with Mr. Leach.

Nerves run from the brain into the antennular tubes, with fibers branching to the bases of many of the setae that are known to be mechanosensory. What is sensed, deflection by relative water flow, is well illustrated by a diagram (Figure 2.12) of an adult *Acartia*, a genus of small, coastal, and estuarine copepods, from a paper by Jeanette Yen and colleagues.[19]

Pairs of the longer setae are set along the antennule at right angles, and they can be tipped by flow speed variations in the water. Tilting pulls a stretch receptor at the base that fires a fiber in the nerve that passes from the brain near the base out to the tip. Variations in the velocity of flow relative to that moving the body as whole can be distinguished according to source direction because the pairs of setae are at 90° angles, and because the antennules themselves extend along the third axis. As already discussed, the delay from stimulus to response can be measured in small aquaria using high speed video and various mechanical sources of flow pulses. Pulses can mimic an onrushing attacker (say an arrow worm) or the rapid suction with which fish try to capture small prey. According to Yen et al., a fish attempting to slurp in a copepod from 1.5 mm or closer will capture it; the copepods response latency is too long for escape. From a little more than 3 mm, the odds for the copepod are much better.

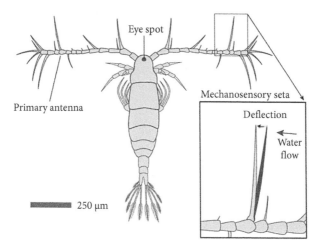

Figure 2.12 Drawing of a copepod showing the paired antennules bearing the mechanosensory array of setal hairs. Deflection of the hairs by flowing water alerts the copepod to a disturbance in its fluid environment. From Yen et al.,[19] Permission, *Integrative and Comparative Biology* (Oxford University Press).

I have done a lot of experiments with live copepods (*Calanus, Neocalanus, Metridia, Acartia*) that I collected in the field and sorted from bulk collections in beakers using glass pipettes. The pipette tips were cut off and heated to provide a smoothed opening around 3 mm. If you hold that a few mm, even about 5 mm, directly before a copepod's head and release the squeeze bulb, the copepod will shoot right into captivity. If you approach from the side or rear, they usually escape. The information about approaching accelerations cannot be equally good from all directions. Video estimates of escape latency and accelerations have mostly been made with the attack pulse or suction applied on a horizontal axis, so the simulated attacks are from the side. On the other hand, fish often attack from below. An experimenter lowers her pipette into her sorting dish from above.

Not all the setae on the antennules are flow sensors. Another type, termed *aesthetascs*, are usually stocking-like with very thin and porous outer skin. These, too, have nerve cells attached, and in their case the fibers extend way to the inside. Like those in the frontal organ and rostrum, these nerve terminals are ciliary dendrites, so these antennular aesthetascs extend olfactory capability out into the surrounding water. It turns out that the algae constituting the phytoplankton leak into their surface boundary layers some of the amino acids and other small odorants copepods are likely able to detect. Moreover, when any living thing is chewed by neighboring animals, small soluble molecules spew out. So their presence is a signpost for the presence of food or enemies. As suggested for dendrites in the frontal organ and rostrum, possibly the antennular aesthetascs can detect algal toxins and stop feeding. Female copepods release pheromones to attract males, and the male antennules of many pelagic copepods have more and larger aesthetascs than do those of the females. It is definitely proved that the antennular aesthetascs respond to those pheromones. I cover that in a chapter on mate-finding.

Slow Swimming: The Antenna

Professor Cannon, the artist of Figure 2.1, was trying to work out how the limbs around the mouth move water, so that food can be captured, particles like algal cells or protozoa. He learned that the antennae, the second limbs from the front, (Figure 2.1) with their bases articulated beside the anterior bump of the labrum, extend their two branches upward and downward beside the body (as drawn). They are rotated in a plane beside the body, the power stroke below the body, the return stroke above. When copepods, say 2.5 mm long, are free in the ocean this can drive steady forward swimming at speeds of about 10 cm s^{-1}. This is most readily observed in males running searches for the pheromone tracks laid down by females seeking to mate. In this searching mode of swimming, the antennae whirl in a blur beside the body. Fans of muscle fibers to drive that rotation, particularly robust in males, attach broadly to the inner top of the carapace and at a sequence of points around the edges of the exoskeletal casing of the basal segments of the limb. The muscles contract around that edge in a sequence that rapidly repeats. The whole limb rotates as the muscle fibers flex the soft arthrodial membranes connecting it to the body. This limb-movement mechanism is quite different from the personally familiar antagonistic muscle pairs of vertebrates with internal skeletons (e.g., the biceps and triceps of your upper arm).

Maxillule, Maxilla, Maxilliped

Behind the attachments of the mandibular palps sit the maxillules (Figure 2.13, right). Flattened out, as in the drawings, they look like random stacks of blobs. The homologies to primitive exopods and endopods are obscure, though some experts are confident of them. Most of those with long, flexible setae extend out to the sides and participate in generating feeding currents. The thick-based bristles of the maxillules, shown at the top left, extend toward the midline and through the almost parallel setae of the uniramous maxillae (Figure 2.13, middle) seated just behind. These fans of setae on the maxillae curve inward from the limbs,

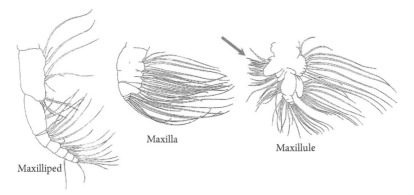

Maxilla

Maxillule

Maxilliped

Figure 2.13 Posterior mouth limbs of *Calanus finmarchicus*. From Anraku and Omori.[20] The red arrow points to the stout setae on a "basal endite" (or expansion) that can reach back to comb through the long setae of the maxilla, part of the particle feeding movements (Chapter 6). Permission, *Limnology and Oceanography* (Wiley).

pointing to and almost reaching into the posterior opening of the mouth between cone like lips on either side, the labial palps. They bear stiff, close-spaced setules that help hold food particles between the two fans directing them toward the mouth. The basal bristles of the maxillule rake forward through those fans, clearing them forward.

The flexibly jointed, tubular maxillipeds (Figure 2.13, left) come last in the tangled array of mouth limbs. They stroke back in generating feeding currents, and they can swing out and down to capture small prey, even in species that mostly feed on very small, single-celled algae or protozoa.

The whole layout of head limbs is shown in a scanning, confocal, laser micrograph by Jan Michaels and Stanislav Gorb[21] (Figure 2.14) of a small, primarily coastal species, *Temora longicornis*. Also prominent in this ventral view is the broad, articulated plate, the *labrum*, which covers the mouth. It is a feature that copepods have in common with all arthropods. The pattern is similar in many particle feeders. Predominantly predatory copepods have the same sequence of limbs, but they are modified for grabbing or stabbing. Both the maxillae and the maxillipeds have evolved into baskets of stiff sabre-like setae that impale relatively large animal prey.

Back to Jaws, Opal Teeth, and Dr. Sullivan

Particularly among calanoid species, many of the teeth along the gnathobase edge are the hard, siliceous crowns that Barb Sullivan rediscovered. They are,

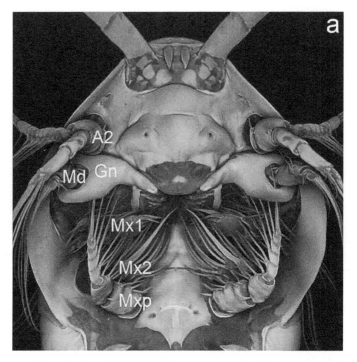

Figure 2.14 Ventral view of the mouthpart arrangement in *Temora longicornis*. Abbreviations: A2, antenna; L, labrum; Md, mandibular palp; Gn, mandibular gnathobase; Mx1, maxillule; Mx2, maxilla, Mxp, maxilliped. This is a confocal laser scanning micrograph of a specimen stained with fluorescing dyes. From Jan Michaels and Stanislav Gorb.[21] Compare this to the side view of *Calanus* in Figure 2.1. From *Beilstein Journal of Nanotechnology* (open access).

almost literally, opal teeth with a Mohs-scale hardness likely to be 5.5–6.5. That matches the hardness of the frustules (shells) of diatoms that are foods for many copepods. Fracturing tests of frustule hardness, like those by Ghatu Subhash and colleagues[21] using "nanoindentation" instruments, produce fragments (halves and pie-pieces) just like those found in copepod guts. Subhash et al. point out that the frustules are arrayed with lines of pores that guide the cracking. The teeth (Figure 2.10) sit on a thick and solid ridge constructed of the thickest chitin anywhere in the exoskeleton. That transfers the force for this rock crushing.

Japanese (e.g., Katsutoshi Itoh[23]) and Russian (e.g., Elena Arashkevich[24]) workers had long since considered the implications of different tooth shapes for feeding habits. Barbara Sullivan's SEM pictures showed similar distinctions between animals that are primarily particle feeders (Figure 2.10) and others that

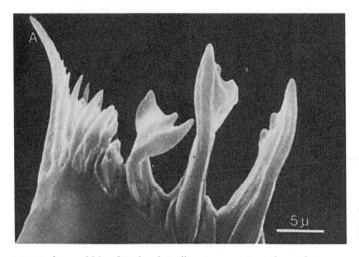

Figure 2.15 Left mandible of *Scolecithricella minor*, a primarily predatory copepod. SEM from Sullivan et al.[3] Permission, *Marine Biology* (Springer Verlag).

are carnivores. Teeth of the former species tend to feature dull, pyramidal points; those of the latter are long spines or blade-like (e.g., Figure 2.15).

Much recent work on copepod teeth has been done in Germany by Jan Michels and colleagues. They have found evidence[5, 21] that much of the crown volume is 2–4 nm (~3 *billionths* of a meter) nanocrystals, which they characterize as α-cristobalite doped with aluminum (1 part in 37). The crystals suggested to them that the teeth are not proper opal, but they also say there is much amorphous silica. It turns out that opal gems also are not entirely amorphous at the 2 nm scale. Michels's group has also shown[5] that the bezel-like sockets on the edges of the gnathobase are lined with the elastic protein *resilin*, which they believe provides some "give" when the teeth are compressed together against food.

Barbara Sullivan did not pursue studies of opal teeth; I did that. They became a fascination of mine. The obvious question in the 1970s was what process forms these mineral teeth, somehow getting them attached on the *outside* of the gnathobase. The partial answers (science is never fully finished) are reviewed in the chapter on molting and development.

Completing her dissertation in 1977, Barbara, by then divorced, moved to Harbor Branch Oceanographic Institute (HBOI) on the Indian River Lagoon in Florida. Experiences there included a ride in HBOI's Johnson Sea Link submersible. That was a double-ended oddity, with the pilot forward seated in a transparent sphere and scientists looking out through a small dome aft while prone—not comfortable. She got down 350 meters into near darkness, except for impressive arrays of bioluminescent lights all about. Other work included some plankton collecting in Indian River Lagoon. She then got an offer to move to the University of Miami (the one *in* Miami, Florida). She accepted and then worked on very large, enclosed, and somewhat controlled experimental water column enclosed ecosystems (CEEs) with a catalogue of famous marine planktologists. Among those were her advisor Michael Reeve, who was also interested in soft-bodied, predatory plankton, the copepod systematist George Grice, and the plankton ecologist Roger Harris.

The group worked with CEEs contained in polyethylene sacs suspended in bays. I suspect, though she denies it, that Barbara was particularly important in the last experiment, termed Food-I. Its CEEs were deployed in Saanich Inlet on Vancouver Island, Canada.[26] Three bags were involved: they were 9.5 m diameter, 23.5 m vertical length (1,335 m³) suspended from doughnut-shaped floats. The bottoms were closed with polyethylene funnels to collect sediment in a weighted central bucket. The bags were wadded into a circle, sunk to depths equal to their wall lengths, and then the upper edges were pulled upward with lines attached to heavy counter weights passing over pulleys on the surface float (Figure 2.16). The result was an enclosed plankton community.

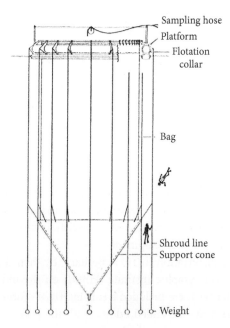

Figure 2.16 Sketch of a contained environmental ecosystem bag as deployed in Saanich Inlet. From Grice et al.[25] Permission, *Journal of the Marine Biological Association, U.K.* (Cambridge University Press).

The experimental goal was to compare a community supported by small flagellated phytoplankton to one supported by diatoms. Nutrients were added to both, but one received no additional silicate and received a bulk culture of diatoms to take up the initial silicate. Those grew and then sank to the bucket for removal by divers. Without silicate for diatom frustules, it became the flagellate bag. The other was left to grow the diatoms already in the inlet. The third bag was for mixing and other physical studies. The bags, sampled from July 9 until October 17, started out with abundant copepods, suitable food for small fish, which might have allowed a good food chain comparison as intended. However, and maybe fortunately for Barbara, two species of ctenophores were present and developed into a sort of bloom. They are soft-bodied, spherical planktonic predators that capture prey, typically copepods, on long tentacles. Eating the copepods promoted ctenophore growth, but pretty soon they had eaten most of them. The food chains lost their middle links and dwindled away. Barbara wrote the paper about that with Michael Reeve.[26] It was a sort of residential experiment: work much of the day, every day, all summer long. She credits Michael Reeve for providing much valued education.

A somewhat similar program, the Marine Ecosystems Research Laboratory (MERL, funded by the Environmental Protection Agency) had been in progress in the 1970s at the Graduate School of Oceanography of the University of Rhode Island (GSO-URI). It had experimental tanks on a wide platform next to Narragansett Bay that were filled with water pumped up from that wide-mouthed coastal estuary. The tanks could have sediment on their floors. That mattered because water in bays interacts intensely with the bottom. Barbara got a temporary position there at the time when multiple postdoctoral jobs became frequent in all of science. Barbara liked the place and people, wrote a proposal for more work, and was hired. She has been working there ever since, but in recent years is also teaching at Providence College. There she emphasizes courses on climate change. She has made valuable contributions to plankton ecology: from tank experiments, from the bay offshore from GSO, from over Georges Bank, and in the Gulf of Maine. She continues to work on the seasonal trade-offs of *Acartia tonsa* and *Acartia hudsonica* populations in Narragansett Bay and the roles in that of predation by ctenophores. For that work she maintains a field-population time series by sampling those small neritic copepods.

Not long after moving to URI, Barbara married a physical oceanographer, Randy Watts, and they raised daughters together. She is birth mother to one of them and speaks eloquently about the complexity and stress of combing child-rearing duties with sustaining a scientific career. Interested in this recurring issue for professional women and dual-career families, she conducted a survey of related institutional policies at universities prominent in aquatic sciences. It covered leave granted for childcare, consideration of dual and split appointments, and modified evaluation procedures for professional staff with parallel family duties. Her report [27] was published in 1995.

References

1. Cannon, H. Graham (1928) On the feeding mechanism of the copepods *Calanus finmarchicus* and *Diaptomus gracilis*. *British Journal of Experimental Biology* **6**: 131–144.
2. Green, J. D., & R. J. Shiel (1999) Mouthpart morphology of three calanoid copepods from Australian temporary pools: evidence for carnivory. *New Zealand Journal of Marine and Freshwater Research* **33**: 385–398.
3. Sullivan, B. K., C. B. Miller, A. Soeldner, & W. T. Peterson (1975) A scanning electron microscope study of the mandibular morphology of boreal copepods. *Marine Biology* **30**: 175–182.
4. Marlow, C. J., & C. B. Miller (1975) Patterns of vertical distribution and migration of zooplankton at Ocean Station "P." *Limnology & Oceanography* **20**: 824–844.
5. Michels, Jan, J. Vogt, P. Simond, & S. N. Gorb (2015A) New insights into the complex architecture of siliceous copepod teeth. *Zoology* **118**: 141–146.
6. Beklemishev, K. V. (1954) The discovery of silicious formations in the epidermis of lower crustacea. *Doklady Akademia Nauk SSSR* **97**: 543–545. [Original in Russian.]
7. Sullivan, B. K. (1980) In situ feeding behavior of *Sagitta elegans* and *Eukrohnia hamata* (Chaetognatha) in relation to the vertical distribution and abundance of prey at Ocean Station "P." *Limnology and Oceanography* **25**: 317–326.
8. Elofsson, R. (1971) The ultrastructure of a chemoreceptor organ in the head of co-pepod crustaceans. *Acta Zoologica* **52**: 299–315.
9. Derby, C. D., P. Steullet, A. J. Horner, & H. S. Cate (2001) The sensory basis of feeding behaviour in the Caribbean spiny lobster, *Panulirus argus*. *Marine & Freshwater Research* **52**: 1339–1350.
10. Teegarden, G. J. (1999) Copepod grazing selection and particle discrimination on the basis of PSP toxin content. *Marine Ecology Progress Series* **181**: 163–176.
11. Hong, J., S. Talapatra, P. A. Tester, R. J. Waggett, & A. R. Place (2012) Algal toxins alter copepod feeding behavior. *PLOS One* http://dx.doi.org/10.1371/journal.pone.0036845.
12. Cohen, J. H., & R. B. Forward Jr. (2005) Photobehavior as an inducible defense in the marine copepod *Calanopia Americana*. *Limnology and Oceanography* **50**: 1269–1277.
13. Elofsson, R. (2006) The frontal eyes of crustaceans. *Arthropod Structure & Development* **35**: 275–291.
14. Frost, B. W. (1974) *Calanus marshallae*, a new species of Calanoid copepod closely allied to the sibling species *C. finmarchicus* and *C. glacialis*. *Marine Biology* **26**: 77–99.
15. Park, T. S. (1966) The biology of a calanoid copepod, *Epilabidocra amphitrites* McMurrich. *La Cellule* **66**: 129–251.
16. Leach, W. E. (1819) Entomostracés. *Entomostraca*. (Crust). *Dictionnarire des Sciences Naturelles* **14**: 524–43.
17. Marshall, S. M., & A. P. Orr (1955) *The Biology of a Marine Copepod*, Calanus finmarchicus, *Gunnerus*. Oliver & Boyd, Edinburgh.
18. Damkaer, D. (2002) *The Copepodologist's Cabinet, A Biographical and Bibliographical History*, Vol. 1. American Philosophical Society, Philadelphia.

19. Yen, Jeanette, D. W. Murphy, L. Fan, & D. R. Webster (2015). Sensory-motor systems of copepods involved in their escape from suction feeding. *Integrative and Comparative Biology* 55: 121–133.

20. Anraku, M., & M. Omori (1963) Preliminary survey of the relationship between the feeding habit and the structure of the mouth-parts of marine copepods. *Limnology and Oceanography* 8: 116–126.

21. Michels, Jan, & S. N. Gorb (2015B) Mandibular gnathobases of marine planktonic copepods, feeding tools with complex micro- and nanoscale composite architectures. *Beilstein Journal of Nanotechnology* 6: 674–685.

22. Subhash, Ghatu, S. Yao, B. Bellinger, & M. R. Gretz (2005) Investigation of mechanical properties of diatom frustules using nanoindentation. *Journal of Nanoscience and Nanotechnology* 5: 50–56.

23. Itoh, K. (1970) A consideration on feeding habits of planktonic copepods in relation to the structure of their oral parts. *Bulletin of the Plankton Society of Japan* 17: 1–10.

24. Arashkevich, Ye. G. (1969) The food and feeding of copepods in the north-western Pacific. *Oceanology* 9: 695–709.

25. Grice, G. D., R. P. Harris, M. R. Reeve, J. F. Heinbokel, & C. O. Davis (1980) Large-scale enclosed water-column ecosystems, an overview of FOODWEB-I, the final CEPEX experiment. *Journal of the Marine Biological Association, U.K.* 60: 401–414.

26. Sullivan, B. K., & M. R. Reeve (1982) Comparison of the predatory impact of ctenophores by two independent techniques. *Marine Biology* 68: 61–65.

27. Sullivan, B. K. (1995) Balancing work and family: Report of survey on institutional policies. *Limnology and Oceanography Bulletin* 4(1): 3–6.

3

Let's Go Inside

Who Has Looked Inside? Esther Lowe and Tai Soo Park

There may be only three comprehensive studies of the anatomy of free-living copepods. One, by the British worker Esther Lowe,[1] is based on specimens of *Calanus finmarchicus*. She illustrated her 1935 paper with beautiful pen-and-ink drawings of organs and the spatial relations of organs. The byline of her remarkable paper reads, "Esther Lowe, M.Sc., Lately Honorary Lecturer in Marine Biology, Department of Oceanography, University of Liverpool." At the end of the introduction she says that the "greater part of the investigation was carried out in the Zoology Department of the University of Manchester," and she thanks Professor H. Graham Cannon, whose side-on drawing of *Calanus* is Figure 2.1. Dr. James Peters, Archivist at the University of Manchester Library, found her staff card for me. It tells us that she took her M.Sc. at Liverpool in 1925, and her thesis is in deep storage (at the Brunswick Library Store) in the university library there. Its title is "Variation in the Histological Condition of the Thyroid Glands of Sheep and Cats with Regard to Season, Sex, Age and Locality." While the subject animals are on a wholly different branch of the evolutionary tree, training for that work would have provided exactly the skills implied by her work on copepod anatomy.

Lowe was employed as an assistant lecturer in zoology at the University of Sheffield 1925–1926, and she held the same position at University College, Bangor from 1926 to 1929. Then she moved on to University of Manchester as an assistant lecturer in zoology between September 29, 1929 and September 29, 1933, when her appointment "expired" (the card is precise there). It must have been during those four years that the work on *Calanus* anatomy was done. Finally, the card in the Manchester archive states that Lowe was working at the Convent of Mercy High School in Liverpool in 1958 under her married name of Drane, or possibly Drave. The card was probably annotated in handwriting. I did not see it; it was read for me by Dr. Peters. So we know that Esther Lowe survived the Second World War, married at some point, and was still teaching biology three decades after receiving her advanced degree.

That's all that British friends have found for me, but likely prowling through Liverpool newspaper obituaries or similar detective work could establish more. Perhaps an adequate immortality can lack copious personal details. I picture her

Oar Feet and Opal Teeth. Charles B. Miller, Oxford University Press. © Oxford University Press 2023.
DOI: 10.1093/oso/9780197637326.003.0003

in a lab at Manchester staring for long hours down the almost vertical tube of a monocular microscope at Mallory's trichrome-stained sections of *Calanus*. Her ability to translate the circular images of 5 μm-thick slices into three-dimensional drawings of organs was remarkable. I wonder if the unfortunate disrespect women often suffered in Esther's day (and often enough still) could have played a role in her leaving research. But there are many other possibilities. Maybe she accompanied Mr. Drane to Liverpool, where she found great satisfaction in teaching.

A 1966 monograph by Korean American Tai Soo Park[2] reported a thorough examination of *Epilabidocera longipedata* (Sato, 1913),[*] a species with some uncommon elaborations, particularly eyes more developed than typical in copepods (Figure 2.11). It is reasonably large (total length, 3.2–4.0 mm), which helped with anatomizing, and it is common in parts of Puget Sound. The study was a dissertation project for Park, who is now over 90. He met with me in early April 2018 in his newly purchased house outside Olympia, Washington. In addition to telling me about his career path, he has published a memoir, describing that in a commemorative issue of *The Sea: Journal of the Korean Society of Oceanography*, translated by the editors into Korean. Fortunately, he reached out to his computer, put in my thumb drive, and downloaded the original English language version. When we met, Tai Soo had just moved to Olympia with his wife to be near his sister's daughter, a concession to old age. He was born in 1929 to a Korean farm family, and despite the nearness of the sea to both sides of the country, he did not see it until the age of thirteen. He studied mechanics at a technical high school, and was then influenced by neighbors who were attending Busan Fisheries College to join them there. He lost a full year of study to the displacements of the Korean War. Resuming study at Busan in 1951, he first saw plankton samples at the nearby National Fisheries Experimental Station, noticing the diversity of the copepods in particular. At the time there were no facilities (either keys or experts) for learning to identify them, so he studied diatoms from net samples. After graduating he joined the faculty for a time.

In 1958 a fellowship from a US agency promoting international cooperation provided Tai Soo an opportunity to study for a year with the Pacific Ocean Fisheries Investigations of the Bureau of Commercial Fisheries, located at the University of Hawaii in Honolulu. There he met Paul Illg, a professor of zoology on sabbatical from the University of Washington (UW). Illg, a prominent student of parasitic forms, taught Park how to examine copepods and dissect them to reveal identifying features. This led later to Park's first systematic publication

---

[*] Sato described this copepod as *Pontella longipedata*, and the species name is recognized to have precedence, so it is now known as *Epilabidocera longipedata*.[3] The parentheses show that he had placed it in some other genus, *Pontella* of course. Before that recognition, Park, I, and many others identified specimens as *E. amphitrites*.

in 1968, a study of the calanoid fauna of the north central Pacific.[4] The relationship with Illg continued, and Tai Soo was invited to study for a PhD at Seattle. Much of the UW course work repeated earlier studies at Busan, but there must have been benefits from studying the terminology of his profession in English. As already stated, his thesis was the study of *Epilabidocera longipedata*. Also studying copepod anatomy under Paul Illg were Wolfgang H. Fahrenbach[5] and Patricia Dudley,[6] both of whom started a little earlier. Fahrenbach anatomized a highly modified harpacticoid, which burrows into algal fronds with antennae modified as cutting tools like dental burrs. Patricia Dudley studied larval development in the parasitic family Notodelphyidae. All these studies, including Park's, were partly done at UW's Friday Harbor Marine Laboratory. Tai Soo and his wife moved to the United States for these graduate studies that occupied five years, advancing him to his middle thirties. He has remained here since, though with visits home to Korea. He became a US citizen around 1970. More of his career path will be covered in Chapter 5, about finding and describing new species.

There are many studies of one copepod organ or another, but for the large and most commonly studied calanoids, it is Lowe's and Park's work that lets us see how everything fits together. Perhaps it is the excellence of these studies that has discouraged more such work. Just as likely, it is the fiendish difficulty of getting reasonably continuous sets of serial sections, viewing and imaging them in detail, and finally reassembling them conceptually into a three-dimensional whole. Tai Soo emphasized that in one way his task was simpler than Lowe's. She sectioned copepods infused with wax, while he could use the epoxy plastic developed for electron microscopy. She had to dissolve the wax to stain her sections, which may have released the tissue pieces to move about inside the blood spaces (haemocoel) of the copepod. Maybe that explains why some of her drawings of organs seem to float free on the page. It's clear she did know where the organs are located. Park could stain without dissolving the epon, so muscle, gut, and gonads retained their spatial relations. On the other hand, Lowe could get serial sections in a ribbon, while Park's sections floated off his glass microtome knives one-by-one, and he agrees he had a devilish time getting their sequences right. Assembling three-dimensional understanding of microanatomy from slices has gotten somewhat easier by using computer assembly of slice images. Programs similar to MRI and CAT scan outputs have replaced constructions of wax, clay, and cardboard models. Maybe somebody is doing that for copepod anatomy.

Blood and Guts

Geoff Boxshall[7] provided a longer summary in 1992 than the one that follows here. Copepods do have blood, but it is not red. Blood with respiratory pigment, like our red hemoglobin or a crab's blue hemocyanin, would be opaque, ruining the near transparency that makes copepods more difficult for visual predators (think fish) to see and eat. I can find no references to blood cells of any sort, though there must be some amoebocytes moving about among the tissues. They would be part of a copepod's immune system, and at least a rudimentary one must be present. Amoebocytes have been found in brine shrimp.[8]

The blood is circulated through the body by a small heart located up against the mid-dorsal exoskeleton of the prosome. Its pulsing is obvious in a live co-pepod viewed with a microscope. According to Lowe, the heart of *Calanus finmarchicus* (Figure 3.1) is a simple, muscular sack that squeezes blood into an aorta through a very simple valve in the ventral side of the pointed anterior end. An electron microscope study[9] of the heart of *Anomalocera ornata* showed the muscle is constructed of rather unexceptional, striated fibers.

The thin-walled aorta directs the blood forward, where it turns against the forehead and drops through passages into a blood sinus (Figure 3.2) above the nerve cord running along the ventral body wall. Lowe called this channel the "supraneural sinus." As you can see in the figure, she gave formal anatomical names to everything. This sinus opens at the back and blood flows into the back of the prosome, then up and around into a thin membranous sac (pericardium) that directs it again toward the heart.

The esophagus extends dorsally from the mouth into a long bag with an inner lining of secretory and absorptive cells, the stomach. That extends forward over the brain and back to the middle of the thorax. It narrows there to an intestine that runs into the urosome and empties between the caudal furcae from the anal segment. A fifth copepodite of *Calanus finmarchicus* whose gut is stuffed with green algae is pictured in Figure 3.3. Using "whose" exposes my tendency, shared with others who see this world through microscopes. We tend to personify these small critters. We come to think of a specimen like this as about 15 cm long, when really it is about 2.5 mm. When someone asks, "How big is it?" we hold up our hands a hand-length apart.

The gut lies below a storage gland, the oil sac, which in the pictured spec-imen is starting to fill with the liquid wax, which is the nutrient storage com-pound of many marine copepods. The red dot behind the antennae is pigment surrounding the mouth area. Many marine crustaceans have red areas around the mouth. Those are believed to shield from view the bioluminescent algae or other bioluminescent prey that light up when touched (Chapter 9). In a world filled with predator eyes, it is best not to be holding an uncovered candle. As I show various anatomical sections, try checking this figure to see where they cross through the body. All of this anatomical diagramming leads to an experi-ence akin to that of looking at *Gray's Anatomy* with no human body before you.

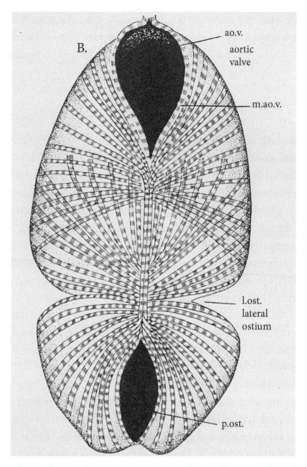

Figure 3.1 *Calanus finmarchicus.* Heart in ventral view. Drawing from Lowe.[1] The striped bands are striated muscles. The top of the picture is anterior. Muscles closing the anterior aortic valve and posterior *ostium* ("mouth") are designated by paler striations. Permission, Royal Society of Edinburgh.

The diagrams and the three-dimensional wet and messy body are at best vaguely related.

Below the gut and above the central nervous system is a webbing of tendons that holds in the sides of the thin and flexible exoskeleton (Figure 3.2). It is termed the *endoskeleton* (inner skeleton). It also provides insertion points in the center of the body for some of the limb muscles (like the mandibular muscles, m and md in the figure). The chitinous, double-walled *apodemes* are struts from the exoskeleton extending internally to provide strong attachment points for endoskeleton fibers and for muscles reaching both above and below

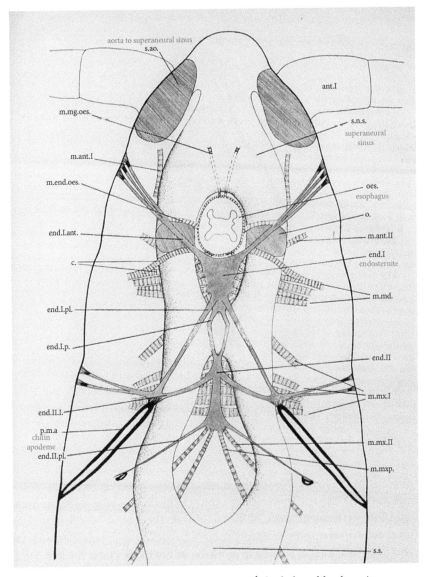

Figure 3.2 *Calanus finmarchicus*. Lowe's Figure 1,[1] shaded tan like the aging pages of her paper, a diagram of the anterior structure below the gut and above the nerve cord. A few of her abbreviated labels are explained in red. Some of the internal strapping of the copepod "endoskeleton" is shown. Lowe's label **s.s.** (bottom right) indicates the superaneural blood sinus just anterior to the heart. Remember, the body is less than a millimeter wide, so these are tiny details. Permission, Royal Society of Edinburgh.

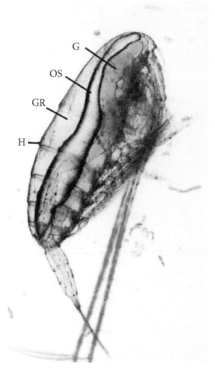

G

OS

GR

H

Figure 3.3 *Calanus glacialis* fifth copepodite, showing the position of the gut thanks to it being filled with algae. Indicated organs: G, gut; OS, oil sac; GR, gonad rudiment; H, heart. Photo by Anette Wold, Norwegian Polar Institute, with her permission.

the plane of the figure. Notice the small muscles extending from the central triangle of the endoskeleton (endosternite) to the front and possibly from the body floor to the outer lining of the esophagus. These must be able to pull the throat tube open, maybe suddenly enough to suck in food particles from the mouth. We will look at feeding in more detail, but likely that is the last step in food capture.

Excretion

Just internal to the maxillae, and termed "maxillary glands," copepods have paired sacs of thin membrane stretching across the ventral haemocoel and opening between some "guard cells" (V in Figure 3.4). Those are in the same location as muscular sphincters in some other crustacea. They allow some flow

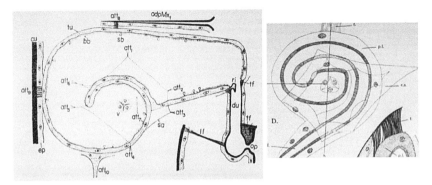

Figure 3.4 Left: Maxillary gland of *Epilabidocera longipedata* drawn by Park.[2] You are looking toward the head and the maxilla extends just below. A membrane (**sa**) shaped like a cone is suspended from attachments (**att$_1$** to **att$_6$**) and intercepts blood flow, directing some of it to the valve (**v**) with three "guard cells," which pass it into the lumen of the gland. Villi (**bb**, just visible) on the gland surface resorb water and some molecules; the rest passes to the duct (**du**) near the midline and out at an opening (**op**) on the base of the maxilla. The gland is suspended above from an internal extension of the exoskeleton (an apodeme, **apdMx1**) and attached at the side (**att$_9$**) to the body wall. The suspending fibers are contractile "tono fibrils" (**tf**). Permission, Éditions Nauwelaerts. Right: Lowe's drawing[1] of the more rounded excretory coil in *Calanus*. Permission, Royal Society of Edinburgh.

into coiled ducts lined with a dense layer of absorptive villi. The coils end in thin membranous tubes that exit the body on the medial surfaces of the basal segments of the maxillae. The function of these structures is not particularly well studied, even in lobsters, which also have them, but they are kidneys of a sort. The urine carries nitrogenous waste as ammonia, based on analysis of water collected from around copepods held in small containers.

Some Muscles

Looking at a side-view photograph of a living copepod (a macrophotograph, taken with a lot of magnification leading the camera: Figure 1.2, Figure 3.3), you do not see muscles. They are transparent. If you have ever picked the meat off a boiled chicken to make soup, you know how complex muscle systems can be: big strips for big power, smaller strips to move limbs back to be pulled again by the big ones, tiny strips for postural adjustments. Copepod muscles are simpler, yet remarkably complex for such small beings. Lowe did not consider them. There are three main sets: long strips to flex the back of the body (sectioned in

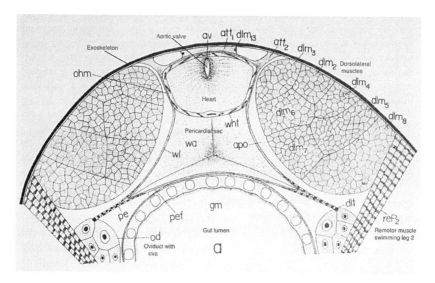

Figure 3.5 Diagrammatic cross-section through the heart and dorsal-longitudinal muscles of *E. longipedatai* drawn by Tai Soo Park.[2] A few of Parks's codes are written out in red. Permission, Éditions Nauwelaerts.

Figure 3.5), strong fans to pull the thoracic legs back then forward again, and swirls of long, narrow strips to swing the antennules and to rotate the antennae and other mouthparts. There are also small muscles between most of the limb segments allowing elaborate bending back and forth. The antennules, for example, can roll into circles. I have no idea what use that has.

Park drew the longitudinal muscles of the body in sideview (Figure 3.6). Separate strips attach to apodemes at the edges of the body segments and run back to surfaces on the insides of segments farther back. Our bones also have side projections, apophyses, for attachment of long muscles that pull on the sides of the next bones along our limbs. In copepods, these longitudinal thoracic muscles allow the body to flex, primarily up and down, swishing the tail fan. The dorsal muscles are more numerous and larger than the ventral ones, suggesting that the down-swipe of the tail is its power stroke. The figure also shows the extensive lacing together of the body by tendons.

Large fans of muscles insert on the sides of the thoracic segments (labeled re in Figure 3.7) and then extend to apodemes inside the posterior edges of the basal segments of the thoracic legs, pulling the whole leg powerfully back. Smaller muscles, also originating on the segment walls, attach on the anterior edges of the leg segments and return the legs forward. They also control the feathering of these oar feet.

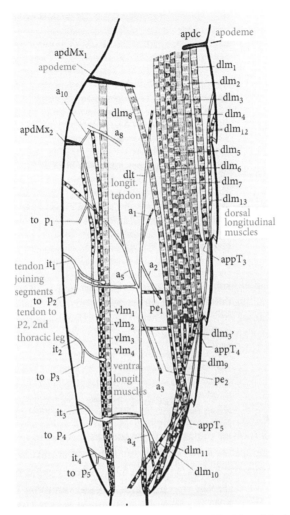

Figure 3.6 Diagrammatic sagittal section drawn by Park[2] of the of the longitudinal thoracic muscles. These are bundled together and form thick masses at the level of the heart, as shown in cross section in Figure 3.5. Some of Park's abbreviations are decoded in red. Permission, Éditions Nauwelaerts.

The basal segments of the antennules and mouth limbs all attach at their edges to circles of thin muscles originating on the surface of the anterior carapace (Figure 3.8). As those muscles pull in sequence, the arthrodial membranes flex and the whole limbs swing from side to side, or can rotate in sweeping circles, particularly the antennae.

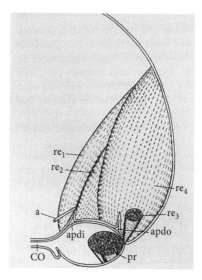

Figure 3.7 Section showing the attachments of thoracic leg muscles: large multislip remotor (**re**: for pulling the leg back during escape jumps) and smaller promotor (**pr** - for pulling the leg forward) are shown. The depiction is looking toward the tail. The "arm" (**a**) of an endosternite and the leg-to-leg coupler (**co**) are also labeled. From Park.[2] Permission, Éditions Nauwelaerts.

Nerves and Senses

In the early evolution of multicellular organisms (and even before), selection of characters produced ways to sense many habitat features carrying information: direct pressure from objects (touch); motion of surrounding fluids (water, later air); sound waves; levels, direction, and spectral variations of light; specific kinds of molecules as tastes or odors; direction of gravity relative to the body; positions of body sections and limbs relative to the main body axis and to each other; temperature and internal body-fluid pressure; concentrations of oxygen, carbon dioxide, electrolytes, and nutrients; and damage to tissues (pain). Moreover, many of these senses are actually systems of multiple sensor types.

The copepod repertoire of senses does not include all of those just listed, but many of them. Sensations are simplest to study in people: "Tell me, Jack, does that hurt?" "So, Jack, what shape is the blue dot in the middle of the red field?" Jack can answer these questions. We can ask copepods, but they say nothing. We can examine likely organs and identify their likely functions from those of systems in other animals. I discussed eyes evaluated that way in Chapter 2. There are also informative experiments, like those already discussed, determining what stimuli induce copepod escape jumps. The intense frenzy of activity in a plankton

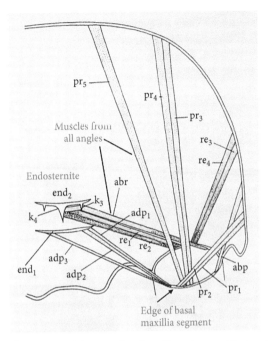

Figure 3.8 Park's[2] diagram of the muscles that allow the maxilla to rotate. Some attach to the top and sides of the carapace, some to the endosternites shown in Figure 3.2. The flexibility of the arthrodial membrane between the ventral exoskeleton and the basal maxilla segment is the key to this rotatory capability. Permission, Éditions Nauwelaerts.

sample when formaldehyde preservative is added suggests that copepods (and arrow worms and krill) feel pain. It must closely resemble, at least as an experience, the pain I felt in the field when handed the formalin squirt bottle, not the one containing alcohol, to sterilize some cuts. I sometimes hope when I preserve plankton samples that copepods and krill have no souls; otherwise, I may be assigned to their lives in my next incarnation.

In any case, appropriate (or even frenzied) motor responses require that sensory information be gathered from scattered sensors, interpreted as a situation so that instructions can be sent to muscles and glands. Even for animals evolved only to an intermediate level of sophistication, like copepods, this integration occurs in a central nervous system and at least partly in a brain. Copepod brains sit just in front of the esophagus, with bilateral nerves emerging to the frontal organ, rostrum, antennules, and eyes (as shown in Chapter 2). As in insects and annelids, there are fat nerve tracts running back from the brain that divide around the esophagus and then join again behind it (Figure 3.9). A double

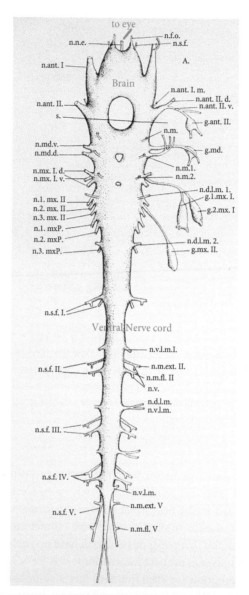

Figure 3.9 From Lowe[1]: Layout of the central nervous system in *Calanus*. It is similar to those of many invertebrate animals (insects, annelids). A few of the small labels on nerves: **n.s.f.** – superior frontal nerve, **n. ant. I** – antennular nerve, **n. ant. II** – antennal nerve, **md.** – mandible, **mx.** – maxillule or maxilla, **msp.** – maxilliped, **s.f.** – swimming foot. Permission, Royal Society of Edinburgh.

central nerve cord then runs all along the floor of the body, close to the sternites (ventral exoskeletal plates) all the way back to the fifth thoracic segment. At each segment there are sizable ganglia (clusters of nerve bodies and branching nerves) on each side of this double cord, ganglia are associated with each leg. Probably some responses of the mouthparts that have large ganglia operate almost independently. Impulses from sensors along the legs stimulate motor reflexes that originate from their ganglia. The paired nerve cords include giant axons that carry very fast impulses to the muscle complexes driving the escape responses of the thoracic legs, the oar feet.

The giant axons of squids are very large indeed, and many aspects of basic nerve signal transfer along axons, from sensors to brain and brain to muscles, were initially studied using them. Some similar studies are possible with insects large enough for implanting tiny electrodes. There is every reason to suppose copepod giant axons work in the same way, which is extremely fast. Note, for example, the stimulus to reaction intervals discussed in Chapter 1. An electrical potential is maintained by channel molecules in the cell membrane. They use energy to move potassium ions into the axon and sodium ions out, both against their concentration gradients. A stimulus at the membrane, say stretching by a setal motion receptor, opens the channels locally, and the ions rush though, down the gradient, creating a small electric impulse that opens adjacent channels. Those do the same all along the axon, and from neuron to neuron, until muscle cells are stimulated to contract, a sphincter to open, or a gland to release a secretion. If the sensory data are good, the specific response will, sometimes at least, be appropriate to the situation. Lowe's[1] diagram of the giant fiber system (Figure 3.10) shows the left branch originating from sensors innervating the front of the right side of the brain, crossing over in a chiasma just ahead of the esophagus, then extending back to the limbs. Strong interaction with impulses from the right antennule is also suggested. The right branch is a mirror image of the left.

Many, maybe most, multicellular organisms have similar sensors for the direction of gravity. In crustaceans these are termed *statocysts*. The most obvious statocysts are in the tail fans of the shrimp-like mysids. In those a spherical calcareous ball rests in a tissue shell at the ends of nerve fibrils. At rest, the direction of gravity is indicated by which fibrils are compressed by the ball. In motion the dominant force may no longer be Earth's gravity, but it can swing to any direction as the tail swings and the animal turns. Copepods have statocysts in the anterior part of their brains, first described by Carl Claus in 1863, who examined the brains of *Eucalanus*. While they probably are functionally statocysts, in copepods

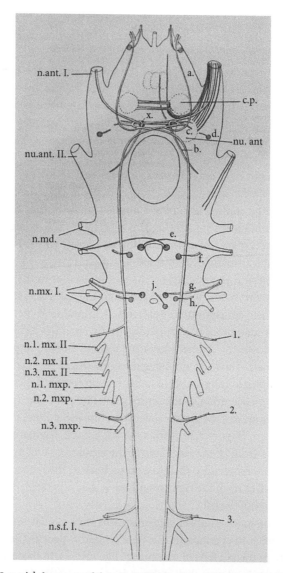

Figure 3.10 Lowe's[1] diagram of the giant fibers of the *Calanus* central nervous system. The contributing fibers appear to include the superior frontal nerve (n.s.f in Figure 3.9) and an antennular nerve (n.ant.I). Lowe gives a thorough description, having traced most of the fiber connections. Permission, Royal Society of Edinburgh.

they are called "Claus's organs." Lowe[1] shows their locations (Figure 3.11) in the brain of *Calanus*, anterior to the eye center and the ganglion associated with the frontal organ.

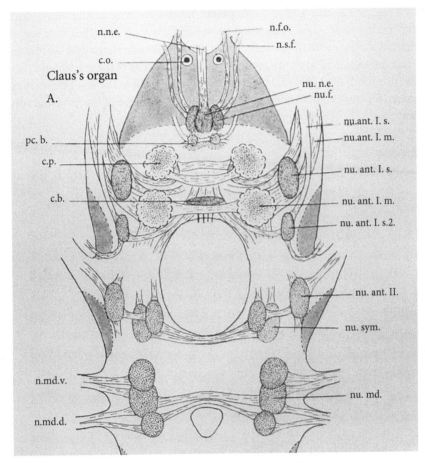

Figure 3.11 Lowe's[1] diagram locating Claus's organs (**c.o.**) from her Figure 11. She examined these in "oil-immersion" detail: "They consist of a pair of small cavities hollowed out in the neuron mass at the extreme anterior end of the brain, each cavity containing a single granule of concretion of about the same size as the nuclei of the surrounding neurons. Fine nerve fibrils can be seen in the walls of the cavity and the nerve tract to the frontal organ passes close to it ventro-laterally, but a direct connection has not been seen." The layout of the many ganglia (Lowe termed them "nuclei," e.g., **nu. ant. II**) of the anterior central nervous system are also diagrammed. From Lowe.[1] Permission, Royal Society of Edinburgh.

Reproductive Systems

In different species, and even different stocks, the gonads and reproductive ducts begin developing in different copepodite stages, but often are enlarging in the third or fourth copepodite. More discussion of development is offered in Chapter 12. Full gonad development follows molting to the adult, sixth copepodite stage. Lowe drew good figures of *Calanus* gonads, but they sort of float on the page with no reference to location in the body. Park's figures are more diagrammatic, and his body outline shows the locations of the gonads and their ducting. Both the ovary and the testis are above the gut in the first thoracic segment (Figure 3.12). Oocytes push out of the anterior end of the ovary into left

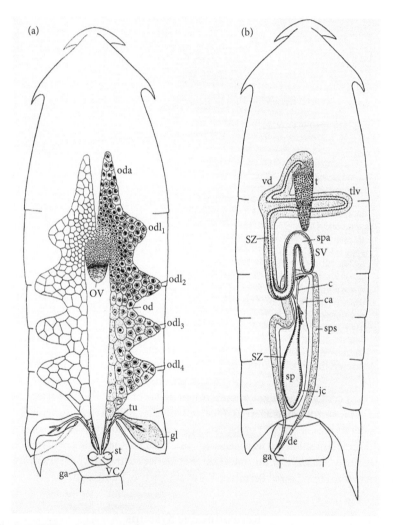

Figure 3.12 Reproductive systems of *Epilabidocera longipedata*, a female, b male, as drawn by Park. Some of the labels from a: **ov** -ovary, od -oviduct, **odl** -oviduct lateral diverticulum, st -spermatotheca, **ga** -genital aperture, **gl** -gland, **t** -tubule. Labels from b: **t** -testis; **vd** -vas deferens; **sv** -seminal vesicle; **sp** -spermatophore; **sps** -spermatophore sac; **ca** -coupling apparatus. From Park.[2] Permission, Éditions Nauwelaerts.

and right oviducts, which extend forward to above the esophagus and then turn down and extend to the genital segment in the anterior urosome. As they move the oocytes enlarge, filling with cytoplasm and yolk, and they bulge down into side pockets adjacent to the articulations of the body segments.

The sperm, smaller of course, are also pushed forward by pressure from spermatocyte division behind them. The vas deferens is a single tube, not double, and it turns left, then takes different turnings in different species delivering sperm to a seminal vesicle, where they are packaged into spermatophores. Those are thin-walled bags with a tubular duct, apparently one bag at time. Those are passed for further processing to a large spermatphore sac, which secretes an outer coating and, in *Epilabidocera* at least, an elaborate, gelled coupler to fasten the spermatophores to the female genital segment. Park examined all this processing in extraordinarily fine detail. In talking to me, he recalled with considerable pride having established that the coupler would seal over the connection between the spermatophore and the female genital aperture. He had also discovered paired glands in the "wings," at the back of the female prosome, with ducts to the space above the flap closing the genital aperture. His theory is that the secretion from those glands dissolves a path through the coupler for the sperm to move from the spermatophores into the small storage chambers called "spermatothecae."

Mate-finding and mating in the vastness of mostly empty water-space will be covered in a separate chapter. But briefly, once a spermatophore is attached, some specialized sperm (Quellenspermien) swell up and push the "regular" sperm through the spermatophore's duct and into the female's spermatothecae. After a mating or two, the female has a stock of sperm in those small sacs that fertilize her eggs as they pass from the oviducts through a narrow genital atrium and out under a flap. There are copepod families in which the females do not have spermatothecae and require several matings to produce more than a few egg clutches.

That covers the basics. It may be more than you imagined you ever wanted to know. But if you want to know more, read Lowe, read Park, read Boxshall.[7]

References

1. Lowe, Esther (1935) On the anatomy of a marine copepod, *Calanus finmarchicus* (Gunnerus). *Transactions of the Royal Society of Edinburgh* 58: 561–603.
2. Park, Tai Soo (1966) The biology of a calanoid copepod, *Epilabidocra amphitrites* McMurrich. *La Cellule* 66: 129–251.
3. Sato, T. (1913) Pelagic Copepoda, 1. *Reports Fisheries Investigations Hokkaido Fisheries Experimental Station* 1: 1–79. [In Japanese.]
4. Park, T. S. (1968) Calanoid copepods from the central North Pacific Ocean. *Fishery Bulletin* 66(3): 527–572.

5. Fahrenbach, Wolf H. (1962) The biology of a harpacticoid copepod. *La Cellule* 62 (3): 301–376.
6. Dudley, Patricia L. (1957) The development of the Notodelphyid copepods and the application of larval characteristics to the systematics of some species from the Northeastern Pacific. PhD dissertation, University of Washington.
7. Boxshall, Geoff (1992) Copepoda. In *Microscopic Anatomy of Invertebrates*, Vol. 9, *Crustacea*, ed. F. W. Harrison & A. G. Humes, 347–384. Wiley, New York.
8. Lochead, J. H., & M. S. Lochead (1941) Studies of the blood and related tissues in *Artemia. Journal of Morphology* 68: 593–632.
9. Howse, H. D., R. A. Woodmansee, W. E. Hawkins, & H. M. Perry (1975) Ultrastructure of the heart of the copepod *Anomalocera ornata* Sutcliffe. *Transactions of the American Microscopical Society* 94(1): 1–23.

4

Alpha Taxonomy I: Distinguishing Species

The meaning of "species"

Alpha taxonomy is the work and record of distinguishing among the species of animals or plants. Almost every group of organisms has specialists (taxonomists). They learn the history of work on their group's variant forms, examine specimens for distinctive characters, and describe those and assign species names to closely related types, which they think constitute closely related populations in nature. That has to be done despite an overlay of individual differences (not all chickens look alike). Taxonomists also initiate the search for patterns of relationship among those types. Once evolution became accepted by biologists, soon after Alfred Russell Wallace and Charles Darwin offered natural selection as an explanation of its process; that search became the attempt to reconstruct the pattern of evolutionary diversification, the "tree of life." Occasionally "beta taxonomy" is used as a term for that work of evaluating the patterns of variation across sets of similar and apparently related species. It can be conceptually separated from just distinguishing species. An "alpha taxonomist" needs a reasonably clear notion of what she means to imply when she assigns a new species name to a group of probably related specimens, and by implication all of its living kin still out in nature. So what, then, is a species?

Andrew Bennett was a longtime colleague in the Oregon State oceanography program who recently asked me that question. What, he asked, is the implied meaning when a species is given a name? I came up with an attempt, in a letter abbreviated here, at explaining what is philosophically, a rather irksome concept:

Dear Andrew,

You asked about the meaning of the term "species" as employed by biologists. Possibly this should be prefaced by an essay on the philosophy of nomenclature in biology—why we name organisms at all, and the utility of ranking named groups in hierarchies. Mostly I'll spare you. But much of it comes down to the notion that if an organism can be placed in a category on the basis of some characters and that category given a name, then other things about it can be predicted (always within limits) from the characters of the other members. If an animal can be identified as a mammal (a named category) because it has hair (a shared character), then its females will almost certainly produce milk

Oar Feet and Opal Teeth. Charles B. Miller, Oxford University Press. © Oxford University Press 2023.
DOI: 10.1093/oso/9780197637326.003.0004

(also a shared character). That mammal category would not tell you whether its bite is likely to make puncture wounds, though knowing that it is some sort of cat would. All of this philosophical underpinning can be translated into jargon, of course. As for "species," an oft quoted phrase from Charles Darwin[1] is still valid: "No one definition has as yet satisfied all naturalists; yet every naturalist knows vaguely what he means when he speaks of a species." Naturalists in Darwin's time, with marvelous exceptions, were men; hence "he."

The definition of species that you recalled learning, Andrew, says that populations kept by biological mating barriers from mixing their distinctive inherited traits are separate entities, species, and worthy of distinct names, so are given different Linnean binomens: a genus name and a species name. For example, the binomen *Calanus finmarchicus* is given to possibly the most studied marine copepod. Being certainly capable of interbreeding, or thought likely to be capable of it, is the so-called biological definition of species. This biological species concept (BSC) was agreed upon by what basically was a US and British committee in the 1930s, and strongly promoted by Ernst Mayr and Theodosius Dobzhansky. Books were written about it, many by Mayr and a couple by Dobzhansky (e.g., his *Genetics and the Origin of Species*, 1937). Those books were prominent in both my undergraduate and graduate reading lists.

For taxonomic purposes the BSC works for very few species, mostly just those subject to direct crossbreeding experiments. And even when those experiments can be performed, they are difficult and take massive and informed replication. The outstanding example is *Homo sapiens.* Humanity includes distinctive subspecies, and those have well recognized geographic origins. Given the levels of distinctive characters between, say Asians from central China and Africans from Nigeria, separate species assignments have often been considered. For the BSC to be applied, we need to cross a suitable number of pairs to see whether there are mechanical, behavioral, or genetic barriers to successful mating. The standard is not that babies would be produced in those experiments, but that the babies would be reproductively successful in multigenerational crosses (and beyond) of the hybrids, and in back crosses with the parent strains. Voila: we (*H. sapiens*) have done these experiments. As you know, not every copulation produces young, and many crosses persistently fail, both at every level of subspecific distinction and at none; many individuals are simply sterile for whatever reason. So what counts is the massive replication of the experiments and the statistical outcome: that very little fertility difference applies to out-crossing compared to "in-crossing," which justifies the notion that "we" are one species in this "biological" sense. In fact, it can be argued that out-breeding produces some level of hybrid vigor; many excellent people emerge from cross-fertilizations.

Perhaps we have almost equal experience with species of fruit fly, domestic plants, and animals; perhaps a few organisms chosen explicitly to evaluate the BSC. However, for the vast array of distinctive organisms, often distinctive only in extraordinarily refined characters, we have not and cannot apply the test. Deep-sea copepods, for example, only mate in almost absolute darkness at great pressures, and they soon die when brought into captivity. When we could even find males and females in the same samples, they would not deem the confines of a mating experiment conducive to copulation. So we cannot run the experiment, and the BSC is usually not *operational*, a term from science generally. For a term to be operational, its definition should include an explicit rule for determining whether it applies to a particular situation. A quantity should be defined by an explicit rule for how to measure it. "Biological species" has a rule, but it can only rarely be applied. There are catalogues of other objections to the BSC.[2]

So most named species are in fact "typological," a tradition going back to before Linnaeus, and to which he added the binomen formalism. He was basically a Platonist, and he thought that sets of evidently quite similar organisms (species) were realizations of Platonic "forms" or ideals. The belief was that it should be possible to catalogue the diversity of creation by finding and naming each of God's creatures, a list expected to be on the order of 10,000. For each set of organisms with closely recurring features, a dead, but otherwise undamaged and insofar as possible, typical specimen would be selected, preserved as well as technique allowed, catalogued, named and stored in a safe place. That specimen would be the "type," the *holotype*, for the species. That practice is still very much in play, with great national and other museums devoted to storage of type specimens and other vouchers (paratypes and other syntypes). After the 1930s, the goal became to name new species and revise the nomenclature for previously recognized ones so that they would probably correspond to the interbreeding (or potentially interbreeding) populations of the biological definition. Geographic distributions, knowledge of behavior, everything that comes to hand was and is invoked in trying to establish this correspondence. The latest data, of course, are DNA-sequence comparisons.

It should be added that a more Aristotelian interpretation prevails, a sort of essentialism in which individuals in a species share enough characters that they are in essence the same, or at least more closely related than those not sharing so many. Genus names are assigned to groups of species that seem certainly to be close relatives. Divisions among those are somewhat more arbitrary, with breaks in a near continuum of variability assigned in part to give genera numbers of species convenient to remember and apply. Genera do provide useful correlations by grouping species not only likely to be related through evolution, but likely to feed, swim or fly, rear their young in similar ways, and that have modest genomic differences.

The impossibility of fully establishing the correspondence between forms distinct in body shape, color, or habitat with interbreeding populations, particularly in the face of morphologically indistinguishable "sister species," led in the 1960s and 1970s to the numerical taxonomy (NT) movement.[3] The idea was that, since most "species" are just defined by commonalities among sets of organisms (they called individuals OTUs, operational taxonomic units) that have very high levels of similarity in their characters (anything: length, ratio of skull volume to big toe length, eye color) should be identified by explicit methods. To that end, algorithmic clustering methods were applied based on character scoring. Named sets were to be called "morphemes," not species. That made some sense, but NT was not adopted. However, NT clustering algorithms are one basis of techniques used to evaluate relationships with quantitative morphological data and DNA sequences.

There are other notions of what should constitute a species. There are the phylogenetic species concept (PSC)[4] and the monophyletic species concept (MSC),[5] and yet still more. If you care, you can chase those down. To an extent seeming almost inevitable, they are still typological, but are applied with the paleohistorical reality of an evolutionary tree held in mind.

Reasonably easy DNA sequencing has produced several species definitions based on specific levels of counted base-pair, or amino-acid coding, differences in modestly long bits of sequence. Gene sequence differences are just characters, as for copepods are: "endopods of leg 1 with 3 segments versus endopods of leg 1 with 2 segments," or long versus short rostral filaments.

For many purposes, particularly in ecology and biogeography, species defined by essentially typological methods are what we use to study distributions, habitat associations, interactions among populations, prey in gut contents, and on and on. For those purposes a species is recognized because a competent taxonomist says it should be (Darwin's opinion). If he or she does a good job of morphologic, behavioral, genetic, or other description, a great deal can be said about nature by examining the biology of such entities: *Cyclops strenuus, Diaptomus kincaidi, Calanus finmarchicus, Acartia californiensis,* and *Temora stylifera* are all such "species" among planktonic copepods. The copepodologists introduced in this book and I have spent much of our careers examining aspects of their life and times.[*]

Thanks for the question, Andrew. Best regards, Charlie

---

[*] Actually, at this point, I told Andrew about the naming of species among unicellular organisms and the division of fossil sequences into species over long stretches of time: trilobites, ammonites, horses. Neither is much of a concern for multicellular copepods with a very limited fossil record. Species names for organism populations are to an unavoidable extent arbitrary, however explicitly defined. They are shorthand, codes by which to talk about entities that are partly real, partly conceptual.

Bruce Frost an Alpha Taxonomist With Other Dimensions

Bruce Frost (Figure 4.1) and I arrived at the Scripps Institution of Oceanography (SIO) to begin graduate studies in September of 1963. We were both newly married and moved, as vertical neighbors, into the university's married-student apartments. The Frosts were upstairs, the Millers were down. Bruce had a long-standing interest in zooplankton, having started to study them in high school in the vicinity of his home at Brunswick, Maine. His science fair project there had been a study of the species captured in nets lowered from a bridge to sample from the flow in a tidal channel. Copepods (*Acartia* species) were dominant in the catches. He did not go far for undergraduate studies, majoring in biology at Bowdoin College in Brunswick. He remembers fondly the mentoring received from a biology professor, James Moulton. Many years later, Bruce named a newly recognized species in his honor, *Pseudocalanus moultoni*. Moulton welcomed Bruce into his lab, gave him access to microscopes and other gear for individual projects, and he recommended him for a summer-break internship with Arthur Humes at Boston University. Humes was an important student of parasitic copepods, and Bruce worked at sorting specimens from the tissues of preserved marine hosts of all sorts, such as anemones. That, he says, was not the best part. Humes encouraged him to spend some of the time at the Museum of

Figure 4.1 Bruce Frost, when he was director of the Oceanography Department at the University of Washington. With his permission.

Comparative Zoology on the grounds at Harvard. Bruce did so, reading in its extensive collections of invertebrate literature.

We learned (maybe from Helen Frost, because Maine men do not volunteer much about themselves), that Bruce was a football lineman for Bowdoin and a nationally ranked shot-putter. It was and is possible, though demanding, both to play sports and study seriously at small colleges in the United States. That is very different from Division I and II schools, where players basically major in sports, and very few study, say, thermodynamics or medieval history.

Max Dunbar, an oceanographer and planktologist at McGill University in Canada, learned of Bruce's potential, probably from Moulton, and asked him to apply for graduate work in Montreal. Bruce did, and Dunbar, who worked in Arctic Ocean areas, invited his participation in a summer sampling project in a high-latitude Canadian fjord. However, Bruce was hedging his bets, as everyone does, and applied also to Scripps. He got a call from a professor there, accepting his application and urging him to come. For reasons he says are no longer clear (maybe they never were), he decided on Scripps. After graduation from Bowdoin, Bruce and Helen were married, and then he headed north to participate in Dunbar's fjord sampling. He says he just took and preserved the collections with no time to look at them. Helen returned to the University of Maine at Orono to finish her degree in English literature. Together again in August, they drove the corner-to-corner distance to La Jolla, California in an old Ford Falcon in time for Bruce's first classes.

I might tell my "getting to grad school story" somewhere else. But after different experiences establishing relationships with Scripps faculty, Bruce and I both acquired John McGowan, then an assistant professor, as our dissertation advisor. After a year-long sequence of basic courses on the physics, chemistry, geology, and ecology of oceans, we both participated in Ursa Major, a long summer cruise organized for 1964 by McGowan. On the RV *Alexander Agassiz*, a converted army freighter, we sailed west to 45° W then north to Alaska, led by Dan Brown. Dan was the engineer who provided a subsurface opening-closing mechanism that we tried on the cruise for the earliest version of the bongo net (Figure 1.1). Other Scripps students aboard were Peter Wiebe, Elizabeth Venrick, and Willis Hayes. Both Peter and Elizabeth became prominent plankton ecologists. Hayes turned to work on shoreline ecology. There were several technicians along from the Scripps ocean data group. They taught us methods like reading reversing-thermometers and data recording.

The object was to study the distribution of zooplankton species along 45° West, from the North Pacific Central Gyre, north through the North Pacific Drift, and then across the Gulf of Alaska to the port of Kodiak. We deployed vertical flights of that first opening-closing version of the bongo net at stations to 400 meters about every 40 nautical miles on the way north. McGowan and some others joined us in Kodiak. After some beers, we went back south all the way to Hawaii. Southbound stations were more closely spaced in the transition zone between subtropical and subarctic waters. We had identified that zone from samples taken going north, based on the small number of krill and chaetognath species endemic to its narrow east-bound flow. That analysis was done at the dock in Kodiak. The training was hands-on all the way, and a great learning experience. It included Nansen-bottle profiles of temperature, salinity, phytoplankton nutrients, and chlorophyll concentrations. Electronic profilers (conductivity-temperature-depth sensors, or CTDs, and rosettes of bottle samplers were still to be invented).

In the winter of 1965–1966 Bruce, Elizabeth, and I, together with some marine chemists, repeated the northbound leg on Ursa Minor, generating data about the seasonal contrast. Waves were huge while crossing the Gulf of Alaska. We could run the RV Argo, a converted submarine rescue vessel with a massively heavy keel, along the troughs and crests for net tows, sinking into great valleys then rising to scan across surrounding hills of water under steel-gray clouds. Argo just rocked a little as g-forces varied, our bodies chasing the decks down into troughs, then the decks shoving us skyward. The bosun strung a heavy line from the deckhouse across the fantail to a post at the stern. To put nets on the towing cable and drop the net-opening messengers from the open stern, we wore harnesses roped to that overhead line. Bruce and I got an astounding lot of bongo profiles under those conditions. Back at Scripps I proposed as a thesis project to identify and count the species in those two north-south sections. Professor McGowan did not agree. I'm no longer sure why, except that he had another idea he wanted pursued. It felt odd then and still does now, but the original ideas for dissertations are only occasionally the choice of their authors. The sections never were fully analyzed. Bruce and I learned to work from ships, to do oceanography, on Ursa Major and Ursa Minor.

Bruce Frost Reworks *Clausocalanus* Taxonomy

Bruce accepted a project suggested to him by Abe Fleminger, a very prominent copepod systematist, on the SIO staff but not tenure-track faculty. Abe would convert copepod genera, groups of species already recognized as obviously related, into extensively detailed projects, particularly including their geographic distributions and morphologic relations to one another. Part of the goal was to recognize likely processes leading to divergence of single original species into two or more new ones, termed "speciation." An example of his work is discussed in Chapter 19. He had been examining specimens of the genus *Clausocalanus* (Figure 4.2) from all over the world oceans, and he had found several clusters of morphologic characters suggesting new species. Abe turned the lot over to Bruce, who then during several years repeated all those observations, gathered more samples from collections at Scripps and Woods Hole, from institutions and museums all over the world. The

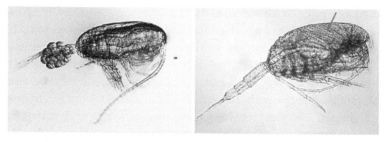

Figure 4.2 Photos of *Clausocalanus furcatus* (Brady). Left: female (ca. 1.1 mm) carrying an egg mass. Most *Clausocalanus* species are egg carriers, though *C. lividus* is not.[6] Photo by Albert Calbet, Institut de Ciències del Mar, Barcelona, with his permission. Right: male (ca. 0.8 mm). Notice both its very long, very thin fifth thoracic leg that is involved in spermatophore transfer and the great mass of muscles (arrow) in the cephalosome that are used for steady, antenna-driven swimming during searches for female sex-pheromone tracks. Photo by Stéphane Gasparini, Observatoire Océanologique de Villefranche, with his permission.

total tabled in their monograph counts out at 438 sorted samples. *Clausocalanus* species live in the readily sampled upper few hundred meters; Bruce could assemble a globally distributed sample set including all latitudes and from coast to transocean coast, except for the Arctic where no *Clausocalanus* are found. For the already named species, Bruce borrowed the type specimens when they still existed.

Frost's work focused on adults, both male and female. Until that full development many of the distinctive characters have not yet appeared. Examining a preserved copepod involves painstakingly dissecting all the limbs and laying them out in order on microscope slides. A usual technique is to soak a specimen in a drop of glycerin-water solution on a slide. That slows drying and provides viscosity so the "pod" can be held. Then, working with a stereomicroscope and using very thin needles (the smallest, stainless steel insect pins, *minuten nadeln*, are good) held in the chucks of microbiological loop holders serving as handles, you steady the animal with one hand, then with the other hand insert a needle into the membrane between the coxa and the body, and then pull the limb away. With one of the needles you lift the limb out of the drop and place it in a much smaller drop at the other end of the slide (or on another slide). Too often a limb can fly off somewhere in the process, but this occurs less and less often with practice. Each limb is covered with a suitably small chip of coverslip and is ready for examination with a compound microscope.

Among body-shape details and somewhere on all those limbs there will be features that characterize the genus and its species. The specimens are dead, so these will be the typological sorts of species defined above. *Clausocalanus* adults are small; *C. jobei* females (Figure 4.3) are 1.0 to 1.6 mm from forehead to caudal

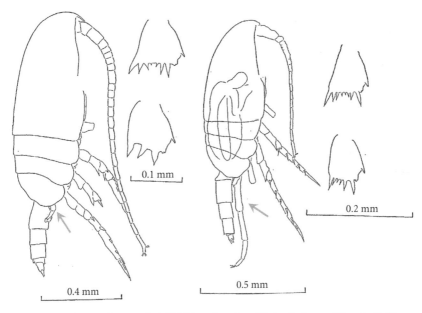

Figure 4.3 *Clausocalanus jobei.* "Habitus drawings," lateral views of female (left), and male, each with P2 (above) and P3 basis segments showing the distal edge projections. Female and male P5 indicated with arrows. From Frost and Fleminger.[7] Permission, Bruce Frost.

furca, and the males only about 1.0 mm. So, pulling the legs neatly off, preferably keeping the swimming leg pairs attached by their couplers is tricky. Like everything, it gets easier with practice, especially given really good stereomicroscopes (science can be expensive).

All species of *Clausocalanus* share the basic shape shown in Figure 4.3. All the females have simple and paired tubular, three-segmented fifth legs ending in two spines. They also have three strong, pointed projections along the edges of the bases of the 2nd and 3rd thoracic legs (P2 and P3). Just one projection between the first joints of the exopod and endopod is usual in most other copepod genera. The males have very long and tubular right fifth legs (P5) terminating with two tiny and seta-tufted branches. Somehow that is involved in spermatophore transfer. The male drawn here is shown with its testis and spermatophore-forming organ visible in its prosome. Indeed, with a microscope light showing through a specimen from below, that reproductive system is visible.

Frost and Fleminger[7] raised the number of recognized species from seven to thirteen, with five new names; for the sixth addition, *C. mastigophorus* was brought back from being considered to be a synonym of *C. arcuicornis* by Wilhelm Giesbrecht (in 1892). The rules of zoological nomenclature declare that the first (or senior) name given to an adequately described species should be kept (have precedence) when somebody else later gives it another name. So, later taxonomists will consign the junior name to synonymy lists. However, greater precision in comparisons by still later workers may reveal differences they consider do justify separating as species the animals described by the two names. All this can require as much time at the library as at the microscope.

Clausocalanus jobei Frost and Fleminger, 1968, which all together is its full, formal name, belongs to a group of five species within the genus sharing more features than characterize the genus as a whole. To illustrate the extent to which Frost took close observation to distinguish types reliably, see Figure 4.4. He

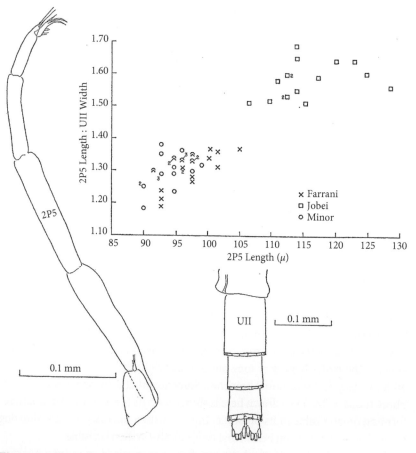

Figure 4.4 Typical graph from Frost and Fleminger's *Clausocalanus* monograph[7] showing variation among closely related species (*farrani, jobei, minor*) in ratios of segment dimensions as a function of a body dimension, here the length of the second segment of the male P5 (2P5). Permission, Bruce Frost.

measured the lengths of male fifth leg segments of *C. jobei* and its closest relatives *C. farrani* and *C. minor*. That is done by overlaying an eyepiece-image of a ruler over views of the legs in a compound microscope. He divided the length of the second segment (marked 2 in Figure 4.4) by the width of the second urosomal segment. Then he compared for 16, 14, and 23 specimens of the three species (in that order) the ratio of the leg segment length to the urosome segment width. In *C. jobei* the leg segment was longer, and longer relative to the width of the urosome (1.6 times longer vs. 1.3 times in the other two). If you have occasion to attempt identification of a male *Clausocalanus*, you could well come to distinction 4 in Bruce's key:

4. 2P5 length more than 1.45 times as great as UII width (Figure 4.4 here) ....*jobei*
4. 2P5 length less than 1.45 times as great as UII width (Figure 4.4 here) .... 5.

Key statements at 5. involve a similar distinction in lengths of the fused antennule segment
20–21 and segment 22 between the antennules of *C. farrani* and *C. minor*.

Use of graphs to illustrate these subtle but definite differences between groups of specimens was a practice used by Fleminger in his Harvard Museum of Comparative Zoology dissertation on copepods from the Gulf of Mexico. Frost later used it extensively for other genera, and I used it to illustrate consistent quantitative differences in the anatomies of *Neocalanus plumchrus* (Krøyer) and *N. flemingeri* Miller. (Yes, I named the latter in honor of Abe Fleminger, because he taught us so much.)

There are various schools of thought about choices of names for species. Sometimes the same authors apply different philosophies to different species, even Frost and Fleminger. For example, *Clausocalanus lividus* is bluish or lavender ("livid"), in life and soon after preservation, and was named for that. The danger is that another species discovered later might be even more intensely blue or purple. For *Clausocalanus ingens*, "The large size of females of this species suggested the name *ingen*"[7] (Latin for "great"). Fortunately, no larger (or any other) species fitting the genus description has been found since 1968. Names for species termed "patronymics," based on the names of prominent (or not so prominent) scientists, friends, or beloved beings, does not have that danger, and at least when used for persons it allows some distribution of credit. In the choice of *C. jobei*, drawings of which are reproduced in Figure 4.3, it was "in homage to a now-deceased companion, who, true to his species, contributed a lifetime of devotion, loyalty, and good cheer to one of another species (A. F.)."[7] Job was Abe's beloved dog. Many other names, particularly for species of *Calanus*, refer to the localities (often the "type locality") where the animals are found: *Calanus finmarchicus, C. pacificus, C. sinicus,* and so on.

Bruce and Abe's original monograph on *Clausocalanus*[7] included maps with dots for stations examined for specimens and circles for stations with particular species. Figure 4.5 shows the maps for *C. jobei*. It is unusual among plankton animals in that it lives in patches in different sectors of relatively warm surface waters all over the world's oceans: the eastern tropical Pacific and Atlantic, spots along 35°S in the Pacific and around the Cape of Good Hope, and especially the Gulf of Mexico and Gulfstream. A few other zooplankton species have similar zoogeography. The other *Clausocalanus* species have more usual patterns in areas with persistently returning (gyral) circulations. That keeps major parts of their populations from drifting out of survivable habitat. Traversing these long and closed routes can take many generations, but they apparently do allow (or require) adaptation to conditions encountered all along the way.

Bruce did not finish his doctoral studies with publication of the extraordinary 1968 monograph on *Clausocalanus* taxonomy. It did have distributional maps, but there were gaps, particularly across the Atlantic (see Figure 4.5, right), which he then proceeded to fill by searching through even more, and more appropriate, samples; more appropriate because he selected those taken with finer mesh nets. They provided a more fully developed view of the zoogeography of the genus. He sent a paper based on that study and its maps to a journal. It was rejected by that journal because of the existence of the inferior maps in the monographs (imagine the horror of an author publishing an improved version of something already available). Rather than pursue publication in other journals, he just shelved the paper and went to work at his new job. Later, he allowed me to teach from the dissertation maps, and eventually he let me publish them in a textbook, *Biological Oceanography*,[8] in which the detailed story is summarized of what they tell us about distributions of upper ocean zooplankton. Most of the patterns stretch all the way around the Earth between particular latitudes. Except for *C. laticeps* in the circumpolar drift around Antarctica and *C. ingens* in the circumglobal band of subantarctic water south of 40°S, all of the distributions are tropical or subtropical.

Moving to Seattle and New Directions

Leaving Scripps, Bruce became Assistant Professor Frost in the Oceanography Department at the University of Washington (UW). He, Helen, and their two children, who had been born in La Jolla, moved to Seattle. Now as Emeritus Professor, he's still working there. He conducted extensive studies of feeding rates and mechanisms in copepods, of their vertical migrations, of their mate

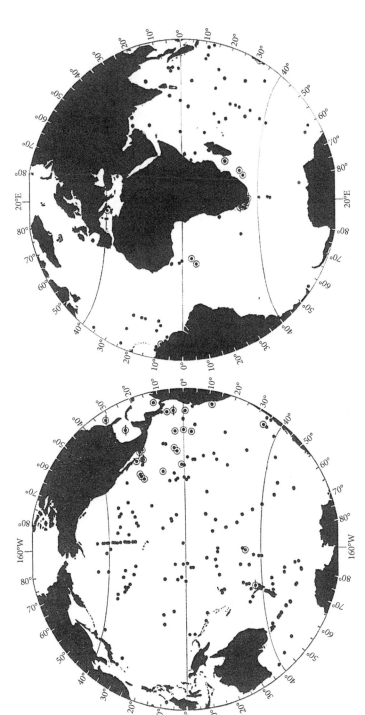

Figure 4.5 Distribution pattern of *Clausocalanus jobei*. Dots show stations for which samples were examined. Circled stations produced *C. jobei*. Specimens were measured from stations with a bar across the circle. A detail map for the Gulf Stream, Gulf of Mexico and Caribbean had 60 examined station samples, 24 with *C. jobei*. From Frost and Fleminger.[7] Permission, Bruce Frost.

attracting pheromones. In a more purely intellectual vein he worked, initially during a sabbatical in Scotland, with the theoretician John Steele on description of the general function of upper ocean ecosystems using difference-equation numerical models, and he extended that to theoretical description of the iron-limited ecosystem of the Gulf of Alaska, an understanding that also extends to vast areas in the Southern Ocean from around 45°S to the Antarctic continent. At the same time, he trained a large and talented group of oceanographic copepodologists. Some of those people and their work will be discussed as we go along. Eventually he served a few years as Director of the Oceanography Department at UW.

On top of all that, Bruce kept some taxonomy going: he described *Calanus marshallae*, from the northeast Pacific, distinguishing it from its more northerly cousin *Calanus glacialis*.[9] He carried out a years-long study of the genus *Pseudocalanus* that inhabits northern waters from the Arctic Ocean southward in both Atlantic and Pacific to about 40°N.[10] He had important help in that study from fellow copepod taxonomist Gayle Heron, working at UW in her retirement from the National Marine Fisheries Service. She drew many of the figures, including the extraordinarily detailed habitus drawings. I used one as an example of such "flattened fauna" sketches as (Figure 1.4). Nobody has questioned or added to the four *Pseudocalanus* species that Bruce redescribed and the three new species he described in 1988.

Some of these *Pseudocalanus* were again distinguished by very refined morphologic observations. In some of the species only the males have prominently distinct characters. Again, he used graphs of relative proportions, in some cases of features it takes a high-drive lens to see at all. However, early enzyme comparisons by J. M. Sévigny (then a graduate student working at Dalhousie in Canada with Ian McLaren) and later gene sequencing by Ann Bucklin and others have confirmed the distinctions. An application of gene sequencing to routinely distinguish *P. moultoni* from the very similar *P. newmani* is told in Chapter 18. Those two species are both found together over Georges Bank east of Massachusetts and in Puget Sound, Washington, but apparently not at present across the Arctic Ocean in between. Distributions of copepod species can be surprisingly disjunct, even separated by continent-sized barriers.

By now you will have noticed that outline drawings of bodies and of limbs flattened under coverslips are a basic mode of communication among copepod systematists about shapes and distinctions. The drawings are preferred because they are oddly more explicit than photographs (when those are possible), because copepod limbs (and sometimes those of other crustaceans) usually flatten well, and drawing is a powerful aid to noticing variations of shape and the presence of microscopic edge and surface details. In addition, of course, drawing has been an available means of communication since ancient times, and capabilities for etching and detailed lithography have been available for several centuries. Thus, drawings began to reach modern quality when microscopy had attained sufficient power and image definition to reveal very minute structural detail. Betty Edwards has pointed out, in her instruction book *Drawing on the Right Side of the Brain*, that persons focused on drawing accurate likenesses tend not to talk and can be helped by choosing not to. Some aspect of keeping track of time also can become suspended. Bringing a drawing of, say, a mandibular palp and its opal-toothed gnathobase to completion can suddenly produce the odd sense that virtually no time has passed, though the clock says you missed lunch or that your late afternoon bus has left. Time suspension can be rather lovely and possibly some students of copepods become addicted. I once thought I was for about a year.

The SUPER Project

Bruce and I shared the theorizing, organizing, and field work of a National Science Foundation supported effort in the 1980s to understand the production ecology of the Gulf of Alaska. We called the project Subarctic Pacific Ecosystem Research (SUPER). The object was to explain why the whole oceanic (well out from land) sector never supports phytoplankton blooms, despite abundant levels of the so-called major algal nutrients: nitrate and phosphate. Our initial hypotheses, which we and our crew showed were wrong, involved grazing by large copepods. The work kept us funded for over a decade. Some of the copepod results appear later in the book.

More Copepod Systematics and Ecology

Once Bruce retired from classroom teaching and department directing, he started new work with new graduate students and some postdocs on more of the open questions in planktology, mostly copepods. He and Mikelle Nuwer worked on questions about the "*helgolandicus*" species group of the genus *Calanus*. The range of *Calanus helgolandicus* itself is the temperate North Atlantic, from the Mediterranean north toward the Gulf Stream and mixing with the more northerly stock of *Calanus finmarchicus*, especially in the North Sea and along the Norwegian coast. Other species in the group are distributed in high temperate waters throughout the world. Mikelle,[11] working also with Virginia Armbrust, applied DNA sequencing to issues going back to a prominent Russian copepodologist. Konstantin Brodsky named three subspecies of the group member *Calanus pacificus* based on minor geographic variations in morphology: *C. pacificus pacificus* in the western subarctic Pacific, *C. pacificus oceanicus* in the vicinity of the international dateline, and *C. pacificus californicus* along the eastern shore (Alaska down to northern Baja California). Bruce, started with hints left in Fleminger's files and specimen cabinets at Scripps and reviewed those distinctions with particular care.

Nuwer, Frost and Armbrust[11] showed, based on a mitochondrial gene, that *Calanus pacificus* most likely has only *two* distinct variants: a coastal subspecies found along the North American coast and in the Sea of Japan, and an oceanic subspecies found in the Gulf of Alaska and west to the Oyashio Current off Hokkaido. They added that their result confirms work of Erica Goetze (Chapter 19) showing that planktonic animals with high dispersal capacity can develop genetically distinctive populations in the absence of obvious geographic barriers.

In 1973 Sifford Pearre,[12] who was studying arrow worms (chaetognaths). suggested that vertical migrations down away from daylight to avoid visual predators (Chapter 9) were more complex. Once arrived in surface layers with abundant food, they tended to leave immediately once they found a meal. Bruce and I discussed that paper when it came out, and it never left our minds. Testing was needed. So, decades later, Bruce and a group of students and postdoctoral scholars decided to look for evidence of what Pearre had come to call "Eat and Run"[13] among copepods. Their strategy was to deploy plankton nets as traps. A net suspended vertically from its cod end, with its mouth well below the surface, would catch copepods headed up. Another with its mouth at the same level, but its bag and cod end hanging below would catch copepods headed down. The trap for upward bound plankton was only opened once it reached depth, catching nothing passing through the upper layer. That for down-bound animals would flush when lowered. Deployments were in the quiet water of Dabob Bay, Washington, a fjord with plentiful plankton like that in the ocean offshore. Mouth depths were 10 or 15 meters and deployments 45 minutes. The results, written up by James Pierson and others,[14] were a resounding yes. The gut contents of upward-bound *Calanus* and *Metridia* females were nearly zero; the guts of downward-bound specimens were variably but mostly filled. Repeated trials showed consistent results. Question answered.

One other major project was checking in the field, again in Dabob Bay, the notion that copepod females eating diatoms often produce embryos failing to divide properly.[15] This had been powerfully demonstrated to be possible in lab experiments,[16] and shown[15] to be the effect of a toxin. However, it had not been shown to be important in field habitats. Bruce, working from a small ship loaded with grad students and postdocs, generated with them a whole book[17] reporting studies showing that sometimes, indeed, there are toxic effects of a diatom diet on copepod egg development in nature.

All of these projects were a tour-de-force, the capstone of a great oceanographer and copepodologist's career. Bruce has told me he is fully retired, busy with Helen enjoying their children and grandchildren.

References

1. Darwin, C. (1958 [1859]) *The Origin of Species*. Signet Classics Edition, New York.

2. Tobias, J. A., N. Seddon, C. N. Spottiswoode, J. D. Pilgrim, L. D. C. Fishpool, & N. J. Collar (2010) Quantitative criteria for species delimitation. *Ibis*. doi: 10.1111/j.1474-19X.2010.01051.x.

3. Sokal, R. R., & P. H. A. Sneath (1963) *Principles of Numerical Taxonomy*. W. H. Freeman and Co., San Francisco.

4. Wheeler, Q. D. (1999) Why the phylogenetic species concept? Elementary. *Journal of Nematology* 31(2): 132–141.

5. Rieppel, O. (2010) Species monophyly. *Journal of Zoological Systematics and Evolutionary Research* 48(1): 1–8.

6. Saiz, Enric, & Albert Calbet (1999) On the free-spawning reproductive behaviour of the copepod *Clausocalanus lividus* (Frost & Fleminger 1968). *Journal of Plankton Research* 21(3): 599–602.

7. Frost, B. W., & A. Fleminger (1968) A revision of the genus *Clausocalanus* (Copepoda: Calanoida) with remarks on distributional patterns in diagnostic characters. *Bulletin of the Scripps Institution of Oceanography* 12: 1–235.

8. Miller, C. B., & P. W. Wheeler (2012) *Biological Oceanography*. 2nd edition. John Wiley and Sons, Oxford, UK.

9. Frost, B. W. (1974) *Calanus marshallae*, a new species of calanoid copepod closely allied to the sibling species *C. finmarchicus* and *C. glacialis. Marine Biology* 26(1): 77–99.

10. Frost, B. W. (1989) A taxonomy of the marine calanoid copepod genus *Pseudocalanus*. *Canadian Journal of Zoology* 67(3): 525–551.

11. Nuwer, M., B. W. Frost, & E. V. Armbrust (2008) Population structure of the planktonic copepod *Calanus pacificus* in the North Pacific Ocean. *Marine Biology* 156(2): 107–115.

12. Pearre, S. Jr. (1973) Vertical migration and feeding in *Sagitta elegans* Verrill. *Ecology* 78: 300–314.

13. Pearre, S. (2003) Eat and run? The hunger/satiation hypothesis in vertical migration: history, evidence and consequences. *Biological Reviews* 78: 1–79.

14. Pierson, J. J., B. W. Frost, B. Thoreson, A. W. Leising, J. R. Postel, & M. Nuwer (2009) Trapping migrating zooplankton. *Limnology and Oceanography: Methods* 7: 334–346.

15. Ianora, A., A. Miralto, S. A. Poulet, Y. Carotenuto, I. Buttino, G. Romano, R. Casotti, et al. (2004) Aldehyde suppression of copepod recruitment in blooms of a ubiquitous planktonic diatom. *Nature* 429: 403–407.

16. Uye, S. (1996) Induction of reproductive failure in the planktonic copepod *Calanus pacificus* by diatoms. *Marine Ecology Progress Series* 133: 89–97.

17. Frost, B. W. (2005) Diatom blooms and copepod responses: paradigm or paradox? *Progress in Oceanography* 67: 283–285.

5

Alpha Taxonomy II: Values of Named Species

An Encounter with Copepods from Below Sunlight

In the early 1990s, I was invited by marine physiologists Jim Childress and his then student Erik Thuesen at the University of California-Santa Barbara to participate in some deep-sea trawling well offshore from Southern California. We sampled zooplankton at various depths, from about 1,000 m down to near the 4,000 m bottom, using an opening-closing trawl with a 10 square-meter mouth. The objective was to bring specimens of everything caught to the surface and pop them into respirometers of various sizes to measure their oxygen consumption. Deep ocean layers are cold, around 3°C, so to keep specimens cold on the way up through the much warmer surface layers, the insulated "cod end bucket" of the net was closed with ball valves when the net's mouth closed at depth. The bucket also had a stout layer of plastic tube to retain pressure. The bucket was opened and drained into a big washtub. Then all the recruited experts on different groups would swarm around selecting specimens. Yes, the pressure was lost, but the animals were not exposed long to warming before being placed in refrigerators. At least the copepods had no gaseous inclusions to expand and kill them, and they were alive and kicking. I used white-plastic soup ladles (Figure 5.1) to remove copepod specimens from the tub.

My job was to select matching sets of the same copepod species for oxygen consumption measurements (done for copepods in syringes), for freezing in liquid nitrogen for enzyme-activity analyses ashore and for identity vouchers. I preserved the vouchers and took notes on collection depths, color patterns and likely family identifications. Yes, color patterns. About those I wrote this later:[1]

Copepods in the deep sea are diverse relative to those in the epipelagic. Common at the tiny end of the size scale are numerous species of *Oncaea*, which crawl about on bits of mucus and submarine snow, eating attached particles. The larger mesopelagic copepods are diverse calanoid genera with many species over 5 mm and even 16 mm in the case of *Bathycalanus sverdrupi*. Some genera (*Bathycalanus*, *Megacalanus*, *Lophothrix*, *Scottocalanus*) are robust-bodied

Oar Feet and Opal Teeth. Charles B. Miller, Oxford University Press. © Oxford University Press 2023.
DOI: 10.1093/oso/9780197637326.003.0005

Figure 5.1 The author sorting large copepods and a University of California-Santa Barbara student sorting something else, from a very deep net haul drained into one of Jim Childress's galvanized tubs. That had been coated with dried mucus from earlier catches, but probably was not trace metal-free. Photo taken with the author's camera

detritus-feeders that mostly eat fecal pellets sinking through their vicinity from above. These copepods and some others of less certain feeding habit (*Gaetanus, Gaidus,* and others) are mostly dull orange or bright red, opaque, and heavily muscled. In captivity they alternate zooming about their aquarium with spells of stillness and fairly rapid sinking.

Lurk-and-grab predators form another large group of deep-sea copepod genera: *Augaptilus, Euaugaptilus, Haloptilus, Disseta, Paraeuchaeta,* and *Arietellus.* Except for *Paraeuchaeta,* these mostly have very thin muscles and elaborate sprays of long setae to inhibit sinking. In aquaria they hang still in the water, tail down with antennules extended, scarcely sinking or rising. *Paraeuchaeta* specimens behave similarly in the ocean, but they tend to rest on the bottom in containers. Unlike the red detritivorous copepods (and large red shrimp, for that matter), these predatory copepods have elaborate color variations. *Euaugaptilus* species come in brilliant yellow, lavender, orange, bright green, and other colors. Some have pale tints over the exoskeleton, through which their gut is visible, brightly colored with a completely different part of the spectrum. Species of *Disseta* come in pale orange and bright white.

Paraeuchaeta have chromatophores, pigment cells that look like a decorative pattern of tiny brittle-star tattoos just under the exoskeleton, and they come in colors and patterns that are species-specific. While dead specimens can be distinguished by experts from very subtle shape differences (following Taisoo Park's monograph[2]), live specimens wear their identities as distinctively colored uniforms. Females of the genus *Euchirella* tow their purple, green, or black eggs along behind them in long zig-zag rows. The adaptive value of this riot of fancy dress in the near dark, with only blue photons to show it off by, is hard to imagine. If Annie Dillard is right, and "we exist so that creation won't play to an empty house," perhaps the color riot of deep-sea copepods is just part of the show.

There have been large improvements in the keys that help with copepod identification since the early 1990s, with much of the best help coming from Geoff Boxshall and Sheila Halsey's *An Introduction to Copepod Diversity*.[3] That allows anyone who knows the body parts to work quite efficiently toward family and genus identifications. But in the 1990s there were various earlier keys that could get me to the right genera for my large and colorful deep-sea specimens. From there I could go to descriptions of species, and I made good progress. However, I often ran aground, gaining a general idea but no species name. Thuesen, Childress and I wanted to have our copepods identified to species. So I recruited help, and particularly helpful were two outstanding experts on calanoid families, Taisoo Park, then working at Scripps, and Janet M. Bradford-Grieve in New Zealand. Dr. Park was introduced in the chapter on internal anatomy, and we'll come back to him later. Now let's meet Janet.

Janet Bradford-Grieve

Janet (Figure 5.2, right) and I talked at the 13th International Conference on Copepods, in July of 2017 at Los Angeles. She told me she developed an interest in marine life on camping trips with her family, studying seashore organisms and gut contents from fish. Both her college biology major and her graduate studies were at the University of Canterbury in Christchurch on the northeast shore of New Zealand's South Island. Like Bruce Frost and me, her graduate training was as a biological oceanographer, that is, as a student of ocean ecosystems far from the intertidal. Her advisor was George Knox, best known for his text *The Biology of the Southern Ocean*, and his work on international committees. He was a serious student of polychaete worm taxonomy.

Knox also established a Canterbury University marine station on the Kaikoura Peninsula looking out toward Chile, some 9,000 kilometers away. That is where

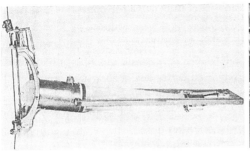

Figure 5.2 Left: the Clarke-Bumpus net used by Janet Bradford-Grieve to sample zooplankton offshore of Kaikoura Peninsula in the mid-1960s. Permission, National Library, New Zealand. Right: Janet still at the peak of her taxonomic work in the twenty-first century. Photo credit: NZ Institute of Water and Atmospheric Resources.

Janet developed her dissertation on the oceanography and zooplankton in the nearshore ocean. From reading the published version,[4] I suspect she was left largely on her own to obtain equipment and get access to the sea. The slope off Kaikoura is close to shore, so depths of 200 m are only 8 km out, and Janet selected a station over that contour for 13 months of fortnightly sampling trips. Needing a capable boat, she recruited a Mr. R. Baxter to take her out on his 12 m fishing boat, *Virgo*-LN168. It needed a winch for lowering and retrieving her homemade water-sampling bottles ("Van Dorn" bottles, to the initiated), her bathythermograph (a brass tube with a fluid-filled coil attached to a stylus that, when lowered into the water, scratched a temperature profile onto a smoked glass slide), and a Clark-Bumpus net (Figure 5.2). The picture of the winch in her 1972 paper[4] shows a sort of welding-class project: a cable spool with no level wind, a gearbox turned by small gasoline engine attached with a belt. Work on station must have required throaty yelling over the roar of that engine. The Clarke-Bumpus net was kindly loaned to her by David Tranter, an Australian planktologist.

With those arrangements, Janet made a strikingly full oceanographic analysis of the waters passing Kaikoura at different times of year. She measured water column temperature and salinity profiles through the seasons, nitrate concentration, chlorophyll in phytoplankton, zooplankton biomass, and vertical distributions of major groups (copepods, euphausiids, chaetognaths, appendicularians, and larvae of shrimp and fish). She estimated primary production (phytoplankton growth) using the formulae of the day. The sampling region is a variable confluence between inshore water of Subantarctic origin flowing from the south and subtropical offshore water from the north.

Clearly, however, what interested Janet most were the copepods. She began by translating the 1933 key to *Copépodes Pelagiques* produced by Maurice Rose for a French series of keys, *Faune de France*.[5] Her version in English with many revised figures is in the New Zealand Oceanographic Institute publication of her dissertation work.[4] I extensively used Rose's key at about the same time in my own graduate-school studies of plankton off Baja California. I learned just enough French to use the key (*avec* vs. *sans*, *gross* vs. *petite*, etc.), never thinking to do the work of typing up a translation. After training herself to identify them, Janet recorded 55 species from the Clarke-Bumpus samples.

To ocean ecologists some of the interest from Janet's first professional study is her demonstration that alternation of the species captured at her standard station matched the seasonal sources of the inshore waters, from the north in winter and the south in summer. That matched the observations of pioneers of Atlantic oceanography like Fredrick Russell. He showed that the shifting geographic sources of flow into coastal zones can be identified from the changes in plankton species taken in net tows. Something must be known of their distribution patterns in the surrounding ocean, and there was enough plankton data from New Zealand waters in the 1960s to make the assignments reasonable.

In her dissertation studies Janet was already recognizing possibly undescribed species but exercising caution about naming them based on the small numbers of specimens in her collections. She did provide good drawings of distinctive characters for types she thought distinct. One example (Figure 5.3) shows the spirit of these recognitions. She termed that apparently undescribed species *Bradyidius* sp. ("sp." stands for "a species not definitely identifiable beyond the genus level"). She had caught one specimen in a net that hit bottom at about 600 m, which was closest in form to a species described by G. O. Sars in 1903, *Bradyidius bradyi*. The published version of her thesis[4] concludes about her animal's distinctions that

> although there are several differences. The distal borders of urosome segments 1, 2, 3 have blunt spines, but only the genital segment is totally bordered. The endopods of the first, second and third legs are densely covered with very small spines. The external spines of the second, third and fourth leg exopods are very stoutly formed. *As only one specimen was captured, a new species is not proposed.* (italics mine)

That statement reflects Janet's awareness that *intraspecific* and *interspecific* differences both occur, while distinguishing them is a statistical issue that often cannot be settled from a single specimen. Some workers will immediately assign new names based on single distinctive characters, or one apparently distinctive specimen. In any case, it is good practice to describe those specimens

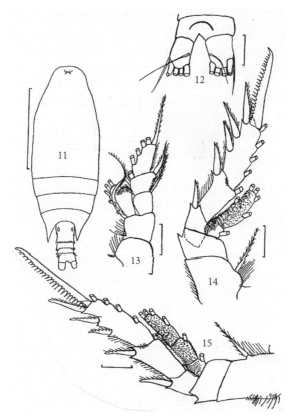

Figure 5.3 *Bradyidius* sp. Details: 11-female dorsal view, 12-caudal furca ventral, 13-first thoracic leg (P1), 14 -P2, 15 -P3. Assembled from Bradford (1972).[4] Scale bar for sketch 11 is 1 mm. That for the other sketches is 0.1 mm. Permission, National Library, New Zealand.

and characters. So Janet drew simple sketches of the distinctions she found on that one *Bradyidius* sp. Quite soon, however, she did provide names for single specimens so radically distinctive they clearly matched no extant descriptions.

At the time she wrote her dissertation, there were only four named species of *Bradyidius*. Clearly there were more, and soon she had an opportunity to work on that. She had married, and not long after she received her degree in 1968, her husband moved to a job in Washington, DC. Janet followed and was accepted in a visitor position with the National Museum of Natural History (NMNH), at the Smithsonian Institution in Washington, DC. It ran a warehouse-like facility, for storage and work on biological specimens. It was in a district that originally was a navy shipbuilding and supply center on the Anacostia River, the Navy Yard.

Janet worked there for years in a temporary position, often uncomfortably because Washington can be very hot in summer and there was no air conditioning. Her principal assignment was supervising the sorting of large plankton samples collected in Antarctic waters by the USNS *Eltanin*; copepods from them were later studied by Taisoo Park, as detailed below.

At the Navy Yard Janet had access to many plankton samples. She had brought her own *Bradyidius* specimen along and could establish that it was not unique, that its characters were shared with a population. She found the likely associated males and described those. Choosing a name, she published a description,[6] so it is now known as *Bradyidius spinifer* Bradford, 1969. Its morphologic characters are presented there in sharply defined drawings. Named species have a type locality, the place where the described type specimens were first collected. For this animal it is offshore from Kaikoura, New Zealand. That paper was also the start of strong interest in the many species of copepods in waters very near the bottom (often called the benthic boundary layer). She included descriptions of three new species of *Aetideopsis* (a family-level relative of *Bradyidius*) from off South Africa and New Zealand.

Again, outline drawings of bodies and of limbs flattened under coverslips are a basic mode of communication about shapes and distinctions among copepod systematists. They are oddly more explicit than photographs (when those are possible). Janet is talented at such drawing and deployed it beautifully for decades.

Working on bringing species of *Aetideopsis* and *Bradyidius spinifer* (all aetideids) "into science," as a saying used to go, Janet's interest in the group expanded. She realized there must be many yet to be recognized species in deep waters, particularly close the seafloor. In collections available to her from an epibenthic sampler deployed to 1,697 m off north New Zealand by colleague D. E. Hurley, she found four more new *Bradyidius* species. She described and named them in a separate paper: *robustus, brevispinus, dentatus, spinibasis*.[7] She wrote at least three other papers while at the Smithsonian, all of them based on cold-water and deep-living southern species. She was a very busy young woman in those years. Also important was that American copepodologists working in the Washington, DC area became her colleagues, particularly Tom Bowman of the NMNH and Dave Damkaer, who worked with the Bureau of Commercial Fisheries (now the National Marine Fisheries Service, part of the National Ocean and Atmospheric Administration, or NOAA).

In 1972, the New Zealand Oceanographic Institute (NZOI) published Janet's dissertation study,[4] that had been submitted to Canterbury University in 1968. By 1972 Janet had returned New Zealand and taken a job with NZOI, locating in Wellington, where she worked the rest of her career. Her interest in *Bradyidius* did not end when she left NMNH. As new species are added to them, genera become cumbersome, at least from a mnemonic standpoint. So systematists chop them up and assign new genus names. Also, on extended study, they may recognize that species assigned to one genus may fit better in another described someplace in the literature, perhaps in papers they had overlooked. Something like that happened. As a sort of coda to a paper describing an estuarine *Bradyidius*,[8] *B. hirsutus*, Janet moved some of her *Bradyidius* species to *Pseudotharybis*, a genus defined in 1909 by Thomas Scott. Thus, we now have *P. brevispinus*, *P. robustus*, and *P. dentatus*. Much later, in 2000, cooperating German and Russian copepodologists Knud Schulz and Elena Markhaseva described *Parabradyidius angelikae*. They reviewed Janet's descriptions and recognized that *B. spinibasis* is a good fit to *Parabradyidius*, so they moved it there.[9] Notice again that taxonomy involves as much library and literature study as field or lab work.

Janet's second 1969 paper[7] also included new genera and species from Dr. Hurley's samples. The mesh size was 1.2 mm, so these were large copepods. Apparently, her confidence regarding copepod identities had increased, and she described five new genera and eight new species, mostly based on one or very few specimens. The paper has the astounding total of 215 habitus drawings and projections of limbs with identifying characters. Again, the level of work during Janet's NMNH years is remarkable, opening a mine of taxonomic gold excavated by her and many others since. Most of the new species and genera were "epibenthic," living very near the seafloor. Her interest in *Bradyidius* did not subside, and she described *B. capax* in 2003.[10] She has contributed descriptions of many other species of Aetideidae.

Settling down at NZOI, Janet with colleagues undertook examination of the plankton in waters spreading around New Zealand, sorting and describing the copepod fauna in exquisitely drawn detail. This resulted in a series of books covering families of the marine calanoid fauna, starting in 1980 with *The Marine Fauna of New Zealand: The Pelagic Calanoid Copepods: Family Aetideidae*, written with John Jillett.[11] Janet followed that with volumes in 1983, 1994, and 1999; the total is 574 pages of descriptions and thousands of drawings. You do not have to be looking at a copepod from near the Chatham Rise to make these useful. If you know its family, some member of its genus is likely in these volumes to get you close to its name (if it already has one). It was a two-decade tour de force of skills and intelligence applied to describing and understanding the diversity of ocean life.

The "bradfordian" Families of the Order *Calanoida*

Taxonomists do not necessarily record what sparked their interest in the specifics of one or another group of organisms. Janet tells me that she in particular was "driven by the ambition to put the southwest Pacific calanoid copepod fauna into global perspective." She placed particular emphasis on certain families. For example, from New Zealand in 1973, Janet published[12] a paper pointing out that several long-defined calanoid copepod families share clusters of olfactory setae (yes, *aesthetascs*) that are thin-walled, sometimes worm-like, on their maxillae and maxillipeds. She noted that their shapes are a source of characters (an example is shown in Figure 5.4). Initially, in that paper Janet called attention to inconsistencies in the classification of four families originally defined by G. O. Sars in 1902: *Phaennidae*, *Scolecitrichidae*, *Diaixidae*, and *Tharybidae*. Sars had split *Phaenna* and its relatives from *Scolecithrix* and its relatives, because he thought *Phaenna* had the thin-walled setae on their maxillae, while *Scolecithrix*, *Diaxis*, and *Tharybis* (respectively classified as families *Phaennidae*, *Scolecithrichidae*, *Diaixidae*, and *Tharybidae*) did not. Janet, however, observed that all four families did have them. So some redefinition of the families and

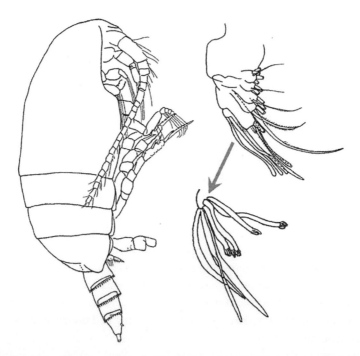

Figure 5.4 Lateral habitus of female *Xantharus cryeri* Bradord-Grieve, 2005,[18] with enlargements of its maxilla and maxillary endopod (red arrow) with its several types of olfactory setae. The total length of the specimens varied from 0.98 to 1.02 mm. Look at the finest divisions on a metric ruler. One of these copepods would fit between them, so removing and mounting a limb took impressive microdissection skills. From Bradford-Grieve[18] Permission, Taylor & Francis.

genera (that is, character lists) were needed, and she proceeded to provide them. The reordering is complex, a matter I leave to any who want to work through it.

Much interest has followed in the likely implications of these bunches of distinctive setae, leading eventually to terming the families with olfactory setae on their maxillae "bradfordian families." Prominent Japanese copepodologists Shuhei Nishida and Susumu Ohtsuka, for example, produced a study[13] using electron microscopy of such setae in the scolecitrichid genera *Lophothrix* and *Scottocalanus*. They found very large numbers of ciliary fibers extending from the nerve dendrites at the base of the setae, with 100 and even 700 fibers in cross sections of single setae. That compares to only a few in the aesthetascs on male copepod antennules (see chapter 10), the detectors for female pheromones. Some of the maxillary aesthetascs are open at the tips, seawater washing directly against the ciliary fibers. Because the copepods in bradfordian families live close to the seafloor and have particulate gut contents, detrital stuff, it is believed these setal systems on the mouth parts help them to select particles likely to be nutritious (having substantial organic content) from others not likely to (say, mineral bits).

References to bradfordian families are a regular feature of current copepodology. For example, Komeda and Ohtsuka[14] are not quite sure to which bradfordian family their new genus and species, *Pogonura rugosa*, belongs. Frank Ferrari of the NMNH in the United States and Russian taxonomist Elena Markasheva have been particularly active in expanding the description and species lists of bradfordian families.[15] Interest, including from Janet, has extended to all epibenthic species. Vladimir Andronov and Nina Vyshkvartzeva of the Shirshov Institute in Moscow also have been describing numerous epibenthic (and bradfordian) species (*Tharybis, Diaxis*, etc.) since the 1960s.[16] Over time, they described a catalogue of new genera and species from near-bottom tows worldwide. Recently Silke Laakmann, Elena Markasheva, and Jasmin Renz[17] have used DNA sequences, trying to make sense of the relationships among bradfordian families, noting that three more families (all so far known mostly from females: *Parkiidae, Rostrocalanidae*, and *Kyphocalanidae*) and at least 30 genera have been added to the list since 1973. They concluded that genes they studied did not as yet help much (see Chapter 20).

Janet has been a lively participant in expanding these lists of epibenthic genera and species. She described new species from near-bottom waters near New Zealand in (at least) 1969, 1999, 2001, 2001 again, 2003, 2004, 2005 (probably more), and 2014. She sorted the 2005 example[18] from a collection with a net attached to the towing warps of a bottom-trawl that sampled 6 cm over the seafloor. It was a new species of *Xantharus*, a bradfordian genus defined in 1981 by Andronov. In her resulting article, Janet's species, *X. cryeri* (Figure 5.4), is depicted with details of every seta available for checking. *Xantharus* species belong, under Janet's definition, to the bradfordian family Scolecitrichidae.

This communal and global "epibenthic copepods" project has far to go for many families, with distinctive arrays emerging in different geographic areas and depth ranges. Epibenthic habitats will provide the alpha taxonomists of copepods with antennules, maxillary setation, and fifth legs to examine and draw for decades to come. In a 2014 paper[19] Janet and coauthors described an epibenthic copepod, *Pinkertonius ambiguous*, that did not fit any already recognized species, genus, or family. I will come back to *Pinkertonius* in Chapter 20.

My own connection to this is that an Oregon State graduate student, Diego Figueroa, began work on the epibentic (though not bradfordian) genera *Ridgewayia* and *Pseudocyclops* as a student at Oregon State University. He worked with specimens collected (I got to help him a few times) in anchialine waters (below sea level, in caves and crevices connected to the sea) of the Galapagos and in nearshore coral rubble elsewhere.[20] Again, it remains to be seen what will emerge from the epibenthic mine of new copepod species. See more on Dr. Figueroa in Chapter 21.

Janet's interest in family relationships and evolution in deep-sea and all copepods persisted over decades, leading in 2002 and 2004 to summaries and speculation[21, 22] about the evolutionary history of the invasion of near-bottom habitats by multiple families of calanoid copepods, of epibenthic species moving into more strictly pelagic water layers. Teaming with Geoff Boxshall, Shane Ahyong, and Susumu Ohtusuka in 2010,[24] Janet participated in deriving a statistically based phylogeny of 40 calanoid families, plus a hypothetical calanoid ancestor. Chapter 20 includes an essay about that paper.

An Oceanographic Aside

Throughout her career, while mostly focused on copepod systematics, Janet Bradford-Grieve has contributed oceanographic insights. An early example is another 1969 paper.[25] She had studied a paper published 40 years before by the Irish zoologist George Farran about the copepods of the British Antarctic (*Terra Nova*) Expedition of 1910. Sampling on the way had occurred along a transect across the long, northwestward extending seafloor ridge above 250 m depth on the shelf off North Island, New Zealand. In the vicinity of Three Kings Islands quite shallow sampling had caught ten species of notably deep-living copepods: *Undeuchaeta plumosa* and others only expected in tows from below about 500 m. Curious how they came to be near the surface, Janet checked through more recent physical oceanographic data, some of it in obscure technical reports, to show that the sampling stations were located in a band of surface water consistently colder than that on either side of the ridge.

Reviewing the physical oceanography, she could explain both the cold band and its unlikely copepods. The easterly (westbound) trade winds drive flow toward Australia where the coast steers it south as the East Australian Current and then east again across the northern Tasman Sea. Approaching the ridge, the deeper layers of the flow are forced to the surface by the ridge. The upwelling carries deep-sea plankton up and over. Janet has participated in many marine ecological analyses at National Institute of Water and Atmospheric Research, coupling ocean biology, sometimes including whales, squids, and plankton other than copepods, with other aspects of oceanography. She has been for many decades an important coauthor in ocean studies near New Zealand and the Pacific sectors of Antarctica.

Something More About Deep-Sea Respiration

The deep sampling off California by Erik Thuesen and Jim Childress produced many puzzling specimens, particularly among the Augaptilidae and Heterorhabdidae. I sent Janet Bradford-Grieve and Taisoo Park many specimens for checking. One or the other helped with every referral. Their taxonomy appears in one of the significant results from this work:[26] a demonstration that the robust, red-bodied copepods (*Gaidius* and *Onchocalanus*), which we termed "Muscular sinkers," and the slender ambush predators (e.g. *Euaugaptilus*, *Disseta*), which we termed "Thin-muscled floaters," differ in "metabolic poise" (Figure 5.5). The muscular sinkers that actively search for meals have low enzymatic activity for anoxic metabolism and high activity for oxidative metabolism. The floaters are the opposite; the short but powerful moves they make are too fast to operate on the oxygen-dependent energy attainable at the low oxygen concentrations in deep waters off California. In fact, oxygen is low in much of the deep sea.

A third group, deep-sea species over 12 mm long with quite solid bodies (*Megacalanus*, *Bathycalanus*, and *Gaussia*), which Erik Thuesen et al.[26] termed "Giants," are relatively well supplied with both classes of enzymes. Clearly, behavioral complexes are supported by appropriate enzymatic ones. Of course, that is true of all animals from snails to squids, from tree sloths to marmosets.

California's Pacific Produces a New *Megacalanus*

The Giants group at depth off California included a distinctive variety of *Megacalanus* closely related to *M. princeps* Wolfenden, 1905. My specimens from deep waters off California, both females and males, had a thin crest on their

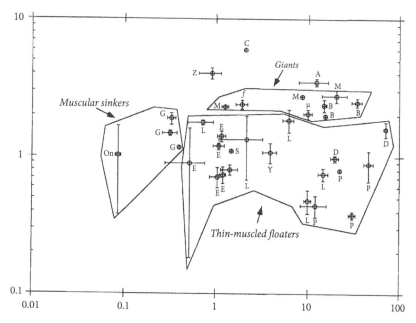

Figure 5.5 Relationship of enzymatic activities for aerobic metabolism (Citrate synthase) versus those for anaerobic metabolism (Lactate dehydrogenase). From Thuesen et al.[26] By family or genus: A-*Arietellus*, B-*Bathycalanus*, C-*Calanus* (surface living), D-Heterorhabdidae, E-Euchaetidae, G-*Gaetanus*, L-Lucicutiidae, M-*Megacalanus*, On-*Onchocalanus*, Y-*Metridia*, Z-*Pleuromamma* (migrates to surface), *f* and µ-*Gaussia* (mesopelagic). Note that *Paraeuchaeta* (all of the "E" species) are more muscular than D-Heterorhabdidae or A-Arietellidae. Permission, Inter-Research, *Marine Ecology Progress Series*.

foreheads (Figure 5.6), a little like the ridge on a Corinthian war helmet. Hoping to describe a new species, I worked very carefully through a reasonably large set of specimens, comparing them to specimens from several old collections from the eastern-tropical Pacific, the North Atlantic type locality and elsewhere. I could not find more than the one distinctive feature, the crest, though specimens with it were only from the eastern Pacific. Thinking of the biological species definition, I did not give it a name. But I did write a paper[27] about it for a volume of papers celebrating Bruce Frost's 60th birthday.

Megacalanus species have some unusual features. The antennules, possibly the prosome surface, some mouthparts, and the thoracic feet bear tiny structures (Figure 5.7, A and B) that Carl With named *maculae cribosae* (sieve spots) in 1915. They are circles, 20 µm or less in diameter, of up to a dozen small spherules embedded in or just under the exoskeleton. Each little ball is attached by a visible (with a compound microscope) fiber to a central body beneath,

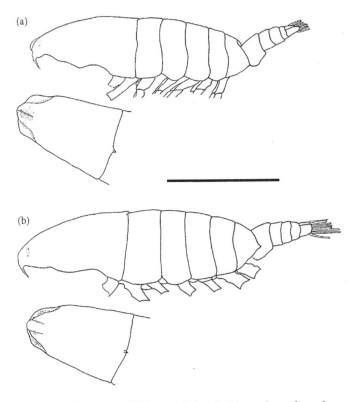

Figure 5.6 Habitus drawings of (A) crested, female *Megacalanus* from deep waters off Santa Barbara, California, and (B) crestless, female *Megacalanus princeps* from off Hawaii. Scale bar is 4 mm (total length is 10. 4 mm). From Miller.[27] Permission, Springer Nature.

and that attaches to what looks like a nerve extending toward the center of the body. These are also found in other genera of the family Megacalanidae, but not all. They could be sensory, points of release for pheromones, or your guess is as good as anybody's. Just proximal to the strong distal spines on the lateral exopodite corners of the second thoracic legs are large glands (Figure 5.7, C) with openings to the water just proximal (closer to the body) to the spines. They are packed with clear secretion granules. Attempts at staining them failed for several likely compositions, but they vanished instantly when a little chloroform-methanol mixture was dropped on them. Apparently they are lipids. The glands are known (Chapter 9) to be the source of bioluminescent secretions. The lipids may be the carrier stuff for luciferin and or luciferase to be released during a swing of the leg. This could result in a sort of visual shock for a small attacking fish.

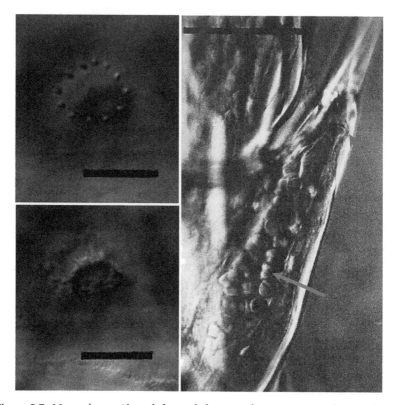

Figure 5.7 *Megacalanus*. Above left: exoskeleton surface expression of a macula cribosa (*mc*). Below left: the "nerve nexus" below the macula cribosa with fibers rising to the spherules at the surface. Right: secretion granules (blue arrow) in the gland that opens (red arrow) just proximal to a lateral spine on exopod segment 1 of P2. Differential interference contrast, light microscopy. Scale bars left are 20 μm. Scale bar right is 0.2 mm From Miller.[27] Permission, Springer Nature.

Janet Bradford-Grieve appears again in this story. She was recruited in 2007 to do what I did on the Thuesen-Childress cruises: examine specimens alive at sea and identify the species from voucher specimens then or later. She participated in a cruise aboard the German research ship *R/V Polarstern* that was part of a Census of Marine Life (COML) project organized by Ann Bucklin (details in Chapter 18). The project intended to expand as much as possible the species lists of marine plankton. On the cruise, Peter Wiebe of Woods Hole deployed depth-separating trawls with relatively fine mesh to depths near the bottom along a station transect down the Atlantic from Spain to South Africa. The goal was to couple high-definition morphological study of plankton, in

Janet's case copepods, with DNA sequencing ("bar coding" with Cytochrome Oxidase I and other genes) done immediately after capture using unpreserved specimens.

Captures of various species of large, very deep-living copepods of the Megacalanidae led Janet, the molecular geneticist Leocadio Blanco-Bercial, and fellow copepod expert Geoff Boxshall to gather specimens from global collections, examine them in detail and revise the family's taxonomy.[28] Janet's drawings of the crested form look like my drawings; that's good. She rated one more morphologic feature on the males captured off California as important: "presence of a fused gripping element on ancestral segment XIX on the right antennule (this element is not present in *M. princeps*)." I'll show some results from the gene sequencing when I come to its copepodological aspects (Chapter 18).

Given another character and some gene differences, Janet decided to name the species *Megacalanus frosti*, respecting its initial description in Bruce Frost's festschrift. Because they differ by a small number of characters, *princeps* and *frosti* do appear as closest cousins among the Megacalanidae in evolutionary trees based on both morphological and a few DNA-sequence distinctions. The full, formal, new name of my crested form is *Megacalanus frosti* Bradford-Grieve, 2017. When patronymic names are given to plants and animals, an important taxonomist can be celebrated by description of an animal captured only by big-budget sampling 3,000 m down in the Pacific, or maybe by climbing up 80 m in trees living only in rain forest six days hike eastward from Peru's Ucayali River.

The entire *Megacalanus* revision[28] is a masterpiece of taxonomic observation, illustration, and argument, much of it clearly by Janet Bradford-Grieve. She named three new species of *Megacalanus*, six of *Bathycalanus* and one of *Elenacalanus*. The last is a new genus divided as distinct from *Bathycalanus*, though the molecular genetics imply a close connection. She assigned five species to it, including the new one, *Elenacalanus tageae*. Drawn figures, 637 of them, for every species in the family, are sharply detailed and accurately depict shapes and critical characters.

Janet Bradford-Grieve's vast array of work has been widely recognized. She is a fellow of the Royal Society of New Zealand, has received an extensive list of recognitions, has been president (2008–2011) the World Association of Copepodologists, and was warmly honored by that group at its 2017 meeting. She told me by email the summer of 2020 that she has now largely retired. She also suggested that I look at the fifth legs of female *Pseudotharybis* (an aetideid), a character not found in any *pelagic* aetideid. I might do that, but Taisoo Park also needs attention.

Figure 5.8 Taisoo Park at his microscope station in 1986. The black compound microscope is a Wild M-5, workhorse in many copepod labs in the 1960s to about 1990. Its arm just below the eye piece prisms is a *camera lucida*, used for making drawings with precisely correct dimensions. Photo from T. S. Park.

Taisoo Park, Another A-list Alpha Taxonomist of Marine Copepods

My other expert helper with deep-sea copepods collected to the west of Southern California was Taisoo Park, who was introduced in Chapter 3. Dr. Park has used three names. He published as Tai Soo Park before he became a US citizen, then Taisoo Park afterward. For personal business purposes, he is Edward Taisoo Park. As mentioned earlier, he was teaching at Busan Fisheries College when he obtained an international grant to study in Hawaii at a National Marine Fisheries Service lab. That coincided with a sabbatical visit there by Prof. Paul Illg of the University of Washington. He taught basic techniques for dissecting, observing, and identifying copepods. Finishing that opportunity, Taisoo returned to Busan and teaching. At age 31, after six years on the staff at Busan, he felt a need for more education. He had remained in touch with Paul Illg, who offered him a chance to study for a PhD at the University of Washington. Taisoo and his wife moved to the United States, where he worked on that degree for five years, much of the time at Friday Harbor Lab in Puget Sound. A majority of his graduate course work was in zoology.

Graduating in 1965 as Dr. Park, he was taken on as a research assistant professor at the University of Maryland, working at its lab in Salomon on Chesapeake Bay plankton. What he really wanted, however, was a job in Korea. To improve his resumé for that search, and after only a year at Salomon, he took an assistant scientist job at Woods Hole Oceanographic Institution (WHOI). That had been offered the day after giving a seminar there at the invitation of the late George Grice, a fellow copepod taxonomist. Taisoo still expresses regret for leaving Salomon early and on short notice. As it turned out, there simply were no appropriate jobs in Korea in the 1960s.

His first study at Woods Hole was identification and description of copepods from the Caribbean Sea and Gulf of Mexico[29] collected by George Grice in 1967. The samples were from vertically stratified net hauls. The mesh size was 239 µm, considered quite fine in the 1960s, before wide attention was paid to *Microcalanus, Oncaea, Oithona, Xantharus* (Figure 5.4 above), and other very small groups. In the 16 samples from six levels (0 to 3,000 meters) Taisoo found 178 species, describing and naming 28 of them as new. All of those came from studies like Fleminger initiated on *Clausocalanus, Eucalanus,* and other genera: examinations in exhaustive detail of numerous specimens then thought to belong to species with well-delimited variations. Taisoo found that previously unrecognized subgroups in twelve genera displayed species-level distinctions. Three new species that he named in the genus *Heterorhabdus* later became part of his global study of the calanoid family Heterorhabdidae. By 1970, when his Woods Hole study was published, and certainly in light of his expertise in respect to copepods in the Gulf of Mexico, Taisoo was offered a job at the Texas A&M University's marine lab in Galveston. He and his family moved there. When the Parks later moved close to La Jolla, California, their son remained in Texas. He serves in Houston as a police officer.

Planning for more work at WHOI, Taisoo had learned about a large collection of plankton hauls taken from the US Antarctic Program ship *USNS Eltanin* that was operated by the National Science Foundation (NSF) in Antarctic waters from 1962 through 1972, a total of 52 cruises. They were the same samples sorted under Janet Bradford's supervision. A navy freighter converted for oceanographic work, *Eltanin* was 81 m length-overall and ice-strengthened. On many of her cruises a so-called 10-foot midwater trawl (mouth area ~7.4 square meters, pulled down by a depressor vane and outfitted with plankton mesh) was hauled through the water column producing massive zooplankton samples that were archived at the Smithsonian Institution. Taisoo got an NSF grant to work on them that transferred to Texas A&M. He published seven monographic studies about particular copepod genera or families. Typical was a 200-page treatise on Aetideidae and Euchaetidae published in 1978.[30] That one focused his interests toward the pair of genera *Euchaeta* and *Paraeuchaeta*. He also went back to

work on copepods living in the Gulf of Mexico.[31, 32] Taisoo advised four doctoral candidates, three of whom produced copepod studies, then became ocean
ecologists with NOAA. He was also on prominent copepodologist Frank Ferrari's
committee. Frank's major professor was Leo Berner at the Galvaston Lab. One of
Taisoo's last Texas papers was coauthored with Frank, who had a long career as a
crustacean curator at the Smithsonian (NMNH).[33]

By 1991 Taisoo was eligible to retire, and he did. Actually, he had another plan.
He was powerfully impressed, he told me, by the Frost and Fleminger studies of
Clausocalanus, the main insight being that to fully understand a related group
of marine copepods it was necessary to study them in all of the oceans, not just
a place somehow convenient like the Gulf of Mexico offshore from his lab, or
Southern Ocean waters accessible via the vast *Eltanin* collections. He applied
to NSF to support a globally comprehensive study of the Euchaetidae. Michael
Mullin invited him to move to Scripps in La Jolla near San Diego to work on the
collection there and with borrowed specimens from institutions everywhere.

Much of the work had already been done, apparently, so Taisoo's monograph[2]
on the family Euchaetidae was published by Scripps in 1993. His book is too vast
to have taken only two years. Look ahead to Figure 6.8. That will help you to see
how beautiful the species of *Euchaeta* and *Paraeuchaeta* can be. Deeper living
ones have different color patterns from species to species. Surely Taisoo could
not see that in the long dead samples he mostly worked on, but the subtle shape
variations in the generally similar species of the family clearly fascinated him.
The monograph is a document for specialists, so its flavor can only be vaguely
conveyed. It deals in the morphologic specifics of the 75 species Taisoo managed actually to see out of the 104 he considered likely valid after declaring some
synonymies. Among those missing, he had no chance to examine Russian taxonomist M. V. Hepner's specimens of a dozen new *Paraeuchaeta* species collected
at all depths over the Kamchatka Trench. There were a few others he could not
obtain.

The monograph includes 89 plates of drawings, almost all of just the details
distinguishing different kinds at the species level. For example, (Figure 5.9) shows
characters for *Paraeuchaeta copleyae*, which Taisoo named for Nancy Copley of
WHOI. She has worked for many years on copepods from sundry collections
by Peter Wiebe. Nancy sent Taisoo specimens Peter and others had collected at
about 1,900 m depth from over a hydrothermal vent field of the Guaymas Basin
in the center (27°N) the Gulf of California. In that relative isolation *P. copleyae*
presumably evolved its distinctions from its more oceanic relations.

Park established that the genera can be subdivided into more closely related
groups: three in *Euchaeta* (plus *E. spinosa* less related to any of those) and six in
Paraeuchaea (plus three others less assignable). He proceeded to speculate on the
likely phylogeny of the genera based on tabulations of the characters shared within
the groups. The resulting relationship charts were defined by *synapomorphies* and

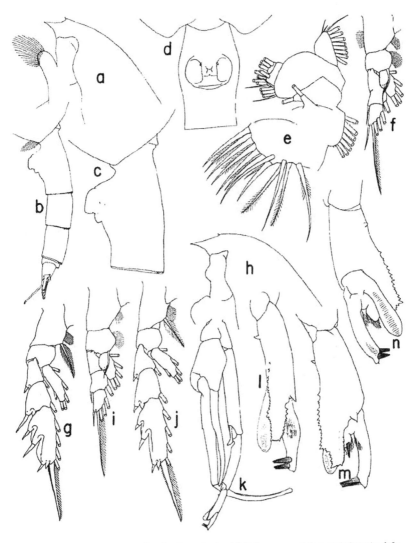

Paraeuchaeta copleyae, new species, female: a, forehead, left; b, urosome, left; c, genital somite, left; d, do, ventral; e, maxillule, first inner lobe separated, posterior; f, first leg, anterior; g, second leg, anterior. Male: h, forehead, left; i, first leg, anterior; j, second leg, anterior; k, fifth pair of legs, anterior; l, distal exopodal segments of left 5th leg, anterior; m, do, lateral, tilted clockwise; n, do, medial.

Figure 5.9 Figure 26 from Park (1993)[2] drawings of significant characters of *Paraeuchaeta copleyae* with original caption. Permission, T. S. Park.

plesiomorphies (Figure 5.10). Those terms were generated by a sort of revolution in taxonomic thinking,[34] an attempt to move closer to an objective basis for higher level classifications (families, order, etc.). Actually, they are new names for long recognized distinctions. A *synapomorphy* is a character shared by members

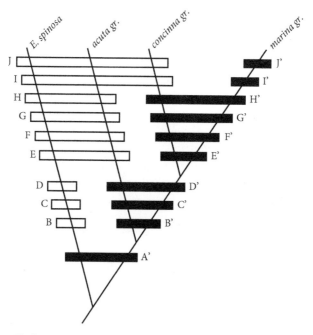

Figure 5.10 Phylogenetic relationships among species groups of the copepod genus *Euchaeta*. Synapomorphies are black bars. Plesiomorphies are open bars. See text for those terms. From Park.[2] Permission, T. S. Park.

Group sizes:

E. spinosa, singly distinctive;

acuta group (5 species);

concinna group (4 species);

marina group (4 species).

of two or more groups (hair in dogs and monkeys) and thought to have been present in the earliest common ancestor population of all the species sharing it. A *plesiomorphy* is a character common to members of two or more groups, but not derived from a common *and quite recent* ancestor (mammals have segmented spines, but so do fish). Park did not list the groups' *apomorphies*, characters unique (and supposed to be novel) that are shared only within a group. It is common to be uncertain about the status of a character's place in this scheme.

Taisoo Park spent the rest of the 1990s working through the same collections to find many representatives of the family Heterorhabdidae; eight genera (*Disseta, Heterorhabdus*, and others), all with left and right caudal furcae of unequal lengths. That was the interval in which he was helping me with puzzling species. He published another extraordinary monograph on that family in 2000.[35] Then, living in north San Diego County, he took up golf. If I have the story right, both he and his wife played often. I cannot make any connection between golf and copepods, except that there are surely diaptomids and cyclopoids in the water hazards. If Taisoo looked for them, he didn't mention it to me.

A General Impression of A-list Taxonomists

From working with copepod taxonomists at the levels reached by Fleminger, Frost, Bradford-Grieve, and Park, I have the sense that they were simply blessed with or worked to develop especially powerful memories for shapes and very subtle differences in shape. For the entire vast array of calanoid genera, if one handed Abe Fleminger a slide with a specimen, he would look at it with his stereomicroscope, promptly announce its genus name, and very often offer a first call for a species name. Those were mostly correct, especially for California Current species. When I sent any calanoid specimen to Janet or Taisoo, an identity would come back by the next internet message. And it is not just a memory for shapes, as those shapes are paired with names from what is a voluminous literature with dates from over two centuries.

As mentioned before, I am convinced that drawing has something to do with seeing details. I also think it has a role in developing memories of forms and in understanding of relationships. The mental quiet usually imposed by drawing, and the determination to find every discernible feature of the animals before them, revealed things I could find when directed by, say, Taisoo, where to look. However, all the A-list copepod taxonomists can see distinctive things I would miss just scanning surfaces myself, even if I were also drawing. Looking at specimens in the thousands from a vast array of sizes and shapes kept filling their extraordinary memory banks, and the safes in those banks seemed to be infinitely capacious.

Taisoo Park was committed to drawing every distinctive copepod he came upon. When I sent him specimens of species he had not seen before, I would get his opinion on its identity. That would be accompanied by a very polite request to stain it, to draw it, and to mount its limbs on slides to be drawn also. These did not lead to descriptions or re-descriptions necessarily. I think he was just using the drawing process to seal the forms into his memory. He eventually contributed his collection of drawings from a very large number of copepod species to the Smithsonian. He told me it is in ring binders and can be viewed there by anyone. Probably only aspiring copepodologists will apply.

There are more such minds at work on copepods (Shuhei Nishida, Geoff Boxshall, Rony Huys, Susumu Ohtsuka, Elena Markasheva, N.V. Vyshkvartseva, and so on). There are some younger talents coming along, and I hope that research systems will persist in supporting them.

A Note on Sources

Details of the development of the taxonomic literature for copepods, the shifts from briefly described distinctive specimens to named species in known genera, to shifts of those into different older or newly established genera, is not my skilled library work. Some has come from publication lists provided by individual scientists or available on websites, theirs or their institution's. For marine plankton the work of Claude Razouls and colleagues at Observatoire Oceanologique de Banyuls is invaluable. They continue a task first undertaken by the Dutch copepodologist Willem Vervoort (1917–2010), who was also a world-class expert on hydroids. He gathered virtually all of the copepod literature ever written and catalogued it. A recurringly updated marine version is maintained at https://copepodes.obs-banyuls.fr.

There are both French and English versions. This catalogue from Banyuls covers marine planktonic copepods. For the oceanographically inclined it is a sort of *vade mecum* ("handbook," from the Latin "go with me"). It includes vast numbers of the original drawings, provides careful work on synonyms, and has detailed listings of naming priorities. A great deal of ecological information is also included, with citations. The World Register of Marine Species (WoRMS) website for copepods also provides access to currently accepted nomenclature and is maintained by T. Chad Walter and Geoff Boxshall. Access to the reliable status of a name can usually be obtained by entering it in Google and opening the WoRMS article for it. Try it for the *Parabradyidius spinibasis* mentioned above.

References

1. Miller, C. B. (2004) *Biological Oceanography*. Blackwell Publishing, Oxford.
2. Park, T. (1993) Taxonomy and distribution of the marine calanoid family Euchaetidae. *Bulletin of the Scripps Institution of Oceanography* 29 (i–xi): 1–203.
3. Boxshall, G. A., & S. H. Halsey (2004) *An Introduction to Copepod Diversity*. The Ray Society, London.
4. Bradford, J. M. (1972) Systematics and ecology of New Zealand central east coast plankton sampled at Kaikoura. *New Zealand Department of Scientific & Industrial Research, Bulletin* 207: 1–89.
5. Rose, M. (1933) *Copépodes Pélagiques. Faune de France* 26: 1–374.
6. Bradford, Janet (1969) New species of *Aetideopsis* and *Bradyidius* Giesbrecht (Copepoda: Calanoida) from the Southern Hemisphere. *New Zealand Journal of Marine and Freshwater Research* 3(1): 73–97.
7. Bradford, Janet (1969) New genera and species of benthic calanoid copepods from the New Zealand slope, *New Zealand Journal of Marine and Freshwater Research* 3(4): 473–505.

8. Bradford, Janet M. (1976) A new species of *Bradyidius* (Copepoda: Calanoida) from the Mgazana estuary, Pondoland, South Africa, and a review of the closely related genus *Pseudotharybis*. *Annals of the South African Museum* 72(1): 1–20.

9. Schulz, K., & E. L. Markhaseva (2000) *Parabradyidius angelikae*, a new genus and species of benthopelagic copepod (Calanoida: Aetideidae) from the deep Weddell Sea (Antarctica), *Mitteilungen aus dem Hamburgischen Zoologischen Museum und Institut* 97: 77–89.

10. Bradford-Grieve, J. M. (2003) A new species of benthopelagic calanoid copepod of the genus *Bradyidius* Giesbrecht, 1897 (Calanoida: Aetideidae) from New Zealand. *New Zealand Journal of Marine and Freshwater Research* 37(1): 95–103.

11. Bradford, J. M., & J. B. Jillett (1980) The marine fauna of New Zealand: Pelagic calanoid copepods: family Aetideidae. *Memoirs. N.Z. Oceanographic Institute* 86: 1–102.

12. Bradford, J. M. (1973) Revision of family and some generic definitions in the Phaennidae and Scolecithricidae (Copepoda: Calanoida) *New Zealand Journal of Marine and Freshwater Research* 7(1 & 2): 133–152.

13. Nishida, S., & S. Ohtsuka (1997) Ultrastructure of the mouthpart sensory setae in mesopelagic copepods of the family Scolecitrichidae. *Plankton Biology & Ecology* 44(1/2): 81–90.

14. Komeda, S., & S. Ohtsuka (2020) New genus and species of calanoid copepods (Crustacea) belonging to the group of Bradfordian families collected from the hyperbenthic layers off Japan. *ZooKeys* 951: 21–35.

15. Markhaseva, E. L., & F. D. Ferrari (2005) New benthopelagic bradfordian calanoids (Crustacea: Copepoda) from the Pacific Ocean with comments on generic relationships. *Invertebrate Zoology* 2(2): 111–168.

16. Andronov, V. N. (2002) The calanoid copepods (Crustacea) of the genera *Diaxis* Sars, 1902, *Parundinella* Fleminger, 1957, *Undinella* Sars, 1900 and *Tharybis* Sars, 1902. *Arthropoda Selecta* 11(1): 1–80.

17. Laakmann, S., E. L. Markhaseva, & J. Renz (2019) Do molecular phylogenies unravel the relationships among the evolutionar[il]y young "Bra[d]fordian" families (Copepoda; Calanoida)? *Molecular Phylogenetics & Evolution* 130: 330–345.

18. Bradford-Grieve, J. M. (2005) New species of benthopelagic copepod *Xantharus* (Calanoida: Scolecitrichidae) from the upper slope, eastern central New Zealand. *New Zealand Journal of Marine and Freshwater Research* 39: 941–949.

19. Bradford-Grieve, J. M., G. A. Boxshall, & L. Blanco-Bercial (2014) Revision of basal calanoid copepod families, with a description of a new species and genus of Pseudocyclopidae. *Zoological Journal of the Linnean Society* 171: 507–533.

20. Figueroa, D. F. (2011) Phylogenetic analysis of *Ridgewayia* (Copepoda: Calanoida) from the Galapagos and of a new species from the Florida keys with a reevaluation of the phylogeny of Calanoida. *Journal of Crustacean Biology* 31(1): 153–165.

21. Bradford-Grieve, J. M. (2002) Colonization of the pelagic realm by calanoid copepods. *Hydrobiologia* 485: 223–244.

22. Bradford-Grieve, J. M. (2004) Deep-sea benthopelgic Calanoid copepods and their colonization of the near-bottom environment. *Zoological Studies* 43: 276–291.

23. Bradford-Grieve, J. M. (2002) Colonization of the pelagic realm by calanoid copepods. *Hydrobiologia* 485: 223–244.

24. Bradford-Grieve, J. M., G. A. Boxshall, S. T. Ahyong, & S. Ohtsuka (2010) Cladistic analysis of the calanoid copepoda. *Invertebrate Systematics* 24: 291–321.

25. Bradford, J. M. (1969) Notes on anomalous British Antarctic (Terra Nova) Expedition copepod records in the Three Kings Islands (New Zealand) region. *Transactions of the Royal Society of New Zealand. Biological Science* 11(8): 93–99.

26. Thuesen, E. V., C. B. Miller, J. J. Childress (1998) Ecophysiological interpretation of oxygen consumption rates and enzymatic activities of deep-sea copepods. *Marine Ecology Progress Series* 168: 95–107.

27. Miller, C. B. (2002) A variant form of *Megacalanus longicornis* (Copepoda: Calanoida) from deep waters off Southern California. *Hydrobiologia*, 480: 129–143.

28. Bradford-Grieve, J., L. Blanco-Bercial, & G. A. Boxshall (2017) Revision of family Megcalanidae (Copepoda: Calanoida). *Zootaxa* 4229(1): 1–183.

29. Park, T. S. (1970) Calanoid copepods from the Caribbean Sea and Gulf of Mexico. 2. New species and new records from plankton samples. *Bulletin of Marine Science* 20(2): 473–546.

30. Park, T. S. (1978) Calanoid copepods belonging to the families Aetideidae and Euchaetidae from Antarctic and subantarctic waters. *Antarctic Research Series* 27: 91–290.

31. Park, T. S. (1975) Calanoid copepods of the family Euchaetidae from the Gulf of Mexico and western Caribbean Sea. *Smithsonian Contributions to Zoology* 196: 37–48.

32. Park, T. S. (1976) Calanoid copepods of the genus *Euchirella* from the Gulf of Mexico. *Contributions to* Marine Science 20: 101–122.

33. Park, T., & F. D. Ferrari (2009) Species diversity and distributions of pelagic calanoid copepods from the Southern Ocean. In *Smithsonian at the Poles; Contributions to International Polar Year Science*, ed. I. Krupnik, M. A. Lang, & S. E. Miller, 143–179. Smithsonian Institution Scholarly Press, Washington, DC.

34. Hennig, W. (2000) *Phylogenetic Systematics*. University of Illinois Press, Urbana. [Original in German.]

35. Park, T. (2000) Taxonomy and distribution of the calanoid copepod family Heterorhabdidae. *Bulletin of the Scripps Institution of Oceanography* 31: 1–269.

6

About Feeding I: Various Modes

What Do Copepods Eat?

Different kinds of copepod eat different things, and there are many kinds of copepods. The most restricted are the predators, usually having just one setup for getting prey. Versions that eat algal and protozoan cells, or pick at plant tissue, are more likely to be omnivores with a somewhat more versatile tool kit. I consigned parasites earlier to somebody else's book. Let's consider free-swimming predators first.

Copepods are small, so they are limited to prey smaller than they are; or, any larger prey must be suitable to grasp, control, and bite bits from. Most small animal meals are mobile, and the first problem is to detect them passing or just nearby. One solution is to modify the naupliar eyes so that prey can be seen, an adaptation of the species of *Corycaeus* (both sexes), *Copilia*, and *Sapphirrina*. In *Corycaeus* (Figure 6.1) the bilateral eyes have clear, convex corneas appearing in front view rather like car headlights (Figure 6.2). These focus light back through narrowing, transparent cones (light guides?) onto not-quite spherical secondary lenses, which are just ahead of the left and right retinas. Those consist of three elongate rhabdomes, the pigmented light-receptor cells typical of sighted arthropods. The field of view is very narrow, perhaps 3°, but the spherical lens is equipped with miniscule muscles that rotate it back and forth across the image at about 15 cycles/second.[1] Thus, some sort of spatial information, perhaps as rasters, like those from the projector in a now old-fashioned TV, can be conducted to the brain. Likely this amounts to two overlapping, albeit narrow, views of the water ahead.

Several laboratory studies of feeding by *Corycaeus* species were done in the 1980s. Moshe Gophen, a young Israeli planktologist, worked with Roger Harris at the Plymouth lab of the Marine Biological Association of the United Kingdom.[2] They put a few 10s to 1,000 brine shrimp nauplii (a sort of miniature, crustacean "lab rat") and 2 to 150 *Corycaeus anglicus*, which they had collected within sight of Plymouth's picturesque offshore lighthouse, into one-liter bottles and mechanically rotated those slowly for 24 hours. They compared the

Oar Feet and Opal Teeth. Charles B. Miller, Oxford University Press. © Oxford University Press 2023.
DOI: 10.1093/oso/9780197637326.003.0006

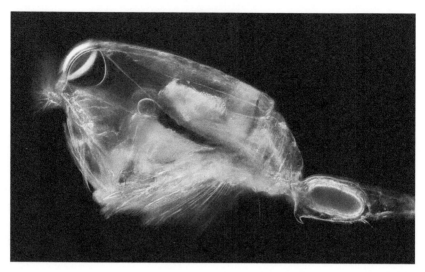

Figure 6.1 Photo of a *Corycaeus* female (about a millimeter long) focused on the left eye. The image-forming convex lens is in the anterior exoskeleton. A light-conducting cone stretches almost half the prosome length to a second spherical lens that moves light across the dark visual pigment of the rhabdomes. The system has been described as comparable to a telescope, with a light-gathering "object" lens and an eyepiece lens. Scanning has definitely been observed in the eyes of the related genus *Copilia*. Photo by Charles Krebs, Krebs Photomicrography, with his permission.

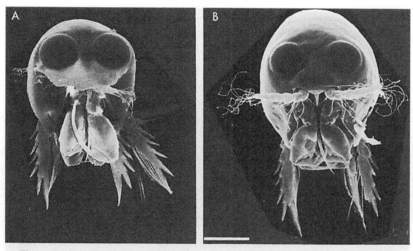

Fig. 1. Scanning electron micrographs of anterior view of male (A) and female (B) *Corycaeus anglicus* showing the paired eyes. Scale bar = 100 μm.

Figure 6.2 From M. Gophen and R.P. Harris[2] (their Figure 1). It emphasizes the likelihood that *Corycaeus* vision is binocular and also shows the arrangement of the antennae converted to grasping limbs. Permission, Cambridge University Press.

average numbers of nauplii eaten each day by the copepods between males and females and between dim light and complete darkness (bottles wrapped in two layers of aluminum foil). Maximum numbers eaten in the light were about 1.7 nauplii per copepod per day for both sexes. Maximum numbers in the dark were about 0.1 per day over a very wide range of availability, from a few to 1,000 per liter. Gophen and Harris reasonably concluded that *Corycaeus* eats what it can see.

A few years later, plankton ecologists Jefferson Turner, Patricia Tester, and Walter Conley[3] ran a more realistic experiment on *Corycaeus amazonicus* collected over the continental shelf off North Carolina. They skipped measuring feeding in the dark, but they used naupliar stages of the coastal copepod *Acartia tonsa* from laboratory cultures. Those are available to *C. amazonicus* in the nearby ocean at about the concentrations tested. They compared feeding rates of *single* female or male adults in four replicate treatments in 500 ml bottles, testing each of 10, 20, 30, 40, and 60 nauplii of each stage (N1 to N6) sorted alive with pipettes using a microscope. There were some ancillary experiments with and without phytoplankton cells present to confuse things (they did not). All in all, one of the authors (or somebody) counted out at least 12,000 individual nauplii that had to be recognized by stage. There is a warning in that: almost any science can involve very boring, detailed work that has to be done with attention and done well.

Plankton samples were collected at sea while the nauplii were counted into their bottles in the lab. *Corycaeus* were sorted from the samples coming ashore and added, then the bottles were slowly rotated with a motor to keep the nauplii close to evenly distributed during 16 to 27 hours (dim lights on the whole time). Predation rates on the N3 as individuals eaten per day (Figure 6.3) were the highest, but they were typical in that more were eaten as more were made available. That is termed the copepod's *functional response*. There is a hint of a limit being reached by females at about 50 nauplii per bottle. Numbers eaten were successively less for the N4 to N6 *Acartia*, especially for males. Turner, Tester, and Conley offered as explanations either that there would be faster appetite saturation by bigger naupliar stages, and/or that there would be better nauplius escape capability as development advanced. They noted in respect to escaping that sucking the older nauplii into pipettes was more difficult.

Corycaeus have one of the several, surely separately evolved, means for capturing their prey after they see them or detect their motion. Their antennae (A2) and maxillipeds (Mxp) have bases expanded for housing large muscles that can snap a long, curved, sometimes pointed, spine closed against that base (Figure 6.4). I have found no description of these limbs in action, but likely a meal is gripped against the mouth, held between the spines and the bases of both

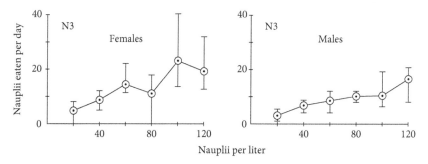

Figure 6.3 Mean and 90% confidence limits (4 replicates) for numbers of *Acartia* (an abundant coastal copepod) third nauplii eaten by one female or male *C. amazonicus* in 500 ml experimental bottles offered 10 to 60 nauplii. The rising rates are the "functional response" of the tested *Corycaeus*. From Turner et al.[3] Permission, Elsevier.

antennal and maxilliped pairs. All the other limbs of the oral area are reduced to a sort of abrading system. Prey must be scraped apart and the bits sucked in. In a separate paper, Jefferson Turner,[4] using scanning electron micrographs of their fecal pellets, has shown that adult *Corycaeus* also ingest large diatoms. Those could be visible at their relative size scales.

There are other copepods with elaborate eye lenses, particularly those in the family Pontellidae. Those may have spherical lenses just below the exoskeleton at the top of the head, as in *Labidocera* (discussed in Chapter 3), or plates on the surface of the head as in *Pontella*. The latter are believed to be image forming, as perhaps are all pontellid eyes. Pontellids are predators, or mostly so. Whether any of them find prey by sighting them is an open question. Michael Landry reported in 1978[6] that *Labidocera aestiva* collected off the California coast captured prey (nauplii and smaller copepodites) equally well in darkened bottles with no light during 24-hour experiments as in clear bottles with illumination during 16 hours daily.

Based on his visual observations and some holographic studies, Landry described *L. aestiva* as both swimming slowly forward by sculling with the antennae (not the antennules) and the maxillipeds, paddling that also pulls a "feeding current" in from in front and toward the body midline. They lunge at and grab prey coming close. Presumably the stimuli they receive are like those attracting *Euchaeta*, explained next. Capture of prey spotted visually appears to be the mode only of *Coryceaus* and the few related genera with similar eyes (*Copilia*[7] and *Sapphirina* females).

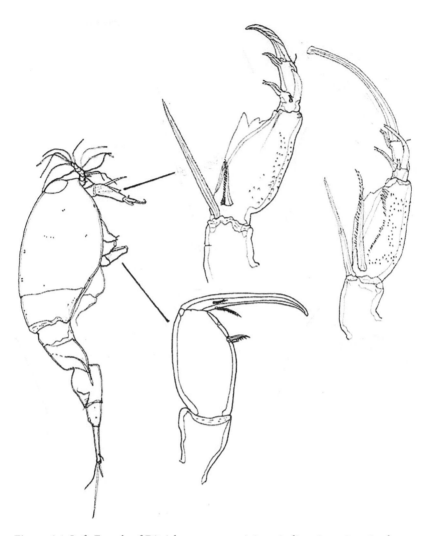

Figure 6.4 Left: Female of *Ditrichocorycaeus minimus indicus*; its swimming legs are not shown. Center above: female antenna; the tip is thought to be homologous with more standard antennal endopods. Right: male A2 with longer terminal spine likely involved in holding females during copulation. Center below: female Mxp, that of the male is similar. Total length of the female is approximately 0.82 mm. These drawings are by Olja Vidjak and Natalia Bojanić [5] of the Institute of Oceanography and Fisheries in Split, Croatia. Dissections of such small copepods with needles and mounting of limbs to prepare them for drawing require superb skills. The tiny mouthparts behind the labrum, visible on the female drawing between the A2 and the Mxp, are miniscule rasps. Drawings of those are in the Vidjak and Bojanić paper. Permission, Oxford University Press.

Figure 6.5 Scanning electron micrograph of central lengths of maxilliped setae from *Euchaeta elongata*. Scale bar (upper left) is 10 μm; the denticles are about 12 μm long. From J. Yen.[8] Permission, Wiley & Sons.

Predatory calanoid families have several modes of setation on the posterior mouthpart limbs, the maxillae or maxillipeds. In the Euchaetidae, forearm-like maxillipeds bear long, somewhat flexible setae that form a curving basket held open to the sides of the body. For example, in *Paraeuchaeta elongata*, 6 to 7 mm long, the "arm" (basis) and setae can stretch out and down about 3 mm. They swing in and forward to surround and hold prey. They are lined with sawtooth denticles (Figure 6.5) for gripping the prey. Stiffer short setae on the more anterior maxillules can then impale the prey and move them under the labrum for chewing and stuffing into the mouth.

In *Candacia* the setae of the maxilla (Figure 6.6) are actually saber-like, a rather Edward Scissorhands-like arrangement. They are flattened, apparently sharp along their concave edges and pointed. How those work is readily imagined; their feeding must be a sort of *World of Warcraft* scenario. The maxillipeds just behind are much smaller with short, brush-like setae on a series of terminal segments.

Jeannette Yen and Prey Finding by *Euchaeta*

In 2005 The Oceanography Society published a whole issue of its magazine, *Oceanography*, devoted to one-page autobiographies of women oceanographers. That was partly in recognition of how many currently active oceanographers are women, partly of the extra difficulties many of them have encountered compared to their male colleagues. Each essay included a self-selected picture, and Jeannette Yen[10] chose, of all things, a baby picture (Figure 6.7). I have left

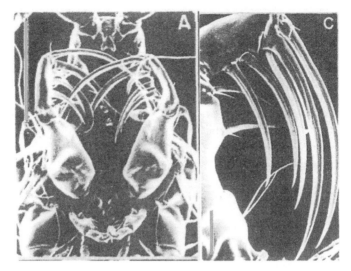

Figure 6.6 (A, left) (scale bar 0.5 mm) scanning electron micrograph of the mouth area of *Candacia bipinnata*, ventral view. Large limbs with scythe-like setae are the maxillae. (C, right) (scale bar 0.1 mm) close-up of those setae: smooth, stiff, capable of opening wide beyond the sides of the prosome. From Ohtsuka and Onbe.[9] Permission, Oxford University Press.

the caption, which suggests what an extraordinary family she was born into. I first met Jeanette Yen when she was studying oceanography at the University of Washington under Bruce Frost. I forget the exact circumstances, but in the 1970s and 1980s I was there often to plan research with Bruce.

Years later, out to dinner during a meeting, Jeannette told me of the pressure to learn and excel exerted by the family in the picture. She said that was quite typical for American-born Chinese children. Her learning included high-level performance on oboe (and glockenspiel), success as a pupil, and wide cultural understanding. She told me in August 2018 that surviving and excelling, despite being both a small person and the only Asian American in her huge New York City high school class, took strength and determination she did not expend gladly. The family pressure, she says, led to a near lifetime of self-doubt and thus rather radical over-preparation. That has lifted off, she says, after a recent health crisis. But it got her into Bryn Mawr to major in biology and biochemistry.

Jeannette did not seem certain during an interview in August 2018 why she chose marine science, but in early 2021 she told me this: "I knew I would do something aquatic because my childhood home was bordered by two sparkling brooks. I would spend hours, playing with the salamanders, crayfish, trout, water striders, caddisfly cocoons, minnows." In any case, she did apply for marine

The Yen family, where I sit in my father's lap, in the summer home of my uncle I.M. Pei.

Figure 6.7 Picture with its caption from Jeannette Yen's essay about herself in *Oceanography* **18** (1), 2005, page 243. The article also shows an example of "the meditative calligraphy of my Dad," a block of perfect 漢字. It is worth looking up the article to admire them. Permission, *Oceanography* is licensed under Creative Commons.

studies, but was turned down by every oceanography program. So in 1975 she took trips along the West and East US coasts, stopping at every marine lab and finally landing a position at University of Washington (UW), one of seven new students that year.

At that first early meeting, Jeannette had told me (I have an often-irritating habit of asking scientists I meet, "What are you working on?") that she was going to become the leading expert on the ecology of the genus *Euchaeta*. I'm not sure who was leading at that moment, but for 20 years or so it was certainly she. Her master's thesis had evaluated how much plutonium could accumulate in growing phytoplankton cells from very dilute concentrations in seawater. She had had some experience with radioecology as a summer intern at Argonne National Laboratory. Plutonium acting chemically is weakly toxic (supposedly less so than caffeine, but its alpha particle emissions are very carcinogenic), so working with any of it is less than fun. That might partly explain the shift to copepods.

Paraeuchaeta elongata
Hopcroft/UAF

2000 µm

Figure 6.8 *Paraeuchaeta elongata,* female, carrying an egg cluster. The scale is 2 mm (2,000 µm = 2 mm). One of the saber-like setae of the left maxilla is in sharp focus crossing the left antennule. Photo by Russ Hopcroft.

She said in 2018 that she first met *Euchaeta elongata* Esterly (Taisoo Park moved it in 1993[11] to *Paraeuchaeta,* the other genus in family Euchaetidae; Figure 6.8 here) when she accompanied fellow UW student Robin Ross on a collecting trip to Puget Sound. Robin, who later became a prominent student of Antarctic krill, was studying the ecology of euphausiids (a more formal name for krill) in that fjord. The females of *P. elongata* are large (for copepods), have bright red markings on the mouthparts, and females typically have a bulging bag of bright blue eggs attached to their genital area. The aesthetic experience of watching these predators with feeding setae like sabers steered Jeannette toward detailed studies of it and related species.

There had already been interest on Jeannette's part in working with Bruce Frost, because, she says, he worked so impossibly hard. She admired that, maybe because extended endurance for working through the endless details of science is characteristic of her as well. His obvious connection to her new interest in copepods led to his being her major professor for her PhD. Jeannette's dissertation work followed earlier studies by Frost on the functional responses of copepod feeding rates to availability of prey. We'll come to his own functional response studies on herbivory later.

Jeanette examined the functional responses of *P. elongata,* which she collected together with suitable prey from a branch of Puget Sound from 1977 to 1980. Oceanography graduate degrees usually take what at the time seem very long years, but in the United States they can also be pleasant years. Reports from her fellow students in that era suggest they recognized her brilliance and work ethic, but she already knew how to enjoy life. While Jeannette followed a line of work like Frost's, she says there was only occasional direct interaction with him. That seems to have been frequent as women started their parade into oceanography: somewhat distant relations with male major profs. Jeannette recalls being a bit afraid of Bruce Frost. That must partly have come from the intense focus that made him interesting, but also just because he "towered over her" (her words). That focus did encourage good results from and strong recommendations for all his students.

Jeannette's dissertation results included a study of gut contents showing that *Paraeuchaeta* mostly eats other copepods. She also ran feeding experiments like Landry's on *Epilabidocera* described above. One or two of these small predators were placed in 1-liter (also 4-liter) jars of filtered seawater with various, counted numbers of smaller copepod prey. In most experiments those prey were adult *Aetideus* (1.4 mm), female *Pseudocalanus* spp. (0.95 mm), or female *Paracalanus* (0.65 mm) from Puget Sound. The result was that the more suitable copepod prey available, the more of them *Paraeuchaeta* eats.[12] Older, bigger stages eat more and at rates accelerating faster with prey availability than younger, smaller stages. A limit is reached asymptotically (Figure 6.9), the typical form of copepod functional responses. For the paper, Jeannette chose

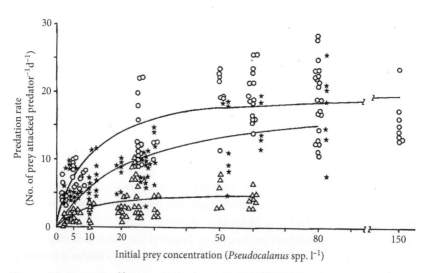

Figure 6.9 From Yen.[12] Rates at which *Euchaeta elongata* females (o), C5s (★) and C4s (Δ) feed on adult female *Pseudocalanus* at different concentrations (numbers per liter). The curves are fitted by Holling's "disc equation." Permission, Springer Nature.

to fit Holling disc equations to the data, but the paper also says Ivlev functions could have been used:

Prey eaten/time = [maximum ration]$[1 - e^{i\rho}]$,

where i is the best-fit Ivlev constant and ρ is prey density.

There you are: it would not be ecology without nonlinearity and asymptotes, *plus* contentious debates about what deterministic functions fit the wildly scattered data best. The full array of experiments showed that different prey are variously attractive to *P. elongata*, or are relatively easier or harder to sense and catch. Starving adults eat more when offered prey than those recently fed. Adult females collected from the field in spring and summer attacked fewer offered prey on average than those collected in winter, but the confidence limits broadly overlapped. It's all in the paper.[12]

Her diploma ready for framing, Dr. Yen obtained Fulbright and North Atlantic Treaty Organization (NATO) fellowships (thanks, along with talent, to those good recommendations) and spent a postdoctoral year at the Institute of Marine Biology of the University of Bergen in Norway. The rain there is even colder than the winter drizzle of Seattle. She worked, among others, with J. B. L. Matthews, one of several British planktologists then working in Norway (copepod ecologist C. C. E. Hopkins was a lead researcher at the University of Tromsø). She set up experiments presenting different copepodite stages of *Paraeuchaeta norvegica* with possible meals of cod eggs and larvae.[13] She had already studied feeding by *P. elongata* on Pacific Hake larvae, work published with fellow graduate student Kevin Bailey.[14] The new work in Norway expanded the number of variables tested, as well as testing a different fish and an Atlantic copepod.

The two *Paraeuchaeta* species are of comparable size, total adult lengths about 6 mm. Fourth, fifth, and adult copepodites of the *P. norvegica* she collected off Norway could capture and eat cod larvae. Adult females ate a maximum of around 8 (average ~6) larvae per day from concentrations of 10–15 per liter in 4-liter jars. They did not eat cod eggs. Eggs, while close in size to larvae, do not swim, so they make only very slight fluid disturbances by rising or sinking, apparently not enough for *Paraeuchaeta* to detect. Detection, Jeannette hypothesized, depends on motion-detecting setae on the antennules. As a crude test, she squeezed the antennules off some females with forceps, learning first that the breaks would scab over, the copepods surviving for two weeks. Captures of larvae dropped to 2 per day per copepod, which to me seems surprisingly great.

Cod have long supported important fisheries along the Norwegian coast, particularly in the vicinity of the Lofoten Islands that cross 68°N, a prominent spawning region for them. *Paraeuchaeta norvegica* is also abundant there, and

Jeannette showed that its predatory impacts on larval cod must contribute substantively to the high fractional mortality of larvae hatching from the massive clutches of eggs from female cod: as many as 2 million eggs at 2 years of age, and 7 million eggs at 6 years. So cod are feeding copepods as well as Norwegians, Icelanders, and fish-and-chips consumers everywhere.

Her European adventure not quite over, Jeannette took a research-faculty post in the oceanography program of the University of Hawaii, but she moved first to the laboratory of the Scottish Marine Biological Association at Oban on Loch Etive. Collections of *P. norwegica* were made from the Loch and photographed in a small aquarium just a centimeter from front to back. That allowed determinations of the orientations relative to their body of setae on the antennules as they are typically held at rest, that is, straight out to the sides. With questions about those setae still to answer, she also obtained a visiting scientist award from the Smithsonian Museum in Washington to do copepod morphometrics (measurements of body parts) with Frank Ferrari, a curator and a prominent copepodologist there. She measured the antennal setae of preserved specimens using a microscope fitted with a *camera lucida*. Scottish scientist Norman Nicoll had a part in the work and is a coauthor on the report.[15] That was published about a decade later. Such delays are common for a scientist living on "soft money," which was the case for Jeannette in Hawaii. They must shift from their finished but unpublished results to writing the next proposal, and then they must pursue the next run of experiments or observations.

The antennular seta results can be simplified. At two points along the antennule, on and near segments 3 and 13, there are groups of setae: some stout and straight, some curved; one points forward, one aft; one points dorsally, one ventrally. Motion of surrounding water will bend any of those away from the source, probably (certainly, as she showed later) firing stretch-sensor neurons attached to the setal bases. With a set near the head and a set well out along the antennule, the combined signals must provide the direction to the source of the motion with considerable precision. Jeannette credited Lauren Haury, then at Woods Hole, and colleagues[16] with suggesting "the setae of the first antennae [antennules] as the most likely receptors involved with detection of fluid disturbances." But with the orthogonal (at right angles) arrangement of the setae established for the Euchaetidae, she went to work on proving it. One step was to demonstrate the innervation of those setae and the mechanics of directional sensing of flow impinging on the antennule.

Finally established in Hawaii, Jeannette had the advantage of working with a talented (and married) pair of neurophysiologists, Petra Lenz and Daniel Hartline. Sectioning the A1 from various species for electron microscopy, they found bundles of axons passing through every segment.[17] Only a very of the few sections must have shown the lever-like neural connections at the bases of the

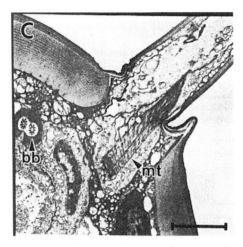

Figure 6.10 Electron micrograph of a setal base on the antennule of *Labidocera madurae*. Scale bar 5 µm. Labels: bb - basal body; **mt** - microtubules. Both of those are aspects of sensory nerves. The dark block of microtubules, would only pull against the nerve extension when the seta is bent toward the exoskeletal cup to the right of the base in the section. From Yen et al.[17] Permission, Oxford University Press.

setae and on one side only of each seta (Figure 6.10). Hitting the exact planes of these features about 10 µm across with a microtome knife cutting on a circular A1 cross-section, all but invisible in the face of a plastic imbedding block, requires the patience of a saint (or determined scientist). A nerve termination at a micro-tubule body (labeled mt in Figure 6.10) that extends through the flexible base of the seta was assumed to initiate depolarization in that nerve. It likely constitutes the sensor for flexing of the seta by impinging flow. The basal body structure (la-beled **bb**) is likely associated with molecular assembly of the microtubules.

The Lenz-Yen team then developed an elaborate scheme, rather like an ultra-miniaturized electrocardiograph, to record electrical signals as nerves depolarized in the antennule. Isolating the set-up from the vibrations in the building, even sounds, was a major part of the effort. Stimuli were applied in various ways, all generating tiny water movements at the antennule. In one ex-ample,[18] a tiny ball on a stick in water near the antennule made series of pulsed jumps, so that little pressure waves moved the setae. An example is shown in Figure 6.11. A pulse series of around 8 µm amplitude at around 3 millisecond intervals generated a roughly proportional series of voltage shifts along the an-tennule. After a sort of start-off chatter, the nerve signals phase-locked to the tiny hydraulic disturbances. The team generated examples from many species of copepod for this: one water-shift impinging on antennular setae produces one depolarization, and those can recur at very high frequencies. Extremely rapid

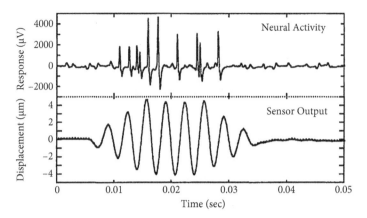

Figure 6.11 Below, displacement in microns of an instrumented ("sensor") solid near the A1 of a copepod (*Pleuromamma xiphias*) held gently in a clamp in an insulating drop of oil, the A1 extending down into seawater. Above, electrodes in the clamp and the water allowed recording (microvolts) of nerve depolarizations in the antennule. Repolarizations took 1 millisecond or less. From Gassie et al.[18] Permission, *Bulletin of Marine Science*.

neural repolarization is characteristic of pelagic copepods. The actual water displacements at a seta on the antennule can be sensed at miniscule scales. In the example shown it was calculated as only 70 nm (nanometers), that is, 70 billionths of a meter. There was other work along these lines, particularly with David Fields, who is still studying aspects of how copepods sense and respond to small-scale hydrodynamics of their habitat. In fact, he and others[19] have learned in molecular detail how the microtubule-triggers of setal motion sensors work.

Let's leave Dr. Yen there in Hawaii for a while (it's nice; she bought a coffee farm in Kona). I will come back to her in the chapter on mate-finding. Now let us consider copepods that feed on particles like protozoa or algal cells that "filter feed." Several somewhat unrelated questions have been asked about them: How much do they eat? How does filtration work?

Functional Responses: How Much Do Copepods Eat?

Ecologists have long been interested in what controls population sizes in plants (or, say, planktonic algae), herbivores (many copepods), omnivores, and carnivores. Early on they classified variations as density-independent and density-dependent. Stock changes independent of population density could be, for example, reduction by an unseasonal freeze that kills a lot of an organism. If conditions are good and a stock not overly exploited by herbivores or predators, then increase, apart from in-migration, is always density dependent: more parents produce more offspring. At the extreme when a population exceeds the carrying-capacity of the habitat, which definitely is density-dependent, starvation kills. In a predator-prey interaction, changes in the stock of prey will depend on its resources and reproductive capacity *and* on both the abundance and feeding responses of the predators. The density dependence works in two directions. The more predators, and the more each of them chooses and manages to eat, the greater the downward pressure on prey. The greater the availability of prey, the more of them that will be eaten, but also at some level the greater the amounts left over when predator appetites are saturated. This does get complicated, and it is very hard to analyze out in nature, say in the ocean. However, at least one part can be evaluated in laboratory experiments.

If you are lucky enough to have a full refrigerator, you do not open it, move everything to stove and table, then eat all of it. It is too much. If your fridge is not full but only offers a half sandwich, when you look in hungry you will surely eat all of it. Your eating responses have come to be called your "functional response," a term first used in a 1949 paper by M. E. Solomon.[20] Many varieties of copepods graze on algal and protozoan cells: diatoms, dinoflagellates, sundry other flagellated algae, and ciliates. They are said to be particle feeders. Bruce Frost, once established at the University of Washington, decided on experimental studies of the functional responses of particle feeding copepods. He worked with *Calanus pacificus*, available in nearby Puget Sound, feeding them single-cell diatoms.[21]

The basic idea was not original with Frost, but he improved the methods. Starting out, he simplified an experimental system from the complexity a copepod encounters in the local fjord. He tested feeding rates on just pillbox-shaped centric diatoms of one genus, *Coscinodiscus* (Figure 6.12). He compared rates on three species of different cell diameters and organic matter content (as carbon): 35, 75, and 87 μm diameter, respectively with 840, 1,644, and 3,334 picograms of carbon per cell. For all three foods, he measured feeding rates for groups of ten to 30 female *C. pacificus* on many cell concentrations in 3.5-liter beakers, very gently stirred with a motorized paddle so the diatoms remained suspended. Then at various intervals he used a hospital-type, electronic cell counter to estimate the numbers of cells remaining. The copepods were fed before and during the repeated sampling, because results from other workers had shown that initial feeding after starvation is much faster. Frost confirmed that for *C. pacificus*.

The numbers from the counts required interpretation. If you used a fine-mesh tea strainer to remove the cells from half the water in a beaker, then returned the filtered water to the beaker, the new concentration would be half the initial one. When a copepod removes a cell from the water in its aquarium (filters it out) and

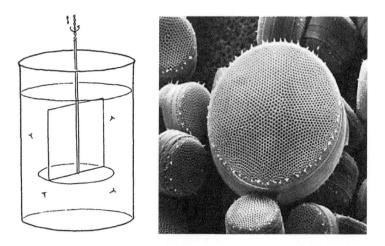

Figure 6.12 Left: Frost's stirring arrangement in a large beaker. Permission, Wiley & Sons. Right: example of *Cosinodiscus* frustule shape. SEM photo by Anne-Marie Schmid, of University of Salzburg, Austria, with her permission.

swallows it, that cell's share of the water remains in the volume, reducing the concentration, C, of remaining cells. That declines as more cells are eaten in a negative exponential fashion:

$$C_{after\ t} = C_{before\ t}\ e^{-gt},$$

where e is the base of the natural logarithms, g is the bulk "grazing rate" (volume cleared per unit time) for all the grazers in the container, and t is the amount of time (say, hours) between cell counts. Frost also took account of the potential for cell growth by running controls without copepods, just algal cells. From the exponential grazing function it is also possible to estimate the average of C over the interval t; Frost wrote it as $\langle C \rangle$. And since the feeding process can be treated as filtering, the rate at which the "average" copepod in an experiment is processing water to remove cells can be estimated, a so-called filtration rate:

$$F = Vg/N,\ \text{milliliters per copepod per time,}$$

where V is the container volume, and N the number of grazers removing cells. You don't have to think it through unless it's fun for you, but the amount of cells ingested by the average copepod in the experiment is

$$I = \langle C \rangle F,\ \text{cells per copepod per time (say an hour).}$$

That can be multiplied by the organic matter content per cell to estimate the rate at which nutrition was provided by the measured feeding.

The results[21] were simple, but news (Figure 6.13). The rate at which diatoms were removed (water was clear of them) went up as their concentration went up, but with a satiation point. That is the functional response of C. pacificus females. Frost's experiments showed also that bigger cells were eaten faster (greater F) than smaller ones, implying that the latter were harder to find or capture. Notice in the figure that to slow eating at cell concentrations above the satiation point, the filtration rate must fall off and does so in a negative exponential curve.

However, the total amounts of nourishment (measured as amount of carbon in organic matter) from smaller Cosinodiscus at the satiation point were comparable to the amounts consumed of large cells. That is why feeding on larger cells satiated faster than feeding on smaller ones. Bruce also compared females steadily fed before the experiment with others that had been starved awhile. The starved Calanus removed cells from the water faster, and they ate much more at first than shown in the "steady eating" data above.

A long list of questions came up at this point. Throughout the 1970s and 1980s, Frost and a legion of other workers experimented with many other aspects of feeding. Filtration rates of many common genera were estimated: Acartia from nearshore and estuaries; Eurytemora from brackish waters; Paracalanus, Pseudocalanus, and others among smaller oceanic species. The feeding frenzy on feeding experiments persisted about 15 years. Experiments were run to see whether the F values various copepods could achieve were sufficient to sustain respiration, growth and egg production at levels of nutritious particles available in estuaries and oceans. They often did not seem to be. However, it is very difficult to be sure exactly what individual copepods are encountering, whether special spots or layers of more concentrated particles could provide enough when the particles in grabbed water samples do not seem sufficient. Mostly there must somehow be enough food, because copepods persist. In all feeding experiments the data are scattered. Not all individual planktonic grazers, especially those brought in from the local fjord or ocean for testing, are doing exactly the same thing or have necessarily had close to identical experiences. At least generalities emerge from enough data, and suitable equations (Ivlev, Holling disc; take your pick or sketch in a line) run lines through the main tendencies.

It was asked whether grazing copepods would stop eating if edible particles were too few to support the energy cost of filtering. Frost's early experiments suggested not (Figure 6.13), but at low cell concentrations, changes in cell counts are only very slightly more than the counting errors. He did find evidence for a feeding threshold in one careful experiment with C. pacificus,[22]

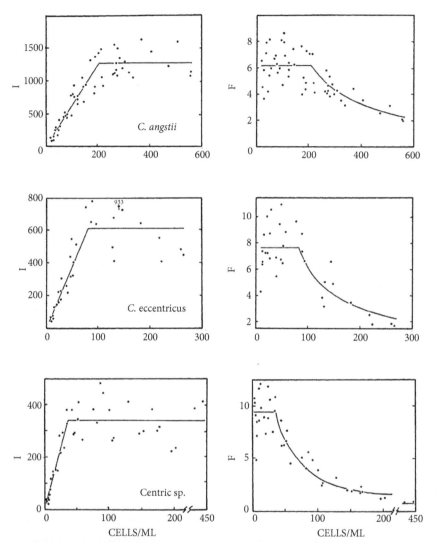

Figure 6.13 *Calanus pacificus* females feeding on different single-cell diatoms. Left: functional response curves: I, cells eaten/copepod/hour. Right: calculated water clearance rates: F, ml swept clear/copepod/hour. Top row: *Coscinodiscus angstii* (35 μm diameter). Middle row: *C. eccentricus* (75 μm). Bottom row: *Coscinodiscus angstii* var. *granulomarginatus* (87 μm). From Frost.[21] Permission, Wiley & Sons.

and many people working with other species have found that S-shaped curves fit best to their data: slow acceleration of ingestion at low particle counts, rising steeply with more particles available, then gradually leveling off. Maybe because people, for example, will often eat much more than required by their metabolic needs when faced with a banquet, it was asked whether copepods engage in superfluous feeding. Given the scatter of the data, it's hard to say. It's fair to guess they mostly do not overeat. However, I have seen *Calanus finmarchicus* fifth copepodites so stuffed with oil before diapause (Chapter 15) that their prosome articulations were all popped out straight. They were undoubtedly obese.

Experiments on selection from mixtures of different particle types were popular, such as diatoms versus dinoflagellates. Filtration rates were estimated for mixtures of particles collected in estuaries or the ocean. That was feasible because as the studies progressed, so did clinical instrumentation. Electronic particle counters were modified to count separate size categories (say, red cells vs. macrophages). One example was a study[23] by Sumner Richman, a professor from a small college in Wisconsin, while he was on sabbatical near Chesapeake Bay. He worked there with Donald Heinle, who by then had been long interested in the ecology of *Eurytemora* and *Acartia* in that estuary. The literature of marine biology is loaded with summer and sabbatical studies by faculty and students from inland areas. Richman and Heinle ran bay water through their Particle Data, Inc. automated particle-sizing and counting system before and after 5 to 25 copepods had grazed from it in bottles (Figure 6.14).

Most of the particles were confirmed to be algal cells. Those from 5 to 10 μm and those from 12 to >30 μm were grazed, but there was some mechanism allowing no filtration of the cells from 10 to 12 μm. What that mechanism is was a question leading to a whole new understanding of the life in water of small and very small organisms (see Chapter 7).

How Are Particles Captured by Copepods Vastly Longer and Wider?

How copepod filter-feeding works was considered long before most currently active copepodologists were born. Esther Lowe's mentor, H. Graham Cannon,[24] described some observations of *Calanus finmarchicus* feeding currents in 1928. He used a version (not well described) of a "compressorium" to hold a copepod on its back on a glass slide with its feet extending up into the bottom of a drop (a dome on the slide) of water containing dyed starch grains. He watched the flow made visible by the starch with a microscope, and he sketched the pattern with respect to the limb positions (Figure 6.15). "Sculling" by the antennae pulled water from in front, driving large eddies in the quadrants behind the extended antennules and the body, and in Cannon's view above the body. Those frictionally drove counter-rotating flow (red arrow in Figure 6.15) that passed in under the setal tips of the anterior swimming legs, then between the maxillipeds

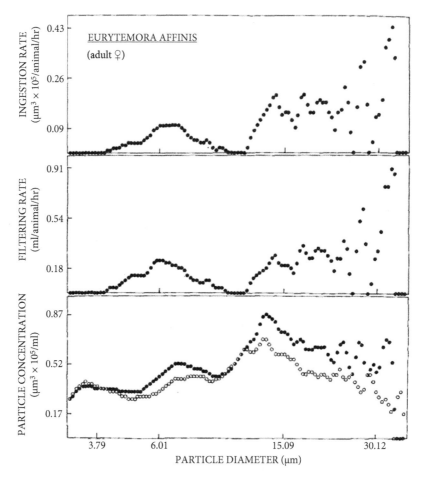

Figure 6.14 Bottom: Abundances of particles in narrow size ranges from 3 to 33 μm equivalent diameter (spheres of the same volume, regardless of shape) in Chesapeake Bay water before (solid dots) and after (circles) grazing by *Eurytemora affinis* females. Middle: implied filtering rates as milliliters filtered/copepod/hour for each counted particle size. Top: Ingestion rates as particle volume/copepod/hour. Plots are semilogarithmic. Note: No changes for particles <5 μm or from around 10 to 12 μm, and thus zero filtering rates and ingestion. From Richman et al.[23] Permission, Springer Nature.

and through the filters of the setal array on the maxilla. Those setae converge toward the mouth. Cannon supposed that the thick, stiff setae on the base of the maxillule, just anterior, would comb collected particles forward toward the mouth and between the mandibular gnathobases (the toothed jaws) meeting under the labrum from the sides.

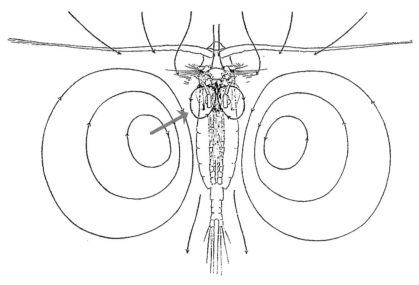

Figure 6.15 Ventral view of *Calanus* with flow streamlines during feeding sketched from dyed particle trajectories. From Cannon.[24] Permission *Journal of Experimental Biology.*

Cannon also drew a close-up of the arrangement of the mouthparts and the flow through the filters. It is worthwhile to copy it here (Figure 6.16), because looking down on *Calanus* from above, the feasibility of the filtering process seems very logical.

This understanding of filtering by the maxilla persisted a long time. It was cited by Sheina Marshall and A. P. Orr (for whom some biography is offered in Chapter 15). They provided[25] a careful drawing (Figure 6.17) of the maxilla's setae, setules, and the variable spaces between them almost three decades after Cannon's report.

The maxilla is a filter, but notions changed about how particles come to contact it. Professor Cannon did get much of it right. His stroboscopic estimates of the frequencies of limb movement are as accurate as measurements, since

> . . . the maxillipeds, maxillules, mandibular palps and antennae vibrate regularly, in the case of *Diaptomus* at the rate of about 1,000 times a minute and in the case of *Calanus* about 600 times a minute. Their movements are synchronous but not in the same phase. . . . The phase differences are such that the maxilliped commences its back stroke (outward movement) just after the antenna commences its forward stroke, so that these two limbs move almost in

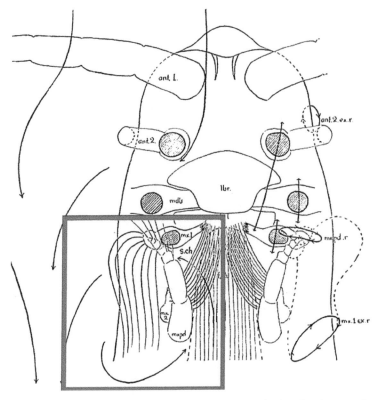

Figure 6.16 Enlarged view, ventral side up of the mouth limbs of *Calanus* and Cannon's concept of the flow pattern (as single-headed arrows) driven by those limbs while feeding. Movements of the limbs on its left side are indicated by double-headed arrows and trajectory ovals. His notion of the flow through the filter on the maxilla is indicated by the blue box. Cannon's codes for the now standard limb names: ant.1. antennule; ant.2. antenna; mdb mandible; mx.1 maxillule (labeled on the comb-like basis, with the palp setae shown on the animal's right); mx.2 maxilla; mxpd maxilliped; s.ch. suction chamber; f.ch filter chamber (closed ventrally by setae from the anterior thoracic leg); lbr labrum. Permission, *Journal of Experimental Biology*.

opposite phase. The phase difference between mandibular palp and maxillule is very small. The back strokes of antennae, mandibular palps and maxillules are faster than their fore strokes.

All that was revealed to Cannon, for the two species, by his strobe illuminator flashing 1,000 or 600 times per minute (17 and 10 cycles per second, or Hertz). He noted that the movements occurred in bouts. He added that the maxillae did

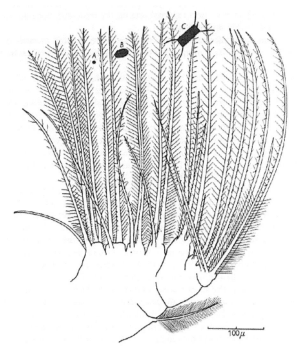

Figure 6.17 Medial-side view of the left maxilla of *Calanus helgolandicus* female, comparing three algal cell sizes: (A) *Nannochloris*, a green alga (1–2 μm), (B) *Syracosphaera*, coccolithophorid (6–8 μm), and (C) *Chaetoceros*, diatom (~20 μm), respectively. This was drawn partly to show that very small algae or other particles could pass through the grating. From Marshall and Orr.[25] Permission, Cambridge University Press.

not move much, unless becoming clogged with particles, they flapped, flinging the mess away from the mouth area.

Cannon left many questions unanswered, and some of his results were artifacts. To divide the old, almost paleolithic understanding from the current one of particle grazing by copepods, I declare a time out and will start a new chapter.

References

1. Downing, A. C. (1972) Optical scanning in the lateral eyes of the copepod *Copilia*. *Perception* 1: 193–202.
2. Gophen, M., & R. P. Harris (1981) Visual predation by a marine cyclopoid copepod, *Corycaeus anglicus*. *Journal of the Marine Biological Association of the U.K.* 61(2): 391–399.

3. Turner, J. T., P. A. Tester, & W. J. Conley (1984) Zooplankton feeding ecology: predation by the marine cyclopoid copepod *Corycaeus amazonicus* F. Dahl upon natural prey. *Journal of Experimental Marine Biology and Ecology* 84(2): 191–202.

4. Turner, J. T. (1986) Zooplankton feeding ecology: contents of fecal pellets of the Cyclopoid copepods *Oncaea venusta, Corycaeus amazonicus, Oithona plumifera* and *O. simplex* from the northern Gulf of Mexico. *Marine Ecology* 7(4): 289–302.

5. Vidjak, O., & N. Bojanić (2008) Redescription of *Ditrichocorycaeus minimus indicus* M. Dahl,1912 (Copepoda: Cyclopoida, Corycaeidae) from the Adriatic Sea. *Journal of Plankton Research* 30(3): 233–240.

6. Landry, M. R. (1978). Predatory feeding behavior of a marine copepod, *Labidocera trispinosa. Limnology and Oceanography* 23: 1103–1113.

7 Gregory, R. L., H. E. Ross, & N. Moray (1964) The curious eye of *Copilia. Nature* 201(4925): 1166. [cited by Downing (1972)[1]]

8. Yen, J. (1985) Selective predation by the carnivorous marine copepod *Euchaeta elongata*: Laboratory measurements of predation rates verified by field observations of temporal and spatial feeding patterns. *Limnology & Oceanography* 30(3): 577–597.

9. Ohtsuka, S., & T. Onbe (1989) Evidence of selective feeding on larvaceans by the pelagic copepod *Candacia bipinnata* (Calanoida: Candaciidae). *Journal of Plankton Research* 11(4): 869–872.

10. Yen. J. (2008) Jeannette Yen. *Oceanography* 18(1): page 243.

11. Park, T. S. (1993) Taxonomy and distribution of the marine calanoid copepod family Euchaetidae. *Bulletin of the Scripps Institution of Oceanography* 29: 203 pp.

12. Yen, J. (1983) Effects of prey concentration, prey size, predator starvation, and season on predation rates of the carnivorous copepod *Euchaeta elongata. Marine Biology* 75: 69–77.

13. Yen, J. (1987) Predation by a carnivorous marine copepod, *Euchaeta norvegica* Boeck, on eggs and larvae of the North Atlantic cod *Gadus morhua* L. *Journal of Experimental Marine Biology and Ecology* 112: 283–296.

14. Bailey, K. M., & J. Yen (1983) Predation by a carnivorous marine copepod, *Euchaeta elongata* (Esterly), on eggs and larvae of the Pacific hake, *Merluccius productus. Journal of Plankton Research* 5: 71–82.

15. Yen, J., & N. T. Nicoll (1990) Setal array on the first antenna of a carnivorous marine copepod, *Euchaeta norvegica. Journal of Crustacean Research* 10(2): 218–224.

16. Haury, L. R., D. E. Kenyon, & J. R. Brooks (1980) Experimental evaluation of the avoidance reaction of *Calanus finmarchicus. Journal of Plankton Research* 2: 187–202.

17. Yen, J., P. H. Lenz, D. V. Grassie, & D. K. Hartline (1992) Mechanoreception in marine copepods; electrophysiological studies on the first antennae. *Journal of Plankton Research* 14: 495–512.

18. Gassie, D. V., P. H. Lenz, & J. Yen (1993) Mechanoreception in zooplankton first antennae: electrophysiological techniques. *Bulletin of Marine Science* 53: 96–105.

19. Fields, D. M. (2014) The sensory horizon of marine copepods. In *Copepods, Diversity, Habitat and Behavior*, ed. Laurent Seuront, 157–179. Nova Publishers, New York.

20. Solomon, M. E. (1949) The natural control of animal populations. *Journal of Animal Ecology* 18(1): 1–35.

21. Frost, B. W. (1972) Effects of size and concentration of food particles on the feeding behaviorof the marine planktonic copepod *Calanus pacificus. Limnology and Oceanography* 17(6): 805–815.

22. Frost, B. W. (1975) A threshold feeding behavior in *Calanus pacificus*. *Limnology & Oceanography* **20**: 263–266.

23. Richman, S., D. R. Heinle, & R. Huff (1977) Grazing by adult calanoid copepods of the Chesapeake Bay. *Marine Biology* **42**: 69–84.

24. Cannon, H. G. (1928) On the feeding mechanisms of the copepods *Calanus finmurchicus* and *Diaptomus gracilis*. *Journal of Experimental Biology* **6**: 131–144.

25. Marshall, S. M., & A. P. Orr (1956) On the biology of Calanus finmarchicus. IX. Feeding and digestion in the young stages. *Journal of the Marine Biological Association of the U.K.* **35**: 597–603.

7

Feeding II: More About Eating

Rudi Strickler Brings Slow-Motion Movies to Copepodology

Data like that selected from Richman and Heinle's paper (Figure 6.14) raised the issue of how a simple filter, even one with several sizes of holes, could capture both 7 μm and 15 μm particles but not catch *any* particles from 10 to 12 μm. And there were many data suggesting other criteria for particle selection. Some species are taken out of mixtures, but not others. Even cells of the same species in different growth phases can be distinguished by copepods for eating, or apparently, rejecting.

This led to a new approach based on *Zeitlupenaufnahmen* (slow-motion movies) advanced by J. Rudi Strickler. He was born in 1938 and raised in Switzerland, trained there as both an engineer and biologist. He told me many things about himself during a day of interviews in August 2018, including that the first Rudi Strickler of his line was born in 1340. His name has a family history of seven centuries. His father was a prominent engineer who worked in Tehran during Rudi's early years. When his family returned to Switzerland permanently during World War II they lived in the small manufacturing town of Mollis, where the factory fumes gave Rudi asthma. To manage that, he spent summers in the late 1940s on high mountain farms near a village in the Engadin doing farm work, including delivering calves. Moving into his teens, he attended a gymnasium school for natural sciences, and then moved on to the Eidgenössische (federal) Technische Hochschule, Zürich (ETH Zürich), a Swiss equivalent of Cal Tech or MIT. His matriculation depended on passing an exam lasting 14 days and covering various subjects. He did that.

Rudi says that one of his initial ETH chemistry professors (the Nobel laureate Vladimir Prelog) suggested to him that he was dyslexic, and therefore that physics would be the ideal subject. Well, many superbly productive scientists were (or certainly have been thought to be) dyslexic. In any case, Rudy plunged into physics, getting opportunities for work with the earliest versions of lasers. That early study at ETH surely is the basis of his mastery of optics and mechanisms, which has informed his studies of copepod swimming and feeding.

The years at ETH were a longish slog through several degrees, with breaks mostly for annual military training and service. He characterizes his part of the Swiss Army as rather like the American National Guard: annual training and

Oar Feet and Opal Teeth. Charles B. Miller, Oxford University Press. © Oxford University Press 2023.
DOI: 10.1093/oso/9780197637326.003.0007

Figure 7.1 Rudi Stickler at one of the optical benches (with high frame-rate video) in his present laboratory in the Global Water Center, Milwaukee, Wisconsin. Photo by GWC photographer. Permission, Rudi Strickler.

some other meetings, service when called up. In 1956, for example, Rudi went as a volunteer to Jena in East Germany to help with Swiss assistance to Hungarian refugees gathered there. During the 1968 Russian invasion of Czechoslovakia, his unit was a one-day march from the Russian Army but did not leave Switzerland. ETH cannot have been all work; he told me he married at age 23 to Eva, a textile artist then 20 years old. Unlike today, he notes, they had to marry to be together.

Attending a lecture on hydrobiology by a Professor Jaag of ETH drew Rudi into plankton studies. Jaag was engaged in work to clean up Swiss lakes and waterways. Through him, Rudi acquired a collection of zooplankton from Lake Lucern, a series of samples in jars collected from the 1870s to the 1960s, and his Master of Science thesis compared those samples still in reasonable condition between the 1930s and early 1960s. Examining numerous *Cyclops* raised the question, "What does it mean to be a pelagic copepod?" Among other interest-arousing observations, he documented that males had many more attached spermatophores than females. He discovered a major problem with the samples, learning by comparing the sample labels to weather records that they were only collected on pleasant sunny days.

Planning for his PhD thesis, Rudi discussed issues for research with experts on lake zooplankton all over Europe: Otto Siebeck in Austria, Arnold Nauweck in Sweden, and Joop Ringelberg in Amsterdam. He gathered ideas on how to learn "what the animals are doing." That led to a study of the changes in swimming of cyclopoid copepods at different illumination levels in aquaria thin front-to-back. He received his PhD in 1969 and published his results,[1] citing similar work by Otto Storch[2] from the 1920s, four decades earlier. The key words in Storch's title are *Mikro-Zeitlupenaufnahmen analysiert*, which translates literally to "microscopic, time-magnifying image analysis." Both of them took and analyzed high frame-rate movies of copepods swimming in aquaria. Rudi used infrared (IR) illumination and movie film sensitive to it. That theme was critical to much of what Strickler contributed in subsequent years, though he and colleagues worked on other aspects of aquatic animal life. Rudi traced the positions of individuals frame-by-frame on large sheets of paper, masses and masses of those that he still has and showed to me. The analysis showed that *Cyclops* swimming behavior varies depending upon imposed light levels.

At this point Rudi needed a paying job and went to work part time designing sewage treatment plants and was supervised by his father, which proved difficult. Meanwhile, his wife Eva, following her work with textiles, had taken an interest in and wanted to work with Inuits. So they emigrated to eastern Canada, initially to Newfoundland and to various odd jobs for both of them. Eva eventually got a position working with Inuit craftspeople in today's Taloyoak, Nunavut, helping them develop markets for their wares. She was a great success at that, winning awards over years from the government for her service. After a few years, Rudi got a chance to attend an aquatic sciences meeting and gave a talk that must have been impressive (he is a talented showman), because it led to a faculty job offer at Chesapeake Bay Laboratory (CBL, of Johns Hopkins University), which lasted four years and allowed more work with movies of copepod behavior. Eva did not join him there but stayed on in Tolayoak. The marriage was sustained for a long time by mail, telephone, and visits.

A paper Rudi published in 1975,[3] in a volume on the mechanics of swimming and flying, contains most of our current understanding of the relative roles of viscous and inertial drag in copepod swimming. Together with his other early papers, it initiated a quantum shift in understanding the mechanics of swimming, which differ as a function of swimming velocity and animal size. Viscous drag dominates for small organisms at low speeds, so sinking (or rising) can be slow. Inertial drag dominates after some acceleration, enabling oar-like propulsion to provide more acceleration and escape jumps. Thus, Strickler provided a basis for understanding the size range found in planktonic copepods. The notion of effective drag modes shifting with changing velocity and body size is summarized in Chapter 1. The text box there explains these two modes of *drag*. Rudi also provided ideas about the proprioceptive and velocity sensing systems needed to control those motions. The four CBL years were very productive.

The CBL work also allowed two years as assistant professor at Yale, where Rudi connected with G. Evelyn Hutchinson and Luigi Provasoli, famous names in ecology, marine and otherwise. Part of Rudi's actual work there was with the Canadian limnologist Frank Harold Rigler. They developed the equipment initially demonstrating the mechanics of particle feeding by copepods. The gear was a tiny aquarium, a "cuvette," with an IR-illuminated, optically clear side facing a horizontal microscope with a movie camera at its eyepiece. They got a movie of feeding in a relatively large marine copepod, *Rhincalanus,* that they showed to Hutchinson and Provasoli. Only Hutchinson was impressed, but Provasoli suggested some partners for further work: Gustav Paffenhöfer of Skidaway Institute of Oceanography and Miguel Alcarez from Barcelona. By the time a partnership was established with them, Rudi had moved to the University of Ottawa, an early move among many that followed.

With an upgrade of the system developed, the three-man team met in Ottawa for some observations. Rudi remembers explicitly that Gus Paffenhöfer brought 37 copepods, surely the *Eucalanus pileatus* used in later studies. Miguel Alcaraz would catch one, set it on a slide, dry off its carapace, and touch to it a hair from Rudi's Shih Tzu tipped with a tiny drop of super (cyanoacrylate) glue. In a few seconds the glue set and the copepod had a handle. That was placed in a clamp and the preparation was slipped down into the narrow aquarium containing a dilute slurry of phytoplankton cells. Then the optical train was brought into focus on the film plane of the camera, illumination was switched on, and a few seconds of the animal's activity were captured at 500 frames per second. Projected at the normal 32 frames per second, time appeared expanded. The movie was shown in August 1978 to an assembly of lake and ocean ecologists at Dartmouth College, many interested in copepods. Those assembled, many of us examining feeding with particle-sizing counters or teaching about feeding mechanics according to Canon (1928) (Chapter 6), were gobsmacked.

Shown at rates reduced by a factor of 16, the head limbs were seen to move out away from the ventral surface, then back toward it, creating water motion toward the maxilla in steps, not as a smooth flow. Most of us present knew some features of viscous flow, but suddenly it was a key part of our understanding about the water habitat of small zooplankton. Very close to zero inertial carry is characteristic of the motion of viscous fluids. Inertial carry is what lets a human swimmer coast into a pool wall. The effects of fluid viscosity depend so strongly on size scales that water at the size of copepod limbs is effectively like honey. Water near them and particles suspended in it only move when the limbs are actively applying force (from muscles). In the films, algal particles approaching the copepod would move and then stop, move then stop. When a particle got close to them, the maxillae would fling outward, surrounding the particle, then squeeze back against the body, removing water from around it. Finally, the particle was sucked into the mouth. All the limb motions were cycling at very high frequencies. Miguel Alcarez was first author of a paper describing the results.[4]

Clearly the alga had to be sensed. Those earliest and many later film observations suggest that cells can be sensed without touching them, probably by olfaction. The boundary layers around cells are surprisingly thick. They surely contain exudates of several kinds, and copepods can sense those. Strickler had anticipated that during his time at Johns Hopkins, and in 1975 coauthored an electron microscope study[5] with Marc Friedman characterizing olfactory sensors on the mouthparts of *Diaptomus pallidus*, a freshwater filter-feeder, as follows (slightly paraphrased; the electron micrographs do not copy well from the journal):

> Distinctive sensilla are found in the maxillule, maxilla and the mandibular palps. More precisely, the setae of these appendages *are* the sensilla. They are characterized by a pore system in their distal regions, permeating the setal wall, which is about 0.25 μm thick. The numerous pores are about 100 Å in diameter.... The setal wall in this region appears distinctly different from the normal chitin in the basal portion of the seta [rather fuzzy, not in continuous in cross-linked sheets]. Each sensillum contains 1 to 5 dendrites, 1 or 2 of which are ciliary and emerge from 9+0 basal bodies, with the accompanying dendrites forming only an unstructured array of doublet microtubules. The presence of several neurons, which can apparently communicate with the outside of the seta through a pore system, and the absence of a "tubular bundle" clearly indicate that these setal sensilla are chemoreceptors. They are structurally similar to the hair chemoreceptors found in other arthropods.

Getting Down to the Details

Microscopic, slow-motion movie and video systems have had several iterations, and observations of particle feeding by copepods have been widely repeated, particularly in Gus Paffenhöfer's lab in Georgia, again using *Eucalanus pileatus* collected offshore from Savannah. Full interpretation required frame-by-frame analysis of the movies, and many people beside Rudi participated. Paffenhöfer[6] continued work on copepod feeding mechanics, and particular attention went to papers by Mimi Koehl,[7] who has had a long career in biomechanics at the University of California, Berkeley, and by Timothy Cowles,[8] who moved in the 1980s from Woods Hole to my department at Oregon State University (OSU). Koehl and Cowles took copepods from Cape Cod Bay to Ottawa to film them with Strickler. Later they did some of their own slow-motion filming at Woods Hole. Once Cowles moved to the Oceanography Department at OSU, we became teaching colleagues for several decades, though taking different research paths. Paffenhöfer guided further studies by Holly Price[9] and others. I must leave my readers to trace the careers of any of those people who interest them.

The Koehl and Strickler version[7] of the story is quite complete, starting with Professor Canon's notion of the feeding "currents" (and why it was wrong), noting the particle-counter based interpretations, and then developing the *Mikro-Zeitlupenaufnahmen* derived explanation of particle capture by *Eucalanus*. They also improved the technique by injecting miniscule streams of dye next to the mouth parts during the few seconds of film run to trace the water motion. Figure 7.2 is their interpretation of the findings. The basic story is in those cartoons and my version of the caption.

Movies by Cowles and Strickler[8] of feeding by *Centropages typicus* show that a particle moved to center (as in Figure 7.2) can be rejected by reversing the stepped flow. A particle steps in (no inertial carry) and then steps back away, uneaten. The whole limb motion and seta orientation has to reverse to allow that. Price et al.[9] compared films made with different algal cell sizes, determining that attention to single particles was limited to those 12 μm and larger, at least for *E. pileatus* (~2 mm long) and the smaller *Paracalanus parvus* (~1 mm). They also showed that for smaller particles the maxilla actually does act as a rather simple filter:

> Continuous low amplitude movement of the . . . maxillae and combing of . . . [its setae are] used by *E. pileutus* to capture cells smaller than the 12 μm sensitivity threshold.

The combing is by the stout setae on the basal endite of the maxillule (Figure 2.13). They can sweep cells from the maxilla and toward (maybe into) the mouth for ingestion. That does sound like H. G. Cannon's notion (Chapter 6) about filter-feeding in *Calanus*, though the flow pattern cannot have been that in the water drops covering the copepods in his compressorium. In addition to confirming the rates of limb cycling during feeding current generation for *E. pileatus* (Figure 7.3, left), Holly Price, working with Paffenhofer and Strickler,[9] counted cycles for *Paracalanus* feeding on 6 μm diatoms (Figure 7.3, right). The limbs of that 1.0 mm copepod swing in and out at a remarkable 90 cycles per second, at least at the 21°C (70°F) experimental temperature comfortable for copepods taken offshore from Savannah. Neural controls for the limb muscles must oscillate at same frequencies.

Long stretches of ratiocination have been devoted to the physics of flow for the different limbs, all of it emphasizing the dominance of viscous (as opposed to inertial) flow. Mimi Koehl[10] later stated a key aspect of that. Even when velocities are great "enough that inertia cannot be ignored, water flow around a feeding

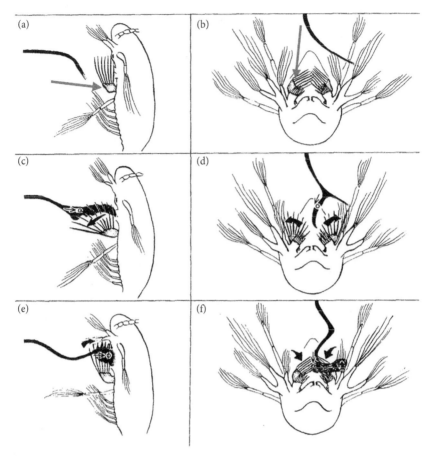

Figure 7.2 Cartoon interpretation of limb movements during particle feeding by *Eucalanus pileatus* from Koehl and Strickler.[7] Body orientations should be obvious, except that the right-hand set are viewed from the head. The urosome has no part in the action, so was not drawn. In (**A**) and (**B**) the antennae and maxilliped, originally bent over the maxilla (red arrows), have swung fore and aft, respectively, and the mandibular palp has swung from over the labrum and out to the side. As indicated by the dye streamer, the flow was stepping toward the nearly closed maxillary setae and then away to the side (it could have gone to either side). In (**C**) and (**D**) the streamer contained an alga, and the maxilla has been "flung" (term used by Koehl and Strickler) open to draw part of the streamer and the cell to the center near the mouth. In (**E**) and (**F**) the maxilla squeezes toward the mouth (black arrows) and the particle is sucked into the mouth and presumably chewed and swallowed. Permission, Wiley & Sons.

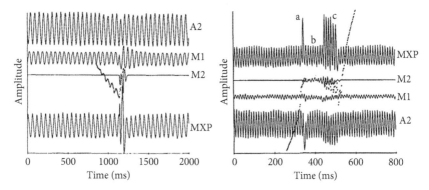

Figure 7.3 Time series of locations of identical parts of four mouth limbs (antenna, A2; maxillule, M1; maxilla, M2; and maxilliped, MXP) of *Eucalanus pileatus* (female, left: for two seconds–2,000 milliseconds) and *Paracalanus parvus* (female, right for 0.8 seconds). The maxilla only moved when a cell (left: a dinoflagellate; right: a small diatom) came near it. The relative cell trajectories are represented by the dots. The "dino" must have been positioned partly by an exaggerated swing of the maxilliped. The diatom was rejected by *P. parvus*, and exaggerated swings of its maxilliped were part of that action. From Price et al.[9] Permission, Wiley & Sons.

copepod is nearly laminar (i.e., the water moves smoothly around the animal and can be considered as moving in layers between which there is no significant mixing)." Once a particle is sensed, that stable layering must assist in predicting where it will be when a copepod reacts to it. At these tiny increments of time and space, water motion is in steps that can be anticipated correctly.

Rather like the particle-counter studies popular through the 1970s, the feeding frenzy of film studies on copepod feeding continued through the 1980s, and late additions still show up. The intense activity produced truly striking additions to general understanding. It was proved quite definitively that particles are evaluated in part by their odor. However, Strickler[11] had shown very soon with a different filming system and untethered *Eucalanus* that the presence of particles distorts the feeding current flow field enough that copepods probably can detect their presence from changed stresses on setae. He has continued to cooperate with others in this work right up to the present. Among the many examples was Jiang-Shiou Hwang, a PhD student who studied with him when he was a faculty member at Boston University in the early 1990s. They continued evaluating results from their films for a decade,[12] while Hwang was working at the National Taiwan Ocean University. Hwang has remained a productive student of copepods.[13]

Strickler as Peripatetic Engineer and Copepod Guru

By the time the 1981 paper with Koehl was published, Rudi had moved to the Australian Institute of Marine Science in Townsville, Australia, where he stayed for four years, much of it continuing work on copepod swimming and feeding.[11] He was still married to Eva, though they had not spent much time together after leaving Newfoundland on different trajectories. They could scarcely have been farther apart than from Townsville to Nunavut.

Rudi then came back to the United States and sought a new job wherever there seemed to be openings. It took awhile, but after giving a talk at the University of South Florida, Richard Dugdale, who was attending, suggested he seek a position at the University of Southern California, in Los Angeles. A dean there agreed to a research faculty position. The meaning of that term is that you can be paid if you can obtain grants from which the university can pay you. However, Rudi took the offer, got a grant or two, and established a lab. It was there that he began an extended research interaction with Jeanette Yen, whom we met in the last chapter. They applied a four-dimensional aquarium vidographic system (cameras on two sides of a rectangular chamber recording at several speeds) so she could pursue her interest in the movement responses of *Euchaeta*. Some of the new machinery involved tracking animals by moving the aquarium or the cameras to track an animal by following its moving image. That was also applied in Yen's later studies (reported in Chapter 10) of copepod mating antics, consistently with Rudi's help.

Open to something other than a soft-money position, Rudi was approached by Jelle Atema of the Boston University Marine Program located in Woods Hole. There were problems there, too, Rudi says, particularly around lab space, which was allocated by the square foot according to counted pages of manuscript production (1 page submitted produced 1 ft^2; talk about publish or perish). In any case, the situation in Woods Hole wasn't particularly happy, and he grabbed an offer from the University of Wisconsin-Milwaukee's Great Lakes WATER Institute (now the UWMilwaukee School of Freshwater Sciences) and has been there since. Somewhere in all that shifting he was divorced, and he remarried. The couple visited Tim Cowles and me in Corvallis in about 2010. Sadly, that marriage did not last long enough. Rudi suffered with his wife through an agonizing illness that was finally fatal to her around 2014.

I found Rudi in Milwaukee about four years later, living alone in a mid-town apartment at age 80. However, he does not seem like an octogenarian. The apartment was small but adequate, and wonderfully decorated with mementos of his long career. He drives a small sporty car at speed through town. While he speaks longingly of losing his wife, he gets around the world often, helping people like

Jeanette Yen and Petra Lenz to engineer equipment for their studies. He and Petra have developed an aquarium system in Hawaii for filming small reef fish from above for studies of their skill development. Apparently even fish must practice to make perfect (or at least to make adequate).

Rudi worked at many things at the University of Wisconsin-Milwaukee, but spoke mostly about helping organize the university for participation in the city's remarkably successful renaissance after falling into postindustrial decay. Milwaukee had been a center of machine-tool and industrial equipment manufacture. At least in part it is now a research center. While still attached to the university (where he has a storeroom for every meter of IR film from all those studies), Rudi has a lab and office in an old factory building refurbished in ultramodern polish as a Global Water Center (Figure 7.4). It is a "water research and business incubator," to quote its brochure, and the university is one of the participating tenants. During many days Rudi can be found there, seven floors up in his spacious lab assembling widgets. His inventiveness proceeds unabated. He proudly showed me a beautifully machined device for scanning an angled sheet of green laser light across any irregular surface. A video camera looking down on the advancing line would record its offsets, providing a computer with pixel-by-pixel data to generate a contour map of that surface. Somebody involved in water research needed those contours, so Rudi invented a way to get them fast when needed.

Figure 7.4 Milwaukee's Global Water Center (the brown building on the left) where Rudi Strickler holds forth these days as a wizard of lasers, cameras and perfectly machined alignment systems. Photo by the author.

References

1. Strickler, J. R. (1970) Ueber das Schwimmverhalten von Cyclopoiden bei Verminderungen der Bestralungsstaerke. *Schweizerische Zeitschrift für Hydrologie* 32: 150–180.
2. Storch, O. (1929) Ueber das Schwimmenbewegung der Copepoden, und Grund von Mikro-Zeitlupenaufnahmen analysiert. *Verhandlungen der Deutschen Zoologischen Gesellschaft, Zoologischer Anzeiger. Supplementband* 4: 118–129.
3. Strickler, J. R. (1975) Swimming of planktonic *Cyclops* species (Copepoda, Crustacea): pattern, movements and their control. In *Swimming and Flying in Nature, Vol. 2*, ed. T. Y.-T. Wu, C. J. Brokaw, C. Brennan., 599–613. Plenum Press, New York.
4. Alcaraz, M., G.-A. Paffenhöfer, & J. R. Strickler (1980) Catching the algae: a first account of visual observations on filter-feeding calanoids. In *Evolution and Ecology of Zooplankton Communities*, ed. W. C. Kerfoot, 241–248. University Press of New England, Lebanon, NH.
5. Friedman, M. M., & J. R. Strickler (1975) Chemoreceptors and feeding in calanoid copepods (Arthropoda: Crustacea). *Proceedings of the National Academy of Sciences* 72(10): 4185–4188.
6. Paffenhöfer, G.-A., J. R. Strickler, & M. Alcaraz (1982) Suspension-feeding by herbivorous calanoid copepods: a cinematograpic study. *Marine Biology* 67: 193–199.
7. Koehl, M. A. R., & J. R. Strickler (1981) Copepod feeding currents: food capture at low Reynolds number. *Limnology & Oceanography* 26(6): 1062–1073.
8. Cowles, T. J., & J. R. Strickler (1983) Characterization of feeding activity patterns in the planktonic copepod *Centropages typicus* Kroyer under various food conditions. *Limnology & Oceanography* 28(1) 106–115.
9. Price, H. J., G.-A. Paffenhöfer, & J. R. Strickler (1983) Modes of cell capture in calanoid copepods. *Limnology & Oceanography* 28(1): 116–123.
10. Koehl, M. A. R. (1983) The morphology and performance of suspension-feeding appendages. *Journal of Theoretical Biology* 105: 1–11.
11. Strickler, J. R. (1982) Calanoid copepods, feeding currents, and the role of gravity. *Science* 218: 158–160.
12. Hwang, J.-S., & J. R. Strickler (2001) Can copepods differentiate prey from predator hydromechanically? *Zoological Studies* 40(1): 1–6.
13. Chen, M.-R., & J.-S. Hwang (2018) The swimming behavior of *Calanus sinicus* under different food concentrations. *Zoological Studies* 57: 13. doi:10.6620/ZS.2018.57-13.

8

Not Being Eaten I:
Diel Vertical Migration

Predator Avoidance Is As Important As Feeding

Like all animals, even elephants, planktonic copepods not only must eat to live, but to keep living they must avoid being eaten. They are ideally sized, as eggs, larvae, and adults, to be eaten by a range of predators from protozoa to fish and even filtering whales. In Chapter 1, the general possession of oar-like thoracic feet for jumping away from predators was introduced. The consistent need for those feet determined the general body form for this entire subclass of crustacea. Copepods also need to know when to hit the accelerator, to scram. So copepods possess sensors on their antennules and elsewhere for very small accelerations of the water along their surfaces. Those, too, were introduced. Mention has also been made of transparency, which cannot be total, given possession of eyes that require pigments and guts filled with pigmented food. However, considerable transparency is typical for copepods living at least some of the time in lighted upper layers. The extreme example of *Metridia* was shown in Figure 2.11 and is shown again in Figure 8.17. In addition to these general features of predator avoidance, various copepods exhibit remarkable adaptations to delay being digested. Many involve being as close to invisible as possible. I will start by describing, and mostly explaining, the predator avoidance adaptation most and longest studied, diel vertical migration.

Diel Vertical Migration (DVM)

In the early days of towing simple plankton nets in oceans and lakes, it was learned that catches are generally greater at night than in the daytime. Explanations were several. Maybe copepods (also cladocerans, chaetognaths, krill, pteropod snails, some fish and fish larvae, and more) could see nets coming at them and dodge during daylight, not in the dark. Possibly, too, they were just sleeping or torpid at night. Victor Franz, writing in 1910 and 1912,[1] insisted those possibilities had to be taken seriously; insistence because migrating down into at most twilight during daytime to avoid being seen by predators, fish among others, had already

Oar Feet and Opal Teeth. Charles B. Miller, Oxford University Press. © Oxford University Press 2023.
DOI: 10.1093/oso/9780197637326.003.0008

become the more standard explanation of greater nighttime catches. It was indeed likely that many animals of many types leave the surface layers in daytime. However, filtration by nets coming through upper layers was included in all early plankton work, so it was difficult to be sure the animals move down during the day. For a long time, the daily up-and-down hypothesis was termed "diurnal" vertical migration. The word "diurnal," however, formally means "during the day," as opposed to "nocturnal," and the previously little-used word "diel" was adopted. According to several dictionaries "diel" first appeared in English in 1934, but they fail to say where. Maybe it came from Dutch, *deel*, meaning "part." In current biology, "diel" describes any process with 24-hour cycling.

Sir Alister Hardy pointed out in his 1956 book *The Open Sea, The World of the Plankton* that opening-closing nets (OC nets) resolve the alternative hypothesis of net-avoidance in daylight. Such nets descend into the water closed, open and filter at the depth to be sampled, then close before being pulled back to the boat or ship. There are many such systems, some of which can be multiply deployed on the same wire, giving samples from, say, five or six depths at nearly the same time. That matters; when plankton move rapidly up or down, time lost to change nets can be too long to show a migration schedule correctly. Hardy's point was that if light assists net avoidance, it should be less effective at depth in near darkness than just below the surface. Thus, if deep-only samples show that abundance at depth increases in daytime relative to night, then surely the decrease in daytime at the surface is due to migration down. It takes repetitive samples to account for the wide variability in plankton abundance estimates, but there have been many stratified-sampling time-series. Some of the older data are as good as the newer. In 1934 George L. Clarke (Figure 8.1), a marine biologist on the Woods Hole Oceanographic Institution staff, sampled plankton in the Gulf of Maine using his then newly designed opening-closing nets (Figure 8.2). He was also on the Harvard faculty and wrote a textbook long popular that was published in 1967, *Elements of Ecology*.

Like many early OC nets, Clarke's sequencing of opening and closing downward was accomplished mechanically by dropping heavy metal sliders ("messengers") down the towing cable. Hitting levers at the net attachment, they released the towing frame, the net's cod end to towing position and released another messenger to open the net below. After a suitable interval another messenger began release of the towing frames, then drag would pull the nets closed with pursing cables threaded through rings sewn into cloth collars behind the frame. Clarke's nets were woven cotton mesh (scrim), 10 threads/cm (mesh openings just less than 1 x 1 mm, explaining the relatively large copepods dominating his samples).

Clarke[3] reported the distributions of *Calanus finmarchicus* from the Gulf of Maine, sampling both shallow and deeper waters (Figures 8.3 and 8.4). He

Figure 8.1 George L. Clarke at about the time of his 1934 paper cited here. From a Marine Biology Laboratory-WHOI library collection. Permission, Woods Hole Oceanographic Institution.

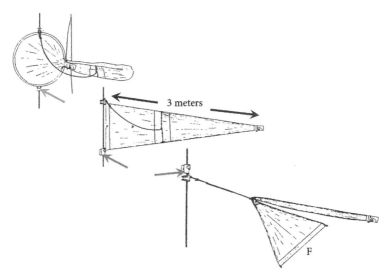

Figure 8.2 "Clarke-Iselin," opening-closing plankton net, 1933. Left: 0.75 m net closed for deployment. Center, sampling. Right: net pursed closed for retrieval. Each net below the shallowest had two messengers to open (blue arrow), then close (red arrows) the net below. Start and then close messengers (metal weights) are sent down from the ship to open then close whole sequence. From Clarke.[2] Permission, University of Chicago Press.

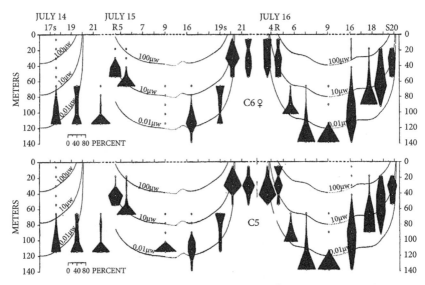

Figure 8.3 Vertical distributions of *C. finmarchicus* C6♀ (above) and C5 (below), over a depth just greater than 140 m in the Gulf of Maine, from July 14 to 16, 1933, a total of 54 hours. Dots represent the towing depths of Clarke's flights of five OC nets. The relative abundances of the copepods at each depth are shown as "kite diagrams" of percentage of the total abundance captured by the five nets. The curves are isolumes of blue light at 100, 1, and 0.01 microwatts/cm[2]. S, sunset; R, sunrise. from Clark.[3] Permission, University of Chicago Press.

plotted the abundances on a background of blue irradiance isolumes measured with a lowered photoelectric cell under blue-passing filters. That has seldom been done since in coordination with net sampling in such an explicit fashion, even though modern optical instrumentation is more sophisticated.

The right half of the deeper-water Figure 8.3 shows the pattern already classic by 1933. Around sunset on July 15 the modes for both stages rose from about 115 m progressively to about 20 m and stayed up overnight. They returned at the next sunrise to below 120 m, moving up some in the afternoon. The left half seems to show failure to rise at night in both stages C5 and C6♀, though some of each were part way up by sunrise.

Clarke compared those mid-July data to those from similar sampling a few days later, closer to Georges Bank and over only 70 m depth (Figure 8.4). At that station *C. finmarchicus* showed no tendency to descend in the water column during daytime, despite the surface irradiance being much greater. Clarke's tables include the fact that the earlier, deeper-water study was under overcast skies, while skies were clear over the station near George Bank. The comparison shows that diel migration behavior is variable, sometimes strong,

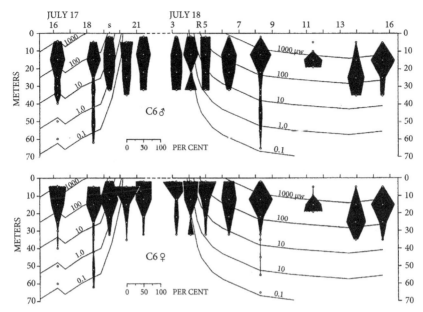

Figure 8.4 Vertical distributions of *C. finmarchicus* males (above) and females ♀ (below), over 70 m depth near Georges Bank, on July 17–18, 1933, one full day. Symbols as Figure 8.3. Clearly here there was no strong migration, perhaps a 15 m descent at mid-day on the July 18. Figure 5 from Clarke.[3] Permission, University of Chicago Press.

sometimes absent. Clarke really had no idea as to the difference between the stations that might have led to the difference in migration behavior. He thought about the small temperature difference and possible distinctions in phytoplankton availability and found them unlikely to be important. William Kerfoot,[4] thinking about Clarke's data later, considered that perhaps the station over a 70 m bottom simply was not deep enough to allow migration. However, the population modes could have moved down 40 meters to 100-fold less light without hitting bottom. Perhaps the modes did move down about 10 m on July 18.

Look again at the graphs. The interpretations of stock distributions depend upon the tails of the distributions, both upward and downward. There were no samples at all in too many depth-time intervals. For examples, there were no upper layer samples the night of July 14–15 (Figure 8.3) until almost dawn. Since the kites at depth are for percentages of what the nets did catch, they could have been based on virtually no specimens. But no, Clarke's paper actually has a table of all the numbers (1934 was in another scientific era) that total to hundreds, plenty to provide a reasonable divisor for percentages.

There were, indeed, no samples below 35 m during the night of July 17–18 (Figure 8.4), so part of the stock might have stayed down around the clock. That can happen with *Calanus* spp.; some of both C5 and even C6♀ can have moved down for diapause (more on that in Chapter 15). Much of the deep-dwelling *C. finmarchicus* resting stock in the Gulf of Maine usually is established by mid-July. At least the actively feeding stock was shallow day and night, the mode moving down slightly late on July 18. Clarke attributed that to the bright sky of the sampling day, likely correctly. Clarke was very taken with the notion that when copepods *do* migrate, they do it following specific isolumes (his curved irradiance lines) down at dawn and back up at dusk. Perhaps they anticipate that the preferred isolume will move (compare morning and afternoon on July 16: Figure 8.3, right). At least some diel vertically migrating zooplankters and fish do follow a preferred light intensity, even following their preferred lighting upward during solar eclipses.

Other hypotheses came forward regarding diel vertical migrations. Alister Hardy[5] suggested that in the sea at least, diel migrations would allow relatively weak swimmers to achieve rather longer horizontal trips across the currents in which they were embedded. Currents tens to hundreds of meters down (some copepods migrate 400 m both down and up daily) are often heading in different directions from those near the surface. If a planker leaves its surface trajectory at dawn and goes down to stay for hours at, say, 200 m, it will be moved across, or even against, the direction of the surface flow. It could possibly have moved by several kilometers when it came back up at dusk. It would be at a new place in the moving water mass, maybe even close to the geographic place it left. Some advantage might come from that, a sort of "station keeping" or maybe exploration for better food supplies. Of course, if it descended in every dawn, it would also leave abundant food it had recently found. Moreover, it would always have less to eat while down, since for phytoplanktivores food tends to be more abundant near the surface.

Ian McClaren[6] pointed out that the daytime depths occupied by migrators are generally colder than near the surface. Colder temperatures change the ratio of maintenance metabolism to growth, favoring growth. It was long known, thanks to Georgiana Deevey,[7] that copepods growing to adulthood at colder temperatures are larger than those *of the same species* that grew in greater warmth. McLaren added that larger individual female copepods produce more eggs than smaller ones; thus, there could be a demographic advantage to spending part or all of most days at depth. The costs would be (1) the nutrition lost from not eating the richer surface rations during daylight, perhaps not eating at all down deep, and (2) slower development and longer generation times. Over even a few generations, the delays could mean more time for mortality and (even without that) less overall population increase.

However, if you ask a dead copepod *why* it had or had not migrated before you captured it, there is no reply, not even if you are reasonably sure that it did migrate. Getting a clean answer requires other comparisons.

Clean Answers: A First One from Maciej Gliwicz of Poland

In the 1980s and later, a number of such comparisons were made, some in lake waters, some in oceans and estuaries. A report was published in *Nature* during 1986 by the Polish hydrobiologist Maciej Gliwicz (Figure 8.5). He is prominent internationally among limnologists, and the Wikipedia article about him explains how richly honored, including the American Society for Limnology and Oceanography's A. C. Redfield prize and membership in the Polish Academy of Sciences. Born in 1939, he trained at the University of Warsaw and became professor of natural sciences there in 1988. His 1986 paper[8] comparing the night and day distributions of *Cyclops abyssorum* among different lakes is one of many significant contributions. The data are simple and direct, presented as figures copied here (Figures 8.6 and 8.7).

As the paper explains, alpine moraine lakes in glaciated Northern Europe did not have freshwater salmonid fishes (trouts) when the glaciers initially retreated. Their drainage streams were too steep for trout to climb, or they percolated through the rocky moraines that held the lakes. However, cyclopoid copepods,

Figure 8.5 Professor Maciej Gliwicz in 2012, talking with colleagues during a meeting in Krakow. Permission, Prof. Z. M. Gliwicz, January 2022.

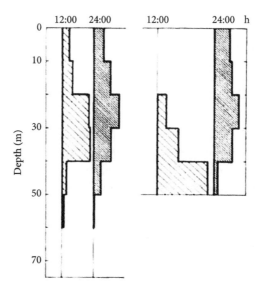

Figure 8.6 From Maciej Gliwicz.[8] Vertical distributions at noon and midnight of *Cyclops abyssorum* copepodites in two lakes: Czarny nad Morskim (left) never with fish and Morskie Oko (right) long with trout. The copepods almost surely vacate the surface during daylight at Morskie Oko to avoid being seen and eaten by the fish. At Czarny Morskim many remain in the upper 30 m through the day. The bars are percentages of the vertical totals and add to 100%. Permission, Nature.

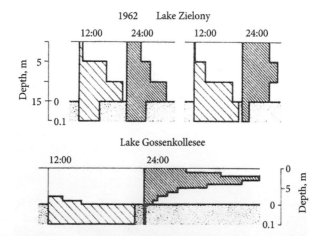

Figure 8.7 Vertical distributions of *C. abyssorum* at noon and midnight in Lakes Zielony and Gossenkollesee. Stipples represent the sediment, from which copepodites were counted to 10 cm depth and added to the sampling total for calculating stock percentages at four (Zielony) or up to 10 (Gossenkollsee) depths. From Gliwicz.[8] Permission, *Nature*.

cladocerans, and the phytoplankton and rotifers they feed upon can arrive airborne in droplets or literally riding on ducks' feet and feathers. Some lakes in the Tatra Mountains between Poland and Slovakia, like Czarny nad Morskim, still have no fish, and they have both cladocerans and several kinds of copepods. One, Morskie Oko, was stocked with trout long ago, Gliwicz says millennia back, and the fish have removed most of the middle-sized zooplankton, all cladocerans and most kinds of copepods. Fortunately for the argument, both of these lakes, with deep points at 76 and 51 m, respectively, have populations of the copepod *Cyclops abyssorum*. Both are reached by hiking in, though carriage rides part way are possible. Gliwicz, assisted by Malgolia Witeska, sampled five or six strata in each of those lakes with OC nets lowered near their deepest points at both noon and midnight on August 12 and 18, 1985 (Figure 8.6). All counts are for copepodites, since the adults mature in winter under ice, and in Morski Oko those are quickly reduced by fish. Apparently, they do have time to produce young that make it to the following summer.

Gliwicz had more than that one demonstration to prove the connection of fish predation to DVM. He also studied some shallower lakes (Figure 8.7). He had sampled Lake Zielony (15 m) in 1962, when it had been stocked with fish for only 12 years, and he sampled it again in 1983. The earlier sampling showed no obvious migration, with some burrowing. The later sampling showed obvious surface evacuation in daytime and proportionally more burrowing. In 1962 at Lake Gossenkollesee (10 m), stocked in 1700, *C. abyssorum* were up swimming at mid-night, and almost all copepodites were burrowed into the sediment at noon.

The three comparisons suggest very strongly that fish, the premier visual predators in water, induce vertical migration. At least this *Cyclops* migrates to avoid being seen by fish. The Zielony versus Gossenkollesee comparison opens a question: Does migration to avoid digestion by fish require genetic selection, or is it consistently available and just skipped when there are no fish? The stocking of Zielony in 1948 did not establish a reproducing trout population until 1965, so there may have been little predation pressure on the copepods during the 1962 sampling compared to that in 1983, particularly not from numerous small fish. Evolution can act rapidly, so likely it did. Diel vertical migration, in fact burrowing down out of the water altogether, was more pronounced in Gossenkollsee, which could have been due to much longer continued selection. So rapid selective evolution of DVM might be possible once fish, particularly smaller fish, become abundant. However, it is also possible that the copepods sense fish hunting: see them, smell them, or sense water motions from them. That could kick in their already inherent tendency to get out of the lighted surface when at risk. Of course, both evolutionary sequences could be active, selection strengthening a tendency already very long present.

More Comparisons, from Steve Bollens and Bruce Frost

Around the time the Gliwicz *Nature* paper came out, Bruce Frost (Chapter 4) had been doing a good deal of sampling in Dabob Bay, a side channel of Puget Sound (now part of what is being called the Salish Sea), northwest of Seattle. Dabob exchanges water with the larger fjord over a shallow sill. It was convenient because a US Navy testing facility maintained a large buoy to which a sampling vessel could be moored in 185 m of water. What's more, slow tidal exchange with the ocean outside sustains a nearly full complement of coastal-zone zooplankton. On the other hand, water residence times are nearly a year. So population processes over periods resembling copepod generation times are more-or-less local. There are two species of *Calanus*, three of *Pseudocalanus*, *Metridia pacifica*, *Euchaeta elongata*, and at depth *Bradyidius saanichi*. All of them reproduce and develop much as outside on the coast. The Navy is generous with use of the buoy unless they are testing torpedoes, which does occur occasionally. When it does, the university's small ship can dock conveniently at walking distance to a store with ice cream. The absence of ocean rollers makes plankton work easier with fewer data gaps than studies well out to sea. Many of Bruce Frost's students participated in projects at Dabob. Among the earliest were Jeannette Yen, whom we met earlier, and Mark Ohman (introduced below). In the mid-1980s Steve Bollens examined several copepod species in Dabob Bay and other habitats.

His study connects directly to arguments raised by Gliwicz. First, let's meet Steve (Figure 8.8). I talked with him in spring of 2020.

Steve grew up in Key West, the island town at the seaward end of the Florida Keys archipelago. His father was a small-boat fisherman, and Steve often worked with him catching grouper and snapper by hook-and-line fishing. It was, Steve says, "one fish at a time." That and snorkeling over the Keys' still healthy reefs opened his interests in animals and oceans. He also claims his interest in DVM started when fishing with his father. The diel migrations of sound scattering layers seen on the fathometer were both obvious and consistent from day to day. Those traces seen on a 1960s fathomer with relatively low frequency "pings" would have been echos from small fish with swim bladders. Steve's high school was not strongly engaged in college preparation for its students, but his parents were readers and he was academically encouraged at home. Thinking college would lead to some sort of biomedicine career, he enrolled at Oberlin College in Ohio and joined the pre-med set. However, he soon found the competitive grade-hunger there ugly and moved over to more environmentally focused courses. One was invertebrate zoology with David Egloff, who has been a significant inspiration to many recognized ecologists. Steve took a

Figure 8.8 Steve Bollens and his wife Gretchen Rollwagen-Bollens. They are the codirectors of the Aquatic Ecology Lab at Washington State University, Vancouver. From the AEL website with permission from Steve and Gretchen.

break working in the Gulf of Mexico on oil rigs with the goal of earning some money to pay for a sea semester at the Sea Education Association (SEA) in Woods Hole. Egloff recommended that to him and him to it. SEA semesters include work from research vessels under sail, and the students learn a range of basic oceanographic topics. Steve emerged primed to train as an oceanographer. Graduating from Oberlin in December 1981, his application was accepted at the University of Washington Oceanography Department for studies under Professor Tom English. Once Steve settled in at Seattle, however, Tom died prematurely. That led to negotiations across the faculty and subsequent work under Bruce Frost. The program they developed took seven years, much of it devoted to the studies of DVM in copepods discussed below. Six or seven years are typically needed to finish PhD programs in oceanography, at least in the United States.

Subsequently, Dr. Bollens has had three long-term jobs. He was a two-year postdoctoral scholar at Woods Hole Oceanographic (WHOI), supervised by Peter Wiebe and Cabell Davis. However, it turned out that he worked mostly with Larry Madin. He continued at WHOI, appointed to the staff as assistant scientist. Some of his projects were with the Global Ocean Ecosystem Dynamics (GLOBEC) Georges Bank program. Steve was part of a study[9] of the unusual blooms of unattached hydroid colonies, *Clytia gracilis*, that develop in the shallows over the bank in spring and early summer. The little polyp clusters (not the jellyfish stage) are greenish-yellow and become dense enough to tint the ocean all across the huge current eddy covering the bank. When that was seen on the GLOBEC cruises, it seemed new. However, it had been observed early in the twentieth century, and then seemingly was forgotten by oceanographers. It was in Woods Hole that Steve first met Gretchen Rollwagen, who was an administrator there at SEA. He remembered her at least. She soon moved to Hawaii for graduate study, after eight years with SEA.

Woods Hole investigators famously (among oceanographers) have to fund their salaries through their research grants, as well as their equipment and technical staff. It's the extreme version of soft-money science allowed by US government agencies and private foundations in astronomy, particle physics, and oceanography. When proposals fail, a lot of nasty pressure appears if the institution has to carry them for a while. I, too, had a role in the Georges Bank study, and Steve reminded me in 2020 that I had advised him to get away from soft money if any reasonable and salaried offer appeared. One did, and in 1996 he moved to San Francisco State University (SFSU), becoming an associate professor.

That opened up teaching opportunities, for which Steve was also looking, and there were excellent facilities for estuary research, including the Romberg Tiberon Science Center adjacent to San Francisco Bay. An example of his training and research at SFSU is an experimental study with Alexander Bochdansky on the effects of thin layers of phytoplankton on copepod distribution and feeding (which is not much, it turned out).[10] Attending a conference in Hawaii led to meeting Gretchen Rollwagen again. That time romance blossomed and led to marriage. Gretchen, then and now G. Rollwagen-Bollens, moved to the Bay Area to make marriage real. During their time there Gretchen obtained a PhD at Berkeley, working first with Debra Penry and later Thomas (Zack) Powell.[11] That effort wasn't enough for the family energy level, which is demonstrably off-scale, so they added a daughter to the family in each of 1998 and 2002. Steve meanwhile began an interest in university administration, becoming an assistant dean for research at SFSU in 1998, and in 2001 he attended a summer institute at Harvard in higher education management and leadership.

A third position was at Washington State University (WSU), involving a move in 2004 to Vancouver, Washington, just across the Columbia River from Portland, Oregon. The main campus of WSU is far to the east in the small town of Pullman, which is also a commercial center to the Palouse wheat farming area. Joining a nationwide trend for state universities to open branch campuses all over their states, WSU decided to see what could be established in Vancouver. Steve, with his new administrative credentials, was invited to become the first director of its School of Sciences and Professor of Biology, and Gretchen was also invited to join the faculty. In the years since, Steve has held other WSU-system-wide director titles: of a School of Earth and Environmental Sciences, of Multi-Campus Planning and Strategic Initiatives, of the Aquatic Ecology Laboratory (shared with Gretchen as codirector), and lately of the Meyer's Point Environmental Field Station near Olympia on Puget Sound. The last has only recently started; he has a bungalow nearby in which to spend three days a week in some seasons. In just 2019 and four months of 2020, Steve and Gretchen, usually together and often with graduate students, published nine papers on environmental issues, some about copepods. Steve is also the WSU representative to the State Legislature, attending a few days a week during sessions, and cochair of the public university system's Council of Faculty.

Steve and Gretchen have also fitted in long sabbaticals in New Zealand (Otago, 2010) and Australia (Perth, 2017). They are active hikers and car campers, and lately Steve has resumed (from high school days) competitive swimming, ranking 14th for fastest time among US master swimmers in 50 m free-style. The daughters are in college, which means they can only phone home for help with schoolwork. I'm exhausted just listing it all. So let's back up to why Steve is considered here, his thesis work on vertical migration:

Bruce Frost had learned by sampling in Dabob Bay on various dates that *Calanus pacificus* of several stages sometimes undertake definite diel vertical migrations, rising near the surface only at night, but at other times they stay shallow around the clock. Teaming with Steve,[12] Bruce tried testing the notion that migrating versus not migrating by *C. pacificus* females could depend, for specific periods, on the presence versus absence of significant numbers of fish that feed on copepods, as shown by their gut contents. They used open-closing net hauls to sample the water column for copepods in sequences of vertical, 25 m thick strata down to 75 m and 50 m strata below that. Two seasonally close dates are shown in Figure 8.9.

With their plankton samples in jars, they pulled a square-mouth (9 m X 9 m) fish trawl through the upper 50 m, filtering volumes equivalent to 9 to 36 Olympic racing pools (25 m x 50 m x 10 m deep). To sample just the surface, a smaller net with floats and an opening down to 3 m depth filtered 0.6 to 2.5 such swimming-pool equivalents. Catches were variously sized fishes, the species dominance varying with season and particular date. They examined fish gut contents and focused on those eating female *C. pacificus*: Pacific sand lance, hake, and several salmon species (on their way seaward), all juveniles. Unfortunately, the data were not perfect. The spring 1985 fish sampling could not occur until June, but small planktivorous fish were few then and throughout the summer. In May 1986, the stock of juvenile sand lance was high, 11 per 10,000 m^3 (0.8 of those swimming pools), so perhaps then the migration was a direct response to the presence of those fish. However, the logic of the test (aiming to show contingency of DVM on abundant fish) was not fully satisfied. Field work is challenging, especially so when it involves competition for ship

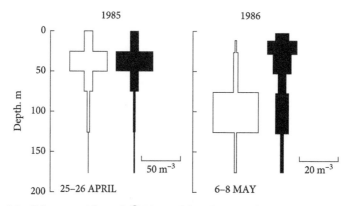

Figure 8.9 *Calanus pacificus*, C6♀. Vertical distributions shown as averages of four noon (white bars) and four midnight (black) vertical, opening-closing net profiles taken on two consecutive days in 1985 (left) and in 1986 (right).[12] Bars are average numbers per cubic meter in each stratum (scales). These are close spring dates. In 1985 there was no evident DVM; in 1986 DVM was pronounced. Permission, Wiley & Sons.

time and other assets. Worse, during all the rest of spring-summer seasons in both years, vertical migration of *C. pacificus* females was strong, even in October when estimated fish abundances were low. Dabob Bay is a great system: variably migrating *Calanus*, variable fish, potential for relatively easy sampling. Eventually more work was done there.

Since the Dabob study did not actually meet the logical requirements of a test, Bollens moved to a smaller system, a saltwater lagoon on San Juan Island in Puget Sound. Jakles Lagoon is nearly isolated by a bar from the open water beyond. Its area is about 2.6 hectares, its center depth is 4.0 m, and it has a numerous stock of the small copepod *Acartia hudsonica*. The development timing and growth of that population had been studied by Michael Landry,[13] one of Frost's early students. There is only one abundant fish, the three spine stickleback (*Gasterosteus aculeatus*). Bollens[14] pulled a small trawl through the upper 1.5 m, identifying dates with many sticklebacks and dates with almost none. The fish have a seasonal cycle: 3 to 20 per 100 cubic meters in spring, 200 to 1,700 in June and August, then back to <200 in October. To examine the relation of *Acartia* distributions to that predator variation, Bollens constructed, with helpers, a sampling raft in the middle of the lagoon to study the vertical distributions of the copepod. Sampling from the raft was done with four one-meter lengths of clear plastic pipe (3.5 liters) spaced along the sides of a square aluminum tube. That was lowered with the pipes open at the ends that were then closed in place with spring-driven stoppers. So sampling around both

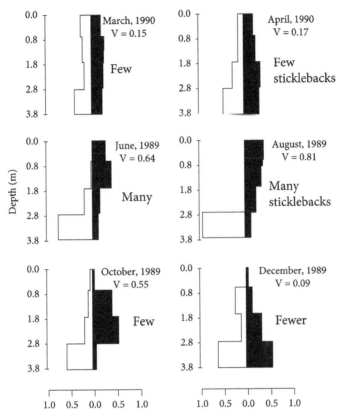

Figure 8.10 Distributions of *Acartia hudsonica* in Jakles Lagoon, San Juan Island, Puget Sound, maximum depth 3.9 m. From Bollens et al.[14] Permission, Springer Nature.

noon and midnight, he could indeed show (Figure 8.10) that when fish were abundant, *Acartia* migrated to near the bottom in daylight. When fish were absent, *Acartia* remained present in the whole water column. Steve calculated a statistic, V, for the strength of migration that was large in summer, small in spring, intermediate in October. Clearly something else was involved in December.

Apparently, *Acartia* can sense the risks from fish and initiate diel vertical migrations in response when those risks are high. There was little vertical stratification of the phytoplankton distribution; suitable food cells were available throughout. Perhaps migrating up in the dark, even when fish are present enables spacing out of food competition when the fish cannot see. Or perhaps it is just what is needed in the more usual, deeper water cases with food above and just safer darkness below: up to eat, down to hide. Let's come back to that, asking first what tells copepods there are fish to avoid. The answer is that it varies.

Bollens[15] also hung long, 1 m diameter plastic bags from one edge his raft. His notion for experiments in such bags came from lake studies he heard about at a meeting. He pumped them full of water, filtering out all the fish and copepods. Then he took vertical net hauls from the other side of the raft and poured *Acartia* from them into the bags, raising their density to about 8 /liter (~21,000 per bag). He did some more trawling for sticklebacks. Some bags had fish added, some also had fish added but in coarse-mesh cages hung in the upper meter, and some had no fish. It was summer, so fish were numerous in the lagoon and the copepods were migrating. Then he sampled the bags with the tube sampler at noon and midnight for a few days.

What did the copepods do? Those with no fish stayed vertically dispersed around the clock, as in the lagoon during the months of spring when there were few sticklebacks (as shown in Figure 8.10). Those with free-swimming fish migrated down to the bag bottom, also like those in the lagoon. Those with fish in cages paid no attention to their presence, finding no reason to migrate. In bags with no fish the copepods did no migrating for five days, but then adding fish initiated their down migration that very day. Apparently, *Acartia* senses the fish and heads immediately for darker layers. No response to fish in mesh cages suggests that *Acartia* senses fish by means other than any odor ("kairomone") the fish emit. Bollens has also shown that fish-shaped lures towed through the water will induce *Acartia* migrations in daylight but not at night. For at least *A. hudsonica*, one cue for getting down and out of the brightest upper layer probably is being tossed a bit in fish-generated eddies.

An Alternative Example from British Columbia

DVM-inducing predators do not have to be fish, other visual predators will serve. One clean example is the interaction of the copepod *Diaptomus kenai* with phantom midge fly larvae, *Chaoborus* spp., studied by William E. Neill in 27 m deep Gwendoline Lake, British Columbia.[16] The predatory larvae can be up to 2 cm long (Figure 8.11) with compound eyes. An example of *Diaptomus* is shown later (Figure 17.1). Neill collected vertical profiles of *Diaptomus* abundance with Clarke-Bumpus nets before and after introduction of fish that removed all the *Chaoborus*. The *Chaoborus* are normal vertical migrators, though Neill does not say what predators they are avoiding. Most likely those would be fish, but *Chaoborus* larvae migrate regardless of their immediate presence.

So, while there were many of them, up to 1 in each 2.5 liters, the *Diaptomus* migrated in reverse (Figure 8.12). That is, the copepods went down at night and

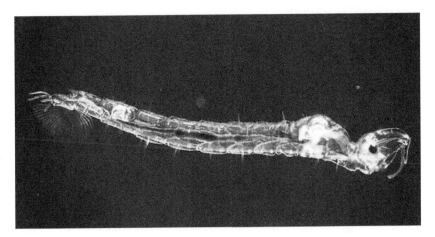

Figure 8.11 *Chaoborus* larva from a Rhine floodplain pool. Note prey clamping forelimb and fish-like tail fin. The dark spot in the head is visual pigment. Photo by *Viridiflavus*. Permission: Creative Commons, Attribution-Share Alike 3.0 Unported.

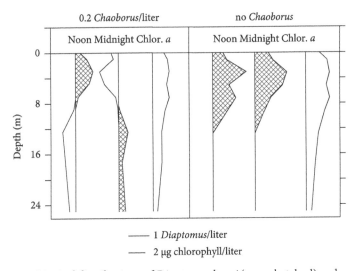

Figure 8.12 Vertical distributions of *Diaptomus kenai* (cross-hatched) and *Chaoborus* spp. (white) in Lake Gwendoline, British Columbia. Left: noon and midnight profiles of *Diaptomus* and *Chaoborus* present. Right: noon and midnight profiles for *Diaptomus* after fish exterminated *Chaoborus*. Chlorophyll was in cells > 5 μm. From Neill.[16] Permission, *Nature*.

came up in daylight to feed. By 1990 Neill had been studying Lake Gwendoline plankton for over a decade. In one year some juvenile trout escaped in Lake Gwendoline, and their growing population effectively exterminated *Chaoborus* in a year or two. Trout gut contents revealed almost *no* copepods. Neill repeated the sampling of *Diaptomus* 28 months after trout exterminated the midge larvae, finding that copepod reverse migration had also disappeared. They stayed in the upper 10 meters of the lake around the clock.

To test the immediacy of the copepod response to midge larvae, Neill stocked *Diaptomus* into 1.5 m diameter x 15 m long plastic bags at approximately 1 per liter, much like Bollen's *Acartia* experiments at Jakle's Lagoon. Some bags he also stocked with *Chaoborus*. That stocking was done around 8:00 p.m., and then Neill sampled all the bags with a pump around midnight, and again the following noon. With no *Chaoborus*, the *Diaptomus* remained near the surface both midnight and noon. In the bags with *Chaoborus* included, the responses were immediate: *Diaptomus* had left the upper water by midnight (reverse migration). *Chaoborus* migrated normally (being down at noon). Finally, during early darkness, Neill siphoned 20 liters of water into the 2 to 6 m depth range of a bag with just nonmigrating *Diaptomus*. That water had held many *Chaoborus* larvae (110 per liter) for 24 hours and then been filtered. Giving the *Diaptomus* four hours to react or not, he sampled again. Having reacted quickly, the copepods were all at depth. So in that case it likely was an odor from the midge fly larvae giving the avoidance signal.

Mark Ohman and Reverse Migration

Mark Ohman (Figure 8.13) also took his PhD at the University of Washington, working with Bruce Frost to extend an earlier observation of DVM in Dabob Bay made by the late Edward Cohen. I interviewed Mark in the summer of 2017 and again in 2020. His initial remarks were about growing up in Berkeley, California, in a politically liberal family. Berkeley did provide some introductory experience suggesting a career in biology. At perhaps eleven, he collected some polliwogs from a park pond, took them home and established them in a jar using newly drawn tap water. He didn't say what he fed them, but it was something. The polliwogs survived: they grew large enough to metamorphose into frogs, but they didn't. He recalls that as requiring explanation and came up with chlorination toxicity of the tap water. Apparently, that was never tested. An inadvertent plunge down along pier pilings at the Berkeley Marina in those early years also made a powerful impression. The profusion he saw of feather-duster worms, mussels, and anemones was a "How can that be?" experience, one he eventually pursued. His

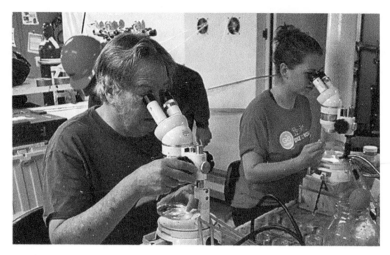

Figure 8.13 Mark Ohman (left) and Stephanie Matthews (a Scripps student) sorting copepods during a Long-Term California Current Ecosystem Research cruise in 2019. Note that the microscopes are tied down to the work bench. The stereomicroscope design (Wild M5) dates from 1958. It is still the best, partly because planktologists are used to what and how we see with it, partly because it is almost indestructible in the field. From *Explorations Now*, an SIO magazine, with permission.

siblings went into nonscientific careers, his brother worked in healthcare finance, and his sister as an administrator for the National Education Association.

Mark attended the University of California, Santa Cruz, on the coast south of San Francisco. He initiated a major in European intellectual history, but science courses derailed that. Of particular significance was an oceanography course taught by Mary and Eli Silver, who had trained at Scripps Institution of Oceanography. Mary taught biological aspects and Mark recalls being radically surprised by her description of larvaceans (*Oikopleura*, *Fritillaria*, etc.). Those distant cousins of the chordates deserve their own book; maybe there's one out there. Plankton, as Mary presented them, had entirely distinct modes for living and remaining always in water. Indeed, those remain surprising, even strange to those long engaged in their study. What also caught Mark's attention was being for the first time in a science course that was partly about what was known or explained, but also about what *was not* explained. Not only did they discuss unsolved problems, but both Mary and Eli emphasized that many of those could be solved with some wit and systems well chosen for study. Up until then, science courses had been either rote learning or solving problems with the tested ideas (and in my experience equations) of a discipline.

Mark realized, as have all these copepodology heroes, that to get into the science game professionally he would need specialized, graduate-level education. He started in the California State University system (distinct from the University of California branches and then offering no doctoral programs), first taking classes with ichthyologist Margaret Bradbury at San Francisco State. He took an invertebrate zoology course at the Marine Biology Laboratory, Woods Hole, learning allozyme genetic techniques (see Chapter 18) and publishing a first note with T. J. M. Schopf.[17] Then he applied allozyme techniques under John Martin at Moss Landing Marine Lab for his 1977 MA thesis examining variation of several esterase genes in the intertidal splash-pool copepod *Tigriopus californicus*. Next, Mark applied to several universities for a doctoral program and chose the University of Washington's Department of Oceanography.

Mark arrived over a decade after zooplankton studies began at Dabob Bay, including a study by Ed Cohen and Bruce Frost. They discovered in 1973 that female *Pseudocalanus* have a reverse migration, at least in summer. In that season they are migrating down at night, up in daylight. Discovery established, Cohen left school for a while, but eventually returned to the University's School of Fisheries and obtained a PhD in 1987. He moved on to a fisheries lab in Woods Hole, where he tracked lobster and fish catches in the Gulf of Maine. Sadly, he died of leukemia after just a few years. His copepod DVM results continued to interest Bruce, who gave a lunch-time seminar about them. Mark found that fascinating. He told me he asked Bruce whether he could take on repeating the observations, and Bruce agreed. At the time, Bruce's study of the genus (Chapter 4) was underway but incomplete, so Mark's counts were listed as *Pseudocalanus* sp., which later work[18] showed were *Pseudocalanus newmani* Frost (1989).

Mark's result from 1979 sampling (Figure 8.14) showed that in mid-summer *Pseudocalanus* sp. indeed moves down at dusk and up at dawn, producing the distributions in the figure at mid-day and midnight.[19] The critical point which Mark added was an explanation. The copepod is not avoiding fish. More likely it is aware of and avoiding the predation potential of larger invertebrates, particularly the predatory copepod (Parae*uchaeta elongata*), arrow worms (*Sagitta elegans* in Dabob), and euphausiids (small krill, *Euphausia pacifica*). Small copepods, including *P. newmani* with mean length of 1.0 mm, can be found in the guts of all of them, and all those predators migrate normally in summer (down at dawn, up at dusk). Obviously, *Pseudocalanus* can reduce that predation both shallow (night) and deep (day) by doing the opposite. A life-table model showed that foregoing the added growth, and thus potentially greater fecundity, from eating at the surface around the clock would be more than repaid by quite modest reductions in predation mortality. Later Mark[20] (in a paper

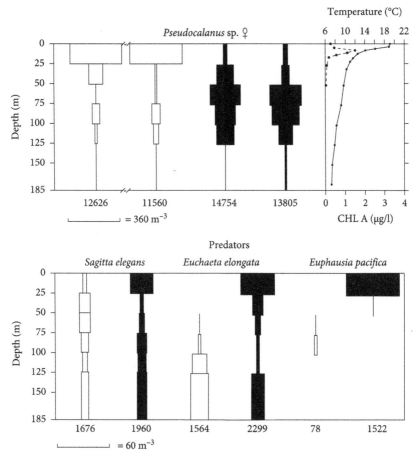

Figure 8.14 Above: day (white) and night (black) vertical distributions of *Pseudocalanus newmani* C6♀in Dabob Bay on July 25 (left stacks) and July 30 (right stacks) 1979. Reverse vertical migration is strongly evident. Phytoplankton (shown as chlorophyll concentration, dashed line) was mostly in the upper 25 m. Below: Principal predators of *P. newmani* in Dabob: an arrow worm, a predatory copepod and a 1 cm krill, all exhibited normal DVM. The krill avoid nets in daytime, but high frequency sonar showed them abundant from 40 to 100 m. From Ohman et al.[19] Permission, *AAAS (Science)*.

with many more sampling iterations) and Frost and Bollens[18] published vertical distributions or *P. newmani* from several times of year, patterns showing both reverse and normal DVM, sometimes none. Sometimes they migrated with fish about and sometimes not. Clearly DVM as a predator-avoidance scheme is very sophisticated: it can be employed only when there are predators to fear. In fact, DVM can be employed against predators detecting copepods with different senses. As we've seen, Bollens extended that to an *Acartia* species, and almost certainly it also applies to more than one species of *Calanus*.

Mark had side interests while living in Seattle. He was part of an Amnesty International Adoption Group 65, one of many around the world working to free individual political prisoners. For example, Mark's group focused its concern on a Chinese dissident imprisoned for speaking out. The groups write letters to encourage the victims, to the incarcerating governments, and to the governments of the United States and other nations alerting them about their adopted prisoner, and pushing for sanctions of various sorts. Of course, adoption groups have a social aspect, and Mark met his wife Cynthia there. They were married when Mark finished a first postdoctoral year at Friday Harbor Lab, and they soon moved to New Zealand where Cynthia pursued dissertation research. She had been studying ancient Greek language and literature at UW.

Mark's dissertation ("The Effects of Predation and Resource Limitation on the Copepod *Pseudocalanus* sp. in Dabob Bay, a Temperate Fjord") is dated 1983 by the university library, which would be a bit after he began the search for a postdoc or an actual job. He landed a one-year position in New Zealand sponsored by John Jillett and Janet Bradford-Grieve (Chapter 5), supported in part by a US/ New Zealand Co-operative Science Program funded by the US National Science Foundation. The work combined field and laboratory work on the populations of large copepods in the family Calanidae available well offshore from the islands. Much copepod rearing from eggs was done with Dr. Jillett at the Portobello Marine Lab located on the Otago Peninsula reaching east from South Island. They studied egg to adult growth and development rates in *Neocalanus tonsus*, collected in winter from subantarctic water, and both *Calanoides macrocarinatus* and *Calanus australis* collected from the Southland Current carrying more temperate species alongshore in summer. Rearing all at 15°C, the patterns of length increase (and probably biomass growth) were similar (Figure 8.15). Development times to C5 were *tonsus* 24 days, *australis* 24 days, and *macrocarinatus* 20 days. For their size, that is relatively fast (take my word for it). Dr. Bradford was first author on the paper.[20] Mark agrees that was because its pages and pages show her drawings of every stage of all three species in lateral and dorsal or ventral views, plus mouth parts and thoracic feet. Specific characters are listed stage by stage.

That dynamic trio also sampled *N. tonsus* to 1,000 m off Otago Peninsula, through the year from October 1984 to September 1985.[21] It, like the *Neocalanus* species of the subarctic Pacific, matures from diapausing C5 and reproduces at depth (Chapter 16). Ohman used the time series to study its wax lipid storage. Those two tours de force of copepod ecology, in addition to demonstrating and explaining an example of reverse diel migration, helped Mark obtain a position at the Scripps Institution of Oceanography in 1985. He has been and is succeeding

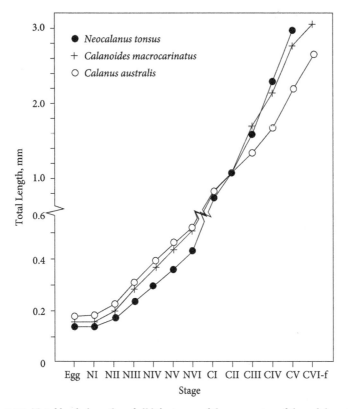

Figure 8.15 Total body lengths of all life stages of three species of three laboratory-reared copepods from waters off New Zealand's Otago Peninsula. The jogging lines between N6 and C1 represent the change from nauplii to copepodites in what is measured. *N. tonsus* would not mature in the lab, stopping at C5. From Bradford et al.[20] Permission, Taylor & Francis.

there with many studies of copepods and other plankton. At the moment he is director of a Long-Term Ecological Research (LTER) program sponsored by the National Science Foundation, connected with time-series investigations of the California Current that began in 1949, a few years before Mark was born. As of 2020 he is advising a postdoc and two graduate students, Stephanie Mathews (Figure 8.12) and Laura Lilly. He continues as the curator of the Scripps Pelagic Invertebrate Collections.

After Mark and Cynthia moved to the San Diego area, they had two children in rapid succession. Their daughter is now finishing training in internal medicine; their son is interested in economics and certified as a public accountant. Cynthia defended her dissertation and taught ancient Greek and Latin for 30 years at

Figure 8.16 The writing studio at Mark and Cynthia's home in north San Diego County. Photo by Mark Ohman, with permission.

the University of California, Irvine. They live in a north San Diego county bedroom community. They have taken two sabbaticals in France. The first was at the Sorbonne lab on the Mediterranean in Villefranche-sur-Mer near Nice, where they put the children in the French public schools. The second was in Paris, spent writing up finished projects and visiting sites like the Marquis de Sade's cell in the Tour de Vincennes (it was there, so they checked it out). Mark has been a strong participant in activities of the World Association of Copepodologists, maintaining an array of connections with colleagues around the world. At home he builds things. Many fathers put a plywood shack on a platform in a tree for their kids, but Mark built his children an elegant playhouse large enough that it now serves him and Cynthia as a writing retreat (Figure 8.16).

Metridinids Take Long Trips Down, Then Up, Then Down, . . .

Those stories show us that some copepod species can choose whether or not to migrate out of lighted layers, some even reversing the migration schedule, depending upon whether predators are nearby at threatening densities and exactly what predators. Thus, DVM must be mostly an adaptation for avoiding predators. However, *Calanus* species only migrate down about 100 m, and *Acartia* species less than that. There are species that take DVM much more seriously, going much deeper. Such migrations are best documented for the marine family Metridinidae, particularly the genera *Metridia* and *Pleuromamma*. There is a family of sea anemones (Metridiidae), and all but one of its members share the genus name *Metridium*. Do not worry about this minor source of confusion; the spelling distinction was set in nomenclatural law in 1902. A Norwegian, Jonas Axel Boeck, published the genus and the species names *Metridia lucens* in 1864[23] with descriptions entirely in words. He did not provide the meaning of the name *Metridia*. It sounded Greek to me, and Mark Ohman's in-house expert in ancient Greek corrected my crude spelling to μητρίδια. Mark wrote to me: "Cynthia says she does not know what the describer (Boeck) may have intended, but 'Metridia' in Greek means 'having a mother; fruitful; filled with seed." My guess is he named it for the masses of eggs readily seen along the dorsal sides of female specimens.

Boeck did mention bioluminescence as a character. That is considered in the next chapter. His use of *lucens* for the North Atlantic species refers to its *stæke blaau Lys de frembringe ved Berøring* (strong blue light elicited by a touch). I will come back to that. The North Pacific species, *M. pacifica*, is very closely related and was called *M. lucens* in much early work. Again, here (Figure 8.17) is the picture of a *Metridia pacifica* female.

Repeating from here briefly from Chapter 2: Species of *Metridia* swim constantly in fairly tight loops, with diameters of about five to eight centimeters. The reason remains unexplained, but it must alert every small, predatory fish,

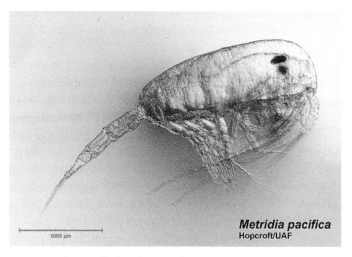

Metridia pacifica
Hopcroft/UAF

1000 µm

Figure 8.17 Lateral view of a female *Metridia pacifica*. Except for the tiny eye just visible here, the genus is entirely unpigmented, as close to transparent as tissue can be. The photo depends on refraction effects. Food in the gut does add one or a few opaque spots. Eggs (arrows) are in the anterior part of the oviduct. Length from head to the ends of the caudal furcae around 3 mm. Photo by Russ Hopcroft.

amphipods, and arrow worms that dinner just swirled past. It must find any cteno-phore or siphonophore tentacle suspended nearby waiting for breakfast. As also explained in Chapter 2, *Metridia* is a dominant component in the gut contents of many zooplanktivores. Presumably to set some upper limit on these popu-lation losses, *Metridia* has evolved every predator-thwarting adaptation known among copepods (apart from hanging quietly by their antennules most of the time). Those adaptations include extreme DVM.

In the 1960s the planktologist and marine ecologist Alan Longhurst worked at the National Marine Fisheries laboratory in La Jolla, California. While there, he developed a plankton-net sampling system that could take large numbers of *separate* samples one after the other.[24] The mouth diameter was 60 cm, followed by a conical net supported by a metal frame. At the back was a box with a mech-anism to wind the catches progressively from each few minutes of tow into a roll of nylon mesh. Individual samples were sandwiched in window-shaped rectangles between two sheets of the mesh. The animals became cross-hatched with fiber marks but were nevertheless identifiable. The basis for the design of that box was the "plankton recorder" developed long before by Alister Hardy to record plankton distributions from freighters crossing the oceans. So Alan's device was soon termed the Longhurst-Hardy Plankton Recorder (LHPR). Several other workers later used and modified them. In the 1970s Alan moved on to the Plymouth Laboratory in the United Kingdom. While there, he and Robert Williams[24] used an LHPR recorder to sample in the Atlantic, producing profiles of many plankton groups. They documented the very extensive ver-tical migrations of *M. lucens* at Ocean Station India (60° N, 20° W, due south of Iceland) (Figure 8.18).

As their graph for March suggests, the bulk of the *M. lucens* population migrates during daytime to depths centered around 420 m, then returns at night to the upper 50 m, its mode very near the surface. The May data look different. Actually, the paper was not about vertical migration; it just shows up in the ex-treme detail of LHPR data. However, reading it line-by-line reveals that the data for March were mostly adults, and the data for May mostly younger copepodites (presumably the offspring of the reproducing stock in March). The graphs are proportions of the estimates summed over the whole water column, so the graphs look as if the total abundances were the same. Indeed, daytime depths move deeper and deeper as *Metridia* copepodites grow.

Migration to 420 m amounts to 140,000 body lengths. Scripps biologist James Enright[26] estimated that adult *M. pacifica* made shorter ascents off Southern California at sustained speeds of 30 to 90 meters per hour. That implies peak sustained speeds of 2.5 centimeters per second (roughly 1.2 body-lengths per second). Yes, that's fast, but also about what they are always doing in a dish under

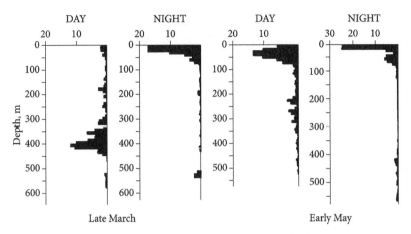

Figure 8.18 March and May vertical distributions of *Metridia lucens* at Station India (60°N, 20°W) in the North Atlantic (Norwegian Sea) as percentages of the water column total in each 5 or 20 meter layer. LHPR data from 1975 collected by Longhurst and Williams.[25] Permission, Springer Nature.

a microscope. To migrate they may just need to straighten it out. I have to note that *Metridia* cannot be readily watched in the ocean, so maybe being in brightly lit buckets and dishes disturbs them into this swirling motion.

To cover 400 m at even 90 m/h would take almost 5 hours, each way. Hiroshi Hattori,[27] based on his LHPR tows, estimated translation speeds from of abundance peaks of *M. pacifica* C5 and females in the Oyashio at around 30 meters/hour. I get those numbers from his graphs, too, using the times of his tows. However, the tows were four hours apart, suggesting that the actual peaks may have moved much faster. His population mode around noon was at 250 m, and at midnight at the surface. Hattori emphasized that not all specimens migrated up from the deep mode; some stayed put at depth. Their gut contents (examined by scanning electron microscopy) contained different foods from specimens collected near the surface. The deeper migrations of *M. lucens* in the Atlantic likely resulted from greater water column transparency.

Another metridinid genus is *Pleuromamma* (Figure 8.19), which includes 16 species for now, most but not all living in relatively warm ocean waters. They have an obvious character making them easy to study, black, knob-shaped organs on just one side of their prosomes. Those do not fade in preservatives, so if you see one, you know you are looking at a *Pleuromamma*. It can be on either side in most species, the proportions of lefties and righties varying. Much of what is known about these knobs was learned from electron micrographs of

Figure 8.19 A female of the larger species of *Pleuromamma*, its transparency lost to preservation. The black dot is the glandular knob characteristic of the genus. Photo by Manuel Elias-Gutierrez, Colegio de la Frontera Sur, Unidad Chetumal. Creative Commons License.

small species by Pamela Blades-Eckelbarger (who comes up again in chapter 10) and Marsh Youngbluth:[28]

> The calanoid copepod genus *Pleuromamma* is easily distinguished from all other copepods by the presence of a rounded, dark-red cuticular structure (pigment knob) that protrudes from the left or right side of the second thoracic segment in both sexes. . . . study of the pigment knob reveals a complex ultrastructure consisting of various cell types within three distinct areas that are bathed by hemolymph from the lateral sinus. The knob is covered by a greatly expanded cuticle through which a pore passes. The pore appears to connect with a centrally positioned pigment cell containing a large mass of darkly staining granules. This suggests that the knob may have a secretory function. Observations of live animals and dissected pigment knobs, however, indicates that the knob does not secrete a luminescent material nor does it luminesce internally.

The authors also note that secondary sexual characters like the males' grasping antennules are on the same sides (left or right) as the knobs. So the knobs likely have some function in mating. I thought you would want to know all that.

Like the *Metridia* species, late-stage *Pleuromamma* make extensive vertical migrations (Figure 8.20). Some of them come up at night into the near-surface layers, but others seem to find nighttime forage between 100 and 200 m. Half a dozen of its species were studied by Scripps planktologist Loren Haury.[29] He had

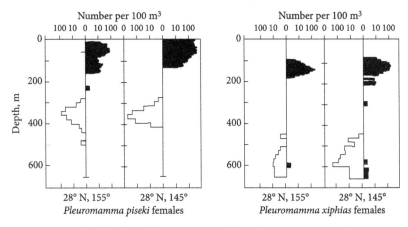

Figure 8.20 Night (black) and daytime (white) distribution patterns of two species of *Pleuromamma* evaluated with in 5–10 m (upper 200 m) and 20 m layers sampled by LHPR from 0 to 650 m in the subtropical gyre of the central Pacific Ocean to the north and east of Hawaii. Collections by Loren Haury[29] were made within 1.5 hours of both noon and midnight. Permission, Springer Nature.

trained as an electrical engineer and followed that with a tour in the US Navy. Then he switched to biology in 1968, entering the Scripps biological oceanography graduate program, studying under John McGowan. His dissertation concerned small-scale zooplankton distributions, both horizontal and vertical, as sampled by LHPR. Taking his degree in 1973, he moved to Woods Hole, where he worked with Peter Wiebe and others. Then he returned to Scripps as a research oceanographer, again sampling with an LHPR. Among the results were the vertical locations of six species of *Pleuromamma* at times close to both noon and midnight on a transect along 28° N in the subtropical gyre of the Pacific from northeast of Hawaii to the edge of the California Current and then on up to Southern California.

Loren Haury's LHPR data (Figure 8.20) for the females of two species at two stations demonstrate just how far these tiny animals swim every day. The smaller of the two, *Pleuromamma piseki*, is 1.3 mm total length. Its night modes were at around 50 m, with some abundance near the surface and down to 150 m (note the semilogarithmic abundance scales). At noon it ranged from 300 to 425 m, its peak at around 350 m. Individuals moving between the modes were swimming 300 m down, then 300 m up, then again down and up again—230,000 body lengths twice each day. *Pleuromamma xiphias* is around 4.5 mm total length, and it swims up and then down daily on the order of 475 m, or about 105,000 body lengths. It seems like a huge amount of work, and various studies have tried to evaluate the energy costs to copepods of swimming. I am not reviewing those here, partly because they largely involve the much faster thoracic-leg-driven escape jumps, not the antennal sculling of steady swimming of only one or so body lengths per second. In order to make the long round trips of *Pleuromamma* species daily, the costs cannot be extreme. Those costs must be fully paid for in improved survivorship.

The study was done in what likely are the clearest natural waters in the world's oceans, so getting to safety in darkness requires going deep. For at least *Pleuromamma* species, very few individuals stay behind at depth and essentially none stay anywhere close to the surface during the day. There do not seem to be days when they do not migrate down at all.

Clearly, two things matter greatly for individual animals, perhaps about equally: getting enough to eat and not getting eaten. For copepods, more than migration down out of well-lighted layers is involved in the latter. In the next chapter, I consider a few more of the copepod safety strategies.

References

1. Franz, V. (1912) Zur Frage der vertikalen Wanderungen der Planktontier (Autorreferat). *Archive fur Hydrobiologie und Planktonkunde* 7: 493–499.
2. Clarke, G. L. (1933) Diurnal migration of plankton in the Gulf of Maine and its correlation with changes in submarine irradiation. *Biological Bulletin* 65: 402–435.
3. Clarke, G. L. (1934) Further observations on the diurnal migration of copepods in the Gulf of Maine. *Biological Bulletin* 67: 432–455.
4. Kerfoot, W. B. (1970) Bioenergetics of vertical migration. *The American Naturalist* 104(940): 529–546.
5. McLaren, I. A. (1963) Effects of temperature on growth of zooplankton, and the adaptive value of vertical migration. *Journal of the Fisheries Research Board of Canada* 20: 685–727.
6. Deevey, G. B. (1960) Relative effects of temperature and food on seasonal variations in length of marine copepods in some eastern American and western European waters. *Bulletin of the Bingham Oceanographic Collection* 17: 54–86.
7. Hardy, A. C. (1935) The plankton community, the whale fisheries, and the hypothesis of animal exclusion. *Discovery Reports* 11: 272–370.
8. Gliwicz, M. Z. (1986) Predation and the evolution of vertical migration in zooplankton. *Nature* 320: 746–748.
9. Madin, L. P., S. M. Bollens, E. Horgan, M. Butler, J. Runge, B. K. Sullivan, G. Klein-MacPhee, & others. (1996) Voracious planktonic hydroids: unexpected predatory impact on a coastal marine ecosystem. *Deep-Sea Research II* 43(7–8): 1823–1829.
10. Bochdansky, A. B., & S. M. Bollens (2004) Relevant scales in zooplankton ecology: Distribution, feeding, and reproduction of the copepod *Acartia hudsonica* in response to thin layers of the diatom *Skeletonema costatum*. *Limnology & Oceanography* 49(3): 625–636.
11. Rollwagen-Bollens, G. (2003) Protozoan-Metazoan Linkages in Planktonic Food Webs: Spatial and Temporal Variability and Trophic Role of Micro- and Nanoplankton in the San Franciso Estuary. PhD dissertation, University of California, Berkeley.

12. Bollens, S., & B. W. Frost (1989) Zooplanktivorous fish and variable diel vertical migration in the marine planktonic copepod *Calanus pacificus*. *Limnology & Oceanography* **34**(6): 1072–1083.

13. Landry, M. R. (1978) Population dynamics and production of a planktonic marine copepod, *Acartia clausi*. in a small temperature lagoon on San Jan Island, Washington. *International Revue der gesamten Hydrobiologie* **63**: 77–119.

14. Bollens, S. M., B. W. Frost, D. S. Thoreson, & S. J. Watts (1992) Diel vertical migration in zooplankton: field evidence in support of the predator avoidance hypothesis. *Hydrobiologia* **234**: 33–39.

15. Bollens, S. M., & B. W. Frost (1991) Diel vertical migration in zooplankton: rapid individual response to predators. *Journal of Plankton Research* **13**(6): 1359–1365.

16. Neill, E. (1990) Induced vertical migration in copepods as a defense against invertebrate predation. *Nature* **345**: 524–526.

17. Schopf, T. J. M., M. D. Ohman, & R. Bleiweiss (1975) Significant age-dependent and locality-dependent changes occur in gene frequencies in the ribbed mussel *Modiolus demissus* from a single salt marsh. *Biological Bulletin* **149**: 446.

18. Frost, B. W., & S. M. Bollens. (1992) Variability of diel vertical migration in the marine planktonic copepod *Pseudocalanus newmani* in relation to its predators. *Canadian Journal of Fisheries and Aquatic Science* **49**: 1137–1141.

19. Ohman, M. D., B. W. Frost, & E. B. Cohen (1983). Reverse diel vertical migration: an escape from invertebrate predators. *Science* **220**: 1404–1407.

20. Ohman, M. D. (1990). The demographic benefits of diel vertical migration by zooplankton. *Ecological Monographs* **60**(3): 257–281.

21. Bradford, J. M., M. D. Ohman, & J. B. Jillett (1988) Larval morphology and development of *Neocalanus tonsus, Calanoides macrocarinatus*, and *Calanus australis* (Copepoda: Calanoida) in the laboratory. *New Zealand Journal of Marine and Freshwater Research* **22**(3): 301–320.

22. Ohman, M. D., J. Bradford, & J. B. Jillett (1989) Seasonal growth and lipid storage of the circumglobal, subantarctic copepod, *Neocalanus tonsus* (Brady). *Deep-Sea Research, Part A*. **36**(9): 1309–1326.

23. Boeck, J. A. (1864). Oversigt over de ved Norges Kyster jagttagne Copepoder henhorende til Calanidernes, Cyclopidernes og Harpactidernes Familier. *Forhandlinger i Videnskabs-Selskabet i Christiania* **1864**: 226–282.

24. Longhurst, A. R., A. D. Reith, R. E. Bower, & K. R. Seibert (1966). A new system for the collection of multiple serial plankton samples. *Deep-Sea Research* **13**: 213–222.

25. Longhurst, A. R., & R. Williams (1979) Materials for plankton modelling: vertical distribution of Atlantic zooplankton in summer. *Journal of Plankton Research* **1**: 1–28.

26. Enright, J. T. (1977) Copepods in a hurry: sustained high-speed upward migration. *Limnology and Oceanography* **22**: 118–125.

27. Hattori, H. (1989) Biomodal vertical distribution and diel migration of the copepods *Metridia pacifica, M. okhotensis* and *Pleuromamma scutulata* in the western North Pacific Ocean. *Marine Biology* **103**: 39–50.

28. Blades-Eckelbarer, P. I., & M. J. Youngbluth (1988) Ultrastructure of the "pigment knob" of *Pleuromamma* spp. (Copepoda: Calanoida) *Journal of Morphology* **197**(3): 315–326.

29. Haury, L. R. (1988) Vertical distribution of *Pleuromamma* (Copepoda: Metridinidae) across the eastern North Pacific Ocean. *Hydrobiologia* **167/168**: 335–342.

9

Not Being Eaten II: Other Defenses

Other Ways to Avoid Being Gut Contents

Beside diel vertical migration, copepods exhibit remarkable adaptations to put off being digested. Many involve being as close to invisible as possible. Let me start with an example of one adaptation requiring another.

The Red Cap of *Calanus*

Much that is known about planktonic copepods comes from studies of the genus *Calanus* and its near relatives (together the Calanidae). In preparing for resting stages they accumulate liquid wax in a sac (Figure 1.2) running above the gut for most of the prosome length. Most copepodites and adults with such a sac have a cone-shaped cap of red pigment in the membrane surrounding its posterior end. During sampling cruises in and around the Gulf of Maine, I photographed small groups (totaling thousands, actually) of *C. finmarchicus* to make geometric estimates of their oil sac volumes. On one occasion, a ship's engineer was looking over my shoulder while the pictures flashed on my computer screen. He asked why all the oil sacs had a cap of red pigment at the posterior end (Figure 9.1). Having no idea, I made up an answer anyway. My wife Martha thinks that is an awful habit. I answered that the refractive index (v) of the oil was greater than that of the surrounding copepod or of seawater. So the sac should act as a light guide, with internal reflections focusing rays entering near the head to the region of the cap. The red pigment, by absorbing downwelling, blue marine radiance, would prevent emergence of a tiny light beam signaling to visual predators that dinner was just above, at least to those close enough to sense the contrast. The engineer himself immediately drew the analogy to the principle of fiber optics: high v core, low v sheath layer. Satisfied with the answer, the engineer departed to check his gauges, leaving me wondering whether that cock-and-bull story could be right.

Could the oil sac really be an oddly shaped lens? Recall (from Chapter 1) that Mark Benfield[1] had shown with images from a towed video camera that

Oar Feet and Opal Teeth. Charles B. Miller, Oxford University Press. © Oxford University Press 2023.
DOI: 10.1093/oso/9780197637326.003.0009

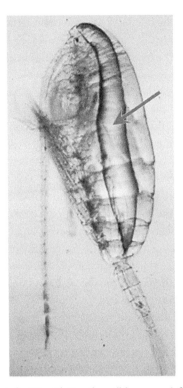

Figure 9.1 *Calanus glacialis* C5 with its oil sac (blue arrow) filling. The red cap is at the very posterior end of the sac's enclosing membrane. Note also the red spot around the mouth. An online photo (uncredited).

Calanus copepodites apparently at rest in the water are head up, tail down, antennules out to the sides. Their oil sacs would gather sunbeams from wide angles overhead and *could* channel them by internal reflections to the vicinity of the cap. With no cap and viewed from below they could appear as a sparkle, a bundle of light rays opticians call a "caustic." Do visual predators like capelin and herring hunt for prey above them? Yes, and many of these "forage fishes" have mouths that can open to the space *above* them to suck prey downward. Thus, possibly it is important for those concentrated sunbeams to be absorbed by the red pigment.

There is often no way to check explicitly a hypothesis like that one, but there can be ways to check its logical parts. I happened to have a vial of the liquid wax from *C. finmarchicus* that had been purified of membrane lipids and triglycerides by the late Margaret Sparrow using column chromatography.[2] Margaret worked in a lab across the hall from my office, an expert in all kinds of chromatography. So I borrowed a refractometer from the chemistry department and measured the refractive index of the oil. At 20°C, with only slight temperature sensitivity, it was 1.4708, which is greater than that of seawater at 1.33. It was probably also that much greater than the other *Calanus* tissue fluids, all of them isosmotic with the ocean.

It was enough different that optical paths to the oil and into its volume could turn at the ultrathin oil sac membrane according to Snell's Law. For a *Calanus* copepodite viewed from the side, a full oil sac would only have a subtle visual effect, maybe not strong enough to say "dinner here" to a fish. Maybe, but you can see the whole sac in the photograph (Figure 9.1) taken with white illumination from the animal's side.

A sort of test was offered by a numerical model of an oil sac viewed from below with a bright sky above the water surface, a model based on Snell's Law of refraction (expanded as Fresnel's equations of reflection and transmission). It suggests that the oil sac *would* act as a light guide. Once a ray is inside the sac and again approaches the oil surface at a low angle, it will not re-emerge; it will be reflected back into the oil. It can then cross the sac again and be reflected down and back several times and into the pointed end.

Across the range of *Calanus*, the spring sun in the middle hours of daylight is typically at about 30° from the zenith. Rays hitting the anterior end of the oil sac at angles around that mostly enter, then cross downward, and are reflected (at incidence and reflection angles <65°) and by that point are trapped and proceed into the pointed tip. Without the red pigment, they would exit at angles opposite to those on entering the oil and together form a "caustic" exiting from the side of the pointed tip (Figure 9.2). The strongest caustic would be approximately at the opposite angle to that of the sun and brightest sky. There would be contrast with the darker sky away from the sun. Calculations[3] of contrast dissipation in seawater suggest a predator with reasonable visual acuity likely could see that from a meter or so away. The challenge at this point is to think up some alternate reason for the red pigment covering on the conical posterior tips on *Calanus* oil sacs.

The red-caps on calanid oil sacs are an example of the back and forth, push-pull of evolution directed by natural selection. As means evolved to generate and store a good liquid lipid (a wax) for sustaining life at great depth in a resting stage, the calanids presumably experienced predation increases from upward looking predators. That would occur while loading wax in the upper strata. The genetic capacity to make red pigment to absorb the blue light predominant even in the ocean's upper layers was presumably present in all their cells. The antennules are red, as are stripes on the thoracic exoskeleton in many species. Activating that capacity in the posterior oil sac membrane could collect any tiny "eat me" beacons. The resulting reduction of mortality could lead to selection of the red caps as a rather consistent characteristic.

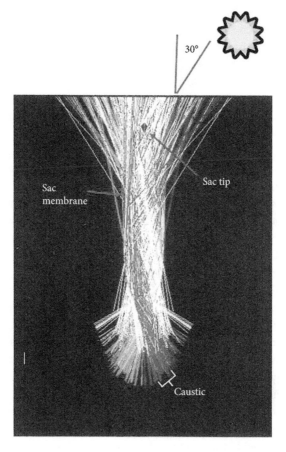

Figure 9.2 Ray-tracing diagram for a *Calanus* oil sac, with light rays impinging from all angles at many points on the upper sac then refracted and internally reflected. This focuses a beam (caustic) at an angle opposite the underwater sun image at 30° from the zenith. Ray colors are arbitrary. Ray tracing model by the author.

This push-pull interaction of predator and then prey evolution is one aspect of the Red Queen Hypothesis of evolutionary theorists. It refers to the Red Queen from Lewis Carroll's *Through the Looking-Glass*; as the Red Queen told Alice, "it takes all the running you can do to remain in the same place." There are aspects of this predatory pressure versus prey countermeasure in all of the predation-avoidance adaptations of copepods. Leigh Van Valen,[4] who coined the term Red Queen Hypothesis, also had a more general phenomenon in mind: the regularity of extinction rates through geologic time in many, perhaps most groups of organisms.

Not Eating During Daylight

I am not sure which planktologists noticed it first, but some species, those not engaging in diel vertical migration mostly stop eating in surface layers during daylight. The mostly likely explanation is obvious in laboratory cultures kept in clear glass containers. If the culture room has been dark, and copepods in their beakers or jars have been feeding on supplied phytoplankton, their guts will be full when you switch on the lights. If your youthful vision is sharp and the walls are pale, you will see their tiny, green-packed guts right through the glass, no magnification needed (especially not for myopic scientists getting up close).

Ted Durbin and the late Anne Durbin were long-time faculty members at the University of Rhode Island's Graduate School of Oceanography. They conducted many and varied studies of fish and plankton in Narragansett Bay and the adjacent Atlantic. Some were investigations of how food, light, and water-quality conditions affect the species of *Acartia* that live in the bay. Much of that work was conducted in large tanks at the National Ocean and Atmospheric Administration's Marine Ecosystems Research Laboratory (MERL), tanks with open tops located above the bay's shoreline. One series of experiments was conducted with a masters student, Eva Wlodarczyk, who finished her thesis in 1988 and then headed off to the University of Texas to study medical imaging, sonography and the like. (Yes, that's more evidence there can be life after copepodology.) Eva's copepod results were published a few years later.[5]

Her October experiments included vertical tows each three hours around the clock with small plastic cylinders having mesh capped ends to collect *Acartia tonsa* females from those 2 m diameter, 5 m deep tanks. The tanks had been filled with water pumped up from the bay with numerous phytoplankton, mostly cells < 10 μm diameter and good food for *Acartia*. New copepods were simply added repeatedly by pumping more water from the bay four times daily. Eva measured the chlorophyll in the guts of captured *Acartia* by macerating them, extracting pigments with acetone and determining the amounts. Chlorophyll fluoresces at red wavelengths when illuminated with blue light. Fluorescent intensity can be measured in an extract with a bench-top fluorometer; intensity is proportional to the chlorophyll concentration. For single small copepods feeding on dense diatoms (more than 10 μg chlorophyll per liter) in beakers, she determined that their full guts held about 7 ng (nanograms, billionths of a gram) of chlorophyll.

In the tanks, adult *A. tonsa* females had much less full guts during daylight than at night (Figure 9.3). Both day and night they could not or did not eat as much as in beakers, but there was a rhythm: less gut content during daylight than in darkness. Eva also measured the gut evacuation rates of copepods transferred to bay water with all the phytoplankton filtered out. The chlorophyll in them

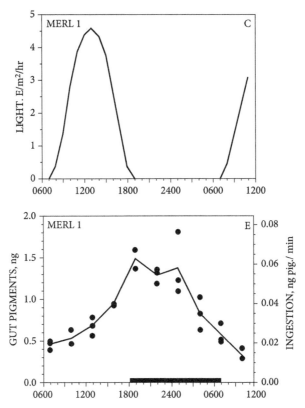

Figure 9.3 Above: light ("irradiance" in Einsteins per square meter per hour) at the top of a 2 m diameter experimental (MERL) tank. Below: average gut content chlorophyll extracted from 3 replicate groups of 20 *Acartia tonsa* females collected each three hours from the tank on October 11–12, 1984. The dark bar on the abscissa represents full darkness. From A. Durbin et al.[5] Permission, Inter-Research, *Marine Ecology Progress Series*.

dropped to zero in about 25 to 30 minutes. From that number and the water chlorophyll level, it is possible to determine the rate of ingestion needed to maintain the chlorophyll level in the gut, provided it is turning over that often. Of course, that turns out to be just a rescaling of the amount in the gut contents (right scale in the figure). The Durbins and Eva then speculated on the reason for a diel feeding rhythm favoring darkness:

If the increased swimming activity and darkened gut associated with feeding make a zooplankter more visible, and hence vulnerable, to visual predators then a reduction in feeding activity during the daytime might enhance survival.

They cited a wide range of papers showing that greater feeding at night than in daylight is very common among nonmigrating zooplankton, with most of the examples from copepods. Apparently, nobody has an alternative explanation. If your body is transparent, the advantage in prevented predation by having less opaque food in your gut must be pretty general. Eva and the Durbins also showed in laboratory containers of several sizes that at very low, near-starvation rations, the diel rhythm was still evident. They also demonstrated[6] that this rhythm is present in *Acartia hudsonica*, the winter dominant copepod in Narragansett Bay.

I find it interesting that gut fullness started to drop by about 3 AM, still in full darkness. The females had probably eaten all they could hold soon after sunset, given the high food levels available. But why stop so soon? If the *Acartia* only slowed feeding at, say 4:30 AM, their guts would have been empty before sunrise. The visibility to fish would have remained minimal. Some things will go unexplained. Maybe they just had absorbed all the food their metabolism could use for a while.

Dangers of Eating at Night: Is Nothing Safe?

The question remained open whether it matters to actual predation losses. Atsushi Tsuda, Hiroaki Saito, and Taro Hirose[7] ran some experiments at their laboratory in Kushiro, Hokkaido with Chum Salmon fry feeding on *Pseudocalanus* sp. and with Pacific herring larvae feeding on *Acartia longiremis*. The data are most convincing for herring larvae. In 20-liter tanks with black or opaque walls filled with filtered seawater, they placed 10 to 15 larvae and 100 each of newly fed (with gut contents) and starved (without gut contents) *Acartia* females under mid-day summer sun. After 20 minutes the fish were taken from the tanks and their guts opened. The copepods were not yet digested, and the fed ones distinguished from the unfed by fluorescing red under blue light, a property of chlorophyll as just noted. See the results in Table 9.1, next page:

It was not an extensive test, but the results[7] are convincing. Atsushi Tsuda is introduced in Chapter 13. He and Hiroaki Saito led the Japanese contribution to international demonstrations that low iron availability limits phytoplankton stock levels in the subarctic Pacific.

Not Eating Bioluminescent Food

Many kinds of copepods "pick" phytoplankton and protozoan cells from around them (Chapter 7). Some cells that are nutritious have the ability when touched

Table 9.1   Results of short feeding bouts by herring larvae offered fully fed *Acartia* and unfed *Acartia* in 20 liter tanks in daylight. Larvae that were eating consistently ate more of the copepods with full guts.

Experiment	No. herring used	No. fish eating eating	No. eaten copepods of 100 with full guts	No. eaten copepods of 100 with no gut contents
1	12	4	15	7
2	11	6	23	19
3	10	5	27	12*
4	13	7	49	26**
5	12	12	67	49*
Total	61	34	181	113***,

Statistical significance by Fisher's exact test of proportions: *($P<0.05$), **($P<0.001$)

or caught in fluid shear to generate brilliant flashes of blue or blue-green bioluminescence. Most important in this respect are the dinoflagellates, sometimes termed Pyrrophyta (fire plants). Some "dinos" are naked, some covered with cellulose plates. Some are bioluminescent, others not. Those that are have tiny organelles called "scintillons," around their periphery that are set up to supply energy (as adenosine triphosphate, ATP) via the enzyme luciferase to a fluorescent molecule, luciferin. Agitation in the water near them will elicit a brilliant (amazingly bright for a cell only 20 μm across) flash from the luciferin in the scintillons. The cells also have a light-receptor organelle, which strongly damps the scintillon response in daylight; bioluminescence from dinos is nighttime only. That is, it is available just in the times of day when zooplankton are most abundantly present in the upper layers of water and most are actually feeding.

Thus, in the literature, dinoflagellate bioluminescence was considered most likely to serve, well somehow, by reducing the population impacts of grazing zooplankton.[6] Look again at the late copepodite of *Calanus glacialis* in Figure 9.1 and notice the red pigment in its mouth area. Not only copepods, but other crustaceans, some shrimp in particular, surround their food-handling limbs with red pigment. Almost certainly that pigment serves to absorb at least some of the wavelengths emitted by flashing meals, because in a realm loaded with visual predators, it is dangerous to hold a lighted candle in your mouth. Nobody, so far as I know, has proved that the dinoflagellate flashes are actually shielded from view by that red pigment, but that is widely believed to be its function.

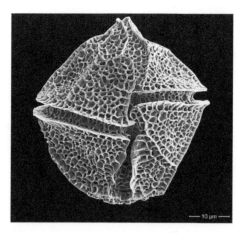

Figure 9.4 *Lingulodinium polyedrum*, about 30 μm diameter. One flagellum lies in the equatorial groove; a second extends from the groove at lower right. This species is usually bioluminescent, and it can form red tides off the US West Coast. SEM by Counsuelo Carbonell-Moore, Botany Dept. Oregon State University, with her permission.

In the early 1970s an Oregon State oceanography student, Wayne Esaias, studied the bioluminescence of dinoflagellates under the direction of Herbert Curl. Among other aspects, Wayne showed that exposure of *Lingulodinium polyedrum* (Figure 9.4), *Gonyaulax catenella*, and *Gonyaulax acatenella* cultures to continuous light for a few days would suppress their diel rhythms of luminescence, and that it would resurge strongly when they were moved into alternating light and darkness. Using that, he could obtain two cultures in opposite diel phases. In the dark, the one in its night phase would flash brightly, the one in daylight phase would only flash dimly. They would remain that way long enough in darkness (12 to 14 hours) to compare copepod grazing rates on cells with strong luminescence to grazing rates on those flashing weakly. Amounts of light available from the cells were measured by stirring the cultures in a uniform way under a blue-measuring photocell (a technique then long in use for dinoflagellates).

Using his phase-shifted cultures, Wayne ran comparative feeding rate tests with females of *Calanus pacificus* and *Acartia hudsonica* (collected from Yaquina Bay at Newport, Oregon) grazing on his *Gonyaulax* cultures.[9] The results were based on cell counts before and after the dinoflagellates were exposed to feeding by the copepods. The experiments were run at night in full darkness and at a wide range of cell concentrations (Figure 9.5) from cultures of both luminescence-suppressed and fully luminescent cells. For both copepod species the suppression of feeding rates by more luminescence relative to less was substantial. In the *Calanus* case suppression of feeding by luminescence was relatively greater

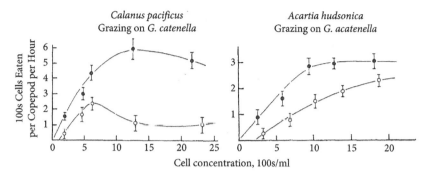

Figure 9.5 Cell ingestion rates of copepod females feeding on weakly (filled symbols) and strongly luminescent (open symbols) dinoflagellate cultures.[9] The ratio of stimulable luminosity between the cultures of G. *catenella* was (dim:bright :: 1.2 : 6.7) and between those of G. *acatenella* was (dim: bright :: ~2.5 : 8.4). Permission, Wiley & Sons.

at higher cell concentrations. Flashing in *Gonyaulax* responds to viscous shear stress in the water along their surface.[10] Evidently more flashing, from applying such shear by their limb motions during feeding, carried an urgent message to the copepods: slow it down! Being at the core of a swarm of flashing dinoflagellates would be like switching on a neon window sign at the copepod café saying "EAT HERE."

Again, just as important as eating is not being eaten. Their relative importance changes, however, depending on how hungry an animal is. Plankton animals have quite elaborate instincts by which to weigh their risks and benefits.

Wayne Esaias was not a copepodologist for any extended time. Most of his career was at NASA's Langley Research Center, NASA Headquarters, and Goddard Space Flight Center, working on evaluating the productivity of oceanic phytoplankton using color-spectrum data collected by satellites viewing the ocean from far, far above. His thesis work[11] on the action spectra (effectiveness of the full range of wavelengths from blue to red) of bioluminescence suppression from dinoflagellates by light probably supplied some of the intellectual training for that. At present he is an adjunct professor of entomology in Maryland, working on evaluation by remote sensing of the foraging potential of different habitats for honeybees.

Copepodamide, The Red Queen Paints Another Rose

More recent research shows that three-level predator-prey interaction (phytoplankton to copepods to fish), has undergone another turn of the Red Queen evolutionary ratchet. It has been known for some decades that copepods grazing on chains of diatom cells tend to break them, one end or both dispersing away in the water.[12] Now Swedish workers have shown that both shorter diatom chains[13] and enhanced dinoflagellate luminescence levels[14] can result from responses of phytoplankton cells to compounds that copepods excrete into the water. Moreover, those compounds have been identified and form an array of detergent molecules (long chain alkyls with terminal sulfate groups, rather like those in the early *Dreft°* and later *Tide°*) termed "copepodamides." Fair enough, they haven't been identified elsewhere; they are all versions of the structure in Figure 9.6, left,[15] with different fatty acid attachment points (carbons 3 or 5), chain lengths, some double bonds, or none. Also shown is a graph of the increasing amounts of luminescence produced over time by cultured cells of *Lingulodinium polyedra* (Figure 9.6, right) when exposed to different concentrations of one specific copepodamide for different time periods.[14] Much of the work has been led by Eric Selander of the University of Gothenburg, who published a good review with more details in 2019.[16] He also presented the observation that copepodamides degrade in seawater in about a day, then suggested that their concentration in the upper ocean must vary from vertical migrations and other aspects of copepod feeding schedules.

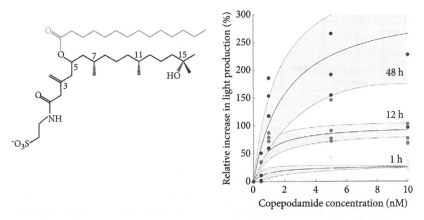

Figure 9.6 Left: An example of a molecular structure for a copepodamide.[15] Black carbon chains are the copepodamide backbone. Variation comes from different fatty acids (blue) attaching at carbon positions 3 or 5. The one shown has 14 carbon atoms and no double bonds. The nitrogen attached to carbon 1 accounts for "amide" in the compound name. The sulfate anion allows solubility in water. Right: the shift in luminescence production by cultures of *L. polyedra* (when shocked with acid) after different intervals of exposure to a range of concentrations (nM, nanomolar) of a copepodamide derived from *C. finmarchicus*.[14] Shadings are 1 SD from fitted curve (N = 4 at each copepodamdie level). Permission, Wiley & Sons.

Many Copepods Generate Their Own Predator-Scare, Bioluminescence

I keep coming back to *Metridia* (μητρίδια, Figure 8.17), because they have all the recognized predator defense schemes evolved by copepods, *except* for doing nothing to keep their eddies too small to alert the nearest arrow worm. Next on the list is bioluminescence. The *Metridia* scheme is a step beyond the luminescent defense of dinoflagellates. Scintillons flash when a dino cell is touched, or a micro-eddy washes their cell membrane. Those flashes are probably too late for the touched cell, already in the incurrent pattern of an herbivore's feeding circulation. Most of the help goes to the stock of its siblings. Since most of the dinos in a swarm, a red tide for example, have identical genes, saving all the "sisters" by flashing as one dies, has a pretty good payoff, at least to those genes.

The likely benefit/cost ratio for *Metridia* isn't quite so favorable. What they do resembles the anti-predator strategy of squid: blast out a roughly squid-shaped ink cloud, shift skin colors instantly, then jet away backward. *Metridia* has an array of gland pairs, some at the back of the urosome (Figures 9.7 and 9.8), some directed backward on the prosome, some around the head. Each gland in each pair has a reservoir, one loaded with a viscous, *luciferin*-rich secretion, the other one loaded with *luciferase* in a carrier fluid.

Placing a specimen of *M. pacifica* (abundant in northern Pacific waters I have studied) in a drop of seawater on a depression slide and then letting it rest awhile, you can observe its luminescent response by poking it gently. The glands squirt their viscous contents, and those mix together in whorls, the luciferase activating the luciferin to emit a copious shower of bright blue photons. Luciferins are also fluorescent, so you can light up one gland in the pair with short blue wavelengths

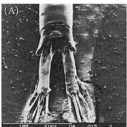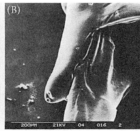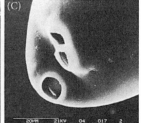

Figure 9.7 (A) anal segment and caudal furcae of *Metridia princeps*, showing the openings of luminous gland sets on either side (arrows); (B) enlargement of left openings; (C) the three pores opening at its tip. The little stripe inside the larger opening is the meeting edges of a valve. I have not found a source telling whether there are actually three glands, or just a divided port on one of them. From Peter Herring.[17] Permission, Springer Nature.

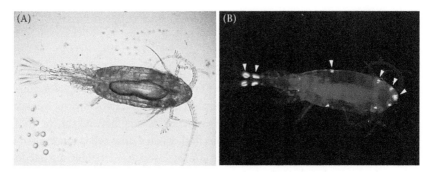

Figure 9.8 *Metridia okhotensis.* Left: dorsal view, brightfield illumination.
Right: same specimen using darkfield epifluorsence, revealing locations of
luminescent organs (arrows: caudal furcae, anal segment, thorax and head).
Takenaka et al.[18] Permission, Oxford University Press.

(epifluorescence microscopy), showing their presence (Figure 9.7, B). Another
demonstration of the intensity of this luminescence occurs frequently when
bringing net hauls aboard at night. As the *Metridia* in the haul rub against the
sides of the draining net, they release their glowing secretions, painting the net a
glowing blue. It isn't necessary to shut the deck lights off to see it, the luminosity
is that strong. However, the "Wow!" sensation is stronger if you do darken the
ship (leaving the running lights on).

Direct observational or experimental tests of the squid-ink hypothesis for the
use of luminescence by *Metridia* are difficult to arrange. Takenaka et al.[18] say it
hasn't been done. An attempt was made, however, some 60 years ago by Charles
David and Bob Conover.[19] They recorded flashes with a photo multiplier tube
from groups of *M. lucens* in a 600 ml beaker held in darkness with one predatory
euphausiid (krill, *Meganyctiphanes norvegica*). Flashes (10^{-7} to 10^{-3} µwatts/cm^2
at 18 cm from the beaker center) were both numerous and in some cases very
strong. Probably many of the *Metridia* were encountered by their potential predator and flashed, as shown in Table 9.2:

Just one flash in the no-krill control might indicate that *Metridia* encountering
each other in the dark and confined space sensed each other as potential
predators. Maybe one individual bumped head-on into the beaker side once
in almost 16 hours. The substantial numbers eaten (8 or 10) may imply that
flashing doesn't always work, or the smaller numbers eaten (3, 4, 5) with 47, 62,
and 24 flashes (over varied intervals) may imply that they usually do. The one
krill might also have had enough to eat in some of the runs. The experiment was
not massively replicated, but it did not eliminate the predator-scare hypothesis
that held up pretty well. David and Conover did run it four times with other

Table 9.2 Experimental Set ups and Flash Counts from David and Conover[19]

No. Metridia	No. Eaten	Flashes above 10^{-7} μW/cm^2	Flashes above 10^{-4} μW/cm^2	Duration Hours
9	8	30	15	15
10	10	17	3	8
10	3	33	14	16
11	5	41	21	13.5
10	4	23	1	10.5
10 (no krill)	0	1	0	15.75

potential predator species. Only one *Metrida* was eaten (at least it was gone after 15.5 hours) by a different species of krill, and only 4 weak and no strong flashes occurred over 53 total hours. Somebody could repeat this test with, say, adult *Euchaeta* (a predatory copepod), large arrow worms or fish larvae.

All the known instances of bioluminescence in copepods (Table 9.3) were listed by Peter Herring,[17] a British oceanographer and planktologist. He studied the phenomenon in marine crustacea generally through an extended career, much of that at the Southampton Oceanography Centre in the United Kingdom. He also wrote a book-length and readable account of deep-sea biology.[20]

Bioluminescence is usually an adaptation in species of the calanoid family Augaptilidae.[17] Its function is not known explicitly for all of its genera, but luminescence consistently occurs as mixing swirls outside of their bodies. *Megacalanus* species have large glands on the lateral edges of the exopods of just their second thoracic legs (P2, Figure 5.7). They do come up alive from 500 m or greater depths. Peter Herring has checked his field notes at my request, and reports that he definitely observed their luminescence directly in 20% of many live specimens. The source was the glands on their second thoracic legs, which could also be induced to fluoresce. *Oncaea conifera*, in contrast, do not secrete their luminescence externally. Their bodies light up like billboards, but Peter Herring and colleagues[21] nevertheless think that predator defense is the purpose of the light show. Herring's 1988 paper[17] says that he saw luminescent sites on the head, swimming legs, and abdomen of the marine, pelagic harpacticoid *Aegisthus mucronatus*, adding nothing specific about them. In addition to scaring predators, there are other possible functions for copepod luminescence. Mate attraction, as in fireflies, is a candidate. All of the certainly luminescent genera do have eyes with some directional capability, though none for image formation.

Table 9.3 Occurrence of Bioluminescence in Marine Copepods

Family	Bioluminescent genera
Metridinidae:	*Metridia, Pleuromamma, Gaussia*
Lucicutiidae:	*Lucicutia*
Heterorhabdidae:	*Heterorhabdus, Hemirhabdus, Hetrostylites, Disseta*
Augaptilidae:	*Euaugaptilus, Centraugaptilus, Haloptilus,*
	Heteroptilus, Pachyptilus
Megacalanidae:	*Megacalanus*
Oncaeidae:	*Oncaea*
Aegisthidae:	*Aegisthus*

Summarized from Peter Herring (1988),[17] who listed the names of species checked by him personally or by experts he trusted.

The metridinids, lucicutiids, heterorhabdids and augaptilids are all in the calanoid superfamily Augaptiloidea. Megacalanidae is a more distantly related calanoid family. Oncaeids are poecilostomatoids and *Aegisthus* an harpacticoid; both of their bioluminescence schemes are distinctive from those of the Augaptiloidea. More on all this higher order taxonomy in Chapter 20.

Coelenterazine, From Oceans to Clinics

For reasons outside copepodology, the biochemistry and molecular genetics of metridinid bioluminescence are amazingly well known. Those reasons are that, like green-fluorescing proteins from the jellyfish *Aequorea*, luciferins and luciferases of the Augaptiloidea have biomedical applications. They can be modified as "reporter" molecules for the presence of specific, biologically active substances, such as cortisol. That steroid hormone is produced by the adrenal glands of mammals under stress, upregulating some functions, downregulating others. Sung Bae Kim et al.[22] modified a *M. pacificus* luciferase (Mpluc) to be inactive unless "unmodified" by combining with cortisol in a fluid like urine or saliva. Then the level of cortisol in that fluid can be quantified by the amount of light emitted when the luciferin is added. The actual technique is more complex, the cortisol meeting the modified Mpluc in a tissue culture system. Because of the interest from biotechnology, the proteins of more than 40 luciferases have been fully sequenced. All of those from the many Augaptiloidea species have some sequence similarity, implying common ancestry.

One advantage of the luciferin from *Metridia* and *Gaussia* for all that biotechnology comes from it *not* being protein, but rather a fairly stable organic

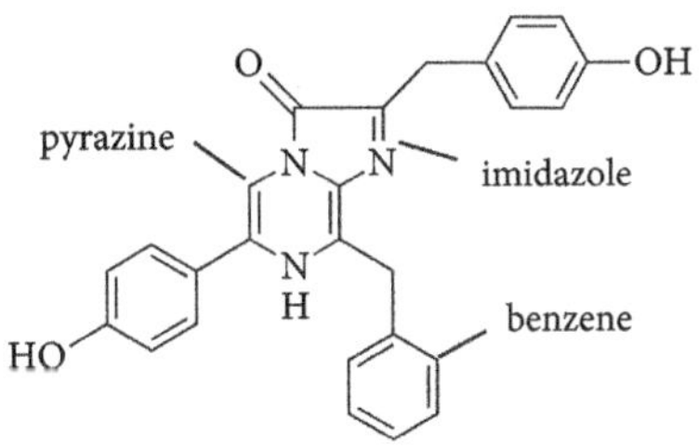

Figure 9.9 Molecular structure of coelenterazine. Each node in the lines is the locus of a carbon atom supporting four bonds. Where only two or three bonds show, the others carry hydrogen atoms.

molecule. In biological terminology it is called *coelenterazine* (Figure 9.8), and it has several names from organic chemistry that are more specifically descriptive. Most luciferins, other than proteins, have structures like that of coelenterazine in one way. Namely, they have multiple benzene and pyrazine (or similar) molecular rings bearing clouds of resonant electrons shifting about over alternating single and double bonds. Such structures can temporarily acquire energy from an enzyme-mediated reaction in them or near them, then emit that energy as a photon, in the case of coelenterazine a blue photon.

Japanese and Russian biologists and biotechnologists have been very active in study of bioluminescence. The premier example was the late Osamu Shimomura, who actually worked most of his career in the United States at Princeton and the Marine Biology Laboratory in Woods Hole. His studies[23] of the variations among luciferins were awarded the Nobel chemistry prize in 2008. Many Japanese scientists came to the United States to study and work with him, including Satoshi Inouye. Later Inouye, working with Yuichi Oba and others,[24] demonstrated by rearing *Metridia pacifica* on diatoms with deuterium-labeled amino acids (exchanging hydrogen atoms for isotopes with two neutrons at various positions) that two tyrosine and one phenylalanine amino acid molecules can be combined by the copepods to produce similarly labeled coelenterazine. Those are two of the three amino acids with structures including a benzene ring:

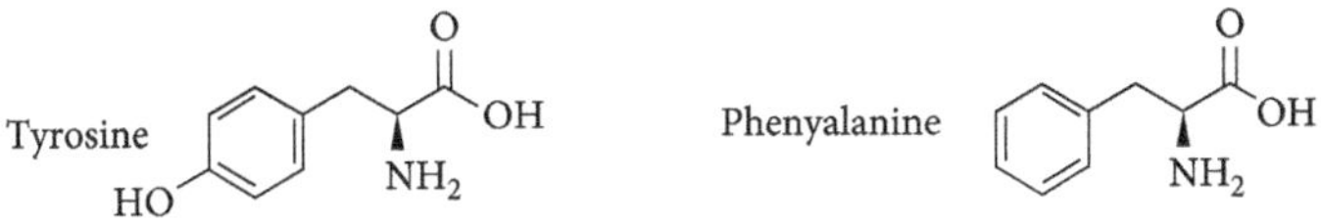

They form the pyrazine-imidazole core of coelenterazine. That is, the copepods biosynthesize their luciferin; it is not from luminescent bacteria or anything else they must eat in order to flash. Animals do not synthesize the amino acids with benzene rings; all of us either digest them from food, or enteric bacteria make them for us.

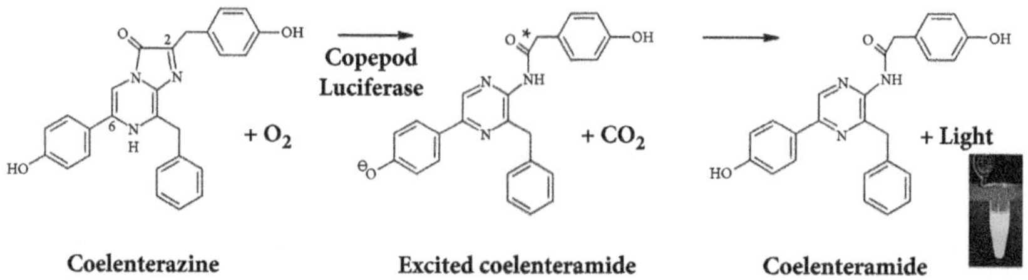

Coelenterazine **Excited coelenteramide** **Coelenteramide**

Figure 9.10 The luminescent reaction of coelenterazine (luciferin) with oxygen in the presence of copepod luciferase. Diagram from Svetlana Markova et al.,[25] with their permission. Permission, Wiley & Sons.

Prominent among the Russian contributors have been Eugene S. Vysotski and Svetlana V. Markova, who work with colleagues at Krasnoyarsk in Siberia, far from the ocean. In a recent paper,[25] they review the chemistry of luciferase action when it mixes with coelenterazine in oxygenated seawater, providing a reaction diagram they kindly allow me to show (Figure 9.10).

Because of their roles in biomedical reporter systems, the genomes of augaptilid species have been extensively searched (using cDNA, chapter 15) for their luciferase sequences. It turns out that all the studied species, except *Gaussia princeps*, have more than one luciferase gene. Table 9.4 shows the 2019 numbers from Markova et al.:[25]

Table 9.4 Numbers of significantly distinctive DNA sequences for the Luciferase genes of bioluminescent copepods

Species	Distinctive Luciferase Sequences
Metridia longa	11
M. pacifica	7
M. curticauda	2
M. asymmetrica	2
M. okhotensis	3
Gaussia princeps	1
Pleuromamma scutullata	2
P. xiphias	3
P. abdominalis	2
Lucicutia ovaliformis	2
Heterorhabdus tanneri	4
Heterostylites major	2

The enzyme lengths run from 169 to 223 amino acids, the most extreme range for isoforms in one species being 169 to 219 in *M. longa*. One homologous end of most sequences is the same (conserved), and it codes for extracellular secretion (presumably into the gland reservoir). The other end is variably shortened. Gene sequence comparisons suggest that the multiple isoforms are not alleles of the same gene, but fully separate (paralogous), and at least two or more can be functional. The proteins coded by some of these have now been synthesized by inserting the genes into cell cultures from insects, then purifying the luciferases. They turn out to have different temperature optima for oxidizing coelenterazine. Specifically, MLuc2 and MLuc7 from *M. longa* have maximum luminescent efficiency at 5° and around 15°C, respectively. The temperature optima for those proteins in their natural habitat (viscous suspension in a copepod's gland reservoir) could be even more different. Larinova et al.[26] propose that having multiple luciferase proteins may allow for luminescent function both at depth in the cold and at the surface in much warmer waters. Remember, the Metridinidae are the long-distance champions of diel vertical migration (Chapter 8).

Retention of such elaborate equipment for producing bioluminescence suggests its great importance for survival in augaptilid copepods. Like Peter Herring, I think that value is almost entirely attributable to improved chances of not being eaten.

References

1. Benfield, M., C. S. Davis, & S. M. Gallager (2000) Estimating the in-situ orientation of *Calanus finmarchicus* on Georges Bank using the Video Plankton Recorder. *Plankton Biology and Ecology* 47: 69–72.
2. Miller, C. B., C. A. Morgan, F. G. Prahl, & M. A. Sparrow (1998) Storage lipids of the copepod *Calanus finmarchicus* from Georges Bank and the Gulf of Maine. *Limnology & Oceanography* 43:(3) 488–497.
3. Aksnes, D. L., & J. Giske (1993) A theoretical model of aquatic visual feeding. *Ecological Modelling* 67: 233–250.
4. Van Valen, L. (1973) A new evolutionary law. *Evolutionary Theory* 1: 1–30.
5. Durbin, A. G., E. G. Durbin, & E. Wlodzrczyk (1990) Diel feeding behavior in the marine copepod *Acartia tonsa* in relation to food availability. *Marine Ecology Progress Series* 68: 23–45.
6. Wlodarczyk, E., A. G. Durbin, & E. G. Durbin (1992) Effect of temperature on lower feeding thresholds, gut evacuation rate, and diel feeding behavior in the copepod *Acartia hudsonica. Marine Ecology Progress Series* 85: 93–106.
7. Tsuda, A., H. Saito, & T. Hirose (1998) Effect of gut content on the vulnerability of copepods to visual predation. *Limnology and Oceanography* 43(8): 1944–1947.
8. Burkenroad, M. D. (1943) A possible function of bioluminescence. *Journal of Marine Research* 5: 161–164.
9. Esaias, W. E., & H. Curl Jr. (1973) Effect of dinoflagellate bioluminescence on copepod ingestion rates. *Limnology & Oceanography* 17(6): 901–906.

10. Maldonado, E. M., & M. I. Latz (2007) Shear-stress dependence of dinoflagellate bioluminescence. *Biological Bulletin* 212(3): 242–249.
11. Esaias, W. E., H. C. Curl Jr., & H. H. Seliger (1973) Action spectrum for a low intensity, rapid photoinhibition of mechanically stimulable bioluminescence in the marine dinoflagellates *Gonyaulax catenella, G. acatenella,* and *G. tamarensis. Journal of Cellular Physiology* 82(3): 325–372.
12. Deason, E. E. (1980) Grazing of *Acartia hudsonica* (*Acartia clausi*) on *Skeletonema costatum* in Narragansett bay (USA)— influence of food concentration and temperature. *Marine Biology* 60: 101–113.
13. Bergkvist, J., P. Thor, H. H. Jakobsen, S.-Å. Wängberg, & E. Selander (2012) Grazer-induced chain length plasticity reduces risk in a marine diatom. *Limnology & Oceanography* 57: 318–324.
14. Prevett, A., J. Lindstrom, J. Xu, B. Karlson, & E. Selander (2019) Grazer-induced bioluminescence gives dinoflagellates a competitive edge. *Current Biology* 29: R551–R567.
15. Grebner, W., E. C. Berlund, F. Berggren, J. Eklund, S. Harðadóttir, M. X. Andersson, & E. Selander (2018) Induction of defensive traits in marine plankton: new copepomide structures. *Limnology and Oceanography* 64: 820–831.
16. Selander, E., E. C. Berglund, P. Engstrom, F. Berggren, J. Eklund, S. Harðadóttir, N. Lundholm, et al. (2019) Copepods drive large-scale, trait-mediated effects in marine plankton. *Science Advances* 5. doi:10.1126/siadv.aat5096.
17. Herring, P. (1988) Copepod bioluminescence. *Hydrobiologica* 167/168: 183–195.
18. Takenaka, Y., A. Yamaguchi, & Y. Shigeri (2017) A light in the dark: ecology, evolution and molecular basis of copepod bioluminescence. *Journal of Plankton Research* 39(3): 369–378.
19. David, C. N., & R. J. Conover (1961) Preliminary investigation on the physiology and ecology of luminescence in the copepod *Metridia lucens. Biological Bulletin* 121: 92–107.
20. Herring, P. (2002) *The Biology of the Deep Ocean.* Oxford, New York.
21. Herring, P. J., M. I. Latz, N. J. Bannister, & E. A. Widder (1993) Bioluminescence of the poecilostomatoid copepod *Oncaea conifera. Marine Ecology Progress Series* 94: 297–309.
22. Kim, S. B., Y. Takenaka, & M. Torimura (2011) A bioluminescent probe for salivary cortisol. *Bioconjugate Chemistry* 22: 1835–1841.
23. Shimomura, O. (2011) *Bioluminescence: Chemical Principles and Methods.* World Scientific Publishing Co., Singapore.
24. Oba, Y., S.-I. Kato, M. Ojika, & S. Inouye (2009) Biosynthesis of coelenterazine in the deep-sea copepod *Metridia pacifica. Biochemical and Biophysical Research Communications* 390: 684–688.
25. Markova, S. V., M. D. Larionova, & E. S. Vyotski (2019) Shining light on the secreted luciferases of marine copepods: current knowledge and applications. *Photochemistry and Photobiology* 95: 705–721.
26. Larionova, M. D., S. V. Markova, & E. S. Vysotski (2017) The novel extremely psychrophilic luciferase from *Metridia longa*: Properties of a high-purity protein produced in insect cells. *Biochemical and Biophysical Research Communications* 483: 772–778.

10

Meeting and Mating

Sex in Wide-Open, Watery Spaces

Free-living copepods in lakes and oceans can be both few and dispersed in vast volumes of water. Not all species are extremely diluted, but based on sampling data, many are or appear so from sampling. Females of many species outnumber the males by substantial factors. Moreover, only a few copepods have eyes likely to be effective in mate-finding, and even for those the line-of-sight distances, say to a bioluminescent flash, are certain to be short. Thus, finding mates must require special tactics. "What are those?" is an obvious question.

Harvard graduate student Steven Katona worked on answering it under Professor George Clarke (the Clarke of the Clarke-Bumpus net) in the late 1960s. Katona published his thesis results in 1973.[1] I met Steve in 1965 when we both took the marine ecology course at Marine Biological Laboratory (MBL) in Woods Hole. I did a project on the spatial distributions of a polychaete (*Diopatria cuprea*) living in tubes on a tidal flat. I have no idea what Steve's project was, but I remember that he was there. After graduate school he got a teaching job at College of the Atlantic in Bar Harbor, Maine (far Down East), eventually becoming its president (1993–2006). He garnered some national fame by leading the entire student body (now 350 young adults) on a cold, quarter-mile swim from Bar Island across Mt. Desert Narrows to campus at the beginning of every school year. He also shifted his science to the other end of the size spectrum, becoming an expert on whales and writing a significant paper (I used to teach from it) titled "Are Cetacea Ecologically Important?"[2] His answer was a well-documented *yes*. For example, Right Whales (*Eubalaena glacialis*), particularly those in the Gulf of Maine, eat tons of resting stage *Calanus finmarchicus* from late spring until autumn. For a time Katona headed up a branch of Conservation International, maintaining an Ocean Health Index.

Katona started on the copepod mate-finding question from some previous reports of search swimming by male copepods, one as early as 1901,[3] and he must have seen a paper by D.T. Gauld from 1957.[4] Katona certainly also had two very suggestive clues. The first is that many kinds of female insects (crustacean relatives) attract male partners by secreting smelly chemicals that male

Oar Feet and Opal Teeth. Charles B. Miller, Oxford University Press. © Oxford University Press 2023.
DOI: 10.1093/oso/9780197637326.003.0010

insects detect with olfactory sensors on their antennae. Such excreted chemicals bearing information were first termed *pheromones* in 1959 by the German bio-chemist Peter Karlson and Swiss entomologist Martin Lüscher.[5] They also cited the earlier term "ectohormones," but that has not stuck. Some such chemicals signal for specific behaviors from other individuals sensing them, like "approach for mating!" Other pheromones mark trails for sensing animals to follow, as is common in ants. Actually, as will be detailed below, female mating pheromones signal both for male copepods.

Second, Katona knew that male copepods have more olfactory sensors on their antennules than do females. Those sensors are in rather baggy-looking setae, with very thin chitin, termed *aesthetascs*. Inside are neuron extensions called ciliary dendrites, a type of nerve ending present in the scent sensors of many kinds of animals. Janet Bradford-Grieve's antennule and mandible drawings (Figure 10.1) for both sexes of *Bradyidius hirsutus*[6] show that sexual

Figure 10.1 Comparison of female (left) and male (right) antennules and mandibles of *Bradyidius hirsutus* Bradford.[6] Journal ceased publication. Permission requested from the South African Museum

dimorphism. In that species the difference is striking, a rather extreme example of bushy olfactory setae in male planktonic copepods. *Calanus* males only have one additional, modestly larger, aesthetasc on every other basal antennule segment. In *Bradyidius*, the female mandibular gnathobase (center of the drawing) has teeth; she presumably eats. The male's jaw does not. Almost certainly the *Bradyidius* male does not eat (or in any event chew) but spends nearly all of its time searching for mate-attracting pheromones and, with luck, mating. Males of some other species, including most classed as *Calanus*, are similarly occupied, but in many species (e.g., *Temora, Acartia*) males do take time out to eat. Katona noted Fleminger's 1967[7] suggestion that the function of such sprays of aesthetascs was sensing female pheromones.

Katona worked with three species of estuarine copepods, *Eurytemora affinis*, *Eurytemora herdmani*, and *Pseudodiaptomus coronatus*. His paper does not say where he collected them, but probably it was Vineyard Sound off Woods Hole. They are common there and elsewhere in New England. His observations were in very small volumes, mostly 5 ml but also 1 ml. He made fine loops at the ends of steel wires 30 µm in diameter, slid females into them and suspended them in his experimental vials. He added males and watched the action. Males would swirl around, propelled by their antennae, not their oar feet, often finding and grabbing the females using their geniculate antennules and then attaching spermatophores. He interpreted the search pattern and its success as following a pheromone gradient to the females.

Katona also watched swimming behavior of males in aquaria also containing females. Males, but not the females, spent much of the time cruising at less than escape speeds, sculling forward with the antennal rami swirling beside the head. Katona's sketch (Figure 10.2) of two male trajectories relative to a captive female shows one farther away not finding a path to her, the other moving in and taking hold. The dots are his conception of the pheromones that must have mediated the success of the one, but Katona supposed them too dilute to direct the other to the goal. A problem was that those pheromones were not actually detected, just quite reasonably hypothesized.

Katona did anticipate a question that had implications regarding the separation of species (in the biological sense): Did pheromones of one species attract related species? He found that they appeared to, but with reduced effectiveness. The data were simply that males of one species would attach spermatophores to females of another, though with reduced success relative to intraspecific matings. More sophisticated work in that regard was done by Erica Goetz,[8] working as a postdoc in Denmark. Indeed, mate-attracting pheromones of copepods can induce some attempts at breeding between species. Erica and interspecific mating will get more attention in Chapter 19.

Figure 10.2 Male search patterns in *Eurytemora affinis*. From Katona.[1] Dots represent possible pheromone molecules. The fan of long arrows is the diffusion trajectory Katona considered likely, given the male swimming patterns around the female. Permission, Wiley & Sons.

Pheromones Demonstrated, Though Not Chemically Identified

Arthur Griffiths, a graduate student working with Bruce Frost at the University of Washington, also asked how copepods find each other to mate. He based his approach on the same combination of two obvious clues: many and larger aesthetascs on male antennules compared to those of females and the likelihood of sex pheromone signaling among arthropods. He must have read Katona. Later he cited him when he wrote up the study.[9] Arthur has disappeared into the mists and cannot, for the moment, be found.

To prove that pheromones are involved, Griffiths decided to work with *Calanus pacificus* and *Pseudocalanus* sp., both readily available in Puget Sound, which was accessible from the laboratory. He tried watching males swim in a tank 19 cm long, divided into sections of 6 and 13 cm by a small-mesh screen. He collected males and fifth copepodites from the Sound, put the males in the long end and females recently molted from fifth copepodites in the short end. The idea was that pheromones, if any, would diffuse to the males and initiate search behavior.

Without females on one side, the males were mostly still, suspended below their antennules. With females, the males swam steadily using antennal sculling in slow figure eights and loops. The paper does not say what happened when the screen was pulled out. The same fashion of swimming was observed when males were placed in water that had previously held females for a while. Males of both species placed with females in beakers would then make a series of rapid zig-zag jumps. Only the *Pseudocalanus* males succeeded in placing spermatophores to complete their role in mating. Confirming that success, Consuelo Carbonell and I have raised multiple generations of *Pseudocalanus mimus* in beakers. However, almost everyone who tries for cultures of various *Calanus* species fails even when using large beakers. In the 1960s Michael Mullin did succeed in getting *Calanus pacificus* to mate and females to produce fertile eggs and offspring in 19-liter carboys. Already, by then, it was clear that container size matters to mate-finding in *Calanus*.

Next, Griffiths went nuclear. Using cultures of several algae raised with ^{14}C-labeled carbonate, he raised females from eggs produced by females collected from the Sound. That produced radioactive females: all of their carbon was labeled, including their pheromones. When they molted into adults, he let some males spend quality time with them. Then, he measured the relative radioactivity of the male bodies and the male antennules by scintillation counting. Both were hotter than controls, and significantly the antennules were hotter than the bodies. Some of the antennules were dissected and mounted on slides, coated with photographic emulsion and left in the dark 70 days for the beta decays of ^{14}C to expose the silver halide particles in the emulsion. When the slides were developed (standard darkroom processing), there were shadows of silver particles on the aesthetascs (Figure 10.3). Radiocarbon from the females had accumulated most strongly at the sites of those little olfactory bags. (I cannot see the aesthetascs, but the sites are in the correct positions). It was Katona's little dots of imagined pheromone made visible. The transfer had to be the suspected pheromones.

Copulation Described, Pamela Blades and Life Beyond Copepodology

At this point the issue seemed settled, mate-finding is mediated by females emitting pheromones that elicit male search behavior. It was not yet clear exactly how the search process works in the field. I will come back to *Calanus*, but first let's examine the mating process from the point at which the male latches on. Katona[10] reported observations for *Eurytemora* in 1975, citing observations by D. T. Gauld from 1957[4] mentioned earlier, who had cited others in 1903 by G. O. Sars for *Euchaeta* and E. Wolf in 1905 for *Diaptomus*. The first of several

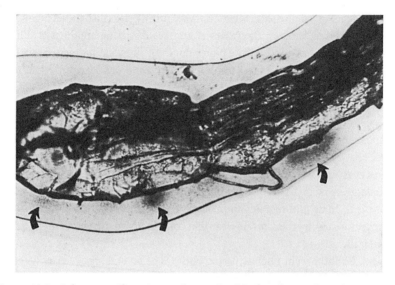

Figure 10.3 *Calanus pacificus.* Autoradiograph of the basal part of a male antennule, with groups of ^{14}C decay tracks shown by the gray clouds in the photo emulsion. The clouds are at the positions of the additional aesthetascs, not (well usually not; see Chapter 13) found in females. The ^{14}C had to have come from females with which they had shared water for a while. This is Figure 3 from Griffiths and Frost.[9] Permission, Brill, Netherlands.

effective and well-illustrated studies of copepod mating by Pamela Blades was, published in 1977.[11] Thus, the interest has been scattered over many decades. It is still scattered, and for many groups there are no observations.

All of Blades's observations were for species from the large group of calanoid families termed Heterarthandria in which the males have one "geniculate" antennule. That is, it has a joint in the middle that can bend like a knee (the Latin *genu* means "knee," hence geniculate). The group name, coined by Wilhelm Giesbrecht, refers to the difference ("heter") between their geniculate right and simpler left male antennules. That name has been parked, replaced by Rhyocalanoidea in a revised phylogenetic analysis,[12] but Giesbrecht's distinction from Amphascandria for families with both male antennules similar, neither of them geniculate, does make a relatively even division of the Calanoida into recognizable groups.

These antennular joints can clamp shut with powerful (maybe locking) muscles, and they can be serrated along the inner opposing sides (Figure 10.4). The males also have complicated fifth legs that strongly differ in form between the left and the right. One side, generally the right, will be another sort of clamp, the other a spermatophore delivery arm. The reason Blades's more thorough studies

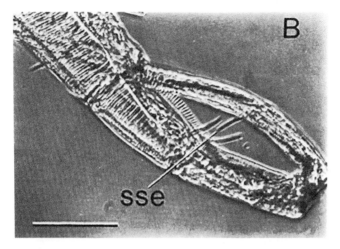

Figure 10.4 Segments 17, 18, and 19 of the right male antennule of *Centropages typicus*. Segment 17 is serrated and terminates with some sensory setae (SSE). Segment 19 folds over to hold a female caudal ramus or antennule against segment 17. From Blades.[11] Permission, Springer Nature.

worked is that the males in these genera will approach and clamp onto females in dishes small enough that the action can be watched with a dissecting microscope.

Pamela (Figure 10.5) made the *Centropages typicus* observations as part of her doctoral dissertation at the University of New Hampshire and extended them to other heterarthrandrids during her tenure as a Research Assistant at the Harbor Branch Oceanographic Institute in Florida.[13] Upon learning electron microscopic techniques, she soon shifted her research focus to the functional morphology of the male and female reproductive systems including the detailed processes of oogenesis and spermatogenesis in Calanoids. In 1982 she married, becoming Pamela Blades-Eckelbarger, and in 1991 she moved with her husband to Maine. Working there as a Research Associate at the University of Maine's Darling Marine Center, she deployed her extraordinary talents as a microscopist and electron-microscopist, delineating aspects of egg formation and egg structure in copepods, and possibly identifying the copepod equivalent of the male gland that signals for sexual differentiation of males in other crustaceans. Weary of writing grant proposals and wanting to spend more time at home with their young son, she combined her knowledge of photography and horses (having grown up in a horse-active family) and established Hoof Pix Sport Horse Photography. From 1993 to 2012 Pamela was the official photographer at numerous equine competitions throughout the Northeastern United States and returned to riding and competing with her own horses.

Figure 10.5 Pamela Blades-Eckelbarger, in her Equus-Soma era, demonstrating how bridles can press on a horse's facial nerve by showing its emergence point on the skull. Photo provided by Pamela Blades-Eckelbarger.

From 2008 to the present, winter months have been spent in the warmer climate of Aiken, South Carolina, a popular community for horsemanship of many types. Frustrated that horses were increasingly showing physical problems related to the stresses of competition, problems that veterinarians are often unable to diagnose, Pamela closed Hoof Pix and ventured into equine physiotherapy and anatomy. With a desire to educate herself and other horse owners, she began to exhume horse skeletons (starting with her horse Petey) and studied their bones for pathology. In February of 2018, she opened the Equus-Soma Osteology and Anatomy Learning Center on a small farm in Aiken, demonstrating with bones from nearly 30 horses. In addition to hosting anatomy clinics for small groups, Pamela and a partner are conducting research to correlate the (dangerous) physical and behavioral issues of humanely euthanized horses to specific genetic mutations: from invertebrates to vertebrates.

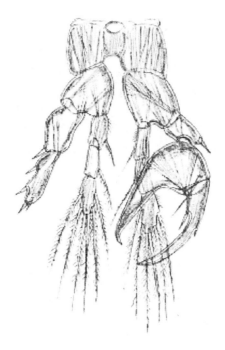

Figure 10.6 Fifth pair of thoracic legs of *C. typicus*, posterior view. The right exopod terminates in a tiny (but *relatively massive*) clamp for holding the female by her urosome. From G.O. Sars.[14]

As observed by Pamela, *Centropages* mating occurs in a series of steps. In a dish the male does not need to search long for the source of pheromones, and once in close proximity he grabs some part of the female with his geniculate right antennule, usually at the base of her caudal furcae or her antennule. He clamps down, and her rather strenuous attempts to shake him loose only work occasionally. This connection can persist for a considerable time, possibly so that things can just calm down, possibly so the male can get a spermatophore ready to extrude. Sooner or later, the male swings around, grabbing the female's caudal furcae with the chela terminating the exopod of his right fifth leg (P5, Figure 10.6).

In many Heterarthrandrids (I just prefer the sound of that), the next-to-last exopod segment of the P5 chela has a lump that fits between the female's furcae from one side, and its terminal segment can swing across the other side in a locking grip. Once a hold is established, the geniculate antennule lets go. The left P5 exopod, which has a sort of abrasive tip, takes a ready spermatophore from the male's genital pore (very obscure, but on the first urosomal segment) and

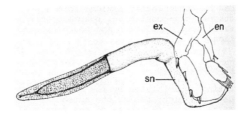

Figure 10.7 Left fifth thoracic leg of *C. typicus* male carrying a spermatophore ready for attachment to the female. There is a visible demarcation between the Quellen Spermien (outer layer near tip) and the fertilizing sperm (central mass). sn labels the spermatophore neck. From Blades.[11] Permission, Springer Nature.

attaches it to the female. Blades offers a diagram (Figure 10.7) of how the spermatophore is held by the P5.

The spermatophore is an elongate bag of sperm with a delivery duct that is inserted under the flap-like operculum of the opening to the female's genital tract. Along the duct away from the bag there is, in many species including *C. typicus*, a complexly shaped blob of sealant to hold the duct in place. Once the spermatophore is outside the male, specialized "swelling" sperm (usually referred to in German: *quellen Spermien*) at the back of the sac absorb water, swell up and push the reproductively effective sperm along a thin duct into the female's seminal receptacles. At this point the male may or may not let go. They sometimes stay put awhile, adding another spermatophore later.

In at least some copepods the spermatophore duct extends all the way into the receptacle on one side, so he may be hanging on to supply the other one; or, the whole thing may be either a male idea of fun or a tactic for keeping other males away. Females collected from the field and those mated in lab dishes can accumulate substantial numbers of spermatophores, most of them not properly installed to deliver sperm. Spermatophores also are found on C5s, on other males, even on bits of detritus. Apparently when sperm packages are ready, they have to be delivered somewhere.

In many heterarthrandrids and some amphascandrids, the females have small fifth thoracic legs. If they bend their urosomes ventrally, the tip or spines on those legs are in an exact position to scrape off empty spermatophores.

Atsushi Tsuda Watches Pheromone tracking in *Calanus*

We need to head back out into open water to see how the pheromones demonstrated by Griffiths and Frost work in the Amphascandria, specifically

Calanus. We are not likely to be able to view the action by descending into the ocean, but we can provide some collected specimens with a much larger aquarium configured to allow both males and females latitude for longer swims. An opportunity for that emerged in my laboratory somewhat by coincidence. The US Coast Guard has quite frequently sent officers to Oregon State University to get a masters degree in oceanography. Usually the Coast Guard insists they study physical oceanography to develop skills appropriate to predicting iceberg trajectories in the Atlantic and oil spill trajectories anywhere along our coasts. Our physicists did not want to take on Tom Heitstuman, who had applied, but they thought his biological background and good records would match my interests. I corresponded with him and the Coast Guard, and they made an exception. Since jellyfish and jellyfish blooms were already a practical concern in coastal areas in the mid-1990s, a thesis project involving them did seem to have Coast Guard relevance.

So, Tom decided to figure out where the bottom attached polyps of the brown sea nettle (*Crysaora fuscescens*), a species with problem blooms off Oregon, might be found and what conditions would set those polyps to shedding larval jellyfish. He was both a qualified diver and an accomplished aquarist (a long-time hobby). He collected all sorts of likely underwater objects from the downstream bays and offshore reefs of central Oregon. He found numerous polyps on the undersides of overhanging surfaces and discovered that they would shed larvae (ephyrae) during the winter when food was scarce, a result found at about the same time by Shinichi Uye in Japan and by others around the world. To identify the type of jellyfish, it was necessary to rear the larvae. For that Tom designed a plankton "Kreisel" and obtained a student-project grant to have a plastic fabricator build it.

These circular aquaria were developed originally in Germany, hence the name, and Tom's was big: 30 cm front to back and 1.0 m diameter. The circumference sits with braces on a stand. Pumps and filters establish flow around the perimeter. Water moves in at the top, sinks very slowly to the bottom, then circulates up around the sides and is pumped out gently through a screen just below the top, where it is filtered and returned. Tom's ephyrae grew well in this, and all of them turned out to be one or another version (indistinguishable to us) of the Moon Jelly (*Aurelia aurita*). He never did find any sea nettle polyps. Tom wrote his thesis, returned to active duty, and took command of a Coast Guard buoy tender.

Back to copepods. In 1988 I had spent three months at the Ocean Research Institute (ORI) in Tokyo and shared an office with Atsushi Tsuda (Figure 10.8), a very bright doctoral student working on zooplankton. Among other talents, Atsushi is an enthusiastic bird watcher. We got acquainted while I worked through preserved ORI plankton samples on variants of *Neocalanus* spp. life histories that occur on the west side of the subarctic. In 1997 Atsushi obtained a

Figure 10.8 Left: Atsushi Tsuda in 1997 returning from a collecting trip near Newport, Oregon (photo by the author). He is a dedicated bird-watcher, so the binoculars seemed permanently attached. Right: Atsushi in 2017 as director of the University of Tokyo's Atmosphere and Ocean Research Institute. After four years he became a university vice president, and is still in that role. Extreme competence can derail a scientist temporarily from brilliant research. Photo by AORI, Permission, A. Tsuda.

one-year fellowship and spent half of it in Louisiana working with planktologist Michael Dagg. The other half year he worked with me in Oregon. I found a small house at walking distance from campus that the recently wed Atsushi and his wife, Chigusa, could rent. She was pregnant, and her time to deliver came during their stay. She went home to Japan to have the baby. She spoke quite good English, but I imagine that delivering a baby where all helpful instructions would be gargled at you was daunting. It went well, and she returned with the baby so Atsushi could start being a dad. Meanwhile, he studied another aspect of mating.

Atsushi and I decided to try to use Tom's one-meter Kreisel for watching *Calanus* adults mate. We filled the Kreisel with artificial seawater (a closely equivalent mixture of dissolved salts) to avoid any pheromones that might be in seawater collected from the field, and then we went sampling off the coast. We caught good numbers of uninjured *Calanus marshallae* Frost (Frost is somewhere in almost everything about copepods in the sea) and sorted out large numbers of males and fifth copepodites (C5). We put the males in the Kreisel and watched using dim back lighting. They stayed up in the middle of the water, pulsing upward a bit as they sank, or as the very slow flow moved them down. The C5 were fed in large beakers, and we resorted them frequently for a few days.

The C5 that molted to females we added to the Kreisel, and again we watched. We knew they were unmated because they had molted without males around, and their seminal receptacles were empty (yes, you can see that; the packed sperm are quite opaque and fill the chambers). Very soon, the males began to swim in looping, mostly horizontal paths across the tank. Speeds were 9 to 20 body lengths/second, about 3.4 to 7.5 cm/s, all driven by sculling beside the head with the antennae. The limb motion was just a blur, but the body's progress was smoothly undulating. We voyeurs found ourselves exclaiming in two languages from our chairs in the dim cold (12°C) at how extraordinarily beautiful that swimming is and how easy to follow with our eyes. They were graceful, though mostly belly-side up, pirouetting like tiny swimmers in a water ballet, antennules held stiffly out at their sides with the feathery setae at their tips like scarves trailing from a hand. Maybe we did not actually see those, just knew they were there.

A diagram (Figure 10.9, A) from our paper[15] suggests the sequence of events. A key aspect is that the females secrete *vertical* pheromone columns and the males search for those on long *horizontal* trajectories, maximizing the odds the two will cross. Perhaps in the field adults of both sexes carry out these functions

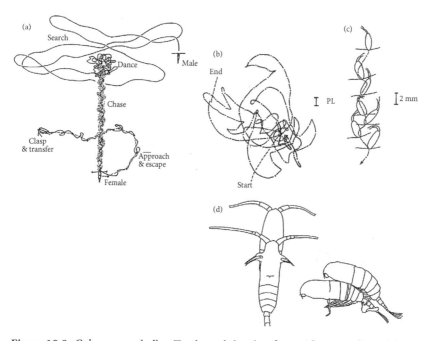

Figure 10.9 *Calanus marshallae.* Tracks and sketches from video recordings. (**a**) the overall sequence of mating events. (**b**) pattern of a grapefruit dance, traced from a 2D video recording. PL = prosome length (~3 mm). (**c**) a male tracking along a pheromone trail. (**d**) two views of the mating position. From Tsuda and Miller.[15] Permission, Royal Society of London.

in a vertically restricted layer they both somehow identify. That remains to be shown.

Quite often (on our best trial there were 22 males and 45 females circulating in the Kreisel), the gentle traverses of a male would stop abruptly, and he would explode into a wild swirling motion that we dubbed a "grapefruit dance." His turning and turning would confine this somewhat faster motion (~10 cm/s) in a spherical volume about that of a grapefruit (Figure 10.9, b). It went on for 5 to 20 seconds then stop. After a moment of stillness, he would head almost straight down at almost the speed (~6 cm/s) of searching, antennules spread to the side, sliding them over and over left to right, right to left (Figure 10.9, c). At the end of the descent he would encounter a female, obviously the source of the signal he had followed, sometimes hitting her head-on. We watched many of the newly molted females, and they were slowly sinking head up, tail down, occasionally taking a few pulsed jumps back up along the same path.

Once we had seen and noted all this, it was time for Atsushi to head back to Japan. He was soon employed at the Kushiro Laboratory of the Hokkaido National Fisheries Research Institute. He and his equally accomplished colleague, Hiroaki Saito, had access to ships for time-series sampling well out to sea, and they ran studies of *Neocalanus* life histories, expanding on earlier work by me and others in the Gulf of Alaska. I discuss the cool aspects of that in Chapter 16 on copepod diapause. However, he took time to return briefly to Oregon before starting his new job, this time with a good video camera. We duplicated the procedures and got enough video of mating sequences to determine speeds and to get some stills of actual mating.

Atsushi and I were not sure how to elucidate the purpose of the grapefruit dances (Figure 10.9, A). The male could be celebrating having found concentrated come-hither scent. He could be doing some kind of preparation for releasing a spermatophore, which Blades thought could take considerable time in *Centropages typicus*. Jeannette Yen suggests, after reading this, that the dances could have the scale of the limit to pheromone diffusion over the duration of the initial search (the Kolmogorov scale). Likely the males are determining the precise location of the pheromone track they must follow.

When the male arrived, the female would use the escape swimming mode to move off a few to maybe 10 cm, and the male would follow. Then jump-and-follow, jump-and-follow proceeded for a few cycles that usually ended with the female blasting off to the far side of the Kreisel. The male would lose track of her, and the encounter was over. Very few times the male was seen to leap onto the top of the female's prosome, apparently clasping her with his extra-long maxillipeds (Figure 10.9, D). The arrangement was such that his left P5, also extra-long, could attach a spermatophore. So we know that much of how *Calanus* females let

Calanus males know where to find them, and we know what the males do to find and follow those instructions.

An obvious question arises: Can a female copepod's vertical trail of pheromone dissolved in the ocean maintain its continuity long enough for a male to find and follow it down to her? Well, we can get into the ocean and create vertical solute stripes, perhaps of chemicals we can see and photograph. If they last a reasonable time, say an hour or so, and do so without just snapping off in shear between layers, then we can be reasonably sure they could serve. Of course, they must; there are always a lot of nauplii appearing. It turns out that a physical oceanographer, J. D. Woods,[16] made the observation long before we knew to ask. He (and a dive buddy) put on scuba gear, swam down into the thermocline in the Mediterranean Sea, about 25 m down, and dropped fluorescein dye pellets that left bright green downward trails as they sank. They photographed them as shear flow installed bends and bends back. The tracks remained continuous (Figure 10.10) for several hours.

If a female has not been found after an hour or so, she can just swim a bit back up and advertise again. If there are males, she will eventually be located. The long continuity found by Woods partly resulted from installing the dye stripes in a thermocline. The density gradient would have stabilized the layering against vertical stirring, and Wood demonstrated that the shearing between thermocline layers was slow enough to just stretch the fluorescein tracks, not shear them untraceably. It could be that thermoclines or haloclines, and very quiet zones deeper in the sea, are the right places to look for copepods mating, at least in *Calanus* and related genera. In any case, almost all captured female *Calanus* and *Neocalanus* have both seminal receptacles packed with sperm. The system works, whatever its details.

Jeanette Yen, Trysts in Coastal Species

Jeanette Yen was introduced in Chapter 6 regarding the mechanics of feeding in predatory copepods. She also has had a major role in understanding mate-finding. In 1997, J. Rudi Strickler (introduced in Chapter 7) organized a series of presentations at a meeting in San Francisco by people working on details of how male copepods find mates in the vast volumes of the ocean. Jeannette Yen and her students (Sean Collin, Michael Doall, and Mark Weissburg) were much of the program, and of course Rudi Strickler were there. I was invited to present the results on *Calanus marshallae* generated with Atsushi Tsuda. Part of the attraction was that Geoff Boxshall of the British Museum, in addition to giving a paper on the anatomy and sensors of copepods relevant to mating, arranged for all of the papers to be published in the *Philosophical Transactions of the Royal Society of*

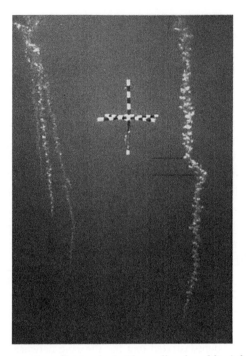

Figure 10.10 Fluorescein pellet tracks (with small wake eddies) from sinking through a thermocline at around 25 m depth in the Mediterranean Sea. The scale lowered by a diver above has 5 cm black and white divisions. A layer of weak shearing is indicated by the black lines. Woods notes that "the thermocline is divided into several rather thick *layers* (ca. 1+ m), of moderate temperature gradient, separated by much thinner *sheets* (ca. 10–20 cm) of steeper temperature gradient." There is vertical shear between the sheets. From J. D. Woods.[16] Permission, Cambridge University Press.

London. That felt like ascending (just once and briefly) to some sort of scientific stratosphere. Jeanette and Atsushi are still up there.

Jeannette, Rudi, and those students had been working on aspects of mate-finding in *Temora longicornis.* That small but wide-bodied copepod is abundant in the waters of Long Island Sound, whereas *Euchaeta* and *Paraeuchaeta* are not. Jeannette had moved to University of New York Stonybrook on Long Island, so she studied the animals she could obtain. However, some background work came from an earlier Yen and Strickler collaboration: her animals and data analysis, his video recording system. Using laser-illumination and recording at 30 frames per second, Jeannette showed[16] that in small aquaria (10 cm long x10 wide x15 tall) *Euchaeta rimana* females, while feeding, mostly move on horizontal paths; males that do not feed move on hop-and-sink vertical paths (Figure 10.11). Yen makes

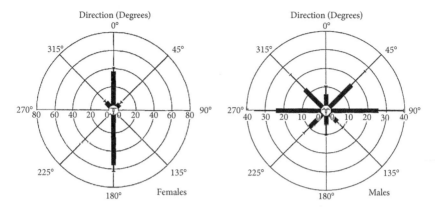

Figure 10.11 Proportions of time that *Euchaeta rimana* specimens, silhouetted against the back wall of a small aquarium, swam toward eight different 45° sectors, from 0° for up to 90° for rightward, and so on around.[17] Recall that copepods have statocysts in their brains. Female results at left on a scale from 0 to 40%; male results at right on a scale from 0 to 80%. Thick lines are average proportions; thinner extensions 95% confidence limits. Permission, *Bulletin of Marine Science.*

no mention of pheromones, and possibly they are not involved in mate-finding by *E. rimana,* and the female does the work of searching. In any case, the notion that efficient pheromone trails and mate-search paths would be orthogonal was already out by 1988.

Work in the Yen lab continued studies of how copepods sense and react to tiny differences in water movement over their bodies. By the mid-1990s, Strickler had adapted a different video recording system for following animals swimming in 10 cm x 10 cm x 10 cm aquaria. Using an arrangement of mirrors, images from two sides of the one-liter volume can be placed side-by-side on a video camera's flat CCD plate and recorded at 30 frames per second. That is frequent enough to follow a copepod moving in three dimensions at less than escape speed. Strickler provided this gear for the *Temora* observations. Initially came a study of groups swimming when attracted into a vertical laser beam in the aquarium's center. Strickler said during our interview that some of the students saw that males were grabbing females for copulation. The reports were shouted, so all the participants, including Jeannette, knew they were onto something useful beside swimming (important as that may be).

They did the obvious: they put more male and female specimens in the aquarium and recorded the action. Then the video was searched frame-by-frame to determine the details. Female *Temora* do not store sperm, so mating must be often, and males must eat and remain a reasonably high proportion of the stock. Also, likely because densities of *T. longicornis* inshore are generally quite high,

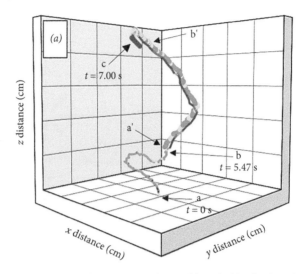

Figure 10.12 *Temora longicornis.* 3D tracking of female (fat dots) and male (narrow dots) locations in a 125 cm³ space. Initially the male rises and turns smoothly at <20 mm s⁻¹ (blue), finding a female trail after 5.47 s. He then accelerates to 20–40+ mm s⁻¹. The female's earlier positions from a' to b' could then be reconstructed. The male tracked those exactly and rapidly (red and orange lines), finding and seizing the female 1.53 s after finding her track. From Doall et al.[18] Permission, Royal Society of London.

the pheromone signal and search patterns need be neither particularly straight nor very elongated, so they are not (Figure 10.12).

There are many more examples in the results of Michael Doall's multiauthor paper.[18] There were 12 males and 12 females in a liter of water in the small aquarium (10 cm x10 cm x10 cm, 1,000 cm³), so a number of such interactions could happen, some simultaneously. Many variants occurred. Males can realize they are going the wrong way along the trail, possibly because the longer diffusion time lowers its pheromone content. They reverse and backtrack, eventually finding the source of the perfume.

Around this time, after about eight years at Stony Brook's Marine Science Research Center (MSRC), Jeannette tackled one of the prominent issues of women's employment: less pay than men with identical credentials and performance receive for identical work. She felt strongly that the discrepancies at MSRC were far out of line. So she found the data and developed a graph with years-since-PhD on the X-axis and salary on the Y-axis. At most academic institutions a PhD is a figurative union card in a nonunion shop, a sort of ticket to faculty status. So the graph represented the normally rising results of professional advancement and promotions, except that the upward trend for men was entirely above that for women, of whom there were by then quite a few. Jeannette's quest for some rectification was met with contempt by the administration: Such discrepancies were everywhere, so why should MSRC, or Stony Brook for that matter, be expected to differ? Other women on the faculty were not supportive, quite uniformly preferring not to rock the boat. As the effort went forward, the time also came to seek promotion. Her request was denied, but she was able to learn that her outside reviews were strong, her teaching reviews were good, and so were her grant-getting and publication records. The explanation came down that despite all those positive aspects, her colleagues at MSRC were unsupportive. This situation was unlikely to improve, so she started looking again for a better situation, and she found it.

In 2000, not long after the copepod-mating symposium, Jeannette was offered a new position at Georgia Institute of Technology (Georgia Tech), a prominent engineering school in Atlanta. A mutual attraction existed there with a group of natural product chemists who were interested in helping with identifying the specific molecules serving as copepod pheromones. They would get a difficult chemical puzzle to solve, and she would help them solve it. She took a faculty position and moved to Atlanta, where she started at double her MRSC salary. Because of the extreme dilutions that can be effectively sensed by chemoreceptors, the partnership has not yet chemically identified copepod sex pheromones. The molecules are probably of modest size, since most characterized pheromones are. She moved her optical and recording instrumentation into Georgia Tech labs and ran new experiments with a new team of grad students.

One suite of those many experiments[19] should be considered here. A new requirement was to have marine copepods shipped in; Atlanta is inland. Working with T. longicornis again, they kept many newly matured females for a while in filtered seawater, then filtered them out. The goal was a sex-pheromone solution. They added some dextran (a short glucose polymer) to the water to make it denser than seawater so it would sink. It would also make artificial pheromone tracks visible by changing their refractive index and so bending light beams. That would make possible video recording of sinking trails using a special optical path to a camera ("Schlieren" imaging, engineered for similar purposes by Rudi Strickler). Then they dispensed it down narrow pipettes into the surface of one of the small aquaria containing some males. Apart from disturbance by the slowly swimming males, the water was still and the mixture sank on smooth, vertical tracks (Figure 10.13, left). Males encountering the tracks recognized their potential and headed up along them. This allowed visualization of the extent to which males stir a trail into its surroundings (Figure 10.13, right). Jeannette even figured out a way to make these visible, artificial trails rise (dextran in the aquarium, not in the pheromone solution), showing that male Temora will track down as well as up.

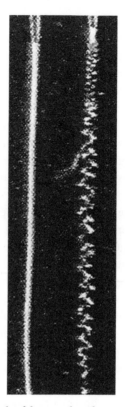

Figure 10.13 Left: vertical path of dextran-laced seawater in which *T. longicornis* females have spent time, width about 1 mm. Right: similar track found by and disturbed by an ascending male of the same species (orange dot) that hit the delivery pipette head-on. Note that the path is distorted in a zig-zag pattern by the male angling back and forth across it. However, the track does remain intact. From Yen et al. (their Figure 10.2, page 74).[19] Permission, Taylor and Francis (Chemical Rubber Company).

Jeannette speculated that the across-path dispersions created by tracking males should substantially increase the diffusion of scent out of phero-mone tracks, helping to obliterate them. Model calculations agreed. In fact, speculations recently modeled[20] extend to the possibility that tracks might in-clude two pheromones with different diffusion rates. That would generate a sort of directional cue for the males: the wider the fringe of smaller (faster diffusing) molecules, the farther it is to the female. When that fringe is getting wider, turn and backtrack. If this two-scent hypothesis is true, then retrieving and characterizing the molecules likely will be even more difficult.

Jeannette's former student, Mark Weissburg, was already on the Georgia Tech faculty when she joined it. Eventually they were married for twelve years. They initiated a program there around 2006[21] that they both work in, the Center for Biologically Inspired Design (BID). She is its director, and Mark one of the codirectors. Gene-based variations developed into designs by natural selection have produced a fantastic array of ways to deal with the world. The classic example is that Velcro was inspired, according to their inventor George de Mestral, by the tiny, springy hooks on the spines of plant burrs. Those hooks make burrs hard to pull from a dog's hair or one's socks. The idea for the Georgia Tech BID program is to extract more design ideas from organisms and find practical applications for them. There are similar programs at Harvard, Cal Tech, and elsewhere. Since not much of the work has taken ideas from copepods, you will need to follow it elsewhere, maybe in the journal *Bioinspiration and Biomimetics*. Mostly Jeannette has directed students and colleagues in this search for the practical in the biologically amazing, but she is also part of a BID consulting firm.

Much of Jeannette's recent and intense work has been to recover from a brain tumor. She showed me an ear-to-ear brain scan in which the tumor, about the diameter of a quarter, sits in the center right atop her *corpus callosum* (thick nerves connecting the hemispheres). It took long months for trials of eight chemotherapies, the last of which simply eliminated the tumor. She refers to the events as her depression, and to her emergence from it as rebirth into a new life. In 2020 she sent me a photo (Figure 10.14), showing the same smile that she has greeted her unexpected gift of added life in late middle-age.

Jeannette is very intent on figuring out how to use this new time well. Some will still go toward identifying copepod pheromones. She has new ideas for that. Her labs are a bit scattered thanks to the long illness and recovery; her optical and recording systems needed realignment when she showed them to me in 2018. But the day after my visit she was off to Montana where a remote lake offered maroon *Diaptomus* that were promising for new swimming and mating studies of copepods, that time done in freshwater. Her energy level is phenomenal.

Along with determination to do science, she is working to show people generally the wonders of marine life. The evening before our interview we went about an Atlanta neighborhood distributing vials of bioluminescent dinoflagellates to some friends, people engaged in many walks of life beside science. In July of 2018 she participated with mixed-media artist Mel Chin and Georgia Tech students in a massive display (titled *Unmoored*) in New York's Times Square.

Figure 10.14 Jeannette Yen working on one of her many art projects. Her shirt is a tiger print, perhaps representing an appropriate totem animal. Photo provided by Dr. Yen.

The scene, viewed through virtual reality goggles, superimposed a stream of pedestrians who were actually moving along the sidewalk at the time, layered under a streaming sequence above their heads of boats floating and marine life swimming. Jeannette contributed photos of plankton: coccolithophores, dinoflagellates, and (how could she leave them out?) copepods, all dramatically enlarged and four-dimensionally arrayed.

Jeannette has moved into an apartment installed under 20-foot ceilings in a repurposed Standard Oil wooden barrel factory. Improvements were in progress when I visited in August 2018. One brick wall was covered with about 25 of her father's detailed paintings. During my visit a new steel porch, welded by a craftsman friend, was being suspended outside and made accessible by a door cut through the second-story outer wall. We visited all the neighbors; Jeannette's new life includes outgoing vigor and a zest for knowing people that few if anyone could match.

References

1. Katona, S. K. (1973) Evidence for sex pheromones in planktonic copepods. *Limnology and Oceanography* **18**(4): 574–583.
2. Katona, S. K., & H. Whitehead (1988) Are cetacea ecologically important? *Oceanography and Marine Biology Annual Review* **26**: 553–568.
3. Parker, G. H (1901) The reactions of copepods to various stimuli and the bearing of this on daily depth migrations. *Bulletin U.S. Fisheries Commission* **21**: 103–123.

4. Gauld, D. T. (1957) Copulation in calanoid copepods. *Nature* **180**: 510.
5. Karlson, P., & M. Lüscher (1959) "Pheromones" a new term for a class of biologically active substances. *Nature* **183**: 55–56.
6. Bradford, J. M. (1976) A new species of *Bradyidius* (Copepoda, Calanoida) from the Mgazana Estuary, Pondoland, South Africa and a review of the closely related genus Pseudotharybis. *Annals of the South African Museum*, **72**(1): 1–10.
7. Fleminger, A (1967) Taxonomy, distribution, and polymorphism in the *Labidocera jollae* group with remarks on evolution within the group. (Copepoda: Calanoida). *Proceedings of the U.S. National Museum* 120(3567): 1–61.
8. Goetze. E. (2008) Heterospecific mating and partial prezygotic reproductive isolation in the planktonic marine copepods *Centropages typicus* and *Centropages hamatus*. *Limnology & Oceanography* **53**(2): 433–445.
9. Griffiths, A. M., and B. W. Frost (1976) Chemical communication in the marine planktonic copepods *Calanus Pacificus* and *Pseudocalanus* sp. *Crustaceana* **30**(1): 1–8.
10. Katona, S. K. (1975) Copulation in the copepod *Eurytemora affinis* (Poppe, 1880). *Crustaceana* **28**: 89–95.
11. Blades, P. (1977) Mating behavior in *Centropages typicus*. *Marine Biology* **40**(1): 59–64.
12. Bradford-Grieve, J. M., G. A. Boxshall, S. T. Ahyong, & S. Ohtsuka (2010) Cladistic analysis of the calanoid Copepoda. *Invertebrate Systematics* **24**: 291–321.
13. Blades, P., & M. Youngbluth (1979) Mating behavior of *Labidocera aestiva* (Copepoda: Calanoida). *Marine Biology* **51**: 339–353.
14. Sars, G. O. (1902) *An Account of the Crustacea of Norway. Vol. IV. Copepoda Calanoida*. Bergen Museum, Bergen, Norway.
15. Tsuda, A., & C. B. Miller (1998) Mate-finding behaviour in *Calanus marshallae* Frost. *Philosophical Transactions of the Royal Society of London*, Ser. B, **353**: 713–720.
16. Woods, J. D. (1968) Wave-induced shear instability in the summer thermocline. *Journal of Fluid Mechanics* **32**: 791–800.
17. Yen, J. (1988) Directionality and swimming speeds in predator-prey and male-female interactions of *Euchaeta rimana*, a subtropical marine copepod. *Bulletin of Marine Science* **43**(3): 395–403.
18. Doall, M. H., S. P. Colin, J. R. Strickler, & J. Yen (1998) Locating a mate in 3D: the case of *Temora longicornis*. *Philosophical Transactions of the Royal Society of London*, Ser. B **353**: 681–689.
19. Yen, J., A. C. Prusak, M. Caun, M. Doall, J. Brown, & J. R. Strickler (2003) Signaling during mating in the pelagic copepod *Temora longicornis*. In *Handbook of Scaling Methods in Aquatic Ecology*, ed. L. Seuront & P. G. Strutton, 149–159, CRC Publishing, Boca Raton, Florida.
20. Hinow, P., J. R. Strickler, & Y. Yen (2017) Olfaction in a viscous environment: the "color" of sexual smells in *Temora longicornis*. *Science of Nature* **104**: doi 10.1007/s00114-017-1456.5.
21. Yen, J., & M. J. Weissburg (2007) Perspectives on biologically inspired design: Introduction to the collected contributions. *Bioinspiration and Biomimetics* **2**: doi 10.1088/174103182/2/4/E01.

11

Reproduction

Free-Spawners versus Sac-Spawners

Copepods, like chickens, do not just appear; individuals start as eggs, or alternatively as female copepods producing eggs. As with chickens, there is no point in
asking which came first; we could start with either. I choose the mothers; mating
was covered in Chapter 10. Among copepod species I know about, mating is about
all that fathers do. Females of some free-living copepod species spawn freely into
the water; those of other species tow their eggs in small sacs until they hatch. Eggs
of the former receive no parental care, apart from some internal provisioning. Eggs
of the sac spawners share the benefit of the mother's escape capacity until hatching.
In both cases, the eggs are single cells, larger in larger species (Table 11.1):

Table 11.1 Some sizes of eggs from free-living copepods

Species	Median Prosome Length millimeters	Egg diameter μm
Oithona davisae[1]	0.3	49
Calanus pacificus[2]	2.5	160
Neocalanus cristatus[3]	7.4	374

Eggs are produced, usually in clutches, by cyclic cell division in paired ovaries
located in the upper middle of the female prosome below the heart (Figure 11.1).

A mass of dividing germinal cells at the back of the ovary pushes new oocytes
forward into an anterior diverticulum over the mouth and brain. Turning back
above the brain, the enlarging oocytes are conducted back, then ventrally into
the oviducts and pass along the sides of the thorax into the segment in the
urosome with the genital opening. In one large group (the Gymnoplea) that is
on the tail section's first segment, in another (the Podoplea) it is on the second
segment. That genital segment has an opening, or two openings, to the water
protected by a flap or valve. They are located ventrally or dorsally in different
major divisions. As eggs (ova) move along that track, they are loaded up with
nutrients. Their surface layers are supplied with granules that upon fertilization

Oar Feet and Opal Teeth. Charles B. Miller, Oxford University Press. © Oxford University Press 2023.
DOI: 10.1093/oso/9780197637326.003.0011

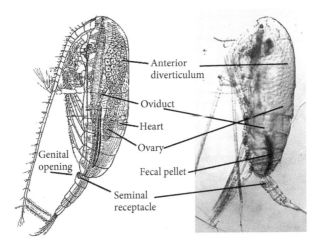

Anterior
diverticulum

Oviduct

Heart

Ovary

Genital
opening

Fecal pellet

Seminal
receptacle

Figure 11.1 Left: a simplified anatomical diagram of an adult *Calanus* female from Marshall and Orr.[4] Those authors are remembered in Chapter 15. Permission, Springer Nature. Right: photo of a live *Calanus glacialis* female (about 3.3 mm) from the Arctic Ocean by Carin Ashjian, copepodologist and oceanographer at Woods Hole Oceanographic, with her permission.

will spread around the cell surface as one or more protective coats or "chorions." A meiotic or "reduction" division occurs starting in the oviduct, and two of the four resulting cells, essentially just cell nuclei with halved chromosome numbers (termed "polar bodies") are extruded to the outside. At least in well studied species, the last polar body emerges just after spawning. The fourth cell produced by meiosis is the egg itself. As it moves through the genital segment, in many species it passes a tributary duct from the seminal receptacle holding the female's store of sperm and is fertilized. Not all copepod groups have seminal receptacles, and their females tend to retain spermatophores (sperm packs) longer. Sperm in those species must stick awhile on the genital surfaces.

Jeff Runge Expands an Era of Egg Production Rate (EPR) Studies

In early June of 2019 I visited marine ecologist Jeffrey Runge (Figure 11.2) at his office in the Gulf of Maine Research Institute at Portland, Maine. He works there on the ecology of that water body, one critical to the maritime interests of New England and eastern Canada. Its westernmost stock of *Calanus finmarchicus* is a keystone species in the transfer of nutrition from phytoplankton to fish and whales in the Gulf. Jeff's interest in the ecology of *Calanus* began during graduate

Figure 11.2 Photo of Jeffrey Runge published in the *Boothbay Register* (July 9, 2017) announcing a talk he would give on changing habitat conditions and plankton in the Gulf of Maine. Photo courtesy of the Bigelow Laboratory for Ocean Sciences.

school in Seattle. Appearing again is Bruce Frost, who was Jeff's advisor at the Oceanography Department at the University of Washington. Raised just south of Boston, Jeff's introduction to the sea was as a surfer. He attributes an early and strong interest in biology, particularly evolution, to excellent teaching in high school. Like Frost, he studied biology at Bowdoin College in Maine, recalling Bruce's mentor, Professor James Moulton, as a major influence. Jeff was a premed student, but like me, he realized (sitting down to fill out medical school applications) that he did not want to spend so much of his life in hospitals. He applied to some oceanography programs and selected the University of Washington (UW), partly because it was far away and likely to be different. An agreement was reached with his then girlfriend that they would stay in touch while she stayed behind a year to continue in a good job. She later joined him in Seattle, and they married. He marvels now how it could have taken eight years to complete the PhD program. I imagine that young wife wondered, too. Jeff's initial research effort included work on feeding by the estuarine copepod genus *Acartia* using methods Frost had applied to *Calanus* (covered in Chapter 6).

Jeff says that Bruce then left it to him to come up with a dissertation project. At that point in our interview Jeff poked around his office and found his copy of Marshall and Orr.[4] He had read that thoroughly while thinking about a dissertation project. Then he opened it to a figure reproduced here (Figure 11.1, left), and he told me, over three decades later, how impressed he had been then with that clean diagram. Marshall and Orr had done some work[5, 6] on how many eggs *C. finmarchicus* females produced both in single spawnings and over time when held and fed in the lab. They had also tried withholding food, which stopped spawning. Jeff decided to take the quantification of fecundity to the field, namely Puget Sound. *Calanus pacificus* is abundant there and can be sampled all year round from small boats, out of Shilshole Bay in Seattle or on short cruises down the fjord to Dabob Bay.

A key to all subsequent work was learning from David Thoreson, a key technician at UW, to collect uninjured specimens with very slow net tows. Jeff (and I and everyone else) then selects only those with all their setae intact, at least apparently uninjured. First, he went to Thoreson's department shop and constructed a multifunnel system[2] to collect spawned, 160 μm eggs. The idea was to place some females in each tall funnel. Their eggs would fall through a 571 μm screen into a narrowing tube at the bottom that was closed by a clamped tube for removing the eggs to count. Since their mothers could not get down there, they could not cannibalize the eggs. This worked, but it also offered a lot of trouble, specifically problems recovering all the eggs.

So Jeff came up with a very simple alternate scheme: a plastic petri dish for each female selected from a sample. The flat, shallow puddle inhibited her from searching effectively and eating her eggs. After loading a stack of dishes, each with a female, that is put in an incubator set at the field temperature. Later the dishes are examined repeatedly, and the eggs, if any, are counted when they show up. Very often they do. Jeff, Barbara Niehoff,[7] Bill Peterson,[8] myself, and dozens of other ecologists interested in copepods have since measured their egg production rates (EPR). Spawning by many genera proceeds despite the narrow vertical confinement of petri dishes.

Most species of *Calanus* spawn in the late hours of the night, some delaying until around sunrise. Marshall and Orr[5] reported that timing for *C. finmarchicus* maturing in late winter from a Loch Striven diapause stock. However, those maturing in summer spawned at random times all day and night. At least in Puget Sound and off Oregon, *C. pacificus* females collected in the afternoon or early evening spawn late at night in all seasons. Females of *Calanus* species that are visibly loaded with large eggs almost all spawn despite the unfriendly imprisonment. A complete review of all the published EPR results would require a book larger than this one.

From March to October, Jeff collected *C. pacificus* from Puget Sound. He found[9, 10] that female size varied with season, which was mostly an effect of food availability: more phytoplankton and certainly more protozoa (when most of the algae are smaller than the preferred foods of *Calanus*) support more growth (Figure 11.3, left). Females are smallest when maturing in early winter, and largest from May to July after developing when algae are most abundant.

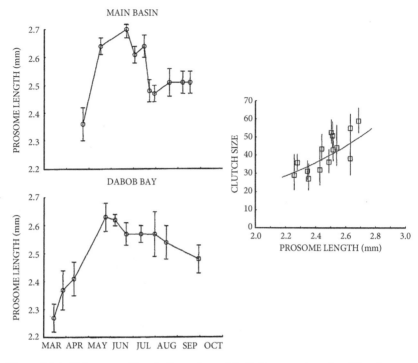

Figure 11.3 *Calanus pacificus* in Puget Sound. Left: Prosome lengths of females from two locations: Main Basin off northern Seattle and Dabob Bay inside Hood Canal. Right: clutch size (eggs/spawning) of copiously fed females. From Runge.[10] Permission, Wiley & Sons.

By August, nutrient minerals are mostly used, algal stocks are down, so newly maturing females are again smaller. In addition to measuring fecundity the day after capture, Jeff brought live females into the lab, kept them in beakers, and fed them all the diatoms they would eat. Given that copious food, he showed that bigger females produce more eggs, the biggest about 60 eggs per clutch, twice the clutch size of the smaller ones (Figure 11.3, right).

For females carrying already formed clutches at capture and spawning within a day, the relationship of clutch size to body size was more complex. The biggest females did produce the biggest clutches in spring and early summer, but also because they had plenty of food in the field in May and June. Females that did spawn when collected from the Main Basin, where food was low from late July through September, produced only about 30 eggs. Those from Dabob Bay spawned only a few eggs in July and August, then bumped up to 45 eggs per clutch in September. Jeff attempted to remove the effect of body size on egg production by

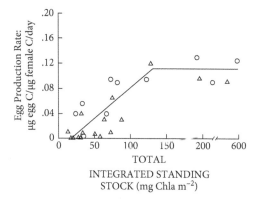

Figure 11.4 Demonstration for females of *Calanus pacificus* collected from Puget Sound that their egg production rate varies with the amount of phytoplankton available (measured as chlorophyll-a summed down the water column).[10] There is a limit of about 12% of female body mass per day. Permission, Wiley & Sons.

estimating each female's biomass, as her bodily carbon content in micrograms, μg C/♀. Then he measured the biomass of her egg output on the sampling day, also as carbon. Finally, calculating her reproductive rate as μg egg C/μg ♀C/day, he plotted that against his field estimate of the standing stock of phytoplankton measured as chlorophyll (Figure 11.4). Voila: the more there was to eat, the more reproduction females achieved, but only up to some physiological limit. That's standard ecology, but it's good to see it proved in any specific case.

Jeff Runge Moves on from Seattle

The last year or so of an advanced degree program gets divided among writing up your research and searching for job opportunities, applying for postdoctoral positions, and wondering if all the work and moderate poverty will produce a way to live and, in Jeff Runge's case, raise a family. An offer did come from Carl Boyd in the oceanography program at Dalhousie University ("Dal"). Carl appears in that role in several of these biographical sketches. Setting aside some doubts about to what going off to Canada might lead, Jeff took the offer and moved with his wife to Halifax, Nova Scotia. He worked at Dal for three years. The first memory of them he recites is that two of this three children were born there; an earlier arrival was born in Seattle. The Dal program was also rich in important ocean ecologists: Boyd, along with Ian McClaren, Bob Conover, and Gordon Riley. Jeff learned from all of them. He worked with the prominent mathematical ecologist Randall Myers, then a graduate student, on a key problem in copepod population dynamics: estimating the variation of mortality rates through the development stages and through the seasons. They produced an early and useful paper on the subject.[11] He also worked with Pierre Pepin and fisheries ecologists from the Bedford Lab of the Government of Canada on the transparent jellyfish *Aglantha*, of interest because they are eaten by Mackerel.[12]

Like advanced degrees, postdocs soon lead to a new job search. But for Jeff a job walked through his office door. Louis Legendre offered him a job at Université Laval in Quebec City, provided he would spend two springs studying under-ice feeding by copepods in Hudson's Bay. It came with an offer of permanent Canadian residency, and it was a job Jeff could do. So he and his expanded family moved. The research was successful, but eventually ended. It did lead to an offer from Canada's Department of Fisheries and Oceans for a job at the new Institute Maurice Lamontagne (IML) in Mont Joli, a 99.9% French-speaking town. French was also the operating language at the lab. His wife got a job in a nearby town where English was dominant, but they both mastered French over time. Stories from Quebec include taking his older son to nearby Rimouski for violin lessons, and the younger kids to practice there with the very good figure-skating coach. There was also driving to skating competitions around Quebec. Both skaters were good; one became novice champion of eastern Canada.

Jeff returned to study of egg production rates, this time of the *Calanus finmarchicus* population in the lower Gulf of St. Lawrence, immediately offshore from Mont Joli. Working with a student, Stéphane Plourde, stacks of petri dishes were again used for incubations. The stock matures from diapause in late April and May but does not begin to spawn until phytoplankton blooms in June, providing enough food to support egg production (Figure 11.5).[13]

They also maintained females in the laboratory, which they had collected from some early maturations in April. This species obtains enough sperm in early matings that re-mating is not needed for prolonged spawning. Fed and unfed groups were both held at two temperatures. Egg production was estimated daily using screens below the females to stop cannibalism. After about 10 days, mean individual egg production of the fed groups rose to about 40 eggs per day, then oscillated while slowly declining to none at about 60 days. With a lot of food at 5.5°C, the average total was 900 eggs. Female mortality in the field is certainly great enough that similar totals are almost never reached. Curiously, unfed females also survived about 60 days. That can only be so if lack of food induces very low metabolic rates. Both Jeff and the German copepodologist Barbara Niehoff[7] have experimented with reintroducing food after a period of starvation and no spawning. Egg production does resume, but at sharply reduced rates, about half those of presumably young females brought in from the field.

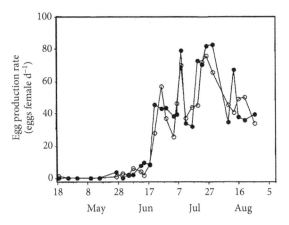

Figure 11.5 *Calanus finmarchicus*. Time-series from 1991 of initial egg production by field-collected females that were fed (●) and not fed (○). Eggs were evidently those supported by feeding in the field. Production did not start until phytoplankton became abundant in June. From Plourde and Runge.[13] Permission, Inter-Research, *Marine Ecology Progress Series*.

Copepod Eggs: What Are They Like?

Let's break off from Jeff here. I'll report on copepods shedding eggs, then look at the eggs in detail. Since females will spawn lying on their sides, even in a shallow puddle on a microscope slide, spawning is readily observed with a microscope. The eggs are pushed past the flap closing the genital opening (operculum), if there is one. Passage elongates them considerably into narrow-waisted, figures-of-eight, not quite barbells. I have watched that many times but have never taken a picture. The last polar body pops to the surface as the egg rounds back up. A clutch of 30, 50, or more *Calanus* eggs can be expressed one at time in a few minutes. In free spawners the eggs do not stick together, at least not for long. Next, a fertilization (or "vitelline") membrane forms, and then a strong, multi-layered coat or *chorion*. In some freely spawning species an outermost layer is a decoration of ridges and points (Figure 11.6, left). The points can extend as rather stout spines on the resting eggs of some kinds of copepods. However, in species of *Metridia* and various other genera, the chorion is just a rather thick layer of cross-linked chitin polymers outside the vitelline membrane (Figure 11.6, right).

    An ultrastructural study of the surface layers of *Calanus sinicus* eggs was done by Euichi Hirose, Hideshige Toda, and Yasunori Saito, directed by Hiroshi Watanabe[15] at the Institute of Biology at the University of Tsukuba. At present, Dr. Hirose puts "Luigi" after his first name on his English-language website. He works at the University of the Ryukus, on a Japanese archipelago in the Pacific

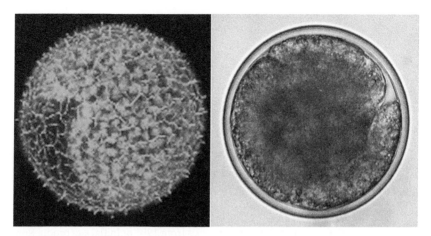

Figure 11.6 Left: *Calanus pacificus* egg, around 160 µm diameter, with individual nuclei still discernible during the gastrula stage (sphere of cells with one side pressing into the center to form a double-layered "thimble"). The outer chorion has a crenulated surface "decoration." PicoGreen staining of both DNA and the outer layer. Confocal micrograph. Right: *Metridia pacifica* egg, around 145 µm, late stage with naupliar limbs visible. Its chorion is a thick chitin layer with a smooth surface. Bright field micrograph, no stain. Both micrographs by Marnie Jo Zirbel, with permission.

far to the southeast of the main islands. His primary interest at present is the strange biology of sessile tunicates. The surface "sculpture" of *C. sinicus* eggs is very similar to that of *C. pacificus* (Figure 11.6 left). The nearly mature ovae (Figure 11.7), still in the oviduct, are surrounded by the shrinking remains of the follicle cells (f) that have fed them, and just inside those is the cell membrane (arrow in Figure 11.7). The outer layer of egg cytoplasm contains some spherical granules (s) that will move to and through the surface after fertilization and spawning to form the chorion layers sequentially in one to two hours, after about seven cell divisions.

At least for *Calanus* species, the chorion's structure starts to form as the initial stages of nuclear division begin. The large secretion granules (Figure 11.7) move to and through the cell membrane, and a sequence of layers is established. Hirose et al. distinguished five layers. The outermost is essentially the thickness of a cell membrane that coats the outside of the surface sculptures obvious in Figure 11.6, left. Layer 2 (Figure 11.8, B) is the rather amorphous "filler" of the sculpture features, and layer 3 is several (2 or 3?) concentric layers of, well, something. Layer 4 they term a "filter" layer. In sections perpendicular to the egg surface (Figure 11.8, B), it is a layer of "tubes" about 0.3 µm long with 0.05 µm "lumens" that may be filled with something or not. In Figure 11.8, B layer 4 has been sliced

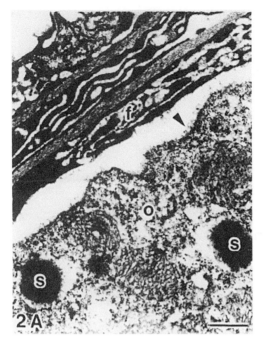

Figure 11.7 *Calanus sinicus*, Transmission electron micrograph (TEM) of the surface of an ovum in the oviduct. Collapsing follicle cells are at the upper left. The arrow points at the ovum's cell membrane. (**f**) necrotic follicle cell; (**o**) cytoplasm (just above a mitochondrion); (**s**) secretion granules ready for chorion formation after fertilization. Scale bar (lower right) 0.5μm. From Hirose, et al.[15] Permission, Oxford University Press.

obliquely showing that the tubes are square in section, sharing sides with adjacent tubes. Layer 5 is multiple layers of curving, cross-linked chitin. They look much like the outer layers of arthropod exoskeleton. What exact purpose each layer serves remains available for speculation (or research).

Egg surface structures vary some among copepod groups (Figure 11.6), but they are consistently tough and permeable only to small molecules (presumably at least O_2, CO_2, and NH_4). Marnie Jo Zirbel and I discovered that by trying to stain nuclei with large, fluorescent molecules (DAPI and PicoGreen) that stain DNA. If we could do that, we could follow cell divisions by counting nuclei. Just flooding formaldehyde-preserved eggs with stain in seawater (to avoid exploding the eggs by osmosis) was useless; no stain got inside. Marnie Jo, however, had the excellent idea to soak the eggs in a freshwater solution of the stains. The eggs did not explode, but osmotic stress apparently stretched the filter layer and the linkages of the chitin layer enough that the stain would pass. Epifluorescence microscopy

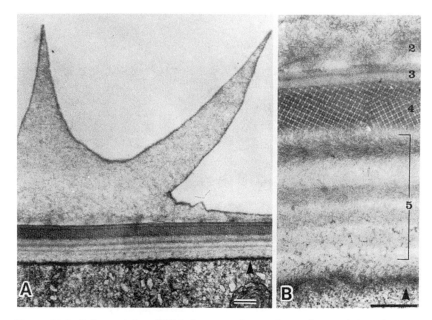

Figure 11.8 *Calanus sinicus*: TEM of egg membrane. A – surface sculpture above a "protein filler" layer, above several layers of polymerized chitin, above the cell membrane and cytoplasm. Scale 0.5 μm. B – the layers close up: 2) amorphous filler of the sculpture, 3) three very narrow layers of undefined composition, 4) a closely regular "filter" layer , 5) cross-linked chitin similar to exoskeleton. Scale 0.5 μm. From Hirose et al.[15] (1992). Permission, Oxford University Press.

then let us count the cell nuclei (Figure 11.9).[14] Sometimes finding a simple solution requires a flash of brilliance, a discernible light coming on above one's head.

Eggs shed directly into the water to make it or not on their own could be subject to sinking if denser than water, to rising if less dense. Selection can shift egg composition to adjust egg density. Basically, with relatively more proteins and less lipid, an egg will sink. Add lipid and an egg will rise. Since copepod eggs are small, considerably less than a millimeter, the velocity will be slow, dominated by viscous drag. And net velocity will also depend upon turbulent displacements up and down. Marnie Jo Zirbel, Harold Batchelder, and I placed *Calanus pacificus* eggs at the top of a tall aquarium with flat sides and its temperature carefully controlled. Its widths were just large enough to avoid viscous effects on drag from proximity to the walls, just small enough to watch eggs move with a microscope looking sideways and moved along the side attached to a pinion gear held on a vertical gear rack. Many people have done that for different species. The results were that all the eggs sank (Figure 11.10). The median of all the rates was 0.04 cm/second. From the density of the water (calculated from temperature

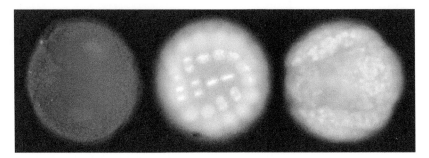

Figure 11.9 *Metridia pacifica*. Field-collected eggs (150 μm) stained with DAPI, a DNA reactive stain. Stages, left: two cells (computer enhanced image, polar body at upper left); center: chromosome clusters of zygote dividing to 64-cells. Right: near hatching, many nuclei, nauplius limbs forming. Epifluoresence micrography from Zirbel and Miller.[14] Permission, Wiley & Sons.

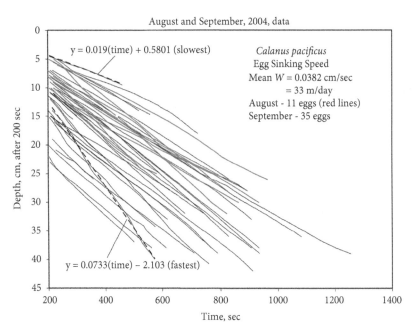

Figure 11.10 Plots of depth versus elapsed time (after 200 seconds) for 46 eggs of *Calanus pacificus* sinking along the center line of an 8 x 8 cm square water column. Depth versus time regression equations are shown for the fastest (0.073 cm sec^{-1}) and slowest (0.019 cm sec^{-1}) eggs. Eggs in the September experiment reached steady sinking sooner than those in the August experiment, so they tended to be deeper after the arbitrary 200 second starting time for data analysis. Illumination for the microscopy induced tiny amounts of heating near the top of the column, and thus some turbulence that only reached down from 5 to 25 cm. Data generated by author with M. J. Zirbel and H. P. Batchelder.

and salinity with density tables) and using Stokes Law of viscous settling velocity for 150 μm spheres (shape makes very little difference for small objects), median egg density can be calculated as approximately 1.07 gm/cm³, a few hundredths greater than seawater.

Everyone else's results for *Calanus* species are similar. Selection for sinking suggests that it is better to risk the bottom (just a little way offshore they are not likely to hit before hatching) than to risk touching the surface. Females must move quite near the surface to spawn. I developed (not very successfully) a pump sampling system ("Megapump") to examine such questions. It could take a profile of two-minute, fine-mesh samples, about 4 m³, at a series of nine depths in around 20 minutes. It was deployed at 7:00 a.m. in April 2008 in Dabob Bay, Washington State (one of Jeff Runge's study sites). Marnie Jo Zirbel sorted *Calanus pacificus* eggs from that early morning profile, when most eggs are in early stages from spawning late the night before. Then, she estimated proportions of embryo development stages using nuclear staining (Figure. 11.11). Thanks to work on *Calanus* embryology published in 1881 by Austrian biologist Carl Grobben,[16] it is possible to determine distinctive stages of embryo development after nuclei can no longer be counted (e.g., Figure 11.9, right).

At 12.5°C, the water temperature on the sampling date, hatching should take about 30 hours. So the expectation is many early stages (blue shades in the figure: from two cells to a blastula stage, say) and surviving late stages from two nights before (brown and then red shades, respectively, for uncountable

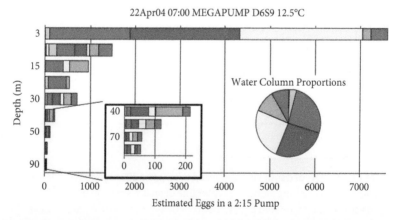

Figure 11.11 *Calanus pacificus* eggs (embryos) in 4 m³ Megapump samples taken in Dabob Bay at 7 a.m. Proportions of early stages are shades of blue to pale green. Proportions of stages likely older than 24 hours at the water column temperatures are in orange and red (near to hatching). Samples collected by Hal Batchelder and M.J. Zirbel and staged by Ms Zirbel. Graph by the author.

masses of nuclei and for distinct naupliar structure). If eggs are spawned near the surface, then those produced late in the night before should be younger and shallower than those from two nights before somewhat deeper. In terms of relative proportions that's correct, at least based on these data. However, the data are messy. There is not only sinking to move eggs mostly down. The bay is a tidal estuary with a lot of stirring most of the time. In fact, eggs spawned two nights back could have been set loose far upstream or down from those spawned last night. However, the expected picture did emerge. Relative abundances of older stages are greater downward.

Jeff Runge Returns to New England

During the late 1990s the US National Science Foundation funded a multiyear, multiship, multi-investigator study of the zooplankton on and around Georges Bank, a submerged sand and gravel pile that was the terminal moraine of the great ice sheet covering eastern Canada and New England in the Wisconsin glacial era. Jeff participated in the study while still based in Mont Joli, seconded by student Stéphane Plourde. German copepodologist Barbara Niehoff, then a postdoctoral scholar at Woods Hole, generated more data on the program's long survey cruises. They measured *Calanus finmarchicus* egg production through the spring and early summer in each of five years. "Cfin" is the largest among the dominant copepods in the deeper waters circulating anticlockwise around the bank, and it is a key food for forage fishes (herring, mackerel) and for cod and haddock larvae and juveniles. So a lot of energy went into understanding its regional ecology. The EPR results[17] can be stated briefly: everything known about Cfin fecundity was confirmed.

Jeff also got a whole new crew of US colleagues. After a divorce he moved to New Hampshire, where he was welcomed at the state university's Ocean Process Analysis Laboratory (OPAL). Over the years he worked his way from grant-supported research work at OPAL to a professorial position at the University of Maine. That was located first at the Gulf of Maine Research Institute in Portland, site of our interview, and now at the University of Maine's Darling Marine Center. He teaches biological oceanography courses, and with his partner Lynn is refurbishing a house in picturesque Rockport between frequent visits to his children and grandchildren in Canada.

With colleagues, Jeff has established a time-series plankton-sampling program at several stations in the Gulf of Maine. In the last few years the Gulf has warmed as dramatically as any place in the world ocean. Surface temperatures were 5°C above average during a 2012 heat spell, a peak repeated in 2018, and have been 2°C above long term averages for over a decade.[17] That is not primarily a local effect. Rather, it is due to changes in relative contributions of warmer waters flowing in from the south and colder waters entering from the Nova Scotian shelf.[18] The dominance order of copepod species has not changed, but their distributions inside the Gulf have.[19] *Calanus finmarchicus* for example are much more confined to the western side. Its diapause stocks (addressed in Chapter 15) have not extended into the lower Bay of Fundy, where they had been dense until lately. So right whales no longer go there to feed on them in late summer. Because the Gulf is packed around with ocean research institutes, very close watch is being kept on the impacts there of climate change. They are substantial and will become more so. Jeff moves all through that institutional structure, struggling to keep it coordinated.

Egg Sac Spawners and an Introduction to Barbara Niehoff

Many genera of copepods do not release their children to make their own way in the water at the egg stage. Instead, their eggs are pushed out into a sort of expanding membrane provided by glands around the genital opening. In *Pseudocalanus* (Figure 11.12, left) and *Euchaeta* (Figure 6.8) that becomes a bumpy blob of eggs, an egg sac.

Some deep-sea copepods carry only a few very large eggs (Figure 11.12, right). In the deep-sea genus *Euchirella*, the eggs form a string in the membrane. I learned from deep sampling off California with Jim Childress and Erik Thuesen

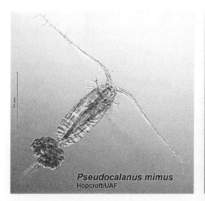

Figure 11.12 Left: *Pseudocalanus mimus* (♀, dorsal view, scale is 1 mm) carrying an egg sac beneath her urosome. Right: egg-carrying female of deep-sea species *Euaugaptilus hyperboreus* (ventral view, total length about 7 mm); she is carrying three eggs; the red dots are oil droplets. *Euaugaptilus* species are "hang and grab" predators. Both photos by Russ Hopcroft. The *Euaugaptilus* photo appears in Kosobokova, Hirche, and Hopcroft.[20] Permission, Springer Nature.

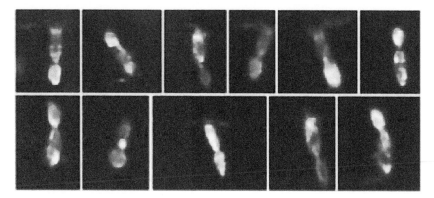

Figure 11.13 Eleven examples of *Pseudocalanus* females actually in the ocean, each with an egg sac, recorded by a video plankton recorder towed at about 8 knots across the south edge of Georges Bank. Compare the shapes to Figure 11.12, left. From Davis et al.[22] Permission, Inter-Research, *Marine Ecology Progress Series.*

(Chapter 5) that different *Euchirella* species place the eggs in different string geometries (like bowling balls in a rack[21] or in zig-zags), and their eggs are of different colors. Egg sacs are often broken or broken off during capture of females. Cabell Davis, Scott Galleger, and colleagues[22] have towed video cameras through the ocean recording frames at frequencies high enough (60 per second) to include all of the tow track at a speed of 4 m/second (~8 knots!). Each frame, then, was about 6.6 cm ahead of the camera, a challenge for precise focus on small objects like copepods. Nevertheless, many frames over many kilometers of track showed identifiable and apparently undisturbed plankton animals. When those images were of *Pseudocalanus* females along the edge of Georges Bank, virtually all of them were bearing egg sacs (Figure 11.13).

Generally, sac spawners are ready to pack up a new clutch immediately after eggs in the previous one hatch, and the emerging nauplii break up the old sac membrane. Thus, almost all the *Pseudocalanus* females would be towing eggs. Time between spawning and hatching generally takes longer than in free spawners, twice the time or more. The spawning strategies differ starting with egg formation itself, different sequences that Barbara Niehoff, an expert on copepod reproduction, characterized in cartoons (Figure 11.14).[23]

Dr. Niehoff (Figure 11.15) is a senior researcher in the Polar Biological Oceanography Section of Alfred-Wegner Polar Institut (AWI) in Bremerhaven, where she is deputy head. Most of her career from graduate studies (supervised by copepodologist Jürgen Hirche) to senior status has been at AWI. She did have a postdoctoral tour at Woods Hole Oceanographic Institution, participating there in the monthly and extensive Georges Bank survey cruises that were part

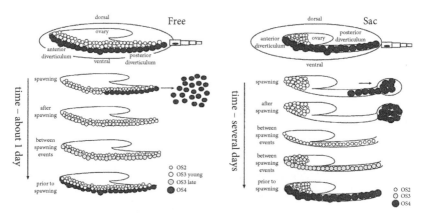

Figure 11.14 Egg formation cycles of free-spawning (*Calanus* type) and sac spawning (*Pseudocalanus* type) female copepods. Reproductive tract and oocyte locations relative to generalized prosome shapes at the tops. Small oogonia in the spaces labeled ovary enlarge to oocyte size and then to stages OS2, OS3, and OS4 in different spatial/temporal patterns. The best studied free-spawners release eggs daily, their ovaries with several stages present at once. Sac spawners prepare eggs for fertilization and towing in clutches more separately in time. Not shown: an *Acartia* type with all stages present all the time, releasing a few eggs throughout the day. From Niehoff.[23] Permission, Elsevier.

Figure 11.15 Prof. Dr. Barbara Niehoff of Alfred Wegner Institute, Bremerhaven. Photo provided by Dr. Niehoff.

of a Global Ocean Ecosystems Dynamics (GLOBEC) project. On many of those January through June cruises in the 1990s we shared a lab table aboard the National Ocean and Atmospheric Administration ship *Albatross IV*, both of us sorting *Calanus finmarchicus* specimens. She sorted females and ran egg production rate studies. I sorted C5s to photograph their oil sacs (see Chapter 15 on copepodite diapause). We talked plankton through many night shifts, working the decks on stations, sorting the catches underway to the next stations.

Since then Barbara has been everywhere in the Atlantic, from Arctic to Antarctic sectors and tropical waters in between. She has measured spawning type, egg production rates, time to hatching, and hatching success for every copepod species coming up in her nets. Many details for those are available in her 2007 summary paper.[23] Her recent research has branched into many aspects of ocean ecology, but emphasis on copepods remains. Her principal hobby is equestrianism. The horse she owns and rides now is named Schiller. A happier, more positive German reference no one could ask for.

More on Free versus Sac-Spawning Modes, Marina Sabatini and Thomas Kiørboe

Three papers by Marina Sabatini and Thomas Kiørboe (Figure 11.16) from 1994 to 1995 started a frenzy of observation and theory about the relative fecundities of free- and sac-spawning copepods. Thomas Kiørboe is continuing a long career studying the ecology of marine zooplankton, originally working from Charlottenlund Slot (Castle). The castle is really a palace (windows, not arrow slits) developed by Danish royalty over several cycles from about 1730. However, from 1935 until 2017 its rooms housed fishery and other marine scientists, which is appropriate since Øresund channel from the Baltic to the Kattegat is visible from the higher windows. Various laboratory facilities were set up in the surrounding buildings. I visited briefly once; it's somewhat sad to think it is now a cultural event venue. Kiørboe works now at the National Institute of Aquatic Resources, part of the Danish Technical University at Lyngby. Much of Danish marine research was relocated there. In both places he has worked with many visiting scientists and postdocs. It seems to be his preferred mode and has generated a vast bibliography. He has a theoretical bent, finding mathematical representations of issues from micro-hydrodynamics to population numbers.

Dr. Kiøboe has promoted, at book length,[24] the notion that ecology as "seen" by plankton organisms themselves occurs mostly in the spaces close around them. The book features events like nutrient molecules diffusing within the capture-range of transport enzymes on cell surfaces, like a copepodite grabbing a meal whose wake tweaked its setae, and like adult copepods finding a mate in the

Figure 11.16 Left: Thomas Kiøboe and Marina Sabatini with their recent books. From internet sources with permission from each of them. Right: *Oithona similis* female with two egg sacs (14–16 eggs total), its prosome with a large oil sac. Note that each oviduct has its own opening. Total length approximately 1 mm. Photo by Russ Hopcroft.

dark (Kiøboe calls that blind-dating). An adult female *Euchaeta* with an egg sac is on the cover of his book. Thinking about such events is in contrast to much of biological oceanography and limnology that focus on time progressions of algal and animal abundance, and on productivity rates (e.g., algal carbon fixation in units of grams C m^{-2} d^{-1}). Moving to Lyngby on the other side of Copenhagen hasn't slowed his output of significant scientific studies, most of them cooperative with others.

For most of her career Marina Sabatini has worked at the plankton laboratory of the Instituto Nacional de Investigación y Desarrollo Pesquero (INIDEP) in Mar del Plata, Argentina. She started there shortly before obtaining her doctorate in 1987 at Universidad Nacional del Sur at Bahia Blanca, also receiving the title Investigadora from the Consejo Nacional de Investigaciones Científicas y Técnicas (CONICET). CONICET sponsors top-line scientists at many Argentinian institutions, paying their salaries. That arrangement was sustained through her whole career until retirement in 2019. Marina's dissertation was titled: "Fito y zooplancton de un sector de la Bahia Blanca: especies dominantes, standing stock y estimación de la producción, con particular referencia a *Acartia tonsa* Dana, 1849." That should not require translation. Over her career Marina characterized the zooplankton of Argentina's southern continental shelves and the waters farther offshore, working primarily from fisheries survey vessels. She was principal editor of a recent and important book; it discusses the planktology seaward of southeast South America.[25] Interviews were not easily arranged as I finished this book, thanks to an international viral pandemic. So here I just want to summarize the impacts of their joint papers.

Starting in 1992, Dr. Sabatini was a visiting scientist in Thomas Kiøboe's group at Charlettonlund. In Denmark, she raised the tiny *Oithona similis*, a copepod often wildly numerous in mid- to high latitudes. It is an egg-carrier. Her results led to speculations on the distinctive adaptations of free-spawning and egg-carrying copepod reproduction. They obtained female *Oithona similis* from local Danish waters.[26] Holding them individually in 50 ml containers, they learned that females collected from the field were probably newly reproducing if they sustained egg production for about five clutches in 24 days at 15°C, which usually ended significant reproduction.

Hatching took *three* days at 15°C (six at 10°C), ending each cycle. The female took a day off, and then she filled new sacs with a new clutch. Marina determined a functional response for egg production rates versus amounts of food she provided as small flagellates. Eating just those, clutch size was 7 to 9 eggs in two sacs. Thus, lifetime totals were about 35 eggs. In just one experiment with 20 females held with the flagellates plus *Oxyrrhis marina*, a protozoan that just showed up in their food cultures, the average clutch size jumped to 14 eggs, which is like the brood being towed by the specimen in Figure 11.16. Lifetime totals (though not stated) would be perhaps 50 eggs. Compare those totals with *daily* spawning by *Calanus* females of 30 to 60 eggs (Figure 11.3, right), running to potential lifetime totals (probably not often realized due to predation) of 900 or 1,000 eggs. At 15°C *Calanus pacificus* eggs would hatch in around 24 hours.

Kiørboe and Sabatini found the distinctions impressive, so they extended the comparison using data for many species from the literature.[27, 28] The difference is not just a matter of the distant relationships among the families. Both reproductive strategies can be found in the Calanoida: species of *Pseudocalanus* (Figure 11.12, left) and some of *Clausocalanus* (Figure 4.2, left) are also egg carriers, while comparably sized *Acartia* species spawn freely. Summarizing their review, the extra development time clearly isn't wasted. Obviously benefitting from the predation protection afforded by their mother's escape capacity, the first larval stages of sac spawners (first nauplii, N1, next chapter) hatch relatively larger, and ready to feed. The new hatchlings of *Calanus* are ready to be eaten but not to eat. They do not yet have open mouths and must

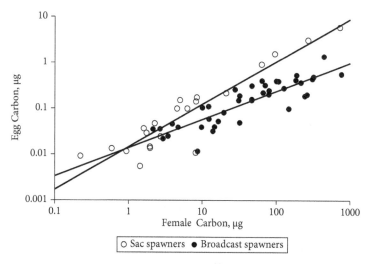

Figure 11.17 Figure 1 from Sabatini & Kiørboe,[28] single egg carbon content versus female carbon mass (note the log-log scales). I removed a few outlier points the authors did not include in fitting the lines. Permission, Inter-Research, *Marine Ecology Progress Series.*

run through two molts on "leftover" yolk before they can feed as nauplius-3. As plankton animals grow larger, they progressively exceed the prey sizes accessible to many small predators, thus sac-spawning copepod mothers provide double predation protection to their young. Free-spawning females must compensate by spawning many times more eggs; they do. The sac spawners must provide more nourishment in each egg, which they do as the review demonstrated with a graph (Figure 11.17).

Nevertheless, the free spawners in compensating must expend about four times more organic matter daily (relative to their own mass) in making clutches as do sac spawners (calculated from Figure 2 in Kiørboe and Sabatini, 1995).[28] Why don't free spawners evolve egg sacs? In a sense, this is because evolutionary inventions are only transferable to new generations, not to separately evolving populations (Well, until the invention of CRISPER).[29]

I need a fresh chapter to discuss hatching, early larval development, copepodite stages, molting, and how opal copepodite teeth are constructed.

References

1. Uye, S., & K. Sano (1995) Seasonal reproductive biology of the small cyclopoid copepod *Oithona davisae* in a temperate eutrophic inlet. *Marine Ecology Progress Series* **18**: 121–128.

2. Runge, Jeffery (1984) Egg production of the marine, planktonic copepod, *Calanus pacificus* Brodsky: laboratory observations. *Journal Experimental Marine Biology and Ecology* 14: 53–66.

3. Saito, H., & A. Tsuda (2000) Egg production and early development of the subarctic copepods *Neocalanus cristatus, N. plumchrus* and *N. flemingeri. Deep-Sea Research I* 47: 2141–2158.

4. Marshall, S. M., & A. P. Orr (1955) *The Biology of a Marine Copepod, Calanus finmarchicus, Gunnerus*. Oliver & Boyd, Edinburgh.

5. Marshall, S. M., & A. P. Orr (1952) On the biology of *Calanus finmarchicus* VII: factors affecting egg production. *Journal of the Marine Biological Association, U.K.* 30: 527–547.

6. Marshall, S. M., & A. P. Orr (1953) *Calanus finmarchicus*: egg production and egg development in Tromsø Sound in spring. *Acta Borealia* (Series A) 5: 3–19.

7. Niehoff, B. (2007) Life history strategies in zooplankton communities: The significance of female gonad morphology and maturation types for the reproductive biology of marine calanoid copepods. *Progress in Oceanography* 74(1): 1–47.

8. Peterson, W. T., J. Gómez-Gutiérrez, & C. A. Morgan (2002) Cross- shelf variation in calanoid copepod production during summer 1996 off the Oregon coast, USA. *Marine Biology* 141: 353–365.

9. Runge, J. A. (1981) *Egg production of Calanus pacificus* Brodsky and Its Relationship to Seasonal Changes in Phytoplankton Availability. PhD dissertation, University of Washington.

10. Runge, J. A. (1985) Relationship of egg production of *Calanus pacificus* to seasonal changes in phytoplankton availability in Puget Sound, Washington. *Limnology & Oceanography* 30(2): 382–396.

11. Myers, R. A., & J. A. Runge (1983) Predictions of seasonal natural mortality rates in a copepod population using life-history theory. *Marine Ecology Progress Series* 11: 189–194.

12. Runge, J. A., P. Pepin, & W. Silvert (1987) Feeding behavior of the Atlantic mackerel *Scomber scombrus* on the hydromedusa *Aglantha digitale. Marine Biology* 94(3): 329–333.

13. Plourde, S., & J. A. Runge (1993) Reproduction of the planktonic copepod *Calanus finmarchicus* in the Lower St. Lawrence Estuary: relation to the cycle of phytoplankton production and evidence for a *Calanus* pump. *Marine Ecology Progress Series* 102: 217–227.

14. Zirbel, M. J., & C. B. Miller (2007) Staging egg development of marine copepods with DAPI and PicoGreen˚. *Limnology & Oceanography: Methods* 5: 106–110.

15. Hirose, E., H. Toda, Y. Saito, & H. Watanabe (1992) Formation of the multiple-layered fertilization envelope in the embryo of *Calanus sinicus* Brodsky (copepoda: calanoida). *Journal of Crustacean Biology* 12(2): 186–192.

16. Grobben, C. (1881) Die Entwicklungsgeschichte von *Cetochilus septentrionalis* Goodsir. *Arbeiten Zoologischen Insitute der Universitat Wien* 3: 243–282, pls. 110–122.

17. Runge, J. A., S. Plourde, P. Joly, B. Niehoff, & E. Durbin. (2006) Characteristics of egg production of the planktonic copepod *Calanus finmarchicus* on Georges Bank: 1994–1999. *Deep-Sea Research II* 53: 2618–2631.

18. Runge, J. A., R. JI, C. R. S. Thompson, N. R. Record, C. Chen, D. C. Vandemark, J. E. Salisbury, & F. Maps (2015) Persistence of *Calanus finmarchicus* in the western Gulf of Maine during recent extreme warming. *Journal of Plankton Research* 37(1): 221–232.

19. Record, N. R., J. A. Runge, D. E. Pendleton, W. M. Balch, K. T. A. Davies, A. J. Pershing, C. L. Johnson, & coauthors (2019) Rapid climate-driven circulation changes threaten conservation of endangered North Atlantic right whales. *Oceanography* 32(2): 162–169.

20. Kosobokova, K. H., H.-J. Hirche, & R. R. Hopcroft (2007) Reproductive biology of deep-water calanoid copepods from the Arctic Ocean. *Marine Biology* 151: 919–934.

21. Ohman, M. D., & A. W. Townsend (1998) Egg strings in *Euchirella pseudopulchra* (Aetideidae) and comments on constraints on egg brooding in planktonic marine copepods. *Journal of Marine Systems* 15: 61–69.

22. Davis, C. B., Q. Hu, S. M. Gallager, X. Tang, & C. J. Ashjian (2004) Real-time observation of taxa-specific plankton distributions: an optical sampling method. *Marine Ecology Progress Series* 284: 77–96.

23. Niehoff, B. (2007) Life history strategies in zooplankton communities: the significance of female gonad morphology and maturation types for the reproductive biology of marine calanoid copepods. *Progress in Oceanography* 74: 1–47.

24. Kiørboe, T. (2008) *A Mechanistic Approach to Plankton Ecology.* Princeton University Press, Princeton, NJ.

25. Hoffmeyer, M. S., M. E. Sabatini, F. Brandini, D. Calliari, & N. H. Santinelli, eds. (2018) *Plankton Ecology of the Southwestern Atlantic, from the Subtropical to the Subantarctic Realm.* Springer, New York.

26. Sabatini, M., & T. Kiørboe (1994) Egg production, growth and development of the cyclopoid copepod *Oithona similis. Journal of Plankton Research* 16: 1321–1351.

27. Kiørboe, T., & M. Sabatini (1994) Reproductive and life cycle strategies in egg-carrying cyclopoid and free-spawning calanoid copepods. *Journal of Plankton Research* 16: 1353–1366.

28. Kiørboe, T., & M. Sabatini (1995) Scaling of fecundity, growth and development in marine planktonic copepods. *Marine Ecology Progress Series* 120: 285–298.

29. Doudna, J. A., & S. H. Sternberg (2017) *A Crack in Creation: Gene Editing and the Unthinkable Power to Control Evolution.* Mariner Books, Boston.

12

Development

Hatching and Early Development

Hatching has been observed by many copepodologists,[1,2] including me. Because the chorion is so resilient, special tricks apparently had to evolve for the first nauplius to escape through it. This is a standard problem with eggs; the young must break out. Birds hatch by breaking their shells with a hard "egg tooth" on their newly formed bill. Copepod eggs must have an inner-most chorion layer, not well characterized, but outside the ready-to-hatch larva. It separates from the others, and then over a few seconds it begins to swell to greater volume, the pressure tearing a slot in the chitin and outer layers. This "hatching membrane" emerges through the slot, leaving the tough outer shell crumpled behind (Figure (12.1). Then in *Calanus marshallae* it rounds up into a transparent sphere about 300 µm diameter, much bigger than the egg at 180 µm.[3] The nauplius is inside. For one to several minutes the nauplius shakes, filling and extending its limbs, which then slice an exit and leave the membrane behind.

Everyone reporting on hatching says the nauplius must release some solute, raising osmotic pressure in the membrane to move water in from outside. It is almost certainly so, but getting some of that substance to prove it has not been accomplished. The mechanism seems so obvious, as you watch it, that maybe its details are unimportant. Maybe you think by now that much of what we know about copepods might not have been important enough to justify learning it. But I wonder whether the chorion has a weak seam somehow built in, what sets the timing of solute release, and what solute is such that suddenly it can be outside the nauplius and then dissolved. Scientists are like that.

Copepod eggs ready to hatch contain a first stage "nauplius," or N1. Nauplii are the larval form common to all crustacea, though some parasitic forms run through those stages inside the egg, including some molting. Nauplii are usually of a somewhat flattened rugby-football shape (Figure 12.2), with three paired limbs: antennules, antennae, and mandibles, the latter two *biramous*. In some species the mandible has a basal spine extending toward the mouth under the fat upper lip covering the mouth, the labrum. In a ventral-side-up drawing by Tage Björnberg (Figure 12.3) those spines look like pickle forks for pushing food bits toward the mouth.[4] Not all copepod nauplii have (or need) such spines. For example, the nauplii of the Subarctic Pacific species of *Neocalanus* are loaded with

Oar Feet and Opal Teeth. Charles B. Miller, Oxford University Press. © Oxford University Press 2023. DOI: 10.1093/oso/9780197637326.003.0012

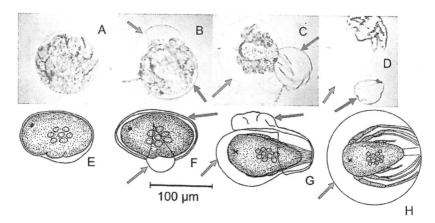

Figure 12.1 A–D stages of egg hatching in *Calanus finmarchicus*, from Marshall and Orr.[1] E-H similar stages in *Diaptomus siciloides*, from Davis.[2] In both species the hatching membrane is indicated by blue arrows, the outer chorion by red arrows. Permissions A–D, Cambridge University Press; E–H, University of Chicago Press.

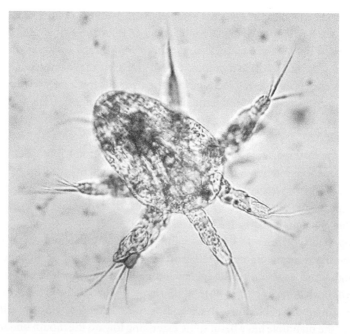

Figure 12.2 Early nauplius of *Cyclops,* a freshwater Cyclopoid copepod. The antennules (A1) are the unbranched limbs at lower right. The naupliar eye is the red spot between the A1 bases. The brown spot is probably food in the gut. Body length about 100 μm. Photo by Jasper Nance. Permission via GNU Free Documentation License.

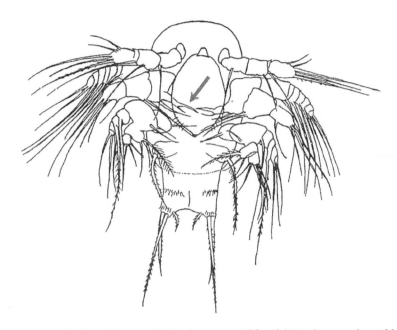

Figure 12.3 Unidentified canuellid (an harpacticoid family) N3 showing the pickle-fork spines on the coxae of the mandibles (red arrow over the labrum) that extend toward the mouth. Tissues were cleared to make the spines visible through the labrum. From T. K. S. Björnberg.[4] Creative Commons License, *Nauplius, Rio Grande*

lipid droplets at hatching and do not feed at all until the transformation to first copepodites.

Larval development in free-living copepods is, with a few exceptions, restricted to eleven molts. There are six naupliar stages (N1 to N6, Figure 12.4) and six copepodite stages (C1 to C6-adults).

There are variations. Nauplii of many genera such as *Pseudocalanus*, whose mothers carried them as eggs in a towed sac, hatch after relatively longer incubations (at given temperatures) and begin to feed immediately. Nauplii of *Calanus* may only start feeding in a somewhat prolonged N3 stage after shorter N1 and N2 stages. So how, exactly, does such a tiny, simply armed, blob-like swimmer feed? Highly magnified video, used much as described in Chapter 7 for copepodites, has allowed some answers. The abstract of a paper by Danish fishery scientist Eleonora Bruno and colleagues[5] provides a description sufficient for our purposes (slightly edited):

We used high-speed video to describe the detection and capture of phytoplankton prey by the nauplii of two ambush-feeding species (*Acartia tonsa*

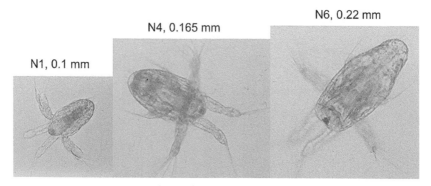

Figure 12.4 Relative sizes of *Acartia californiensis* nauplii, labels are body lengths. Micrographs by J. Kenneth Johnson, with permission.

and *Oithona davisae*) and by the nauplii of one feeding-current feeding species (*Temora longicornis*) . . . The ambush feeders both detect motile prey remotely. Prey detection elicits an attack jump, but the jump is not directly towards the prey, such as has been described for adult copepods. Rather, the nauplius jumps past the prey and sets up an intermittent feeding current that pulls in the prey from behind towards the mouth. A feeding-current feeding nauplius detects prey arriving in the feeding current, but only when prey are intercepted by the setae on the feeding appendages. This elicits an altered motion pattern of the feeding appendages that draws in the prey.

Nauplii presumably sense nearby prey with minute sensors of shear in the water generated by even smaller organisms (protozoa, say), by scent or both.

There are few studies of the internal organs of nauplii. In one of those, Edith Susana Fanta,[6] working in Brazil on the small brackish-water calanoid *Pseudodiaptomus* in the early 1980s, traced their development stage-by-stage. She had done similar studies of nauplii from *Euterpina* (pelagic harpacticoids) and *Oithona*. More and more strips of muscle are added from N1 to N6 to drive and turn the limbs. Otherwise the stages are much alike internally, just progressively bigger. At the very front of the head a frontal (or "naupliar") eye (the red dot in Figure 12.2) sits on the ventral surface, just in front of a tiny brain. The esophagus behind the brain rises into a gut sac that ends, possibly at a sphincter, in a narrower, short intestine emptying at a posterior anus. A ventral nerve cord runs from a ganglion behind the esophagus. There is no heart. Wolfgang Fahrenbach[7] found a tiny excretory gland in the naupliar antennule of a parasitic harpacticoid. Fanta, however, found no excretory glands but rather small excretory concretions in the gut wall of nauplii from both *Pseudodiaptomus* and

Oithona. There are a few secretory glands in the epidermis. The anatomy is just complicated enough to hunt in water for tiny food items.

Growth of nauplii from stage to stage is (to me) surprisingly small, though not necessarily slow. Each stage is progressively a little larger, and the molts add little or no new segmentation to the initial three limbs. Marine biologist Veronica Alvarez Valderhaug, while working on a fellowship in Bombay, measured the lengths of the nauplius stages of *Apocyclops dengizicus,* a brackish water cyclopoid.[8] Within a stage they varied by only a few thousands of a millimeter (microns). Increase in length stage-to-stage was as shown in Table 12.1:

Table 12.1 Stage-by-stage body-length measures for naupliar *Apocyclops dengizicus*[8]

Stage	Mean Body Length, mm
N1	0.107
N2	0.129
N3	0.158
N4	0.189
N5	0.228
N6	0.275

She found that the length advanced by those small steps, a stage-to-stage factor of 1.21. That roughly fits the sequence of the nauplii of *Acartia californiensis* (Figure 12.5). Dr. Valderhaug worked later at the University of Oslo on sediment communities in Oslo Fjord.[9] Drawings and measurements by Frank Ferrari and Julie Ambler[10] of all the naupliar stages of *Diothona oculata,* a species living in mangrove marshes and near coral reefs, also show nearly constant length ratios, but they less than double in length from N1 to N6. All the *A. dengizicus* naupliar stages have the same limb segmentation (Figure 12.5), though a few setae and spines are added (counts from Valderhaug's figures are certainly reliable) stage-on-stage. A few additional spines are usually added progressively to the back end (termed the "caudal armature" by Valderhaug). At least in that species, rudiments of more cephalic limbs begin to appear at N3 (Figure 12.6) but are only slightly elaborated from N4 to N6. In some species, the back of the body becomes extended, as a sort of pre-thorax. It becomes vaguely segmented with flap-like, ventral precursors of limbs to come.

Across the full range of free-living copepod diversity, the early naupliar stages are much alike. However, by N6 there may either be almost no

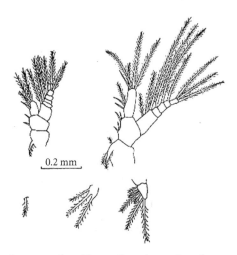

Figure 12.5 Limbs from a cyclopoid nauplius, *Apocyclops dengizicus* Lepeshkin.
Top row: antennae of N1 and N6. Note the changes in setation, but not segmentation. Lower row: rudiments of maxillules from N3, N4, and N5. Drawings by Veronica Alvarez Valderaug.[8] Permission, Brill, Leiden.

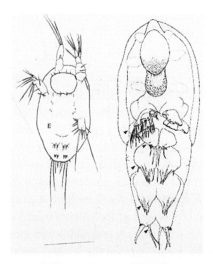

Figure 12.6 Comparison of simple and more elaborated versions of sixth nauplius morphology: *Dioithona oculata* (left),[10] and *Phyllodiaptomus annae*.[11] Only one of each anterior limb pair or simply outlines of their attachments are shown. Permissions, Allen Press and Oxford University Press.

elaboration beside some enlargement (Figure 12.6, left), or there can be quite extensive posterior development of rudiments of soon-to-appear thoracic legs (Figure 12.6, right).

Completing N6, copepods undergo a substantial metamorphosis, molting to a first "copepodite" stage (C1; Figure 12.7) with distinct prosome and urosome sections. All of the cephalosome limbs from antennule to maxilliped are present. The C1 stages of most free-living species have two or three articulated thoracic segments bearing two swimming legs with only a few segments in each. A third thoracic segment may have a limb rudiment (bud). The urosome usually has two segments and small caudal furcae. More molts take the copepodite through five more stages. Successive molts add thoracic segments and swimming legs to a total of four or five. From C1 to C6 articulations can be added between distinct leg segments. Fifth swimming legs can be present or not, and when present they are usually modified in adults, even radically modified, for roles in copulation. Adults are the last stage, C6, usually with males and females strongly distinct, in many species so different it is difficult to figure out from preserved plankton samples which males and females belong to the same species.

For *Diothona oculata*, chosen as an example for Figure 12.7, the increase in length from C1 to C6 is 2.2-fold. In other copepods the growth in length can be greater; in *Calanus* species on the order of 4.5-fold. Recall that volume increases as roughly the cube of length for a given shape, so the *D. oculata* increase in body mass is about 6-fold, that in *Calanus* about 80-fold.[12] This difference in copepodite growth is one of many strong differences between small and large genera. Additions of thoracic legs and urosome segments can occur in somewhat different orders and to different totals.

Growth Rules: Stage Duration and Overall Growth Rates

Because copepods are so dominant in their size-range in the plankton of lakes, estuaries, and oceans, their growth is a substantial part of new animal tissue production in those habitats. Think "food for fish." So large efforts have gone into measuring the rates at which they grow through their stages, into determining the amounts of tissue developed at each stage. It is difficult to weigh one nauplius or copepodite, even with high-tech electronic balances. Weights are usually determined for counted groups. Since most of the tissue is water, and water has no "food value," dry weights or carbon contents are usually measured. There are reasons to estimate wet weights, particularly since copepods can live through being weighed. Carbon contents, to represent amounts of organic matter, are measured with instruments that burn dried tissue in a stream of air or oxygen. The stream is then ducted between a source of infrared light (IR) and an IR

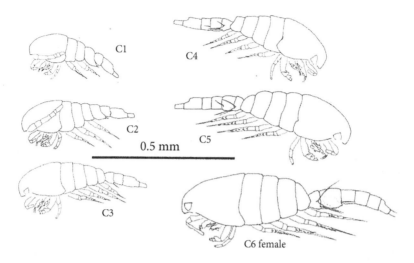

Figure 12.7 *Diothona oculata* copepodite stages.
Numbers of thoracic legs and urosomal segments are:
legs segments
C1: 2 legs, 2 U segments
C2: 3 legs, 2 U segments
C3: 4 legs, 3 U segments
C4: 5 legs, 4 U segments
C5: 5 legs, (5th now on U1), 5 U segments
C6: 5 legs, (5th now on U1), 6 U segments
Ferrari and Ambler.[10] Permission, Allen Press.

sensor. The CO_2 from the combustion is measured by the amount of IR it absorbs as recorded from the sensor. The mechanism is the same as used to measure CO_2 in the atmosphere, to obtain data like the Keeling curve (seasonal peaks at this writing rising past 419 parts per million by volume).

Stage Duration

Most determinations of the time between molting into a stage and molting from it are done by rearing copepods in laboratories, feeding them cultures of phytoplankton, protozoa, brine shrimp nauplii, and so on. The questions asked usually include the effects on durations from temperature, salinity (if marine) and food levels. Temperature and food availability have strong effects. Because laboratory rearing is typical, usually left out are factors like how much work copepods have to do in jumping away from predators or in daily vertical migrations.

Effects of only eating part of the day (sometimes observed in the field) are also neglected. But to some extent the lab data can be checked by comparing results to stage progressions and body sizes in field populations. So how are durations measured? Donald Heinle[13] (and others independently) developed the following method used many times since.

You collect some fertilized females from the field (or from a long-term lab culture) and let them spawn. The eggs are separated from their mothers and incubated at a temperature the species might experience in the ocean, lake, or estuary with food rations already tested for supporting growth. It works best to combine many eggs from a short range of spawning times, say a day or two. Then, as development progresses you repeatedly sample the stock of nauplii and copepodites in the culture containers, and you stage and count them to estimate the stage proportions (hence accuracy depends on large numbers of eggs). Changes of the proportions are then plotted against time (Figure 12.13, an example from Peterson and Painting[14]), with the cumulative proportions of each stage over time connected through time with a regression line (different types may be favored). As the experiment progresses, the lines will cross the 50% mark

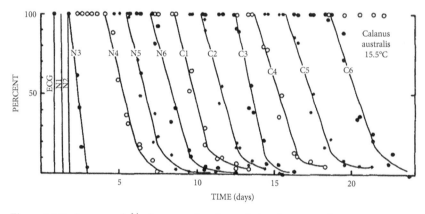

Figure 12.8 An example[14] of progression through development of stage proportions in *Calanus australis*. Starting with eggs of females from a nearshore stock off South Africa, laboratory cultures of 100s of hatchlings (at 15.5 C in 14 hour:10 hour :: light:dark cycles) were fed more than they would eat of a diatom and flagellate. They were staged and counted by stage (alive) daily. Spacings at the 50% level represent the sequence of median development times. (Late advancing specimens (curves at the bottom) were not included in the regression lines. Estimated durations were N2=0.9 days, N3=3, N4=1.2, N5=1.8, N6=1.6, C1=1.5, C2=2, C3=1.7, C4=2.1, C5=3. Permission, Oxford University Press.

successively as the cohort develops. The time intervals between those crossings are measures of the *median* durations of the stages.

This can also be done in the field. It can, provided a situation is found in which a seasonal cohort installed in a body of water develops in a roughly coherent sequence. I tried it far at sea in the Gulf of Alaska with *Neocalanus plumchrus*.[15] That copepod mostly ceases spawning there at the end of winter, about March. On a cruise to Ocean Station P (50°N, 145°W), arriving on May 9, 1988, we found the stock of 5,000 to 15,000 per square meter in surface layers was a mixture of copepodites C1 to C5. There were only a few C1 and no C5 present, and the proportion of C1 declined in the next days to nearly zero, implying that the spawning interval had been adequately terminated. We sampled often for 19 days, seeing over half the stock reach at least C4, with some 15% reaching C5. Graphical analysis of the data (Figure 12.9) suggests the duration of C3 was about 13 days.

However, a problem is imposed, particularly in field applications, by possible differences in mortality between the stages, as pointed out by Hairston and Twombly.[16] If an older stage dies faster (if, say, they are the preferred prey of some fish) than younger ones, their proportion will increase more slowly than determined by stage duration alone. Thus, the younger stage will *appear* to last longer, leading the method to overestimate its stage duration. Conversely, if a younger stage suffers relatively more, say, arrow worm mortality than the next older, its duration will be underestimated. I wrestled with this problem using numerical models. Applying likely, if unknown, mortality rates, the C3 duration might

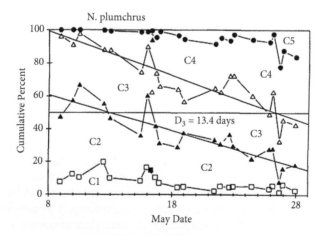

Figure 12.9 Proportions of copepodite stages of *Neocalanus plumchrus* (see C5 in Figure 1.2) at Ocean Station P in the Gulf of Alaska in May 1988. Data from Miller.[15] Permission, Elsevier.

vary ± two days. At least the field estimate provides a rough idea. There are other problems. In an open ocean situation you must assume that stage proportions are more stable than total abundances, which varied in the example by a factor of about three. Year in, year out in the 1980s, spring stage proportions at Station P progressed quite stably (no backing up outside random sampling variation), despite wide abundance variation. My stage proportions graph (Figure 12.14) looks convincing, at least to me.

Development and Growth Rules

Recall the extensive field and culturing study by Sabatini and Kiørboe of *Oithona similis* from Chapter 11. They[17] ran rearings at 15°C and salinities like those typical on either side of Denmark: 20 and 30 g/kg for Kattegat and North Sea, respectively. Nauplii were collected from large numbers of females over two days, then fed two kinds of flagellated algae already proved to support growth for multiple generations. Samples were counted for stage proportions every two days or more often, and stage durations were determined using (basically) the Heinle method. Results were plotted as cumulative median stage durations versus time (Figure 12.10).

Whether the modest mean duration difference for *O. similis* between the two salinities could be a real effect or due to random sampling variations is not clear. Sabatini and Kiøboe also measured the carbon content of many stages, creating a carbon biomass relation to body length, then applied that to the stage durations, to get a biomass versus time relationship (Figure 12.11).

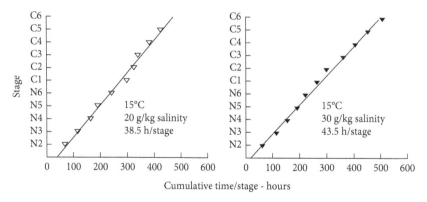

Figure 12.10 Cumulative stage-duration times for laboratory-reared *Oithona similis*. The stages take nearly identical amounts of time from N2 to C6. From Sabatini and Kiørboe.[17] Permission, Oxford University Press.

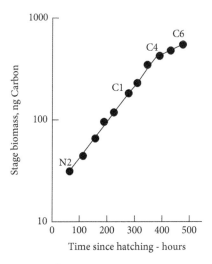

Figure 12.11 Growth in mass of organic matter, as nanograms carbon, as a function of development time. A nanogram is one billionth (10^{-9}) of a gram. The straight line from N2 to C4 on the semilogarithmic (base 10) scale implies exponential growth. That is, each new stage adds the same fraction of the mass of the stage before it. Growth (as change of mass/original mass/hour) slows some during advance from C4 to C6. From Sabatini and Kiørboe.[17] Permission, Oxford University Press.

Most newly hatched young grow exponentially for at least a short time. Even nursing human babies do so for a few months, surprising parents with how they got so big so fast. That phase is a (postpartum) "growth stanza" one. Free-living copepods, especially small copepods like *Oithona*, retain that potential into the last few of their thirteen lifecycle stages (counting the egg).

Long ago I thought a lot about these growth patterns (so allow me more graphs). In Figure 12.12[18] are the development and growth patterns for two *Acartia* species. All of the stages of *A. californiensis* (Figure 12.12, left) are, like those of *O. similis*, completed in roughly equal time. The same is true of A. *tonsa* from Chesapeake Bay (Figure 12.12, right), and at least its copepodite stages add biomass (as dry weight) in consistent multiplicative steps (i.e., *exponentially*). The dependence of rates on temperature is obvious from the rapid *A. tonsa* stage progressions at 25.5° and 30.7°C. Similar temperature dependencies of egg development and growth rates are true of all copepods. Great skeins of data have been generated showing that, and substantial fascination has attended fitting mathematical functions to rate versus temperature relationships.

A similar pattern applies to *Calanus* species, represented here by *Calanus marshallae*, which is abundant most summers near the Oregon coast. It was

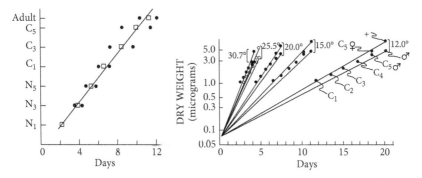

Figure 12.12 Left: Cumulative stage duration times in well fed cultures of *Acartia californiensis*, a stock originating in upper Yaquina Estuary, Oregon (J. K. Johnson's data). Right: Cumulative stage duration and dry weight increase during growth (labeled as C1, C2, etc.) in well fed cultures of *Acartia tonsa* from Chesapeake Bay, Maryland (Donald R. Heinle's data). From Miller et al.[18] Permission, Wiley & Sons.

reared in the laboratory at 10°C by the late William Peterson.[19] He learned (as others have) that N1 and N2 do not feed, their mouths are not open; living on yolk, they molt twice in a few hours. The N3 is prolonged for feeding, but appears only to be replenishing nutriments, and so N4 is about the same dry weight. Then a nearly consistent stage duration moves individuals through N5 to C5, and tissue mass, as dry weight (DW), increases exponentially (Figure 12.13). Maturation to C6 was delayed, with C5 lasting about three times longer than the earlier stage durations. However, Peterson's adults (all female, discussed in Chapter 11) had increased in weight by the same proportional amount (DW_{C6}/DW_{C5}) as at the younger steps (DW_{C2}/DW_{C1}, etc.).

Certainly, C5 must not only grow but must achieve more anatomical construction, in order that new C6 females can be nearly ready to spawn. Also, *Calanus* species often have some delay at C5 for a decision between a long diapause (a version of hibernation) and immediate maturation. No one has figured out how to induce *Calanus* species to enter diapause in culture. The very wide range of sizes of older stages is obscured by the logarithmic scale. They vary in dry weight by factors greater than twofold. Peterson followed development of clutches from different females, but all had similar food and temperature. Much of the variation was between clutches, less of it within.

So, what's the rush? Most animals with continued parental protection, like elephants and people, take really long to grow up. Some copepods, including *Acartia* and *Calanus*, are just tossed out as eggs, getting no protection from the moment of fertilization. Eggs of others, including *Oithona*, *Euchaeta*, and *Pseudocalanus*, are protected as eggs, but their nauplii are on their own from the moment of hatching. Almost certainly that puts a selective premium on getting to adulthood

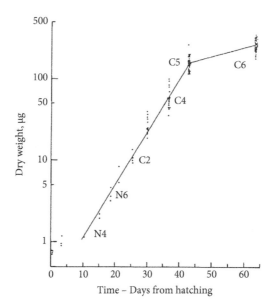

Figure 12.13 Biomass growth (as dry weight) of *Calanus marshallae*. Results from laboratory rearing and electrobalance dry weight determinations by William T. (Bill) Peterson.[19] Permission, Inter-Research, *Marine Ecology Progress Series*.

and new reproduction as quickly as possible. There are exceptions, breaks for diapause that are always spent in a place relatively safer than the feeding habitat. Likely the upper limit is exponential growth during a few stages or at least some, so the means to reach that limit are selected. A very large part of the need for haste must come from predation. One way for very small stages to avoid beasts that could eat them is to grow bigger than grabbing limbs, fangs (arrow worms), and sucking mouths. Growth also adds speed and distance to escape jumps. Copepods just do not linger in youth.

Molting and a Visit with Catherine Johnson

At a given life stage, any arthropod has a nearly fixed size set by its tough, if thin, chitin exoskeleton. Therefore, advancing from one life stage to the next requires (1) separating epidermal cells from the exoskeleton, (2) a new skin to be formed under the old, (3) shedding of the old, and (4) expanding the new skin to a larger size. A transmission electron micrograph (Figure 12.14) of a glancing section through a copepod's exoskeleton shows the interlacing of the chitin fibrils. It also shows distinctive heavy-metal staining between an outer layer and an inner one. Those are laid down and removed at different phases of molting. Before anything

Figure 12.14 Transmission electron micrograph of an oblique section through copepod exoskeleton. The angle spreads the chitin polymer and basal membrane layers. E, epidermal cell cytoplasm; I, inner chitin layer; O, outer chitin layer showing laminae and their curving ("helicoidal") cross links; arrow, three-layered basal membrane of the epidermal cell. Scale uncertain (from my first-ever TEM session). The unspread exoskeleton thickness is about 1 μm (see Figure 12.19, d). By the author.

else the epidermal cells grow some narrow tubes reaching quite far into the tissues. In those, the setae are formed. The tubes first secrete the tips of all the new setae (usually half the length or more). Some sort of separation lubricant is coated on, and then the tubes secrete the proximal sections. Before molting proper comes *apolysis*, the cellular epidermis withdraws slightly but definitely from the old exoskeleton, and the inner layer of chitin is digested off from it. Then a whole new chitin layer, destined to be the outer layer in the next stage, is formed on the epidermis surface. Since the new epidermis must have more surface area than the old, bigger epidermal invaginations into surface tissues, some with complex folds, generate that new area.

Next comes molting proper. In copepodites a crack opens between the upper carapace and the ventral plates bearing the feeding limbs, and a considerable volume of water is imbibed (yes, the copepodite drinks). That raises the pressure inside the new exoskeleton. The antennules and feet are pulled back, and the newly clothed copepod emerges head-first from just inside the rostrum. It sort of shimmies, pulling the antennules and other limbs out of their sleeves. Perhaps there is some lubricant, or perhaps just water, between the old and new exoskeleton. Now free, the copepod is not yet ready to swim. More drinking stretches out the folds down in those invaginated pockets. The new setae, attached at their bases to the new exoskeleton, are extruded (Figure 12.15), another effect of the internal pressure from drinking. During this swelling, considerable shaking can take place. It is suspected that could make molting events obvious to predators

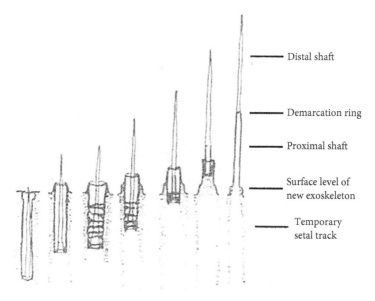

Distal shaft

Demarcation ring

Proximal shaft

Surface level of
new exoskeleton

Temporary
setal track

Figure 12.15 Diagram of new seta extrusion. At left, distal setal shaft and surrounding setal base are ready for extrusion. Shown as a sequence left to right, internal pressure drives the seta out of the epidermis, through the surface chitin of the new exoskeleton to which it remains attached. From Barbara Dexter.[20] Permission, Oxford University Press.

in the neighborhood. It is fairly brief, a few minutes, and the now larger copepod swims off. The epidermis at this point secretes a new inner layer of chitin between itself and the outer layer secreted earlier. Enzymes both polymerize the chitin into long, starch-like chains and cross link them for stiffening.

Timing of all these molting operations is regulated by a series of *ecdysones*, steroidal hormones that direct epidermal cells one by one to initiate and complete the steps of the cycle. Ecdysones have been characterized from insects and large crustaceans. However, more than 70 years ago, extracts from copepod brains were shown by Carlisle and Pitman[21] to include neurohormones that, as do shrimp ecdysones, induce chromatophore activation when injected in shrimp. They claimed in their one-column note to *Nature* to have microscopic (was it a TEM?) evidence of neurosecretory cells in the anterior-lateral part of the brain with microtubular connections to the hemocoel. So far as I can tell, they never published that evidence. Maybe they were citing Esther Lowe.

Catherine Lynn Johnson (Figure 12.16), a Scripps Institution of Oceanography graduate student in the 1990s, thought that perhaps molting hormones could be the source of signals to the tissues of diapausing *Calanus* (more on diapause in Chapter 15) to emerge from rest at depth. Since molting is a key part of ending

Figure 12.16   Photo of Dr. Catherine Johnson on the hill above the Bedford Institute of Canada's Dept. of Fisheries and Oceans. Bedford Basin is in the distance. Photo taken in 2020 by Gordana Lazin. Included here with Dr. Johnson's permission

that resting stage, ecdysones must have some part in its termination. Perhaps Johnson got the basic idea from Carlisle and Pitman's[21] title. Catherine grew up in Arlington, Massachusetts, a suburb of Boston. Her mother loved being at the beach, and organized vacations on Cape Cod every summer through Catherine's youth. Swimming and snorkeling there, she developed an interest in marine life, which was further fostered by her high school biology teacher, Don Bockler. As a student at Harvard (perhaps not wanting to go far from Boston), she included a biological oceanography course taught by Jim McCarthy and one of the Sea Education Association oceanography courses taught aboard a sailing ship departing Woods Hole. Those courses have been initial at-sea experience for several prominent oceanographers.

Rather than moving on after graduation, Catherine worked at Harvard under Joseph Montoya for five years. Montoya was (and still is, now at Georgia Tech) interested in nitrogen fixation by phytoplankton, among other biogeochemical subjects. Catherine was part of his team working on N_2-fixation by the filamentous cyanobacterium (blue-green alga) *Trichodesmium*.

"Tricho" is an important near-surface component of tropical and subtropical ocean ecosystems. Catherine participated in the Caribbean Sea fieldwork and on a cruise from Barbados to the Cape Verde Islands. Eventually she decided to get the credentials to lead her own research, and applying to graduate schools, she was accepted at Woods Hole Oceanographic Institution (WHOI) and Scripps. Tempted by WHOI, she was advised by Doug Capone, another prominent N_2-fixation researcher, to spread her experience outside New England. Scripps tends to accept students as an institution with at least initial fellowships, letting them find advisors when they arrive according to their interests. For Catherine, David Checkley was a good match, and he agreed to advise her.

At the time, Checkley was working with postdoctoral scholar Kenric Osgood on the population of *Calanus pacificus* C5 in diapause and clustering just above the anoxic bottom water in a deep basin between Santa Barbara and Santa Barbara Island in the so-called California borderlands. Fairly soon after Catherine arrived at Scripps, Dr. Osgood moved to a position at the US National Marine Fisheries Service. Catherine joined the work on *C. pacificus* diapause. She found sampling opportunities to locate other stocks of resting fifth copepodites in the borderlands area, stocks closer to Scripps, so it was less difficult to sample and get specimens to her laboratory for study. After some preliminary studies and thinking, she came up with the notion there would be clues to its diapause regulation (see Chapter 16) from study of its late-stage molt cycles. She credits Dr. Checkley, whose own dissertation work was at Scripps on the reproduction of nearshore populations of *Paracalanus*, with giving her great freedom to design studies for work on that notion, and he found funding to enable that.

Along the way, Catherine produced a remarkable demonstration of the timing of elevated ecdysone "titers" in the tissues of laboratory-reared *Calanus pacificus*.[22] One major advantage of cultures, compared to working with collections from the field, is that you can know reasonably exactly how old your specimens are, how much they have had to eat, the rearing temperature and salinity, the lighting regime. From her cultures, started with eggs, she sorted out large numbers of newly molted C5 (age-within-stage was zero days) and maintained them on adequate rations. Then every day or so she froze a sample, and on those she ran radio-immuno assays (RIA) for ecdysone "titers" of numerous individuals. Substantial cultures were grown at 12° and 16°C. Also, following her interest in diapause, Catherine measured ecdysones in *C. pacificus* C5 collected from a stock resting in a deep basin near Scripps, the San Diego Trough.

What is RIA? It is possible for a rabbit or horse to generate antibodies for the steroid core of most ecdysone molecules by injecting those into their bodies. The resulting antibodies can then be purified from samples of their blood and dissolved in a buffer solution. Their capacity to combine with ecdysones generally can be used to quantify those. Biologist Ernie Chang, of the University of

California Bodega Bay Marine Station, was studying crustacean ecdysones, and he had a stock, produced by Walter Bollenbacher in North Carolina, of stable, rabbit antibody to the ecdysone core molecule. Over several decades Dr. Bollenbacher detailed the endocrinology of *Manduca sexta*, a hornworm moth, whose larvae (up to 7 cm long) eat tobacco leaves. Dr. Chang was helpful generally to Catherine, and in particular he shared the antibody. That was the basis of her test system.

To measure total ecdysone in *one copepod* (emphasized because even at peak molting activity concentrations are all but vanishingly small), she ground it up in some buffer, spun out the tissue and skeleton bits, then extracted the steroids (there is a process) from the supernatant. She added those to the system, and she also added a standard amount of synthesized, tritium-labeled ecdysone. After allowing time for a competition reaction, some of the antibody active sites were occupied by copepod ecdysone, others by the ^{3}H-labeled version. Next, she dried the sample (removing the uncombined antibody). Finally, re-dissolving the antibody-ecdysone combination in scintillation fluor (which flashes when a β-particle passes through it), she measured the radioactivity again by scintillation counting (in a machine with photocells counting fluor flashes). The idea is that the *fewer* the ^{3}H-decay counts, the more ecdysone was in the copepod extract. Catherine made a standard curve with amounts of unlabeled ecdysone varying from 5 to 4,000 picograms and her standard amount of ^{3}H-ecdysone. For a copepod sample she compared the ^{3}H counts to that curve. A picogram, by the way, is one-trillionth of a gram (12^{-12} g), a miniscule amount. Sensitivity in that range is the virtue of RIA.

Complicated? A lot of steps? Yes, science rarely renders new information without hard work (I do keep saying that). Here's the result for the C5 raised in culture (Figure 12.17).

In all arthropods, the ecdysones in the signaling sequence are distinguished by the number and locations of hydroxyl (-OH) groups attached at sites on the steroid core (basically, cholesterol). Thus, it would be fun to know when each ecdysone appears as preparation for copepod molting proceeds. For example, which ecdysone (they were once labeled α-, β-, γ-, . . . ecdysones) signals for apolysis in the mandible, the beginning of tooth formation, whatever area is next for apolysis in developing a new, larger exoskeleton. The amount of work required for those determinations would likely make learning them a waste of resources on science fun.

The result for deep-dwelling C5 from San Diego Trough was ecdysone titers on the order of <2.0 pg, which must have been extrapolating below Catherine's calibration range. Low titers are not surprising, but it is good to have them demonstrated. Yes, those *Calanus pacificus* specimens were "unbusy," at the time they were captured, delaying their next molt.[22] Catherine wrote several other significant papers based on her work at Scripps on diapause in *C. pacificus*.

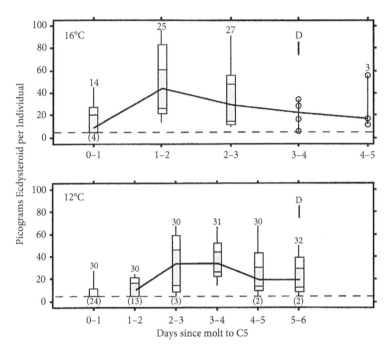

Figure 12.17 Titers of ecdysone in C5 individuals of *Calanus pacificus* from culture. The line follows medians; gray bars are titers in half the specimen numbers above each bar; the white boxes total another quarter; the whiskers show the range. **D** marks the age at molting to C6 at the rearing temperature. The delay in C5 seen in *C. marshallae* (Figure 12.13) was not seen (a difference between species?). Molting hormones appear after approximately 1/3 of the stage duration, peak, then taper off before the actual molt. From Johnson.[22] Permission, Inter-Research, *Marine Ecology Progress Series*.

Catherine had a short tenure at Scripps as a postdoctoral fellow, around 2001. In that interval she initiated a relationship with fellow postdoc and Canadian citizen Claudio DiBacco. His training was with a range of larval forms, including those of copepods parasitic on fish ("fish lice"). Shortly before leaving Scripps they married, but their career tracks diverged geographically. He got a job at the University of British Columbia, and she got a postdoc in New Hampshire to work with Jeff Runge (from Chapter 11) and modeler Wendy Gentleman. That was participation in a data synthesis from a half-decade of cruises and experiments over Georges Bank of New England, sponsored jointly by the National Science Foundation and the National Ocean and Atmospheric Administration in the 1990s. The synthesis actually considered all Gulf of Maine and regional studies. The foci were *Calanus finmarchicus*, other copepods, and larval cod.

A trend through much of her later work started then with a modeling study she published with Jamie Pringle and Changshen Chen.[23] They worked out some of the expected changes in *Calanus* stocks in the Gulf of Maine, based on transfer into it of distinctive stocks from flow along the Nova Scotia shelf. They applied a combination of individual-based modeling of the *Calanus* population life cycle progressions with models of regional flow along the shelves and exchanging with water in the deep gulf basins. Their model of flow into and out of the gulf at in-termediate depths agreed with observations that the size-distribution of late copepodites at rest changed over the winter. The resting stock established in the gulf locally during early summer probably was replaced by smaller individuals that had entered diapause to the northeast outside the gulf.

In 2004 Catherine moved to Vancouver, British Columbia, to be with her husband, and they started a search for work in the same vicinity. That succeeded; they both obtained positions at the Bedford Institute of Oceanography, a Department of Fisheries and Oceans (DFO) laboratory in Dartmouth, Nova Scotia, the entire width of Canada to the east. Dartmouth is on the eastern shore of Halifax Harbour, now governmentally part of Halifax. So they moved to another of Canada's big cities. At present Catherine runs a laboratory group investigating whatever the DFO thinks needs study. She manages to generate wide swaths of good science in addition to completing those assignments. She is working with Dalhousie University's Helenius Laura on the role of essential fatty acids in copepod nutrition, with Bedford's Hui Shen on oil spill movement and internal waves in the Gulf of Maine, on general reporting on ocean conditions in the eastern Canadian coastal ocean, on the shifting of North Atlantic right whale feeding areas (eating diapausing *Calanus*) with Stéphane Plourde of DFO's Maurice Lamontagna Institute (located on the St. Lawrence estuary), Jeff Runge of Gulf of Maine Research Institute, and others. It's a dizzying array of varied subject matter and involves much personnel management. She has joined Dr. DiBacco as a Canadian citizen, but remains a US citizen as well. On interview by internet connection during the 2020 pandemic quarantine, she seemed happy with life and with science as part of it.

Back to Teeth: How Opal Crowns Are Formed

Once Barbara Sullivan rediscovered that copepods have siliceous teeth (Chapter 2), the obvious question was: How do they install essentially stony objects in jewelry bezels on the *outside* of their exoskeletons? We knew how setae form inside the old exoskeleton and are then extruded to the outside on molting (Figure 12.15). However, it was difficult to see how that could work for something not actually part of the new chitinous sheaths being formed before each copepodite molt. Years after Sullivan's discovery, I had a chance in 1988 to work a few months at the plankton laboratory of the Ocean Research Institute in Tokyo. They had an interference phase-contrast (Nomarski) microscope with a camera and many samples of the very large copepods *Neocalanus plumchrus* and *N. cristatus*. With those I was able to examine dissected jaws effectively and see generally how the process (the "morphogenesis") works.[24] Nomarski employs optical tricks to make small changes in refractive index across a microscope image show up as lines and shapes in something otherwise almost transparent.

After a molt, the mandibular gnathobase, a rather long "endite" (side bulge) from the basis segment of the mandible, has a big open extension of the hemocoel beneath the epidermis, proximal to the newly exposed teeth (Figure 12.18, A). Tissue fluid during molting expands out into all of the limbs, and in the jaw the resulting "bubble" persists for quite a long "postmolt" phase of the molt cycle.

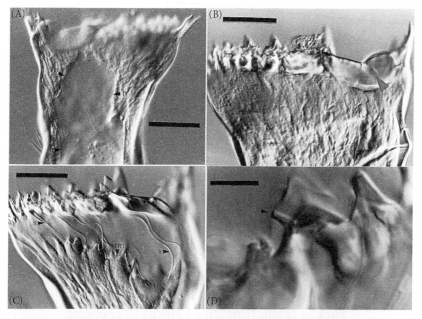

Figure 12.18 *Neocalanus cristatus*, C4, jaw blades. (**A**) postmolt phase; h arrows at hemocoele outline; scale 45 μm. (**B**) molt-cycle interphase, tissue smooth; the red arrow indicates the extra thick chitin supporting some of the teeth and stiffening this diastema; scale 65 μm. (**C**) after apolysis, molds (**tm**) formed for C5 teeth, salivary ducts (**S**) continuous through molds, across apolysis space and opening on C5 teeth; scale 70 μm. (**D**) salivary duct opening (arrow) on side of C5 tooth; scale 10 μm. Originals, previously published.[24] Permission, Springer Nature, *Marine Biology*.

In fact, in C5 *Calanus* and *Neocalanus* species the postmolt lasts through all of the early months of their deep-dwelling diapause (Chapters 15 and 16). Next in the cycle is an "intermolt" phase during which rather undifferentiated tissue fills the space beneath the teeth (Figure 12.18, B).

As preparation for molting begins, that tissue becomes striated, with differentiated columns beneath each tooth outside the exoskeleton. Then apolysis in the jaw comes earlier than over the rest of the body. Presumably one of the distinctive ecdysone molecules signals cells under just that surface to pull back. After apolysis, a tooth-like shape forms in the epidermal surface below the apolysis space (Figure 12.18, C). Looking at many specimens for progressive stages, I found jaws where that tooth shape was extruded well into the space, but with miniscule tubes running across the space, through the old exoskeleton and opening on the old teeth (Figure 12.4). I call them "salivary ducts." Evidently, many kinds of copepods are capable of secreting something into the mouth through the teeth, perhaps into prey. It remains to identify that secretion, but the glands are derived from cilia-bearing cells well back in the gnathobase column (not shown here), and the ducts are of a type formed by hollowing out cilia.

The Nomarski phase microscopy also showed that the tooth molds connect at their dorsal and ventral (see Figure 2.3) sides (Figure 12.19, A) with ducts from the proximal end of the gnathobase. Using a lower power lens showed the origin of those ducts in a cluster of bulb-like "organs" (Figure 12.19, B). I am convinced those supply molecules, yet to be characterized, that can off-load silica (SiO_2) as opal in the obviously tooth-shaped molds.

With the help of Oregon State's electron microscopy team, Al Soeldner and Chris Weiss, I got a better look at some of the structures.[24] Fifth copepodite jaws were identified by Nomarski (I had my own then) as being at likely tooth-development stages of interest, stained with heavy metals, embedded in plastic ("Spurr's medium") and ultra-thin sectioned with diamond knives. My attempts to guide the microtome cuts with light microscopy worked occasionally. Of first interest were the central cells, which produce masses of fibers that disperse into a tooth shape against an outer membrane of some sort, apparently filling it out (Figure 12.20, A and B). The exact character if this "central cell" was not clear in the pictures, but the fibers are delivered by microtubules through a gap in the chitin-secreting, epidermal cells surrounding the center of the tooth mold. As they emerge from the central cell, the fibers appear to be "spraying" into the mold form.

Once numerous fibers are present in the mold (Figure 12.20, A), epidermal cells produce darkly staining blobs of secretion (Figure 12.20, B) apparently open to the base of the tooth mold. Those could be precursors of the resilin protein identified later by Jan Michaels[25] and thought to be a cushion between the rock-hard teeth and the flexible chitin. Adding strength to the tooth-edge, some of the chitin there (Figure 12.20, B, arrow) is much thicker than over the rest of

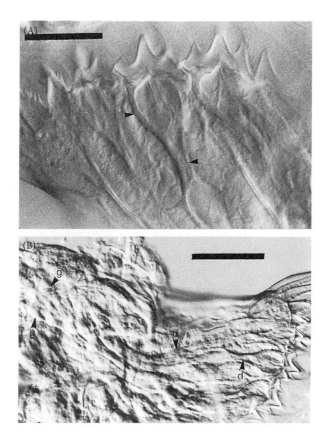

Figure 12.19 *Neocalanus plumchrus,* C4, mandibular gnathobase with new teeth forming. (**A**) The molds for teeth are obvious from their shapes; central cells (red arrow) at the mold bases supply them a fibrous matrix; ducts from the silica glands (black darts) attach at the dorsal and ventral corners; scale 40 μm. (**B**) Silica glands (**g**), ducts (**d**), central cells and tooth moles, left to right, scale 80 μm. Interference contrast micrography. Original, previously published.[24] Permission, Springer Nature.

the body. In finishing stages, the silica glands do their thing and the molds fill in, somehow directed into sequence from the outer surface (Figure 12.20, C) toward the center so that the fully silicified tooth can rest on new chitin secreted by epidermal cells expanding in across the open space at the apex of the central cell (Figure 12.20, D). A scanning electron micrograph (Figure 12.21) shows the edge of a jaw from which one tooth has fully and another partially exploded in the microscope's vacuum chamber (opal is *hydrated* silica). It shows the shape of the bezels that hold the teeth by wrapping around their edges.

There are puzzles left in this tale of bio-mineralogical morphogenesis. I did show, by removing all of the silicic acid from seawater and then raising *Acartia*

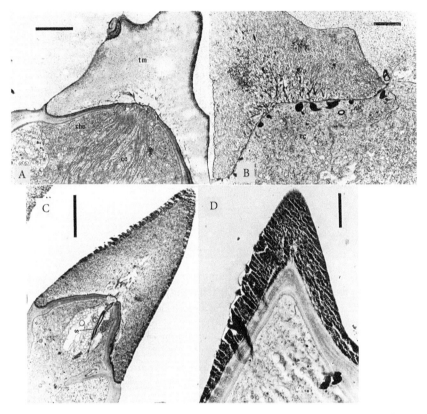

Figure 12.20 *Calanus finmarchicus*, C5. (**A**) tooth mold beginning to fill with fibers; cc, central cell with microtubules carrying proteins for fibers; **chc**, chitin depositing cell; **tm**, tooth mold; scale, 3.5 μm. (**B**) tooth mold with abundant protein fibers, adjacent epidermal cell with dark-staining secretions; **ec**, epidermal cell; scale, 2 μm. (**C**) tooth beginning to silicify (opal is electron dense, so black) from the tooth's outer surface inward; the opal shatters when the microtome knife cuts through; sideways **s**, salivary duct; scale, 6 μm. (**D**) section through the side of a new tooth; silicification is complete down to the new chitin; note opal shattering; scale, 2.6 μm. From original transmission electron micrographs. Original, previously published.[24] Permission, Springer Nature.

tonsa in it, using flagellate food very low in silicon, that copepods cannot make teeth without silicon available.[26] Probably they can get some by eating diatoms, but when there are diatoms, there is also sufficient dissolved silicic acid. Dennis Phillips and Mark Brezinzinski[27] found a way to make useful quantities of a radioactive isotope of silicon, ^{32}Si. Some autoradiography and Si-tracing studies might be done with it; its half-life is 153 years. Kurt Tande leads a business in Norway (*Calanus,* AS) that extracts storage oil from tons of *Calanus* collected by

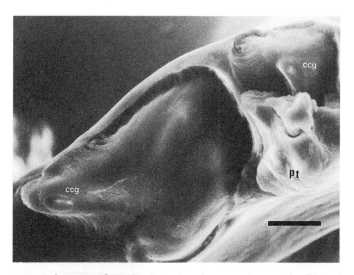

Figure 12.21 *Calanus pacificus*, C5. Scanning electron micrograph of jaw edge from which a tooth and part of a tooth (**Pt**) have exploded in the scope vacuum. **ccg**, central cell gap in chitin surface under a tooth. Scale, 4 μm, Original, previously published.[23] Permission, Springer Nature.

fishermen, producing a dietary supplement for human beings. Someone could obtain the protein matrices of *Calanus* teeth from the resulting, lipid-free, copepod meal. Those could be characterized as done by Nils Kröger, et al.[28] for silifin, the matrix protein in diatom frustules. Silifin is a seriously unusual protein; copepod tooth proteins will be also.

References

1. Marshall, S. M., & A. P. Orr (1954) Hatching in *Calanus finmarchicus* and some other copepods. *Journal of the Marine Biological Association, U.K.* **33**: 393–401.
2. Davis, C. C. (1959) Osmotic hatching in the eggs of some fresh-water copepods. *Biological Bulletin* **116**(1): 15–29.
3. Peterson, W. T. (1980) Life History and Ecology of *Calanus marshallae* Frost in the Oregon Upwelling Zone. PhD dissertation, Oregon State University.
4. Björnberg, T. K. S. (1998) Description of Canuellid nauplii of São Sebastião Channel (Southeast Brazil). *Nauplius, Rio Grande* **6**: 155–160.
5. Bruno E., C. M. Andersen Borg, & T. Kiørboe (2012) Prey detection and prey capture in copepod nauplii. PLoS ONE **7**(10): doi:10.1371/ journal.pone.0047906.
6. Fanta, E. S. (1982) Anatomy of the nauplii *of Pseudodiaptomus acutus* (Dahl) (Copepoda, Calanoida). *Arquivos de Biologia e Tecnologia. Curitiba* **25**: 341–353.
7. Fahrenbach, W. (1962) The biology of a Harpacticoid copepod. *La Cellule* **62**: 303–376.
8. Valderhaug, V. A., & H. G. Kewalramani (1979) Larval Development of *Apocyclops dengizicus* Lepeshkin (Copepoda). *Crustaceana* **36**(1): 1–8.

9. Berge, J. A., & V. A. Valderhaug (1983) Effect of epibenthic macropredators on community structure in subtidal organically enriched sediments in the inner Oslofjord. *Marine Ecology Progress Series* 11: 15–22.

10. Ferrari, F. D., & J. Ambler (1992) Nauplii and copepodids of the cyclopoid copepod *Diothona oculata* (Farran, 1913) (Oithonidae). *Proceedings of the Biological Society of Washington* 105(2): 275–298.

11. Dahms, H.-U., & C. H. Fernando (1993) Naupliar development of *Phyllodiaptomus annae* (Apstein, 1907) (Copepoda: Calanoida) from Sri Lanka. *Zoological Journal of the Linnean Society* 108(3): 197–208.

12. Madsen, S. D., T. Nielsen, & B. W. Hansen (2001) Annual population development and production by *Calanus finmarchicus, C. glacialis and C. hyperboreus* in Disko Bay, Western Greenland. *Marine Biology* 139(1): 75–93.

13. Heinle, D. R. (1969) Effects of Temperature on the Population Dynamics of Estuarine Copepods. PhD dissertation, University of Maryland, College Park.

14. Peterson, W. T., & S. Painting (1990) Developmental rates of the copepods *Calanus australis* and *Calanoides carinatus* in the laboratory, with discussion of methods used for calculation of development. *Journal of Plankton Research* 12(2): 283–293.

15. Miller, C. B. (1993) Development of large copepods during spring in the Gulf of Alaska. *Progress in Oceanography* 32: 295–317.

16. Hairston, N. B. Jr., & S. Twombly (1985) Obtaining life table data from cohort analyses: a critique of current methods. *Limnology and Oceanography* 30: 886–893.

17. Sabatini, M., & T. Kiørboe (1994) Egg production, growth and development of the cyclopoid copepod *Oithona similis*. *Journal of Plankton Research* 16: 1329–1351.

18. Miller, C. B., J. K. Johnson, & D. R. Heinle (1977) Growth rules in the marine copepod genus *Acartia*. *Limnology and Oceanography* 22(2): 326–335.

19. Peterson, W. T. (1986) Development, growth and survivorship of the copepod *Calanus marshallae* in the laboratory. *Marine Ecology Progress Series* 29: 61–72.

20. Dexter, B. L. (1981) Setogenesis and molting in planktonic crustaceans. *Journal of Plankton Research* 3(1): 1–13.

21. Carlisle, D. B., & W. J. Pitman (1961) Diapause, neurosecretion and hormones in copepods. *Nature* 190: 817–828.

22. Johnson, C. L. (2003) Ecdysteroids in the oceanic copepod *Calanus pacificus*: variation during molt cycle and change associated with diapause. *Marine Ecology Progress Series* 259: 159–165.

23. Johnson, C. L., J. Pringle, & C. Chen (2006) Transport and retention of dormant copepods in the Gulf of Maine. *Deep-Sea Research II* 53: 2520–2536.

24. Miller, C. B., D. M. Nelson, C. Weiss, & A. H. Soeldner (1990) Morphogenesis of opal teeth in calanoid copepods. *Marine Biology* 106: 91–101.

25. Michels, Jan, J. Vogt, P. Simond, & S. N. Gorb (2015A) New insights into the complex architecture of siliceous copepod teeth. *Zoology* 118: 141–146.

26. Miller, C. B., D. M. Nelson, R. L. Guillard, & B. L. Woodward (1980) Effects of media with low silicic acid concentrations on tooth formation in *Acartia tonsa* Dana (Copepoda, Calanoida). *Biological Bulletin* 159: 349–363.

27. Phillips, D. R., & M. A. Brezinzinski (1998) Production of high specific activity silicon-32. Los Alamos technical report (LA-UR-97-3612). doi:10.2172/563829.

28. Kröger, N., R. Deutzmann, & M. Sumper (1999) Polycationic peptides from diatom biosilica that direct silica nanosphere formation. *Science* 286: 1129–1132.

13

Sex Determination in Copepods

Sexual Reproduction Is Persistent Despite Its Negative Consequences

Animal and plant cells with nuclei, and with molecular recipes (DNA sequences) wound into chromosomes for enzymes and other proteins, divide by *mitosis*. Any biology book will have a description with diagrams, and the Wikipedia article is excellent. I am going to assume you know the story already or can find it readily. The nuclear membranes disperse. Each chromosome is duplicated and the "daughter chromosome" clusters are moved to opposite sides of the cell. The cell membrane constricts between the two clusters, squeezing two new cells apart, then new nuclear membranes form around their chromosomes.

In making gametes, cells called oögonia and spermatogonia divide by an even more elaborate process termed *meiosis*, producing eggs (oöcytes) and sperm. It is a minuet of chromosomes based on a sequence of two mitoses. Again, if it has gotten foggy since high-school biology, you could look it up. The key point, including for copepods, is that most animal cells, except gametes, have *two* of each different chromosome: one originally supplied in the egg by the animal's mother, another from the fertilizing sperm supplied its father. In the earliest phase of meiosis, producing eggs or sperm, those mother-father pairs (homologues) "recognize" each other, and then they align and attach base-for-base all along their constituent DNA. Finding each other and attaching is termed *synapsis*, which occurs along the nuclear membrane to which the chromosomes' tips attach. Those ends are moved along by contractile microtubules until the right pairings are found.

When the chromosome strands of each gametocyte are duplicated, the meiotic "copying" enzymes can identify missing and broken bases thanks to the synaptic pairing. There are many ways such deletions and molecular damage can happen, from oxidative damage to cosmic ray strikes and poisons. Rather than just skip over these damage points, the meiotic enzymes can check what belongs at the break from the bases on the aligned chromosome. If that has a guanine (G) at the site, the enzymes put a G at that gap in copies of both the good strand and that with the error. Also, A for A, T for T, C for C, as it moves past a gap or damage, then on to the next base. Because there are two copies, the gene sequences can be "proofread" and corrected as gametes are generated. There is

Oar Feet and Opal Teeth. Charles B. Miller, Oxford University Press. © Oxford University Press 2023.
DOI: 10.1093/oso/9780197637326.003.0013

a substantial literature about the rates at which repairable sequence errors and worse errors occur. The rates of damage to DNA are substantial, but many errors can be corrected in gametes.

That's it! Early forms of meiosis-like DNA repair appear in bacteria, even archaea. In 1977 a clinical pathology professor, Carol Bernstein, wrote an article titled "Why Are Babies Young?"[1] Her answer was that unlike DNA breaks in legions of mitotically produced somatic cells, gametes are only ready for mating after their DNA has been proofread and edited. Most of their DNA errors have been corrected. She, her husband Harris (also a pathologist), and biologist Richard Michod have written whole shelves of papers and books about this.[2] Yet biology books mostly emphasize that meiosis also generates genetic variation between one generation and the next, which is the raw material needed for evolution by natural selection. That's true, too, but not so hidden within it is this: once some members of a population chance upon nearly perfect adaptation, it will all but certainly be lost in the next round of random assortment of different maternal and paternal chromosomes to their gametes. By and large habitat conditions do not change so fast that this is not a serious *cost* of sex. With those random assortments of genes from the parental pairs comes a regression toward the population mean for many characters. Meiotic crossovers make this more intense. Thus, near perfection cannot be sustained in the face of the need to proofread the genes at each generation.

There are other drawbacks to the common two-sex setup for meiosis. Some are minor but often cited; for example, reproduction can be delayed by mate-finding and its failures. Moreover, in a sense it is a waste to feed and grow males that do not produce or gestate zygotes ready to take their place in the cycle of development and reproduction. Many plankton animals, *not copepods*, have overcome that through hermaphroditism, all individuals serving both sexual roles as they grow: arrow worms, pteropods, salps, and others. Cladocerans can run off numerous generations by parthenogenesis, but separate sexes develop seasonally or otherwise, and meiosis then gets to repair accumulating gene damage. There are even more drawbacks to reproduction via separate sexes with haploid gametes. A thorough (if obscurely presented) analysis was provided in 1975 by the late George C. Williams.[3] That reads as if he had no clue about what actually goes on in meiotic chromosome pairing (to be fair, until later nobody did).

Accepting the Bernstein explanation for the general persistence of sex, what about sex ratios in adults? There are reasons they can vary from the general expectation that in reasonably large populations at the stage of reproductive maturation, ratios will be close to 1♀: 1 ♂. Many populations, however, have ways around that when other ratios are advantageous. Sex determination in copepods provides examples both of nearly equal numbers of females and males and of skew in one direction or the other.

Sex Determination

At least in serious zoology, sex determination does not refer to determined intent for copulation on the part of either males or females. It refers to the means by which a fertilized egg (a zygote) is put on course to complete development as one or the other of those sexes. Basic knowledge of the human mode of sex determination is widespread: the chromosome pair given number 23 (of 23 total pairs) comes in two distinct forms, X and Y. Female zygotes have two X versions; male zygotes X and Y versions. The Y is quite simplified, but it carries "trigger" genes that direct development toward male characters in babies (gonads as testes, and other genital parts such as scrotum, penis, and some glands) and male secondary sexual characters in adults. In forming eggs and sperm, one chromosome from each of the 23 pairs (two, so "diploid"), is distributed to each of two gametes, then termed "haploid." From the XY pairs of fathers come two versions of sperm, one with X one with Y. Haploid X-sperm fertilizing a haploid X-ovum thus can produce XX zygotes that are female determined, and Y-sperm produce XY zygotes that are male determined. The X chromosomes carry genes for enzymes with various functions, but also including a substantial number of those involved in meiosis, the special cell divisions producing eggs and sperm.

Males in that mammalian system do have an X chromosome, and thus all the genetic information needed for becoming female. In fact, across a wider range of animals than mammals, the basic plan for gonads and gamete delivery is the female one. Development of ovaries and oviduct rudiments is generally the initial basis for any individual's reproductive system. The alternative path to males is to modify those into testes and sperm delivery tubing, the latter setup varies widely across the "tree" of animal kinds.

Although the XX=♀ and XY=♂ "heterochromosome" system is quite consistent in mammals, across the whole array of animals there are many other systems, albeit with a few related genes in common. Moreover, there is little consistency, with wildly different versions even within groups that might be expected to share one scheme or another, flies for example. Heterochromosomes are often either missing or not readily distinguished by microscopic examination of chromosome smears (karyology: a whole subfield of biology). Sex can be determined by genes scattered around the genome, rather than combined in heterochromosomes to assure they "travel together" through meiosis. And which determining genes are translated as enzymes controlling development may depend upon environmental conditions, termed environmental sex determination (ESD). The most studied and cited example of that is the effect of egg temperature during embryo development in turtles and lizards, both reptiles. Most crocodilians determine sex similarly to turtles, but snakes have several different heterochromosomal systems. Even turtles and lizards work it differently:

warmer turtle eggs => female : colder turtle eggs => male
warmer lizard eggs => male : colder lizard eggs => female

Thus, taking turtles for example, the sex for all or most of a clutch of hatchlings will depend upon how high on a beach, riverbank or pond shore the nest is buried. Down low near the cool water, male; up higher and warmer, female. There can be mixed clutches from mid-range temperatures. Triggering temperatures vary among species, and evidently among individuals. Despite extensive studies, the decision processes of females when choosing nest sites remain obscure. There are hypotheses, tests, and modest confusion. I leave it at that, because the subject here is copepods. With ocean warming, more sea turtle nests produce only females.

Environmental Sex Determination in Copepods: In *Acartia*?

Interest in the possibility that sex is environmentally determined in copepods began for me with four more or less separate events. One of my early masters degree advisees was Enrique Carrillo Barrios-Gomez. He came early onto the copepod-culturing scene as scientists, including Edward Zillioux (see Chapter 17), Roger Harris (below) in Plymouth, England, and Michael Mullin with Elaine Brooks at Scripps, learned to maintain live cultures of marine copepods. Enrique cultured *Acartia* from nearshore collections. At the suggestion of another graduate student, he decided to see whether *A. clausi* (later split off as *A. hudsonica*) cultured from the US East Coast could mate successfully with those called by the same name cultured from off the coast of Oregon. Specimens from Vineyard Sound near Woods Hole were air-freighted to him by Peter Wiebe, a planktologist at Woods Hole Oceanographic. Enrique readily got them through several generations, keeping Oregon cultures on separate shelves.

Just as the cultures were ready for the between-oceans cross, there was an emergency in Enrique's family in Mexico, and he headed home to help. I took over to start things. I loaded a cold-room shelf with a 5 x 5 array of beakers with seawater, algae, and one fifth copepodite expected to molt to adulthood. Those would remain unmated until Enrique could return. I did not look closely at the C5, just put them in the beakers and fed them. That set of 25 matured as 24 females and just one male. That struck me as interestingly far from a 1:1 sex ratio. Nothing came of that observation, except that I remember it. Enrique returned. He ran within-ocean crosses that almost all produced young that grew up, and he ran between-ocean crosses that all failed. He wrote his thesis, published a paper about the result,[4] then moved on to further studies elsewhere, a career in marine science and prominence as an educator in Mexico.

Does *Acartia* have ESD? Probably, but it is not clear how it works. A graduate student, Zair P. Lojkovic Burris, working with Hans Dam at the University of Connecticut, now Dr. Burris, reared nauplii of *A. tonsa*, studying what she called their "birth sex ratios."[5] Female *Acartia* do not have spermatotheca, but they retain spermatophores attached by males for a week or more and must retain some sperm after scraping those off. Their eggs are spawned a few at a time (not in big, all at once clutches), and eventually the sperm are gone and females need to mate again.[6] That favors higher proportions of males in the stock than found for many other free-living, marine copepods. For *A. tonsa* during its summer peak in Long Island Sound (off Groton Point, Connecticut) Burris found males averaged 43% of adults. Given that searching males probably have a higher mortality rate, that would seem consistent with chromosomal sex determination insuring 1:1 ratios at maturation. But no. In 2013, Burris raised (in petri dishes) substantial numbers of hatchlings to adulthood from the eggs of 110 field-collected (October) females bearing spermatophores, families of eggs raised separately. Only the 72 families of 20 or more individuals with >60% survival were considered. Sex ratios were all over the range from no males to 100% males (Figure 13.1; all 72

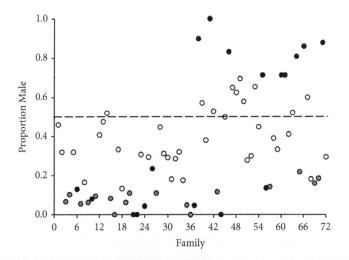

Figure 13.1. Sex ratios of 72 sets of *A. tonsa* siblings, all sets > 20 lab-reared adults. Black and gray dots statistically significant (5%) from 50% males. Gray dots still significant if all rearing mortality would have been male. Apparently sex ratios can take any value. Family number seems to have been arbitrary; there was no actual trend between families before and after no. 36. From Burris and Dam.[5] Permission, Wiley & Sons.

families). Survivorship ranged from 60 to 97%, not correlated with adult sex ratio (I checked her data). The legs of late *Acartia* male and female copepodites differ enough for sex to be determined at C4. At least according to their legs at both stages, the sexes of all adults remained the same as they were at C4. So sex is likely determined by the molt to C4, though not necessarily in the egg.

I have no detailed idea how that would come about. Burris and Dam[5] suggested there could be an effect from female age. Since the females in the experiment came from the field, Burris did not know their age. If there is an aging or other maternal effect, its sex determining mechanism is unknown. Clearly female outcomes are favored, but they can go strongly either way. Sex determination can be very strange indeed.

I had some trouble in the summer of 2019 finding an email address for Zair Burris. So I wrote Hans Dam to inquire whether he knew how to reach her. Other than citations of numerous papers from her time working with Dam, I found only one later reference to her on the web, coauthorship of a poster from California Department of Fish and Wildlife. It was an undated report on the zooplankton of the Sacramento Delta. There had been great changes during a 1972 to 2014, California Fish and Wildlife sampling time-series: mainly shifts to high abundance of invasive, formerly Asian copepods. That was all there was after her University of Connecticut degree. Hans replied with an email that said:

> I am afraid Zair may have dropped out of science recently. Sadly for us, because she is brilliant. She worked on sea spider [pycnogonids] parental care for her MS thesis with Alan Shanks in Oregon. Then, I think because her husband was coming to UCONN, she applied to work with me. Her thesis project was entirely her own. Thus, she educated me on the subject. She finished her PhD in three and a half years.

Maybe Zair is not lost to copepodology (or just marine biology) forever. She eventually replied to an email from me. I had told her about one other found reference, a very old note about score keeping by Zair Burris for women's soccer at Carleton College (my alma mater), so I had assumed she probably attended college there. Her email corrected that. She is a graduate of Grinnell College in Iowa. A soccer team from there might well have played one from Carleton. And yes, she worked for two years in California. She says in an email that: "After I graduated from UConn, I worked for 2 years at the California Department of Fish and Wildlife on the diets of young fishes. That was fun but I missed doing field work and experiments (the job was mostly data analysis)." Indeed, sitting week-in and week-out moving numbers around a computer screen can deaden the interest in almost anything, particularly when the numbers are not really your own.

Figure 13.2 Zair Burris holding some *Laminaria* on an Oregon intertidal flat, about 2010. Photo provided by Dr. Burris.

As of August 2019, Zair was busy raising an infant, her first child, which is an important way to look to the future. From parental care for pycnogonids to parental care. Several pictures of Dr. Burris do show up on internet sites, and she says I can show the cheerful one in Figure 13.2 here.

Environmental Sex Determination in *Eurytemora*?

Shortly after Enrique Carrillo left Oregon, another OSU masters student, Archie Lee Vander Hart, became interested in the general subject of sex determination, and decided to pursue the subject with copepods. So far as he could determine by reviewing the literature, nobody had definitely found heterochromosomes in copepods. The papers dated from well back in the nineteenth century, when the chromosomes of every obscure life form were examined. J. P. Harding, a summer-season colleague of Sheina Marshall and Andrew (A. P.) Orr (introduced in Chapter 15), had claimed in 1963[7] that two somewhat longer chromosomes among the seventeen pairs in *Calanus* were probably heterochromosomes (even though they look alike in his micrographs). Two workers in India, Usha and S. C. Goswami, found them in other genera, including *Acartia*,[8] but Archie was unconvinced. He thought it rather likely that copepods, generally, had genes for sex determination scattered among many chromosomes. He also thought experiments suggested by Curt Kosswig in a 1964 paper[9] might let him demonstrate it. Namely, if sex-determining genes were scattered around, then long sequences of *inbreeding* should sort those genes by multiple rounds of chromosomal crossovers, until those needed for developing as one sex or the other were lost.

So Archie set out to create exactly that: many generations of inbred copepods. Like *Acartia*, the estuarine copepod genus *Eurytemora* (*E. affinis, E. americana*, and so on) can be raised readily in beakers when fed some single-cell algae and perhaps some protozoa or rotifers. They develop and mature rapidly, about three weeks from egg to adult at 16°C, faster at 22°C. *Eurytemora* are egg carriers, so clutches can be counted and followed through development. Archie began with an array of experiments to determine the effects of food availability, temperature, and salinity (8‰, 12‰, 16‰, and 20‰ salinity). All of those salinity levels were readily tolerated with good survival. Effects on sex ratios were never extremely far from 1:1. The C5s can be isolated readily to insure that when they mature, the females can be bred to their brothers. And Archie could add counts from hatchings of several sequential egg sacs to get fairly large families. It was necessary to raise the nauplii to C5 or adult to determine their sex.

He did all that and carried his inbred lines for up to fifteen generations. He examined many family lines with wildly varying sex ratios. A few of them, after nine or ten generations of brother-sister mating, trended to mostly male or mostly female broods, some of them (asterisks in Figure 13.3) with low binomial ("coin-flip") likelihood of occurring by chance. Notice about this experiment that when the point is reached that sex becomes an inbred genetic character, individuals must be "outbred" with a sibling retaining at least enough of the opposite sex genes *to be* the opposite sex.

Perhaps Archie's "Kosswig experiment" on *Eurytemora* did show some inbreeding impacts on sex ratios, indeed suggesting "polygenic" sex determination. Over a decade later, Pauline Vaas and Gerald Pesch published a chromosome study of *Eurytemora affinis*.[11] None of the ten pairs of rather ordinary chromosomes appeared heteromorphic to them. There was no missing member of any chromosome pair in either sex to suggest an XX versus X0 (unpaired X) sex determination mechanism. There were various likely artifacts in some of Archie's results, so he wrote his thesis[10] but did not publish the results. Nobody has repeated his experiments. If anyone is tempted to do so, she should read Archie's thesis first. There are many more pertinent observations there.

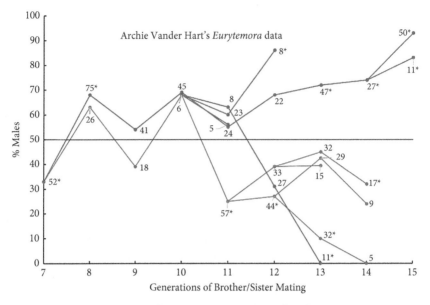

Figure 13.3 Adult sex ratios of successive generations of brother-sister matings in two family lines of *Eurytemora* followed separately after the seventh generation. These are selected as lines favoring female (red) and male (blue) sex ratios. Numbers with the points are the number of adults; stars indicate less than 5% probability of observed %♂, *if actual* ♀:♂ :: 1:1. Redrawn From Vander Hart,[10] which does not list a copyright (Mr. Vander Hart is deceased).

Archie took his MS diploma and left science for theology school in Michigan, becoming a minister. Looking online for Archie's address to ask for permission to publish a few of his numbers, a recognizable picture came up. Unfortunately, the associated article was his obituary. He died in early 2015 after a long career as minister at four Christian Reform Churches (CRC), the last in New Brighton, Minnesota. I had learned at some point that Archie got into difficulties with at least one congregation by insisting on the reality of evolution.

Environmental Sex Determination in *Calanus*?

Abraham Fleminger is mentioned in Chapters 4 and 19. In the mid-1980s Abe's kidneys failed, and he worked hard to finish projects while on intermittent dialysis. He was extraordinary at noticing refined details, one of which is that some (only some, making it harder to notice) female adults in the copepod family Calanidae (*Calanus, Neocalanus, Undinula,* and so on) have male setation

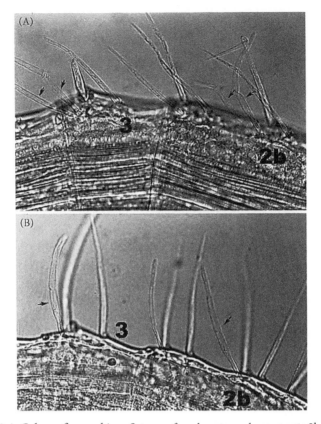

Figure 13.4 *Calanus finmarchicus.* Setae on female antennule segments 2b (2 is two fused segments) and 3. The tiny arrows point to aesthetascs, olfactory sensors.; (a) the quadrithek (male-type) pattern; (b) the trithek pattern. From Miller et al.[12] Permission, Inter-Research, *Marine Ecology Progress Series.*

patterns on their antennules. Males have an additional, larger aesthetasc on every other segment of long stretches of their antennules that females do not have. Abe invented the terms "trithek" (three) for the female pattern and "quadrithek" (four) for the male one. On males those olfactory sensors are certainly involved in detecting male-attracting hormones from females (Chapter 10). Oddly, some females have small but definite versions of these added aesthetascs in the same positions, hence the quadrithek pattern (Figure 13.4).

Abe wrote a paper about this and showed me a draft. So I learned before its publication about his speculation[13] that development in the preceding fifth copepodite stages of quadrithek females with male-type setation had started toward a male outcome, then switched to a female one. The extra aesthetascs remained in place at the final molt, possibly a sign of this sex switching. He speculated that something about habitat conditions was a signal that being female would be more "successful" than being male. That is, sex is likely environmentally determined. Successful would have to be in the sense of natural selection: more matured offspring would be likely to carry a female's genes than would all of a male's offspring, regardless of how many females he supplied with sperm. The switching C5 would not "know" this; it would just be a long-term result of selection based on some habitat clue, whatever that might be.

Environmental sex determination is also likely in species of Calanidae (*Calanus* and near relatives), because when culturing them it is quite difficult to obtain any males at all. Bigger aquaria and particular mixtures of phytoplankton types as food help. For example, Rob Campbell and colleagues raised *C. finmarchicus* in 100-liter tanks in their Rhode Island laboratory in the 1990s, and they did get up to 35% males at maturation.[14] They also made a lot of food available as two species of dinoflagellates. However, families maturing entirely as females have been common in many studies. This suggests an experiment started with *field-collected* specimens. After all, ecology and evolution both occur in natural habitats, and the females with extra antennular aesthetascs suggest sex is determined late in development. Moreover, periods with a lot of available phytoplankton can also have very high proportions of males in net tows, to >70%. What is going on in laboratory rearings? Collect some older copepodites from the ocean, say C3 to C5, sort them by stages into containers, a liter or more, with food surely sufficient, then let them grow and molt to adulthood. Decades back, Sheina Marshall wrote Michael Mullin a letter saying that she tried this, but there was too much mortality overall to conclude that male mortality wasn't responsible for female-biased sex ratios. It was tried at Oregon State in the mid-1990s by undergraduate Michael Green and technician Jennifer Crain. Mortality was again too high for any conclusion. In the late 1990s Xabier Irigoien and Roger Harris succeeded.

Roger Harris and Xabier Irigoien: An Effective Partnership

At the Plymouth Marine Laboratory, on the shore above where Plymouth Sound opens to the English Channel, Roger Harris and Xabier Irigoien (Figure 13.5) worked together from about 1999 to 2000 (and continuing cooperation to about 2006) on questions about planktonic copepods and on other oceanographic puzzles. Roger was a senior scientist, who hired the younger Xabier on receipt of a grant from European Union funds. They were a remarkably effective team, keeping the partnership going from grant to grant, producing many insights.

Roger grew up in Yorkshire coal country, son of two teachers who had met at Oxford. He attended a Quaker-run school through the secondary levels. His

Figure 13.5 Roger Harris (left), and Xabier Irigoien next to Lake Biwa. Both were attending an ASLO Voyage of Discovery meeting in Otsu, Japan, in summer 2012. Photo credit: Marlene Harris. Permission, Harris and Irigoien.

parents taught there, his father chemistry, his mother modern languages. His father's avocation was alpine field botany, so many school holidays were spent in the Alps photographing and collecting plants. The school's biology teacher, V. I. C. Mendham, was particularly influential for Roger, leading field trips to the seashore and various preserves. Roger learned European birds and qualified for bird-banding work. Those interests led to a zoology major at University College London (UCL), which is in the center of the city. Roger says that he studied there at a time when UCL housed research in cell biology, tissue culture and the like, but it also retained a rather traditional course sequence: comparative anatomy, embryology, invertebrates. It was the last of those subjects that captured his imagination. Biological field trips included an Easter-break trip to the rocky intertidal and a visit to the Millport Marine Laboratory, site then of much of Britain's best plankton research thanks largely to Sheina Marshall and A. P. Orr.

Graduating with his BSc, Roger continued at UCL for graduate studies. His advisor, Mary Whitear, suggested that he go where marine biology was strongest in the United Kingdom, the Plymouth laboratory of the Marine Biological Association of the United Kingdom. (the MBA). With some sort of introduction letter in his briefcase, off he went. By means not particularly clear to the recipient, a stipend was arranged and bench space made available. In a 2020 interview, Roger told me that Sir Fredrick Russell, a former MBA director then in his 70s, approached him one day at the library as he tried to get a grip on the current literature. In the course of conversation, Sir Fredrick told Roger that when he had started at the MBA (1924), he could read everything appearing in marine biological publications every month. That might just have been feasible in the 1920s, but not for Roger in the late 1960s; today it would be impossible for anything but a computer indexing system. Later in Roger's tenure at the MBA, Eric Denton became director. Denton was a student of the adaptive ecology of larger pelagic animals: fish and squid, particularly their visual systems. Denton liked to drop in on the scholars at their benches, typically asking questions like "What new things have you learned this week?" I suspect that after the first such bombshell the staff learned to keep an answer to hand.

There was little or no course work involved in the PhD program at UCL, just the requirement to present and defend a dissertation. Roger settled in, selecting a study of the meiofauna of intertidal beach sands. Meiofauna are creatures small enough to live in the interstitial waters between the grains of sediment in beaches, seafloors, and lake bottoms. Prominent among those are harpacticoid copepods, and Roger taught himself how to extract them from sediment, identify them and characterize their ecology. The whole project was completed in about three years. Based on it, he was granted his PhD in 1970 and published the results as five papers in the *Journal of the Marine Biological Association, U.K.*[15]

After those years peering down microscopes at miniscule dead copepods, Roger wanted to work on something large enough to study alive, to work on ecophysiology. He obtained an MBA fellowship to continue on at Plymouth and ran a study of feeding and growth in the tide-pool harpacticoid *Tigriopus brevicornis,* about a millimeter long (Figure 13.6). That is indeed larger than most meiofauna. The work connected him to similar studies in progress at Eric Corner's MBA lab on *Calanus* and other pelagic copepods. Sometime in this interval, a young American woman, Marlene Jenssen, showed up at MBA. She was an undergraduate student at Cornell University in New York, and her professor there had arranged a term of study with intertidal ecologist Allen Southward. She and Roger were introduced and, according to him, moved from that to a wedding in only a few weeks. With their marriage established, Marlene returned to finish her degree at Cornell, and Roger stayed on in Plymouth to finish his *Tigriopus* project.[16]

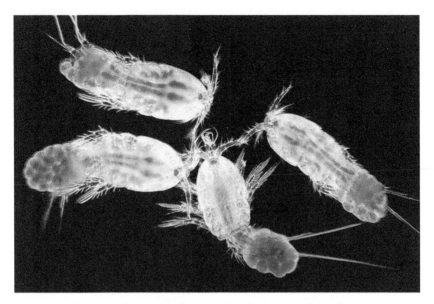

Figure 13.6 *Tigriopus brevicornis* females, all with egg sacs. Shown because *Tigiopus* comes up here and in later chapters. Common in upper intertidal pools, often with elevated (evaporation) or lowered (rain) salinity. Photo by Olivier Dugornay, Ifremer, France, with his permission.

That experience led Roger to ask Donald Heinle whether he could join in his studies of the ecophysiology of estuarine copepods at the Chesapeake Bay Laboratory in Maryland. Heinle agreed, adding him to a study determining whether detritus from seagrasses and benthic algae could be food for estuarine copepods. They worked with *Eurytemora* (a calanoid easy to culture, as noted above) and *Scottolana*, a planktonic harpacticoid. Indeed, the team showed that both species can eat and grow with only decaying bits of marsh plant available, but they only do well when those bits are loaded with bacteria and protozoans that have gotten to them first.[17]

That done, Roger obtained a Royal Society grant to join planktologist Gustav-Adolf Paffenhöfer for some studies at a new German laboratory on the island of Sylt next to the Wadden Sea on the edge of the North Sea. Marlene had joined him by then, and they brought along their first child, Matthew. Culture work with Gus involved feeding and growth studies on species of *Pseudocalanus* and *Temora*, small copepods usually abundant in the North Sea. After a brief spell there, Gus was hired away by the University of Georgia in the United States, and Roger and Marlene returned to Plymouth and settled in.

As for everyone, chance had its part in Roger's career. An oil tanker, the *SS Torrey Canyon* ran aground in April 1967 on a reef near the mouth of the English

Channel, between Lands End in Cornwall and the Isles of Scilly. Beaches in Cornwall and Brittany were oiled. Anticipating more oil spills, labs around the world started to work on how oil dissolving in seawater affects marine life. In the mid-1970s Eric Corner received one of the resulting research grants and invited Roger to join his group at the MBA. Roger put his skills in copepod culturing to work, testing the effects of dissolved hydrocarbon pollutants. Corner brought in other people, too, and Roger began to work with talented colleagues from everywhere. Particularly important in the oil pollutant studies was Viviene Berdugo, trained in Haifa, Israel.[18] As that work wound down Roger, Marlene, young Matt, and younger brother Christopher spent a few months on Canada's Vancouver Island. He worked there with Michael Reeve, George Grice, and Barbara Sullivan on the experimental marine community experiments described in Chapter 2. Roger recalls it as a very pleasant summer outing.

In the 1970s and since, national efforts in marine science in the United Kingdom have been reorganized more than once. Part of the plankton-oriented staff at MBA were moved in 1988 to a newer organization, the Plymouth Marine Laboratory (PML), housed uphill and west of the MBA lab. The move included Corner's group, from which Roger soon became more independent, following his own research priorities. In 2002 PML became a charity, though it was largely funded by grants from the National Environmental Research Council and European Union marine science programs. Roger found resources that way for extensive work on North Atlantic plankton. His astounding list of studies included a 20-year, zooplankton time-series at Station L4 near Eddystone Light, some copepod rearing, feeding, and sex-determination experiments. Eventually he participated in an examination with many colleagues of the balance in the oceans between algal photosynthesis and the respiration of everything from bacteria to whales.[19] Describing all his studies would require another chapter. Numerous postdocs in his lab maintained his education, including a few from the 1990s and later. I can list a few of those from memory: Xabier Irigoien, Catherine Rey-Rassat, Delphine Bonnet, Pennie Lindeque, and Elena San Martin.

Roger and Marlene modified an amazing house in eastern Cornwall, just west of Plymouth. It is located adjacent to the site of a very old, once long-running copper mine. Out behind for a time was the chimney that served the smelter for refining the ore. They started with the shared-wall houses of three miner's families, cutting through the two inner walls to create one comfortable and spacious house with three upstairs sleeping areas. There are gardens, a stone-lined entryway, and a view on the side away from the driveway of a yard filled with birds at feeders. That is bordered by a hedge, beyond which runs the single-line Tamar Valley railroad, dating from 1908. It extends from Plymouth to the town of Gunnislake. Diesel driven passenger cars chug by several times each day. Several generations of planktologists have been guests there, for dinner or brief stays.

The Harrises are warm hosts. If they take you on a nature walk in the vicinity, you learn that the other Dr. Harris, Marlene, is a field botanist who knows the name and ecology of every macroscopic plant. Her dissertation of 1986 examined the ground flora in a selectively harvested Douglas Fir forest near Tavistock (Devon and Cornwall).[20] She found that avoidance of clear-cutting established relatively more diverse ground flora among the big trees, providing both brightly lit and shaded habitats. Her study should be more widely read in the native habitat of Doug Fir on the North American West Coast. There this fir is now found most numerously in tree farms and public forest maintained as tree farms. They are clear-cut in great open gashes visible from space. Martha, my wife, and I have been on forest hikes with the Harrises in coastal Oregon. The comparisons remain fresh to Marlene's mind, and here too she knows every plant by name and provenance. Not long ago, she retired from a career of teaching at a school in the Plymouth area.

The Harris's two sons are long since adults. Matthew is a veterinary pathologist. Roger says that was his ambition from childhood. Matthew and his wife have two sons. Chris became a long-distance hiker on the Pacific Crest Trail from Mexico to Canada, where he met his American wife. Later they hiked the Continental Divide Trail through the Rockies. They have one son and live now in Upstate New York, maintaining a 20-acre farm. Chris is also a software engineer with a scientific computing company. The grandparents are devoted, helping with childcare at Matthew's home well to the north of Gunnislake and traveling often to Upstate New York. Roger retired from PML, then ran off a decade-long tour as the editor of the *Journal of Plankton Research*.[21] In recent years, before and after retiring, he worked a great deal as a British representative to worldwide ocean studies groups (such as ICES and PICES, which have websites). He served two terms as chair of the steering committee of the International Global Ocean Ecosystem Dynamics (GLOBEC) project. The travel opportunities in all that were extraordinary, and the Harrises checked off a very long bucket list of must-see places.

Xabier Irigioen has returned to his Basque origins in northeast Spain, serving at the time of writing as Scientific Director at AZTI-Technalia. He is also a Research Professor supported by Ikerbasque, the Basque Foundation for Science. AZTI seems to be a Basque name, referring to strong capacity for cooperation. Located in Bizkaia (Bilbao) on the Bay of Biscay coast of Spain, AZTI-Technalia is a private, nonprofit organization sponsoring general marine and fisheries research, much of it funded by European Union programs. I have not contacted Xabier for a story of his early years and life outside science. He received a PhD from the University of Bordeaux for studies under Jacques Castel of phytoplankton production and grazing by *Acartia* and *Eurytemora* species in the estuary of the Gironde River. Strong tides and abundant particles coming down the Gironde maintain high turbidity According to Xabier's research,[22, 23] the particle load interferes with feeding in one genus but not the other.

Somewhat like Roger, Xabier gathered research experience at many places, in his case in at least four languages: school days (Basque, Spanish), graduate school (French, finishing 1994), Instituto de Ciencias del Mar in Barcelona (Catalan, 1994–1996), Plymouth Marine Lab (English, 1999–2000) and Southampton Oceanography Center (more English, 2002), AZTI (Head of Biological Oceanography and Living resources, 2004–2011), and director of the Red Sea Research Center at King Abdullah University of Science and Technology (KAUST) in Saudi Arabia (Arabic, 2011–2016). While at KAUST, Xabier established a Red Sea plankton lab in Thuwal near Jeddah. Five years may have been enough of Saudi Arabia, or he had a good offer from AZTI-Technalia to move back and serve as its director. Correspondence with Xabier in 2022 revealed that he recently moved back to a new position at Thuwal. Roger tells me that Xabier is a very quick worker. They would discuss a project or idea, and the next time he saw Xabier he was delivering a draft manuscript.

I am tempted to cite one curious result. Ocean pollution expert Carlos Duarte and Xabier Irigoien, both then working at least part time at the KAUST Red Sea Center, participated in an Arctic Ocean cruise that sampled floating plastic debris.[24] That was done with surface-skimming nets. It turned out that the Gulf Stream carries many plastic fragments from the North Atlantic into the Arctic, particularly along the Norwegian and Novaya Zemlya coasts, leaving it to eddy in the Barents Sea. At some points, as much plastic pollution was found there as in the subtropical gyres. That's a long way from copepods, but please get and use refillable bags and containers. Stop tossing every empty plastic water bottle and yogurt tub into landfills and oceans. Apart from that, I must let Xabier's copious scientific output speak for itself and get on with copepods.

The Successful ESD Experiment: Sex Is Determined in *Calanus* at C5

Back to the experiments finishing rearing in laboratories of field-collected *Calanus* copepodites. The main improvement by Irigoien and Harris was that they held mortality to a few percentage points. For the C4 and younger copepodites, that is harder than it seems. If even one or a few setae are broken off by the quite abrasive net surface during sampling (which does require filtration), tissue fluids clot at the base and involve the epidermis. That scab can cause the new skin to stick to the old, holding the body from leaving it, which is fatal. For the C5 becoming adults, you can still sex the stuck and dying specimens. For C4 and younger, you never get to know how they would have grown up, which in any significant numbers, ruins the experiment. The trick is to closely examine every copepodite, not including those with any broken setae. Xabier and Roger don't say they did that; I just know they had to. Their study[25] was part of an international burst of *Calanus* studies called TASC (Trans-Atlantic Study of *Calanus*)[26] across the North Atlantic in the late 1990s. They worked with *C. helgolandicus* collected at Station L4 out in the English Channel, seaward of Plymouth Sound and near the picturesque Eddystone Light, a navigation beacon built on some wave-washed rocks. They sorted groups of C2–3, C4, and C5 into 5-liter beakers in groups of 20 with replicates. Different containers were given 50 or 300 cells per milliliter of *Procentrum micans*, a dinoflagellate believed to be a good food for *Calanus* to grow on. Water was changed and food renewed regularly. Mortality was 1 to 4 copepodites per beaker; low enough it could not account for the results in Figure 13.5.

Maturation of copepodites collected as C5s in April, May, and June was mostly as males. But nearly all C4 and younger copepodites collected in June and July matured as females. If sex were determined as male in some preliminary way by C4 or early in C5, apparently it could be switched to female by the final molt. In any case, sex appears to be finally determined *late* in development. The difference in food levels appears to have been unimportant, at least in this experiment. However, just being in beakers may have an effect. This experiment screams "repeat this," but of course there's minimal glory in just saying someone else was right.

As part of TASC, May–June rearing experiments were conducted by a team headed by Bent Hygum and Kurt Tande using really big plastic bags suspended in Norwegian fjords at both Bergen and Tromsø.[27] They started with newly hatched *C. finmarchicus* nauplii, letting them feed on nutrient-enriched, but otherwise natural phytoplankton stocks (2–5 mg/m^3 chlorophyll). Maturation produced modest majorities of *males*, about 60% (Figure 13.5). Notice that the data in the figure in respect to sex ratio are somewhat redundant. Once the stocks all matured, by about day 70, the proportion of males could just stay quite high. There were no predators in the mesocosms large enough to eat adult *Calanus*. Even though males have reduced mouthparts, and feed very little, they will live several weeks (even months) if not eaten by chaetognaths, *Euchaeta*, or fish.

Among the TASC projects on the west side of the Atlantic was a five-year, monthly sampling program in winter-spring over and around Georges Bank and out into the Gulf of Maine to the north. A group (known as "the planktoneers") led by Ted Durbin counted copepods in the thousands of samples. Among many

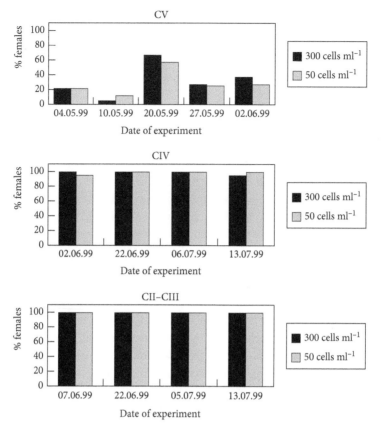

Figure 13.7 Proportions of females from field-collected copepodites C5 (CV), C4, and C2–C3 of *C. helgolandicus* reared to adulthood in beakers with different levels of dinoflagellate cells available as food. Collection dates are in European order: day, month, year. Clearly a majority of the copepodites collected as C5s, matured as males, except on 20 May, while almost all earlier stage copepodites matured as females. That difference is most convincing for the two June collections (last for C5 first for C4), which included all three stage groups. Data from Irigoen and Harris included among coauthors of Irigoien et al.[25] Permission, Oxford University Press.

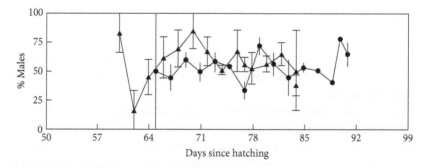

Figure 13.8 Percentage of *C. finmarchicus* males among adults raised from the egg in mesocosms in a fjord near Tromsø, Norway, with abundant phytoplankton for food.[27] Dots are for eggs from females collected at Bergen. Triangles are for eggs from Tromsø females. Confidence bars are standard errors of averages. Figure from Irigoien et al.[25] Permission, Oxford University Press.

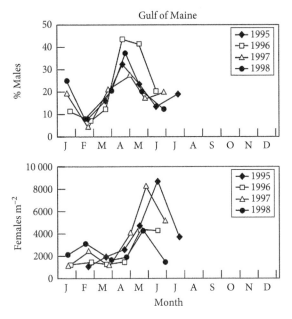

Figure 13.9 Fraction of males (above) in Gulf of Maine waters adjacent to Georges Bank in four winter-spring seasons. The peak of male abundance precedes the peak of female stock abundance (below) by about a month. The diapause stock matures in January as around 12–25% male. In the first new generation, of males make up around 25–45%, then later maturing females and surely predation reduce their proportion in early summer. Data from the Durbin GLOBEC group, shown by Irigoien et al.[25] Permission, Oxford University Press.

other facts, they showed that the *C. finmarchicus* sex ratio varies in a recurring way through those seasons (Figure 13.9).

Using subsamples from the same Georges Bank/Gulf of Maine cruises of 1995 and 1996, Jennifer Crain and I estimated proportions of females with quadrithek aesthetasc counts, and the late Nancy Marcus took some copepodites home to her Florida lab to finish their rearing with different photoperiods. We wrote up all the results together.[12] To get decently estimated proportions of Q and T females, we examined samples of 606 to 1,114 specimens from seven months in 1995 and three in 1996 (Figure 13.8). We discovered that some modified antennules had fewer than the standard Q number (5) of segments with "extra" aesthetascs. We symbolized those as QT, meaning "in between Q and T."

The graph shows that if Q and QT antennules occur on individuals that changed to female development after starting to prepare extra olfactory setae for maturation as males, then the frequency at which that happens changes seasonally. It gets *less* from winter to spring-summer. So we know that much. Sometimes scientists wonder whether anybody else in the world could possibly

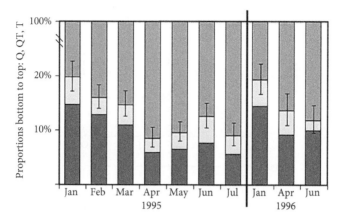

Figure 13.10 *C. finmarchicus* from Georges Bank. Proportions of right antennules with, from bottom to top, Q, QT and T aesthetasc patterns. Bars are 95% confidence limits from the binomial distribution. From Miller et al.[12] Permission, Inter-Research, *Marine Ecology Progress Series*.

care about their discoveries, however definitively proved, no matter how many thousands of dissections and microscope looks it took. And in this case, a young Norwegian worker, Camilla Svensen, had already shown seasonal shifts in proportions of Q females from Grøsundet (a fjord) near Tromsø.[28] However, her Q proportions became *greater* from winter to spring. Maybe there is a clue in that to the questions remaining: What does a C5 gain, presumably in final reproductive output, by changing from male development to female at the molt to adult in any season? And how can that advantage shift seasonally? That is, if change off New England occurs for 20% in January, why then does it only occur for around 10% from April to July? Why does the rate of change shift the other way in Grøsundet? To apply these clues, we need some sort of Sherlock Holmes or Inspector Clouseau to make something of almost nothing.

In 1999 Nancy Marcus, Pamela Blades-Eckelbarger, and Jennifer Crain presented a poster at a final TASC symposium. They had done a copepodite maturation experiment with *C. finmarchicus* from the Gulf of Maine. Survival was not very good, but female maturation was dominant for C2 through C5. This time Crain examined the antennules,[12] and large percentages of the females were indeed quadrithek (Table 13.1):

Do the high values of %Q-females *prove* that switching to female maturation occurs after development of male characters has started? No; maybe many otherwise normal females simply have an undersized male character. However, the high fractions with extra aesthetascs among the very large excess of female maturation suggests that mechanism.

Table 13.1 Sex at maturation and proportions of "Q" females of field-collected *C. finmarchicus* copepodites completing growth in laboratory containers.

Starting Stage	%Surviving to mature	No. males	No. females	%Q females
C2	36	2	45	39
C3	31	0	31	19
C4	45	10	74	41
C5	60	12	42	46

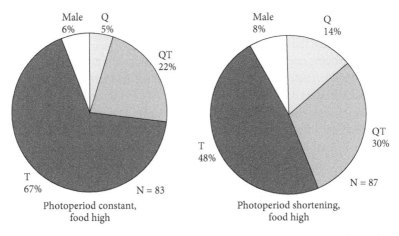

Figure 13.11 Proportions of males and quadrithek (Q and QT), and trithek (T) antennules of female *C. finmarchicus* raised from eggs in the laboratory by the late Nancy Marcus. Photoperiods during rearing differed as labeled. From Miller et al.[12]. Permission, Inter-Research, *Marine Ecology Progress Series*.

Possibly Nancy Marcus's rearing experiment[12] offers another clue. She raised four groups of *C. finmarchicus* nauplii to adulthood at 15°C, in 19-liter jugs, two with more food, two with less, which made no significant difference. Daily illumination for one group of each pair was 15.5 h:8.5 h :: light:dark throughout the 33-day rearing to maturation; the other had its photoperiod shortened progressively to 13.8 h:10.2 h :: light:dark. That is like the progression from spring to summer over the Gulf of Maine. That did make a difference (Figure 13.11).

If quadrithek frequency does relate to that of sex switching, (it must, don't you think?), the effect of photoperiod in this one rearing experiment is the opposite of what might be photoperiod in the field data from the Gulf of Maine

(Figure 13.7). As for Svensen's result in northern Norway, photoperiod goes from 24-hour dark in January to 24-hour light in June. If it really matters, the whole issue could use more experiments and observations. I have written something about the late Nancy Marcus in Chapter 17 on copepod egg diapause.

Tran The Do Shows the Advantage of ESD to One Parasitic Copepod

My colleague Patricia Wheeler at Oregon State decided in the mid-1990s to advise a Japanese graduate student named Nobuyuki Kawasaki (Figure 13.12). He had a master of science degree from Soka University in Tokyo, taken under Professor Tatsuki Toda, where his project concerned sex determination in a copepod, *Pseudomyicola spinosus*, that is parasitic on mussels and clams. He gave a talk at OSU about that work.

Like free-living copepods, *P. spinosus* mates by copulation, with direct male-female contact. Their eggs hatch as nauplii, which move out of their parent's mussel as a dispersal stage. The nauplii are loaded with yolk and molt in the plankton through six naupliar stages without feeding. The first copepodite gets hungry and must find a mussel. But finding a mussel is a chancy thing, and most mussels are found by none. Infected mussels often host only one. That one grows up, and at the maturation molt it might be male or female.

Figure 13.12 Dr. Nobu Kawasaki, as he appears in an internet photo collection, with his permission. Nobu earned his MS at Oregon State University and his PhD at the University of South Carolina. He worked awhile at a university in Malaysia, and now lives in Tokyo, working for DIC Corporation, a fine chemicals business. He extracts and studies specific algal oils.

In either case, it is forced to celibacy like a walled-up anchorite monk or nun. Arrival of another larva of the same species would be welcome, provided it matures as the opposite sex. Usually when two adult parasites are found in the same mussel, indeed they are one male and one female. Nobu Kawasaki had run experiments with adults and larvae in petri dishes, feeding them slips of mussel gill, separating them with diffusion barriers permeable to molecules, so that pheromones could diffuse through. Doing that, he confirmed that sex determination for the new arrivals is by pheromones, as shown earlier by Tran The Do and Takeshi Kajihara in 1984 and 1986[29,30] at the University of Tokyo's Ocean Research Institute (ORI). The nifty aspect of *this* ESD case is that the selective advantage is both obvious and all but impossible to dispute.

I learned about the young scientist who worked on this at ORI by asking Shinichi Uye, a friend, copepodologist, and jellyfish expert. Who, I asked, did the initial Japanese experiments on ESD in copepods parasitizing mussels? He replied on July 29, 2019:

> I can recall that work. It was done by Tran The Do as his dissertation at the Ocean Research Institute of the University of Tokyo. He came to Japan as one of many Vietnamese taking refuge in Japan after the mid-1970s. He became a student at Hiroshima University, completing his masters degree here, working in my laboratory. His excellent thesis on parasitic copepods of fish was directed by Professor Shogoro Kasahara. Do and I worked together for 3 years; I took him on field trips and to local fish markets for sampling fish. Later, at the Ocean Research Institute, Tran The Do worked intensively on a parasitic copepod of mussels in Tokyo Bay under the guidance of Professor Takeshi Kajihara. After his PhD, he established a small trading company in Tokyo handling various goods between Vietnam and Japan. He is still a manager of that company.

Eventually Shinichi Uye established contact for me with Dr. Do, who in September 2019 was in Ho Chi Minh City. Dr. Tran sent some details of his career to Uye in Japanese script, which Shin translated for me. They are remarkable:

> 1973, September: I entered Saigon Agriculture University (currently HCMAU).
> 1973, December: I obtained a scholarship from Taiwanese government, and moved to Taiwan.
> 1974, September, I entered National Taiwan University, Faculty of Agriculture.
> 1975, June, I stop learning in Taiwan, and moved to Japan.
> 1976, April, I entered Hiroshima University, Faculty of Fisheries and Animal Husbandry, as a private foreign student.
> 1978, April, I obtained a scholarship from Japanese government.

1982, April, I entered University of Tokyo, Graduate School of Agriculture (worked at Ocean Research Institute).

1985, March, I completed PhD program (Dr. Agriculture).

On August 6, 2019, Dr. Uye added:

Tran The Do's eyesight was extremely sharp, like old-days copepodologists with only humble microscopes. His artistic sense was also remarkable. He was trained also by the prominent American student of parasitic copepods, Dr. Ju-Shey Ho. If Do would have continued work on copepods, he would still be an excellent copepodologist.

Evidently, Tran The Do is a case of life continuing after copepodology. Toward the end of August, I urged Shinichi Uye to look for a picture of Tran The Do in his files. Some rather formal line-ups of students and professors were there. If Shin had taken them, they would have been in focus, but Dr. Do was in them. There was also a remarkable shot from January 1984, when Do was married in Vietnam (Figure 13.13). He married a high-school classmate, Nguyen Thi Ly. ORI Bulletins based on his dissertation studies came out in the September issues of 1984 and 1986. So he probably married in the middle of his doctoral work. It

Figure 13.13 Tran The Do returned to Vietnam in early 1984 and was married there to Nguyen Thi Ly. Photo provided by Shinichi Uye, printed with permission from Tran The Do.

is tempting to hunt down more details, but the interest here is in Do's copepod studies.

Before working with Do, Professor Takeshi Kajihara had published many papers on parasitic copepods. In 1982, partly cooperating with Ju-Shey Ho, Do and Kajihara conducted a "copepod hunt," collecting over 5,000 blue mussels, *Mytilus edulis*, from Hokkaido to Kyushu, 22 stations in all, and washing parasites from them. They found seven species of tiny gill chewing and sucking copepods (poecilostomatoids). Most prominent was the globally widespread *Pseudomyicola spinosus* (Figure 13.14).

After redescribing the new species they found,[29] Do set to work on a life history study of *P. spinosus*.[30] Collecting live females from Tokyo Bay, he watched their egg sacs in dishes of seawater until nauplii hatched from them. In a little less than two days those molted to C1, and he placed them singly in shallow dishes of filtered seawater with slips of mussel gill as food. They latched on and grew. Of 40 individuals, 18 matured as atypical males (A♂), one as a typical male (T♂). The

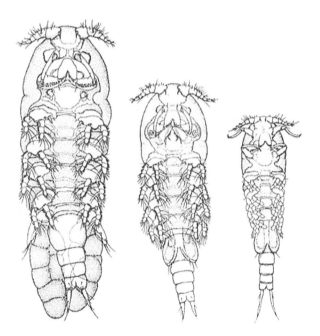

Figure 13.14 *Pseudomyicola spinosus*, ventral views. Left to right: female with egg sacs, typical male with spermatophores and atypical male, also with spermatophores. Mouthparts strongly modified for grasping and chewing mussel gills. All with five thoracic legs, the fifth strongly modified. Female total length about 2.9 mm. From Do, Kajihara, and Ho.[29] Permission, Tomohiko Kawamura, Director, Atmosphere and Ocean Research Institute, University of Tokyo.

other 21 matured as females, an overall sex ratio not statistically different from 1:1. Atypical males are smaller (Figure 13.14, right) and much more active than the typical males. They swim, probably outside mussels, and have plumose setae on the thoracic legs to facilitate that. The antennules have very large olfactory setae, and Do and Kajihara suggest they move about in mussel beds to seek females. Basically, typical males can only crawl and stay on their slips of gill filament in culture dishes. From the numbers (18A♂ vs. 1 T♂) it is likely that copepodites developing alone in a mussel mature as A♂ to facilitate mate-finding.

In two new series Do included either an adult male or an adult female in the dish with the C1 but removed it when the C1 molted. It was the same result: when those specimens matured, the males and females were 1:1. Next, he left the adults in through C2. Leaving males, the C2s matured as one atypical male and 30 females. Leaving the females, the outcome was 19 males (9 A♂ and 10 T♂) and one female. Evidently, the adults can signal for development of the sex that makes its younger fellow mussel-dweller useful, and the signal could not be received until C2.

Taking the adult female out after the larva reached C2, the outcome was 16♂ and 1♀; removing the adult male it was 1♂ and 30♀. Leaving them in until the youngster matured, produced nearly the same result (36♂:1♀ and 3♂:27♀, respectively). So apparently sex is determined in C2, and changes rarely or not at all in later development. Clearly that screamed for another test. Do started that with 20 C1, each with a female, switched the companion to male for C2, then at C3 switched it back to a female. Outcome: 5♂ and 15 ♀. Not perfect, but mostly the male present during C2 directed the sex determination. The opposite "switch and switch back" (adult ♀, then ♂, then ♀) with 40 C1 produced 32 ♂ and 8 ♀. Again, not perfect but clearly adults of both sexes could, more often than not, direct development of mates from arriving copepodites.

Immediately a new question arose: What is the nature of the signal? Did the adults touch or tickle those C2, inject those turning male with masculinizing hormone, or send a message by pheromone? Do did the obvious experiment. He put C1s on slips of mussel gill in his petri dishes and put adults in mesh cages with them. He only reported two treatments with two replicates each from one experimental series: with one caged female (his experiment Y) and two caged females (experiment Z). Summed results of the replicates are in Table 13.2:

Clearly, from experiment Y, one female in the cage provided too low a dosage of stimulus. However, the 16:4 ratio of ♂: ♀ from experiment Z has a probability of around 0.001 (if the true effect of the females were nil, then 10:10 would be expected). Nobuyuki Kawasaki later had similar results to Z, but his experiments were also small. It does seem very likely that the sex determining signal is a pheromone. Do did not report an experiment with males in the cages. So there is more to be done (and maybe it has been). *Pseudomyicola spinosus* does seem

Table 13.2 Maturation sexes of 20 *P. spinosus* C1 placed in a petri dish with 1 caged adult female (series Y) and with 2 caged adult females (series Z). The cages stopped touching of C1 by females.

Experiment	Results		
	A♂	T♂	♀
Y (1 ♀)	9	2	9
Z (2 ♀)	15	1	4

unlikely to provide a system from which the pheromone could be isolated and characterized. That could be helpful in providing hints about the chemical nature of copepod mating pheromones.

Sex Ratios in Lakes and Oceans

A number of fairly recent papers have expressed puzzlement about the sex ratios of free-living copepods in lakes and oceans. Males are often, but not always, less numerous than females. Many have noted, and Thomas Kiørboe[31] summarized the data, that the copepod families with no seminal receptacles (*Acartia*, mentioned earlier, and other genera related to freshwater *Diaptomus*) tend to have sex ratios in the field closer to 1:1 than those with receptacles. The need for repeated mating[6] likely has selected appropriate adaptations: males that feed adequately, and shorter searches with both sexes moving quite actively and leaving hydrodynamic mating cues. In contrast, families (*Calanus* and *Pseudocalanus* species and others) with females signaling via pheromones tend, at least through much of their spawning seasons, to have female-biased sex ratios. In a further and extensive literature review in 2010,[32] Andrew Hirst and colleagues concluded that sex ratios of the latter group tend to shift from near 1:1 at C5 (when that can be sexed) to significantly more females as adults.

They mentioned ESD as sometimes favoring female maturation, even citing Irigoien and colleagues.[25] Then they just dropped it like a hot potato. They did agree that ESD has been clearly demonstrated in *Tigriopus*, harpacticoid copepods that live in small, isolated tide pools, probably clinging to the walls as those pools are topped up by wave splash. They concluded nevertheless that the large sex-ratio differentials of families with seminal receptacles are most likely from greater predation on males due to their space-filling search patterns. Moreover, they recommended developing "a more comprehensive understanding of the importance of feeding preferences in predators." We are free to suspect that most predators tend to lunge at objects stirring the nearby water (which in lakes and oceans will mostly be animals), sometimes swallowing them, sometimes being avoided. Thus, predators will shift copepod sex ratios.

One of the papers Hirst and his team cited is a 2009 review by Luis Gusmão and David McKinnon,[33] published just the year before, who had come to the opposite conclusion. They thought sex-selective predation could not be the whole story, and that ESD was likely one aspect of female-skewed sex ratios. Once they saw the Hirst group's 2010 paper, they added Anthony Ricardson to their team and wrote an over-the-top review[34] titled "No Evidence of Predation Causing Female-Biased Sex Ratios in Marine Pelagic Copepods." You will not be surprised to learn that team Hirst came back[35] with a response, saying that lacking proof of a predation effect is not evidence there is none. You have to live awhile embedded in academic culture to understand the process and importance of such polemics (a wonderful word: polar opposites arguing, rejoindering, pushing back, ink-darkening vast tracts of what used to be journal pages).

The exciting issues, at least in regard to *Calanus* species, are (1) What cue informs their fifth copepodites that male or female maturation is the best choice for maximizing their reproduction (and the numbers of their grandchildren and generations beyond with their genes) in the environment most likely *after* that moment of decision?; and (2) What about those possible environments leads to more offspring for males or for females? The seasonal variation in proportions of quadrithek females suggests that such outcomes vary through the annual reproductive period.

An underlying problem is that evidence must derive from lakes and oceans, not laboratory jugs. Out there in the big waters, all sorts of developmental biology (sex differences in larval mortality, males maturing first and females later, ESD, etc.) and basic ecology (predation, fatal salinity crashes in estuaries, starvation, etc.) goes on between sampling dates. So it is very difficult to prove (absolutely or conclusively) any hypothesis about sex ratios, changes in population size or shifts in mortality rates. That never stops us, and amazingly we have learned a great deal about copepods, their predators, the population effects of flow, weather and food supplies, and even quite a lot about sex ratios.

References

1. Bernstein, C. (1979) Why are babies young?: meiosis may prevent aging of the germ line. *Perspectives in Biology and Medicine* 22(4): 539–544.
2. Bernstein, H., C. Bernstein, & R. E. Michod. (2012) DNA repair as the primary adaptive function of sex in bacteria and eukaryotes. *International Journal of Medical and Biological Frontiers* 18(2/3): 111–146.

3. Williams, G. C. (1975, corrected version 1977) *Sex and Evolution*. Princeton University Press, Princeton, NJ.

4. Carrillo, E. B.-G., C. B. Miller, & P. H. Wiebe (1974) Failure of interbreeding between Atlantic and Pacific populations of the marine calanoid copepod *Acartia clausi* Giesbrecht. *Limnology and Oceanography* 19(3): 452–458.

5. Burris, Z. P., & H. G. Dam (2015) First evidence of biased sex ratio at birth in a calanoid copepod. *Limnology & Oceanography* 60: 722–773.

6. Wilson, D. F., & K. K. Parrish (1971) Re-mating in a planktonic marine calanoid copepod. *Marine Biology* 9: 202–204.

7. Harding, J. P. (1963) The chromosomes of *Calanus finmarchicus* and *C. helgolandicus*. *Crustaceana* 6: 81–88.

8. Goswami, U., & S. C. Goswami. (1974) Cytotaxonomical studies on some calanoid copepods. *Nucleus* 17: 109–113.

9. Kosswig, C. (1964) Polygenic sex determination. *Experientia* 20: 190–199.

10. Vander Hart, A. L. (1976) Factors Influencing Sex Ratio in the Estuarine Copepod Genus *Eurytemora*. Masters thesis, Oregon State University.

11. Vaas, P., & G. G. Pesch (1984) Karyological study of the calanoid copepod *Eurytemora affinis*. *Journal of Crustacean Biology* 4(2): 248–251.

12. Miller, C. B., J. A. Crain, & N. H. Marcus (2005) Seasonal variation of male-type antennular setation in female *Calanus finmarchicus*. *Marine Ecology Progress Series* 301: 217–229.

13. Fleminger, A. (1985) Dimorphism and possible sex change in copepods of the family Calanidae. *Marine Biology* 88(3): 273–294.

14. Campbell, R. G., M. Wagner, G. J. Teegarden, C. A. Boudreau, & E. G. Durbin (2001) Growth and development rates of the copepod *Calanus finmarchicus* reared in the laboratory. *Marine Ecology Progress Series* 221: 161–183.

15. Harris, R. P. (1972) The distribution and ecology of the interstitial meiofauna of a sandy beach at Whitsand Bay, East Cornwall. *Journal of the Marine Biological Association, U.K.* 52: 1–18.

16. Harris, R. P. (1973) Feeding, growth, reproduction and nitrogen utilization by the harpacticoid copepod *Tigriopus brevicornis*. *Journal of the Marine Biological Association, U.K.* 53: 785–800.

17. Heinle, D. R., R. P. Harris, J. F. Ustach, & D. A. Flemer (1977) Detritus as food for estuarine copepods. *Marine Biology* 40: 341–363.

18. Berdugo, V., R. P. Harris, & S. C. M. O'Hara (1977) The effect of petroleum hydrocarbons on reproduction of an estuarine planktonic copepod in laboratory cultures. *Marine Pollution Bulletin* 8: 138–143.

19. López-Urrutia, Á, E. San Martin, R. P. Harris, & X. Irigoien (2006) Scaling the metabolic balance of the oceans. *Proceedings of the National Academy of Sciences* 103(23): 8739–8744.

20. Harris, M. J. (1986) Studies on the Ground Flora Under Selection Forestry in the Tavistock Woodlands Estate. PhD dissertation, Department of Geographical Sciences, Plymouth Polytechic.

21. Koski, Marja (2018) Virtual Special Issue: RPH in JPR. *Journal of Plankton Research*. https://academic.oup.com/plankt/pages/roger.harris.

22. Irigoien, X. 1994 Ingestion et Production Secondaire des Copépodes Planctoniques de l'Estuaire de la Gironde en Relation avec la Distribution du Phytoplancton et la Matiére en Suspension. PhD dissertation, Université de Bordeaux I.

23. Irigoien, X., J. Castel, & B. Sautour (1993) *In situ* grazing activity of planktonic copepods in the Gironde estuary. *Cahiers de Biologie Marine* 34: 225–237.

24. Cózar, A., E. Martí, C. M. Duarte, J. García-de-Lomas, E. van Sebille, T. J. Ballatore, V. M. Eguíluz, et al. (2017) The Arctic Ocean as a dead-end for floating plastics in the North Atlantic branch of the thermohaline circulation. *Science Advances* 3: doi:10.1126/sciady.1600582.

25. Irigoien, X., B. Obermüller, R. N. Head, R. P. Harris, C. Rey, B. W. Hansen, B. H. Hygum, et al. (2000) The effect of food on the determination of sex ratio in *Calanus* spp.: evidence from experimental studies and field data. *ICES Journal of Marine Science* 57: 1752–1763.

26. Tande, Kurt, & C. B. Miller (2000) Population Dynamics of *Calanus* in the North Atlantic: Results from the Trans-Atlantic Study of *Calanus finmarchicus*. *ICES Journal of Marine Science* 57: 1527.

27. Hygum, B. H., C. Rey, B. W. Hansen, & F. Carlotti (2000) Rearing cohorts of *Calanus finmarchicus* (Gunnerus) in mesocosms. *ICES Journal of Marine Science* 57: 1740–1751.

28. Svensen, Camilla, & K. Tande (1999) Sex change and female dimorphism in *Calanus finmarchicus*. *Marine Ecology Progress Series* 176: 93–102.

29. Do, Tran The, T. Kajihara, & Ju-Shey Ho (1984) The life history of *Pseudomyicola spinosus* (Raffaele & Monticelli, 1885) from the blue mussel, *Mytilus edulis galloprovincialis* in Tokyo Bay, Japan, with notes on production of atypical male. *Bulletin of the Ocean Research Institute-University of Tokyo* 17: 1–65.

30. Do, Tran The, & T. Kajihara (1986) Studies on parasitic copepod fauna and biology of *Pseudomyicola spinosus*, associated with blue mussel, *Mytilus edulis galloprovincialis* in Japan." *Bulletin of the Ocean Research Institute-University of Tokyo* 23: 1–63.

31. Kiørboe, T. (2006) Sex, sex-ratios, and the dynamics of pelagic copepod populations. *Oecologia* 148: 40–50.

32. Hirst, A. D., D. Bonnet, D. V. P. Conway, & T. Kiørboe (2010) Does predation control adult sex ratios and longevities in marine pelagic copepods? *Limnology & Oceanography* 55(5): 2193–2206.

33. Gusmão, L. F. M., & A. D. McKinnon (2009) Sex ratios, intersexuality and sex change in copepods. *Journal of Plankton Research* 31: 1101–1117.

34. Gusmão, L. F. M., A. D. McKinnon, & A. Richardson (2013) No evidence of predation causing female-biased sex ratios in marine pelagic copepods. *Marine Ecology Progress Series* 482: 279–298.

35. Hirst A. D., D. Bonnet, D. V. P. Conway, & T. Kiørboe (2013) Female-biased sex ratios in marine pelagic copepods: Comment on Gusmão et al. (2013). *Marine Ecology Progress Series* 489: 297–298.

14

Chromatin Diminution

Grace Wyngaard and Marvelous Mitoses

Scientists often make or extend discoveries by "drilling down." They find an interesting phenomenon, focus intensely on it, and bore long shafts and side tunnels into its details. They watch the rest of science, mathematics, and computing for new tools, new "drill bits," and they closely track others drilling in the same field or nearby. Grace Wyngaard (Figure 14.1, right), who teaches biology at James Madison University (JMU) in Virginia, has drilled for years into the biology and ecology of the freshwater copepods of the family Cyclopidae, *Cyclops* (Figure 14.1, left) and closely related genera. Off and on she is cutting in a side tunnel, learning about *chromatin diminution* (CD), a phenomenon so far thought to be restricted, among copepods, to the Cyclopidae.

Grace has given me a succinct description of cyclopid CD: "When they're embryos, they chop up their chromosomes, throw out anywhere from 30% to 95% of their DNA, glue their chromosomes back together and go on with life. And they do this in all of the cells that are going to become somatic cells: intestine, eye, muscle, nerve. Only the so-called germ cells retain the whole DNA complement of the species. Those cells are destined to divide into eggs or sperm." Let me acquaint you with Grace, then come back to CD.

I met Grace in person at the 13th International Conference on Copepoda at Long Beach, California, in July 2017. We had worked together by correspondence long before on whether or not *Calanus pacificus* embryos have chromatin diminution. At the conference she was stylishly dressed, self-confident, and very active in the discussions. She can be intense, closely attentive to whatever matter is at hand.

Grace places her decision to become a scientist at age 14, about the age when adults ask, "What are you going to do when you grow up?" Sometimes they ask, "What are you going to *be*?" She recalls having decided by then to choose work that was "interesting and not dead end," and that by then she had concluded "scientist, probably marine biologist." She studied biology at the University of Rhode Island (URI), where the work included a study of the campus pond and introduced her to small zooplankton. She recalls that reading studies on *Daphnia* (cladocerans, small crustaceans with a bivalve shell (Figure 14.2, right) by J. L. Brooks particularly extended that interest. The work by Brooks was probably his

Oar Feet and Opal Teeth. Charles B. Miller, Oxford University Press. © Oxford University Press 2023.
DOI: 10.1093/oso/9780197637326.003.0014

Figure 14.1 Left: *Cyclops strenuus*, a copepod common in European temperate lakes and ponds and abundant in Lake Superior where it was apparently introduced. Photo by Dr. Ulrich Hopp, Grafenau, Germany (ulrich_hopp@web.de), with his permission. Right: Dr. Wyngaard at the microscope. Either she has just seen an obviously new species, or someone told her she has just been made a AAAS Fellow. Photo from James Madison University.

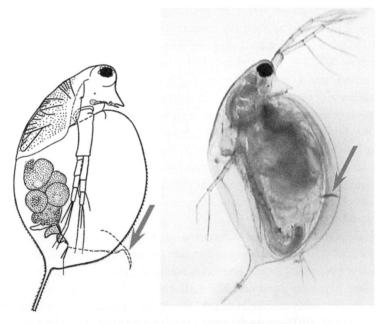

Figure 14.2 Left: Detail from a drawing by J. L. Brooks[1] of a *Daphnia middendorffiana* female with developing embryos. Length about 2.7 mm exclusive of posterior spine. With permission, Connecticut Acad. Arts and Sciences.
Right: Photograph of the same species with an empty embryo sac, its gut full of algae and the posterior claw (arrows) pulled forward between the shells. Photograph from by James F. Haney, University of New Hampshire, with his permission.

monograph on North American *Daphnia* published (in 1957) in the *Proceedings of the Connecticut Academy of Arts and Sciences*.[1]

That journal is (and was) rather obscure, even in Connecticut, so that being inspired by Brooks's paper (and remembering that) implies that her interest in the zooplankton of lakes was at full flood in her college years. My university library has a copy, and what a lovely study it is. Brooks sampled all over North America, checked type specimens and notebook drawings by original authors, redefined all of the then-recognized species of *Daphnia*, drew hundreds of sharply defined figures of daphnid shapes (Figure 14.2, left), and discussed identifying characters. He mapped distributions and from those discerned ten regions across the continent with distinctive species assemblages. The paper could indeed have inspired an ambitious young biologist like Grace to reach for something comparable. In light of DNA sequence data, the species definitions Brooks devised have not held up; the morphology of daphnids is simply too plastic to reflect relationships accurately in the sense of species defined as populations of individuals capable of interbreeding.

Brooks's most famous work was a short paper in the journal *Science*, written with his student Stanley Dodson, about differences in the zooplankton of lakes with and without planktivores, particularly fish. Dodson was a major force in American limnology, who died prematurely after a bicycle accident in 2009. Grace eventually published with Dodson on some studies of cryptic species of *Acanthocyclops*.[2] Such branching connections are usual among academics, although most often, as in this case, they are obscure and remembered only by the participants.

Grace recalls approaching graduate school wanting to study species that always reproduce sexually, unlike cladocera with long intervals of females producing more females without mating (parthenogenesis). URI professor Candace Oviatt suggested copepods as suitable candidates: planktonic, diverse, suitable for experiments and (important to Grace's mind) consistently "dioecious" (two morphologically distinct sexes in separate individuals). Grabbing that notion, Grace headed off to the University of South Florida for a masters degree. She produced a thesis there on a population study of *Mesocyclops edax* in Lake Thonotosassa. In that subtropical lake the copepod had no quiescent phase, unlike its northern populations. Like many other cyclopidae, *M. edax* can shut down metabolism as late copepodites and burrow into their lake's bottom for the winter. Lake Thonotosassa had more *M. edax* in summer than in winter, but reproduction was continuous.

Grace applied an "egg ratio" method to estimate the changing birth rates of the *M. edax* stock through the seasons.[3] She counted the females towing sacs of eggs (Figure 14.1, left), counted the eggs in the sacs, and determined for females with new clutches in beakers how long it took the eggs to hatch at the different temperatures occurring seasonally in the lake. That interval, *D* for duration in days, depends on temperature, so she tested over a temperature range. The female egg production rate is then estimated as:

ER = (eggs/female) / D.

Thinking through that, *ER* is the *per capita* population birth rate (hatching rate): the inputs of new nauplii to the lake from each female each day (Figure 14.3). Birth rate was higher in the warmer summer, when there were also more algae and diaptomid copepodites (*Diaptomus dorsalis*) for food, and lower in winter. Except for some modest pulses in spring of 1977, the copepodite and adult stocks were close to steady at 12 and 8 per liter, respectively. In fact, the population age structure (proportions of nauplii, juvenile copepodites and adults) was close to constant through the year. For numbers to remain the same, the death rate had

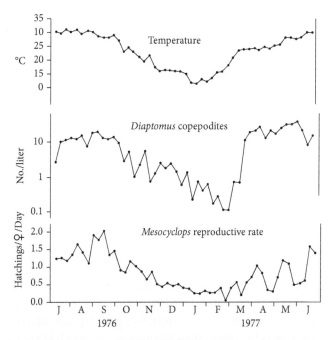

Figure 14.3 Data[3] from Grace Wyngaard's thesis study of copepod dynamics in Lake Thonotosassa, Florida. Temperature has a 15 to 30°C seasonal cycle. Copepodites of *Diaptomus dorsalis*, a prominent prey for *M. edax* varied from a less than 1 *per liter* (L^{-1}) in late winter to about 20 L^{-1} in spring and summer. Wyngaard also counted rotifers, *Diaptomus* nauplii and adults and measured chlorophyll concentrations. Reproduction of *M. edax* increased through spring, peaked in summer, then fell to a January–February low. Permission, Springer Nature.

to decline when the birth rate declined. Grace had no direct proof that death rates dropped in winter, but it was a mathematical necessity. When anyone works in a habitat week after week for a year or more, watching its constituent populations, strong impressions form about the processes affecting their numbers. So she could speculate that fish predation, particularly by Gizzard Shad, on the older stages of *M. edax* decreased in winter because the juvenile fish abundant in spring were both reduced in number and had grown and moved up to larger prey. That allowed the low winter nauplius production and slower growth of *M. edax* to sustain its stock numbers in that colder season (not seriously cold, Lake Thonotosassa is in Florida).

As covered in Chapters 15 and 16 on copepodite diapause, in high temperate and colder habitats, stocks of many species pause at some stage adapted for rest and often hiding. So for a time the whole population is in one stage. When that sleepy stage is aroused, matures, and begins to reproduce, strong pulses are generated in the age structure. That did not happen in Lake Thonotosassa. Sustaining nearly constant age structure, with or without total abundance variation, can continue in such subtropical habitats, because reproduction and mortality both continue all year round. Wyngaard offered one of the first demonstrations of that.[3]

Showing it had prepared Grace for even more graduate school. It made her application attractive to Professor J. D. Allan at the University of Maryland. She had collected, identified, counted, and experimented with animals of interest to Allen. He was studying the heritability of life history characteristics in cyclopoid copepods.[4] She had already, as a masters student, published a review paper on the ecology and cell biology of cyclopoids,[5] no doubt making her even more interesting. Her coauthor on the review was C. C. Chinnappa (working in Calgary, Canada), who had been advised to seek her out to participate in the writing. Evidently Grace knew Allen's subject at his level.

Heritability seems a simple concept: blue-eyed human mothers crossed with blue-eyed human fathers quite consistently produce blue-eyed young. Crossbreeding one of those with a darkly brown-eyed parent often produces offspring with dark brown eyes. Blue-eyes may re-emerge in a fraction of backcrosses of eventual offspring (although generally we do not *design* those experiments for human beings). Heritability of eye color is fairly strong. However, when many genes influence a trait, as is common with body size, crossbreeding experiments produce complicated results, and the level of heritability becomes a statistic with substantial variance. Thus, meaningful measurements of that statistic, usually written as simply H, require very large numbers of crosses, preferably each with large numbers of young from each. Since freshwater copepods can be maintained and crossed in the laboratory and have large egg clutches, such experiments are feasible. However, they are only feasible with repeated field collections, close attention to culture conditions and culture maintenance, counting and recounting at the microscope, and measurements of enough specimens to make comparisons of the variable H values possible. Indeed, long hours of work. What Allen obviously needed, at least in the usual culture of scientific academe, was a patient, determined, smart, well prepared graduate student. Finding one already certified as competently publishing her results was ideal.

Becoming Grace Wyngaard, PhD

So Grace earned an opportunity to get a PhD, and also to work on the heritability of genetically complex characters in copepods: body size, development speed, fecundity, and propensity to undergo resting stages. Her masters thesis subject, *Mesocyclops edax*, became her dissertation subject. It lives in a wide range of lakes, from the tropics to southern Canada. Larger specimens came from her familiar Lake Thonotonassa, a stock with no apparent resting phase in the life cycle. Smaller ones came from Douglas Lake in Northern Michigan, and those do overwinter in the sediments as late copepodites. Some replication was needed to assure that those two lakes were regionally typical, so stocks from Fairy Lake in Florida and Cochran Lake in Michigan were included. Collections were made and transported to Maryland and established as cultures in the laboratory, with due attention to water conditions, algal foods, and temperature control. The Florida stocks in the field were accustomed to seasonal temperatures ranging from 14° to 30°C, reproducing all year. In Michigan they live from a few degrees above 0°C under ice, nearly suspending animation ("diapause") and burrowing in bottom sediment, to 26°C in summer and reproducing from 18 to 26°C. The larger Florida specimens had female prosomes averaging 0.76 mm, those from Michigan 0.7 mm, a 9% difference in length but a 30% difference in likely body volume. Male sizes differed in the same direction.

Grace discovered[6,7] that the relative differences persisted in laboratory rearings of nauplii held and fed at 15° and 25°C (shown for female sizes in Figure 14.4). In her papers she attributed the persistence to a genetic difference, one likely selected by the substantial differences between the Michigan and Florida lakes. There was no difference at 30°C. Perhaps she will forgive me for saying that is not explained, but she noted it as creating an "interaction" of genotype (source population) with temperature in the statistical analysis. Many things are observed that are not readily explained. The differences for males were similar, including larger Michigan-sourced specimens at 30°C.

Look at the numbers over the graph points (Figure 14.4); those are the numbers of specimens that were each raised separately in small dishes to keep them from competing for food or biting one another. *Mesocyclops* nauplii and copepodites

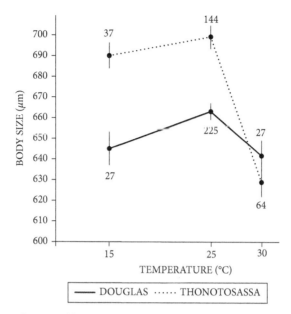

Figure 14.4 Body sizes of female *Mesocyclops edax* reared individually at three temperatures after hatching from egg sacs collected at Lake Thonotosassa in Florida and Lake Douglas in Michigan.[6] Numbers by the points are numbers of specimens maturing at that temperature. Note that body length in microns, μm as labeled, is 1,000 times the same measure in millimeters (mm). Vertical bars through the means of the data are from one standard error above and one below those means, an indication of how far the actual mean might be from the data mean. The small numbers by the data means are numbers of specimens measured in each experiment. Permission, University of Chicago Press.

do not need much space given enough algae to eat. The stacks of six-centimeter dishes with specimens maturing totaled 524, and Grace must have started with more. Not all specimens survived and the 25°C experiments included specimens from a different lake in each state. Probably there were about 1,000 specimens at first, each one moved with a pipette to new water every other day and given new algae to eat. Many who have done similar experiments could tell you about the creeping progress of hours peering at tiny lives through microscopes. An odd form of subconcious caring emerges as the work becomes almost robotic, with little pulses of disappointment when a minute beastie gets stuck trying to molt or turns up dead for no apparent reason.

Again, the modest difference in size persisted between *M. edax* cohorts from northern and southern climes, despite rearing at the same temperatures. Apparently 30°C was outside their usual adaptive ranges for development. In addition, the Florida females produced many more eggs, filling their egg sacs with an average of 36 eggs when reared in the lab and about 50 when collected in the field. That compared to about 21 in both cases for their Michigan cousins.

Perhaps that difference in fecundity is compensated by an opposite difference in egg size. For females collected from the lakes, the contents of Michigan eggs averaged 0.57 cubic millimeters (mm^3), while Florida eggs held only 0.24 mm^3. More substance in the eggs amounts to an increment of maternal care, which might compensate for the enhanced chance of a few more, if smaller, young surviving from larger clutches. Greater predation by planktivorous fish in Lake Thonotosassa than in Lake Douglas was offered as an explanation. Remember that Thonotosassa had Gizzard Shad, which feeds on small zooplankton by sucking in water to filter, whereas Douglas had very few similar feeders. Douglas also had a lot of *Daphnia*, which is unusual where predation is intense. So maybe the Florida stock adapted by raising the nauplius output enough for a few to run the predation gauntlet successfully. Those young could make up for their low hatching size with growth before maturing, supported by their Lake's abundant algae. Times to mature in the lab, starting from the first nauplius, were similar between the stocks, 14 and 16 days at 25°C. It is common for ecologists to find adaptive explanations for all or almost all observations. Some of the time we are right.

With this data in hand, Grace made a further test of whether differences in body size, clutch size, and maturation time are actually based on genes (that is, directly genetic). There are other forms of parent to offspring transfer, which are primarily maternal effects (apart from some teaching by males, though certainly not in copepods). For examples, a difference could be transferred via the mitochondria, almost always supplied only by the egg, not the sperm; or nutritional factors could vary among eggs from different mothers. A test is a comparison of character variations among groups of specimens with some genes for those characters to variations in groups with others. To avoid the maternal effects, Grace compared large groups of half-siblings with different mothers but the same fathers. Half-siblings share a quarter of their distinctive gene alleles, while very distant cousins (of the same species, of course) share very few. Because females of *Mesocyclops* can fertilize multiple egg clutches with sperm stored from a single mating, the stocks of "half-sibs" produced can be numerous and statistically useful.

Raising half-sib adults of both the Michigan and Florida stocks involved, again, thousands of larval copepod transfers: long sessions pipetting and feeding her way through huge stacks of 6 cm dishes. After all the measuring and data tabling, sharing fathers appeared to make a little difference among Michigan females in their time to grow up, but none for any males or for Florida females. In the Florida cultures somewhat more size variation occurred among young not sharing fathers than among those that did, but that was not so for the Michigan cultures. Fathers did not appear to transfer any substantial differences in clutch size to their daughters. Sometimes the results of a well-planned, fully successful experiment are just what happened in the dishes, the cloud chambers, or the crop yields. Were you to read Grace's discussion section about measured heritability of character variation in *M. edax*,[7] you would feel her difficulty in discerning their solid meaning. She is honest about that, while considering every likely evolutionary explanation of the numbers. Integrity is the deepest core value supporting scientific work.

In yet another lab-rearing experiment,[8] Grace crossed the Michigan stock with the Florida one, and then checked the extent to which the fourth copepodite stages lasted long enough, with a pause in feeding, to imply initiation of a resting stage (typical in Michigan lakes, rare in Florida lakes). The offspring had proportions starting rest intermediate between their parents. Backcrossing the offspring to Michigan and Florida individuals (not their parents) produced high and low rates of rest initiation. Almost certainly, diapause, has a genetic basis in *Mesocyclops,* and where it is not needed the relevant genes can become silent.

Grace finished her PhD in 1983. Notice that the papers describing the work are from 1986 and 1988. Such delays in journal publication of dissertation studies were typical up to that time. Graduate students were not expected to publish before finishing their degrees, but to put their introductory essays and results into extended form for submission to their supervising committees for review and approval. With the diploma on the wall, they could then see if anonymous reviewers and journal editors would also accept and publish the dissertation contents. Since about the 1980s most graduate student work is submitted first to journals, and dissertations are stacks of work approved by anonymous reviewers and already published. The change has had significant consequences that have nothing to do with copepods.

Moving On

Her degree in hand, Grace joined the Biology Department at James Madison University in central Virginia, where she still teaches and carries on research. She told me that the location was good in respect to her personal life. She fell in love with a student of amphibians and reptiles who was a curator at the Smithsonian. He was, she says, the first fully satisfactory man who showed an interest and was interesting. JMU is a commute from the Smithsonian, but they have made that work. It turns out that copepodologists (six vowels, nine consonants) are also people with all the complications thereunto pertaining.

As Professor Wyngaard, Grace continued to work on cyclopoids, particularly their life history variations and the role of diapause in the fourth copepodite stages. She has shared the work and publication authorships with many students and freshwater ecologists in the United States and elsewhere. She has traveled to tropical climes, particularly Brazil, to examine the cyclopoids living there and to work with scientists in the region. She established that the family probably originated in the tropics, spreading poleward later. At times she branched out, working, for example, with Barbara E. Taylor[9] on the resting eggs of *Diaptomus stagnalis*, a species of Diaptomidae, a family well apart on the evolutionary tree from Cyclopoida. They demonstrated the extreme capacity of those eggs to survive long intervals in the thoroughly dried sediments of intermittent "vernal" ponds. Moreover, they confirmed earlier work that given some winter chill as a stimulus, they fully develop and are ready to hatch when water returns to the pond in spring (the reason it's called a *vernal* pond).

Many, Maybe Most, Copepods Have Gigantic Genomes

During a sabbatical leave Grace added molecular genetic techniques to her repertoire, running a very complicated protocol to show[10] that the marine copepod *Calanus finmarchicus* has extraordinarily high numbers of the genes that code the ribosomal nucleic acid (RNA) portions of their ribosomes. Ribosomes are miniscule, but very complex, protein and RNA biomachines in the cytoplasm. They assemble amino acids into proteins in the order specified by sequences of messenger-RNA "transcribed" from DNA in the chromosomes. That work was done with Ian McLaren, a marine copepodologist (and also an expert on birds and seals) at Dalhousie University in Halifax, Nova Scotia.

Ian had long taken interest in the relative amounts of DNA in different marine copepods, species of *Calanus* and *Pseudocalanus*. The results must in a way have suggested new directions for work on chromatin diminution. Copepods, particularly the Calanidae, have very large genomes (a lot of DNA in their chromosomes). In fact, there is so much that a complete reading of that DNA has only lately been possible and would require the currently best automated DNA sequencing systems. Those systems chop up an organism's DNA into varied lengths of a hundred or so base pairs and generate multiple copies of all of those pieces. Then they sequence all the pieces, so that "bioinformatics" computer routines can use the overlaps in the "reads" to identify the original sequence. However, with too much DNA, the mass of data can overwhelm the computer programs that make the comparisons and propose the original orderings.

Back in the 1980s, before extended DNA sequencing was possible, McClaren[11] became very interested in simply comparing the total amounts of DNA among related copepod species. He used a technique honed to perfection by Dr. Ellen Rasch[12] that is explained in more detail below.* He opened the copepods, stained all the tissues with Feulgen dye and pressed them out on slides under coverslips and sealed those to the slides. The dye attaches to DNA, which turns the cell nuclei purple, and the concentration of purple can be measured and is proportional to the mass of DNA. Table 14.1 shows some results[10] (means):

Table 14.1 Estimated mass of DNA in seven species of *Calanus.*

Species	DNA per cell, picograms	Diploid number of chromosomes
Calanus finmarchicus	13.0	34[13]
Calanus glacialis	24.2	not known
Calanus marshallae	21.5	not known
Calanus helgolandicus	20.3	34[13]
Calanus pacificus	13.6	not known
Calanus sinicus	17.1	22[14]
Calanus hyperboreus	24.9	not known

The number of DNA base pairs (bp) *per* nucleus has a close relation to their total mass:

$$1 \text{ pg} = 10^{-12} \text{ grams} \sim 0.92 \times 10^{12} \text{ bp}$$

So, *C. finmarchicus* has a genome of about 12×10^{12} bases, 3.7-fold larger than, well, yours. Its relatives *C. glacialis* and *C. marshallae,* the closest relatives according to morphology, behavior and sequences of individual genes, have around *double that.* Similarly large ratios apply to the cousins *C. pacificus* and *C. helgolandicus.* Those differences led McLaren to speculate that back in the mists of time, cells of one or more cousins got a double dose of chromosomes. It seems odd, but we do not have chromosome counts to show that this was actually "polyploidy," two full chromosome compliments per cell. The counts need not be precisely the same to support his speculation. Chromosomes can, if infrequently, both combine and split.

---

* See "In Memoriam: Ellen Rasch," ASCB, September 30, 2016, http://www.ascb.org/newsletter/october-2016-newsletter/october-2016-newsletter-in-memoriam-ellen-m-rasch/. It is stated there that Rasch actually developed the Feulgen-based, microspectrophotometric means of measuring DNA content. The article says that Rasch was introduced to the optical system by her then supervisor, Hewson Swift.

It is difficult to imagine that copepods need genes for more enzymes, more structural proteins, more control of development than people deploy. Copepods are not simple organisms, but neither are we. Much of the extra code is what has come to be termed "junk," among other terms. It is recognized that some of the excess may have roles such as organizing chromosome structure, blocking transcription of potentially active genes or quietly waiting to accumulate mutations to code for modified proteins that could be useful in eventual generations. Almost certainly though, especially with *large* amounts of noncoding DNA, much of it is a sort of parasitic load that must be duplicated at every cell division, incurring expenses in materials and energy. If at some point an "accident" of cell division doubles the full set of all genes, that could allow much faster production of proteins by providing many more DNA templates for all of them. McClaren pointed out that the *Calanus* species with the largest genomes mostly live in the coldest habitats and have the longest development times, multiple years in the case of *C. hyperboreus* with resting phases in each winter. Possibly more genes simply allow protein synthesis to get done eventually, despite the slow rates that low temperatures impose on biochemistry. Grace Wyngaard was very close to all that evaluation and thinking. She says that occurred during a year of consciousness expanding work. Again, her part of the study showed that *Calanus* carries unusual numbers of genes for constructing ribosomes, key cellular devices for protein synthesis.

McLaren and colleagues[15] did similar work on the species of *Pseudocalanus*. That genus, too, has large (if not quite so large) and almost multiplicative amounts of DNA (Table 14.2):

Table 14.2 Estimated mass of DNA in six species of *Pseudocalanus*.

	DNA mass per cell Picograms
P. newmani	4.5
P. moultoni	8.7
P. elongatus	7.7
P. minutus	8.3
P. acuspes	6.9
P. major	3.6

In this case again, no secure determinations have been made of actual chromosome numbers. Note that relations of *newmani* to *acuspes* and *moultoni* to *elongatus* were later shown to be closer based on morphology[16] and DNA sequence similarities in a well-studied gene[17] than the other two-species comparisons among *Pseudocalanus*. Nothing seems suggested by Table 14.2 about those pairs A study by Susan Woods,[18] directed earlier by McLaren, showed the same number of chromosomes, haploid counts of 16, in both large and small species of *Pseudocalanus* sampled in fjords on Baffin Island. Specific identifications only became feasible 20 years later.[16] There was a difference: fatter chromosomes in the larger species, which Woods attributed to "polyteny," not polyploidy. Polytene chromosomes have multiple copies of their entire DNA complement fused in parallel. That is another mechanism providing more gene copies for more transcription and thus more operating proteins. A worthwhile problem in chromosome evolution waits among species of *Pseudocalanus* for a talented microscopist.

People have one of the best-studied genomes, and the human *haploid* genome (in just the chromosomes contributed by the father or mother) runs to "only" about three billion DNA base pairs. A remarkably small fraction of those, about 1.5%, actually code for functional proteins, currently believed to be around 20,000 genes. Some of the rest is codes for the enzyme-like RNA of the ribosomes, codes for "transfer" RNA molecules, and codes that regulate when genes for specific proteins are transcribed. However, a very large fraction probably has no essential function and can be considered "parasitic," gene-like sequences riding along through the replication process generation after generation. This is not what Richard Dawkins famously referred to (in 1976) as "the selfish gene," but the term seems to fit even better than the unconcern he noticed of genes for the welfare of other genes, so long as the organism survives and reproduces. In that view, popularized (not to say it spread far beyond the biological fraternity) by Dawkins, organisms are best interpreted as rather elaborate "machines" produced by natural selection to perpetuate genes. Leslie Orgel and Francis Crick[19] suggested more specifically that some portions of noncoding genes were a better fit to the sense of the word "parasitic." Finally, we are ready to look at *chromatin diminution*, a solution selection has developed in a few freshwater copepods to deal with their excess mass of DNA. Its nearly certain function is to *eliminate* most of it.

More About Chromosomes

Chromatin diminution (CD), which eventually received Grace Wyngaard's fascinated attention, was discovered in the copepod *Cyclops strenuous* during the late nineteenth century in by Valentin Häcker.[20] He was studying the early embryology of that lake-dwelling copepod and reported his observations in 1894, interpreting them as indicating diseased nucleus formation during cell division (mitosis). At that time the mechanics of chromosome duplication and distribution to daughter cells during cell division were being studied intensely, particularly in Germany, and every sort of animal was examined for illuminating details. Patterns of chromosome change similar to those in *Cyclops* had been reported a half-dozen years earlier by Theodor Boveri[21] in embryos of large, parasitic nematodes (*Ascaris*), and it has eventually been seen in a few other animals, including ciliated protozoans, hagfish, and lampreys. The cell division mechanics of CD differ among these groups. It is not found in all copepods, or even all cyclopoids. Grace learned while in graduate school about these early observations from her review of cyclopoid cell biology for that book chapter written with C. C. Chinnappa.[5] She identifies that as the start of her interest in CD.

It may not be familiar that animal chromosomes have undergone substantial diversification in form and function. That has happened over very long time spans, with extensive variations emerging in parallel with the diversification of families, classes, and phyla. For example, the chromosomes of animals other than mammals (e.g., insects, sharks, salamanders, birds) can expand during egg formation into a bottlebrush-like form, an elongate core festooned with long, thin loops to all sides. Those are termed lampbrush chromosomes. Though brushes with bristles arrayed in cylinders are common in modern kitchens and laboratories, we rarely use them now to clean lamp chimneys. Illumination at the time lampbrush chromosomes were discovered was daylight, candles, or oil lamps that slowly smoked up their glass chimneys. Microscopes used by Häcker, Boveri, and all other biologists of their era had a single eyepiece, and usually illumination was reflected through specimen slides from the windows. The resolution was close to that of modern light microscopes, but the image fields were tiny by comparison to our scopes and the focus was not, by later standards, "flat." Images in the nineteenth century were drawings. The extent of the discoveries they made with those stacks of crudely color-corrected lenses seems astounding from our distance.

The chromosomes in the salivary glands of flies become multiplexed into thick tubes, termed polytene chromosomes. Those become banded when stained. Segments being actively transcribed into messenger RNA can expand and stain distinctively. That has allowed specific genes to be located on chromosome maps, contributing greatly to the importance of fruit fly studies to progress in genetics.

Grace's early colleague Chendanda Chinnappa, working in Canada with Reginald Victor,[22, 23] showed that cyclopoid copepods also have distinctive chromosomes. Those results were reviewed in Wyngaard and Chinnappa's book chapter.[5] For example, there is partitioning along the chromosomes of *Mesocyclops edax* between "true chromatin" (*euchromatin*) in their center length and "different chromatin" (*heterochromatin*) toward each end. The centers are thin, and the ends are plump and more intensely staining. In the phase of meiosis

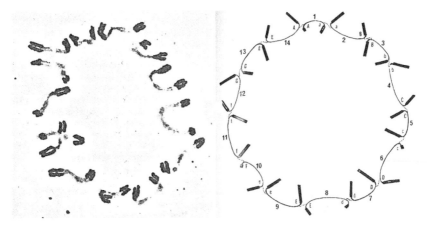

Figure 14.5 Left: *Mesocyclops edax*. Oocyte chromosomes aligned in a circle on the cell division plane during meiosis. Thin curved shadows are euchromatin joining two lengths of heavily stained, plump heterochromatin. Very thin threads, not visible in the micrograph, join the tips of euchromatin lengths on adjacent chromosomes. Right: Diagram of the circle labeled with chromosome numbers. The plump ends are not all of the same length; 1, 6 and 8 are markedly shorter. From Chinnappa and Victor.[23] Permission, Springer Nature.

when the chromosomes have replicated but are still attached in parallel—paternal chromosomes aligned base-for-base with the maternal—the points between thin and fat can be joined from one chromosome to the next by minute threads. They form patterns, at least when spread and stained on microscope slides. In some populations of *M. edax* these can be rings, including the full set of chromosomes with a haploid count of 14 (Figure 14.5). That is very likely an aspect of chromatin diminution.

Together with colleagues, Grace worked during the early 2000s on implications of the chromosomal variations in cyclopoids. The leaders included the Russian biologist Andrey Grishanin, Ellen Rasch, Stanley Dodson, and Dodson's statistical consultant Kevin Gross. They[24, 25] sampled *Acanthocyclops vernalis* from six ponds in two areas of Wisconsin and a lake in Ohio and brought them into culture. Generational lines were established from egg sacs of several females from each of pond. The diploid chromosome numbers (2n) of these strains, that were very close morphologically, included 6, 8, 9, and 10. Some variation occurred in stocks from the same ponds, more variation occurred among ponds. Despite the variation in counts, the total amounts of DNA were very close to the same, as measured by Ellen Rasch's nuclear colorimetry. Amplification and sequencing of particular genes was well developed by the date of the study, and they applied it to a 593-base portion of the genes for ribosomal DNA. There were only differences considered small: 1 to 7 bases. The minimal genetic distinctions raised the issue of whether the differences in chromosome number (and shape) produced barriers to mating success. The gene sequence differences might or might not matter, but the chromosomal ones were likely to generate problems in crosses between stocks during pairing of paternal and maternal chromosomes in meiosis.

The crew tested[25] an elaborate set of crosses between culture lines derived from single females representing the stocks from different ponds, several lines for each pond. Some multigeneration lines had only one chromosome count, others were variable. Nine lines derived from single original females retained high fecundity and viability for at least 60 generations, both for lines comprised of consistent chromosome numbers and lines with variable ones. The papers imply that Andrey Grishanin, during what was evidently a long visit in Virginia, did most of the culture work involved in crosses between the lines, which had to be carried through at least second-offspring generations (affectionately known in genetics as F2). Crosses between parental lines from different ponds succeeded in some cases and failed in others. There were also intermediate cases with apparently successful matings but no hatching, with hatching but weak survival and failure of later generations, and with reduced viability and fecundity but sustained reproduction through multiple generations. Some out-crosses with different chromosome numbers were successful, though mostly they were among the cases with reduced viability. Crosses of one pairing of lines with different chromosome numbers were uniformly unsuccessful.

Paraphrasing all that, successful crosses between lines possessing the same chromosome numbers were more likely than crosses between lines with different chromosome numbers to produce viable and fertile offspring. Examination of chromosomes from a limited number of specimens at generation 20 from four stock lines showed they had persisted in having variable chromosome numbers. Two lines retained their consistent chromosome numbers of 2n = 10 and 2n = 8, respectively. The implications are that moderate differences in chromosome number and structure can have no impact on mating and mating success, *or* they can create a mating barrier. The latter could be one means of speciation, especially if anything else about geographic location or behavior prevents competition between the incompatible stocks from eliminating one of them (termed "competitive exclusion").

What, Again, Is Chromatin Diminution?

Long decades after its discovery by Valentin Häcker, a brilliant insight about chromatin diminution came from the German worker Sigrid Beermann.[26] She was the wife of Professor Wolfgang Beermann, a widely recognized expert on variations of chromosome shape and cell-cycle transformations among different animals, particularly for flies. Her duties as wife, and perhaps as a technician, delayed her own work, but eventually Sigrid, working in Abteilung (laboratory) Beermann at Tübingen, repeated Häcker's observations of *C. strenuous* development, and she realized that the sequence of patterns was not what he called *pathologische* mitosis. It is a distinctive dance in which the fat, heterochromatin segments are removed from all of the cells destined to be parts of the copepod's body. This excision of chromosome ends occurred at the 16-cell stage (during the fifth division of the egg) for 14 of the cells. The cell destined to be the gut shrugged off its excess chromatin at the sixth division. Chromosomes in the cell destined to divide into eggs or sperm (the "germ" cell) did not dance in the same way at all; that cell retained the fatter tips. Those ends must contain genes or other DNA needed for some aspect of reproduction.

The notion that embryonic cells are "destined" for particular developmental roles should be explained. In one branch of the Animalia, development is *determinate*: as early as the two-cell or four-cell stage each progressively smaller cell becomes destined to form a more specific subset of tissues, eventually a particular tissue. If the cell preset to form the gut is removed, the embryo will not grow a gut (eventually fatal). Annelid worms, molluscs, all the varied arthropods and some others share this trait. On a separate branch of the evolutionary tree, the echinoderms (starfish and sea cucumbers), vertebrates, and others delay such cell specialization until somewhat later phases of development, usually to after the eight-cell stage. If a vertebrate embryo is split at the two-cell stage, identical twins result, at the four cell stage quadruplets. Delayed specialization into tissue types is called *indeterminate* development. This explanation can, of course, be wrapped in Latinate terminology.

Sigrid Beermann's micrographs (Figure 14.6, left) of dividing *Cyclops furcifer* embryos[27] clearly show the thin chromatin strands bending at their attachments to the threads of the division spindles. The plump heterochromatin pulls along behind and, at later stages, breaks off and is left at the cell's division plane. The resulting chromatin blobs are divided among the daughter cells by their cell membranes constricting across the division plane.

Building on Beermann's results, Grace, first author Michelle Clower, and other colleagues[28] published a detailed examination in 2016 of chromatin diminution in *Mesocyclops edax* as that species reached its fifth decade of contributing to the Wyngaard opus. The work remains demanding. The copepods must first be cultured. Times of extrusion of eggs into the sacs towed by their mothers must be determined within narrow limits. The times to successive divisions must be determined, and embryos must be preserved at specific stages of their fourth, fifth or sixth cell divisions. The embryos or their cells must be flattened on slides such that the chromosomes are obvious when stained, a procedure involving as much chance and good luck as skill. Only those cells flattened in the right orientation and at exactly the right point in the cell cycle provide useful results. Figure 14.6 shows how whole embryos look at the time of chromatin diminution. When that embryo was preserved, the thin euchromatin was moving along toward the tips of the division spindles, and the heterochromatin was being left behind, about to be approximately divided between the daughter cells by the constricting cell membrane.

Clower and her coauthors propose that the eliminated DNA retained in the embryonic cytoplasm is likely used later to provide DNA bases (the A, T, G, and C) and maybe the ribose sugar needed to construct DNA in the dozens of cell divisions that come after chromatin diminution. That bit of "waste not, want not" parallels the notion that chromatin diminution saves somatic cells the expense of duplicating masses of the likely parasitic DNA needed for some reason to create the eventual sperm and egg cells.

Fortunately, such elaborate protocols are not the only way to determine whether chromatin diminution occurs in any particular species. The fraction of

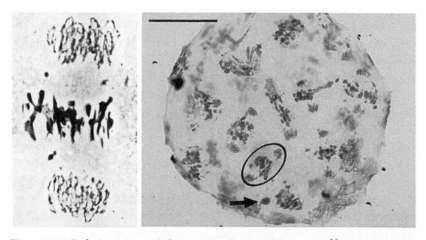

Figure 14.6 Left: A micrograph from a paper by Sigrid Beermann[26] written in 1966 shows one cell at the sixth division of an embryo of *Cyclops furcifur*. Her caption reads, "The eliminated chromatin collects in the equator of the spindle." Permission, US Govt. publication (*Genetics*). Right: Micrograph from Wyngaard's laboratory (Clower et al.)[28] of a partly squashed embryo of *Mesocyclops edax* at its fifth cell division. The nucleus in the oval has the euchromatin at its ends and the mass of heterochromatin in the center that will not be included in the nuclei of the daughter cells. Blobs of eliminated chromosome (arrow) are scattered in the cytoplasm. The scale bar is 40 μm. Permission, Oxford University Press.

the DNA in embryonic chromosomes can be estimated by measuring the ratio of DNA in germ cells to that in somatic cells. The technique, used by McLaren to determine the amount of DNA in *Pseudocalanus* nuclei, was deployed for *M. edax*: microspectrophotometric (sometimes in science the syllables do pile on) estimation of DNA mass.[12] The idea is simple; the practice takes skill. A tissue sample is disaggregated into a slurry of cells and spread to dry on a microscope slide. A whole copepod egg or embryo can also be examined. The dry patch is flooded with Feulgen stain to make the DNA in the cell nuclei reddish purple. Then a light-intensity sensor compares the red wavelengths passing through spots of unstained cytoplasm on those slides with the lesser intensity passing spots within the stained nuclei. Suitable standard curves can be developed for absorbance versus the mass of DNA in the spots. Nuclei of red blood cells from chickens, with known DNA content, are used as standards by including them on the same slide. The area of the nucleus is measured, easily done at present with image analysis software, and an averaged absorbance result for, say, a copepod nucleus is multiplied up to its area, and that is converted to an estimate of the DNA amount in picograms (e.g., Tables 14.1, 14.2).

Dr. Ellen Rasch,[12] who worked at East Tennessee State University, pioneered studies of DNA amounts in nuclei of plants and flies. For progress on chromatin diminution, Grace Wyngaard needed to measure DNA per cell in copepods, comparing somatic cells with germ cells. By working with Ellen Rasch living in Johnson City, Tennessee, just to the southwest of Harrisonburg, Virginia, along Interstate 81, Grace had access to the best technique and to Ellen's world-class expertise. They established a partnership: Grace prepared specimens, Ellen ran the densitometry. Over several years they documented and quantified CD in several species of cyclopoid copepod, including the work with Grishanin and Dodson on *Acanthocyclops vernalis*.

Grace acquired the needed and complex FIAD equipment after Ellen Rasch became seriously ill after about 2010. She died in 2016 at age 89. When FIAD was applied to CD for *M. edax*, as reported by Michelle Clower et al.,[28] the nuclear DNA content of somatic cells went from 15 pg in the earliest divisions to 3 pg after CD. Cells preparing for CD at the beginning of the embryo's fifth division doubled their DNA content to produce two somatic cells. Thus, there was 30 pg in the chromosomes before division, 6 pg retained afterward, divided between two daughter cells. Blobs in the cytoplasm amounting to 24 pg were left in the cytoplasm roughly divided between the two daughters. Thus, 80% of the DNA was eliminated (a ratio of before/after for somatic cells of 5.0). In his Doctor of Science habilitation, Andrey Grishanin[29] reviewed all the literature, including particularly that from Russia, showing a wide range in that ratio among the Cyclopidae with CD: from around 2 to around 16, as little as half to more than 90%.

What Genes Are Kept in the Germ Line Genome?

Something in the plump ends of the chromosomes that are retained in the germ-line cells must be needed for those cells to function, something not needed by any other cells in the body. Obvious candidates would be genes for the enzymes only active in meiosis, the specialized cell divisions producing gametes. The utility of meiosis in protecting a population's DNA from accumulating damage from generation to generation was discussed in Chapter 13. It provides DNA corrections by copy-reading during the formation of eggs and sperm. So what DNA sequences are eliminated in CD, and of what use are those to cells in the germ line?

Not surprisingly, Grace Wyngaard has been a key player an effort to find out what DNA sequences are removed by CD. Some work was started by Grishanin's group in Moscow working with the development of *Cyclops kolensis*. In North America it started with a visit to Grace's lab in Virginia by Guy Drouin[30] of Université d'Ottawa. *Mesocyclops edax* were sampled from Lake Shenandoah and identified by Grace, who also sorted females with egg sacs containing pre-CD embryos. Drouin extracted DNA from those and from whole copepodites, then amplified the DNA using 47 different PCR primers, each with 10 randomly selected bases (PCR, polymerase chain reaction, is described in Chapter 18). Of the 47 sequences from the pre-CD embryos, 45 contained tandemly repeated sequences, including two with CACACACA . . . -repeats, eight with CAAATAGA . . . -repeats, and nine with CAAATTAAA . . . - repeats. That echoed a widely made observation about all large genomes: they tend to have multiple repeats (a few hundred base pairs) of rather short sequences of DNA that are part of so-called transposable elements, also called "transposons." They are transposable because they readily move around in the chromosomal structure, an observation made early for maize by Barbara McClintock (subject of Evelyn Keller's book *A Feeling for the Organism*). Drouin's observation opened an initial answer to the obvious question: What does CD remove from the somatic genome? A few years later Drouin and a colleague[31] showed that some nonrepetitive sequences, likely codes for significant proteins, are also part of both the discarded DNA and somatic cell DNA.

Because we now have total DNA sequences for many organisms, detailed analyses have allowed transposons to be identified and classified from many organisms. They have noncoding ends of various lengths and often some central code that transcribes into an enzyme allowing new copies to insert at new locations in the genome. Some transposons cycle from storage DNA to replication as RNA, then back to DNA by "retrotranscription," then reinsert for storage. Other transposons replicate during regular DNA copying for cell division, then move to new locations using reinsertion enzymes derived from their coding part. Those moves often involve adding copies of a few terminal sequences to each end, usually just one to six bases. With multiple moves these become repeating sequences, from which such transposons can be identified in complete genome maps or sequences of long random fragments. Transposons are large and variable fractions of most organismal DNA, commonly more than half. It is very likely that those are the DNA removed during CD.

Next, Grace needed to obtain sequences long enough for detailed searching from large fractions of the DNA removed at CD, and also sequences from the DNA of somatic cells. Sequencing sufficiently extended to reveal whole genomes rapidly became routine around 2005, and the methods involve laboratory manipulations to produce long and overlapping fragments in massively multiple, PCR-generated copies. Those are then submitted to university or commercial laboratories that maintain automated sequencing equipment. Back comes a library of all the tens or hundreds of thousands of sequences, which are then analyzed by bioinformatics computer routines. Getting involved in this work is not trivial, so Grace recruited a specialist, Rachel Lockridge Mueller, who works at Colorado State University using a system called Shotgun 454 sequencing. Together with Rachel's staff[32] (Cheng Sun, Brian Walton, and Holly Wichman), they compared masses of sequence from somatic cells (207,451 reads averaging 183 bp, covering 0.76% of their genome) to those from pre-CD zygotes (612,470 reads averaging 218 bp, 1.6%). Quoting the results, stated in Gb or billions of DNA Bases, if those

> shotgun reads are representative subsets of the whole genomes, ~13.14 Gb of germline DNA [15 Gb total] and ~2.15 Gb of somatic DNA [3 Gb total] are composed of the identified repeats, accounting for ~11 of the 12 Gb of DNA eliminated from the presomatic cell lineage during *M. edax* development.

So, most of the DNA sequences in both cell types are transposons, but much more is retained in germ cells.

Based on the Bernsteins's "Why are babies young?" argument about the importance of meiosis, it is reasonable to suppose that something needed for that DNA proofreading must be saved in the sperm or egg precursor cell. Something in that eliminated DNA is needed for making sperm and eggs. It might be a gene or genes for the enzymes that carry out the meiotic checking and repair. One might check the two classes of cells for genes for RAD51, one of the essential meiotic DNA-repair enzymes in essentially all animals. Of course, it might be found in both the somatic and germ line cells, which would prove nothing, since the bulk of DNA removed by CD has been shown by Lockridge Mueller's team and Wyngaard to be transposons. And in fact, proteins important in meiotic DNA repair, like RAD51, can also operate in mitosis and other cell cycle phases.

It might not even be genes for specific proteins that must be retained in germ cells, because transposons can operate as regulators of which enzyme-coding genes are transcribed and translated. Another possibility is simply that masses of transposons offer some protection from all the oxidative and cosmic-ray sources of DNA damage, making retaining more of them in germ cells useful.

Grace and Rachel Lockridge Mueller's team did push on.[32] Transposons can be classified into distinct types, and many of those are widely distributed across the whole spectrum of living forms. Next-generation sequencing by Rachel's lab provided a list of the most common transposon types in *Mesocyclops edax* and also sequence data that could be used to select PCR primers. That allowed them to use a quantitative version of PCR (qPCR) to measure roughly the proportions of different transposon classes. The more of a given sequence is present in a qPCR run, the more of it will show up in the amplified DNA. If you are fascinated, you can look up exactly how the quantification works. I am not sure whether it is a surprise, but the proportions of particular and dominant transposon classes differ strongly between the somatic cells and the pre-CD zygotes (Figure 14.7). What that shows is that a lot of selection (*natural* selection; thank you Messrs. Wallace and Darwin) has sorted different DNA sequences into the heterochromatin removed from the somatic cell nuclei by chromatin diminution and the euchromatin retained after it.

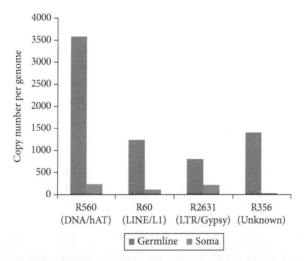

Figure 14.7 From Sun et al.[32] elative proportions of four dominant transposon types in germline and somatic DNA of *Mesocyclops edax*. Four types, three of them common (standard coded names in parentheses, R numbers are laboratory indices) were both more abundant in germline DNA *and* in different proportions from somatic-cell DNA. Results from qPCR analysis. Permission, https://creativecomm ons.org/licenses/by/2.0 .

Cataloguing the genes active in adult cells of both the germline and the rest of the body remains to be done. A molecular biologist living down the street from you can explain how that could be done.

A Personal Note on Grace Wyngaard's Generosity

I was once privileged to participate in the collaboration between Grace and Ellen Rasch. Marnie Jo Zirbel, Hal Batchelder, and I were working on a project to determine the likely age of *Calanus pacificus* embryos at each stage in their development. Zirbel[33] had invented a nuclear staining technique for the form-aldehyde preserved eggs (see Chapter 12). Thus, we (mostly Marnie Jo) could determine an embryo's development status: 1 cell, 2 cells, 4, 8, 16, . . . , cells, and then stages of the complex folding and organ formation that all embryos undergo). Together, we developed a sort of carousel to carry each of many dishes with one live, healthy, field-collected female past an electronic camera to be photographed every six minutes. That was set up with strict temperature control (12°C), and then we sat in a warm room and waited for eggs to show up on our computer screens. Female *Calanus* only take a couple of minutes to release all of the 30 to 50 + eggs in a clutch. When we saw a newly spawned clutch of eggs on a dish bottom, we knew their age to within six minutes. We worked with both *C. pacificus* and *C. marshallae* from Oregon waters. Groups were held for many different times, creating the time sequence in Figure 14.8.

At first the embryo's cells divided close to every 45 minutes, but there was a long pause at the 16-cell stage, which was unexpected and screamed for an explanation. One possibility was chromatin diminution, so an exchange with Grace was arranged. We sent her some embryos at different stages and some adult *Calanus* females. She and Dr. Rasch prepared specimens of both embryos less than three hours old and adult cells for DNA estimation by FIAD. The result was that there was no difference; chromatin diminution is very unlikely. That is a negative result, but it is also a hypothesis eliminated, which is often just as good (though I did not properly publish it; imagine a fact here with no peer review).

There is another possibility. Embryos of many (all?) animal groups pause awhile between one or another of the early cell divisions.[34] In that interval the embryo stops using the ribosomes supplied in the egg by its mother (decidedly *her* ribosomes), and some time is devoted to transcribing its own ribosome genes and assembling the ribosomal RNA and ribosomal proteins as new protein-assembly organelles. That was discovered about four decades ago and has come to be called the embryonic genome activation (EGA), or alternatively the maternal-to-zygotic transition (MZT). Maternal ribosomes are destroyed

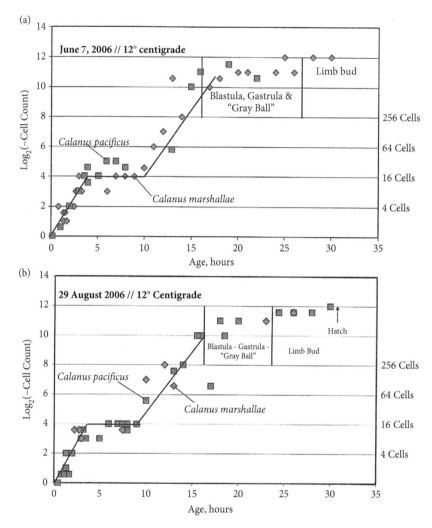

Figure 14.8 Zygote division times in two species of *Calanus*. Graphs (**A**) and (**B**) are from collections on different dates in 2006. Spawning was in the laboratory carousel and camera system. Both spawning and hatching were at temperatures very close to 12°C. The development timing lines were fitted by eye. Both experiments and both species exhibited a roughly six hour period at the 16-cell stage. Graphs by the author.

more or less rapidly at this developmental phase. In most metazoans, this occurs after one to several divisions, eight in fruit flies.[35] There apparently is an early development pause for at least two *Calanus* species, and maybe it is their MZT. This possibility can also be tested; somebody will come along to do that. Somebody might also suggest that the timing of CD very likely coincides in copepods with the embryo switching to translation of mRNA to proteins by ribosomes coded by its own genes.

The sheer amounts of DNA and of repeating sequences in *Calanus*, and some other copepods, have foiled automated sequencing systems until lately. Nobody has moved to take advantage of the latest advances to sequence the really large copepod genomes. Unlike, say, salps,[36] there were no complete sequences at the time this was written for most ecologically significant copepods. However, there is no doubt that much of their huge strings of sequence is junk DNA. There must be a substantial metabolic cost to carrying this parasitic load through every cell division of development and gamete production. Adenine and guanine (purines), and thymine and cytosine (pyrimidines), must be made available to duplicate that load over and over. Chromosomes vastly more complex than needed to support basic functions must be wrestled through every mitosis and meiosis. Very few animals are known to have found ways around paying substantial parts of these parasitic-DNA costs. The cyclopoid copepods have, and the mechanism is chromatin diminution. Andrey Grishanin[29] surveyed the literature from Häcker[20] to 2013, finding reliable reports of chromatin diminution in 23 species of cyclopoid copepods, species distinctive enough to have been assigned to eight different genera and living in widely dispersed areas. Clearly evolution discovered this partial solution to parasitic DNA long ago.

There's More to Come

Grace Wyngaard remains very vital; she has expansive perspective. Her interests continue to spread across many aspects of the lives and the evolution of copepods. Only a sample of those were covered here.

References

1. Brooks, J. L. (1957) The systematics of North American *Daphnia*. *Memoirs of the Connecticut Academy of Arts and Sciences* 13: 1–180.
2. Dodson, S. I., A. K. Grishanin, K. Gross, & G. A. Wyngaard (2003) Morphological analysis of some cryptic species in the *Acanthocyclops vernalis* species complex from North America. *Hydrobiologia* 500: 131–143.
3. Wyngaard, G. A., J. L. Elmore, & B. C. Cowell (1982) Dynamics of a subtropical plankton community, with emphasis on the copepod *Mesocyclops edax*. *Hydrobiologia* 89: 36–48.
4. Allan, J. D. (1984) Life history variation in a freshwater copepod: evidence from population crosses. *Evolution* 38: 280–291.

5. Wyngaard, G. A., & C. C. Chinnappa (1982) Biology and cytology of cyclopoids. In *Developmental Biology of Freshwater Invertebrates*, ed. F. W. Harrison & R. R. Cowden, 485–544. Alan R. Liss, New York.

6. Wyngaard, G. A. (1986a) Genetic Differentiation of Life History Traits in Populations of *Mesocyclops edax* (Crustacea: Copepoda). *Biological Bulletin* 170(2): 279–295.

7. Wyngaard, G.A. (1986b) Heritable life history variations in widely separated populations of *Mesocyclops edax* (Crustacea: Copepoda). *Biological Bulletin* 170: 296–304.

8. Wyngaard, G.A. (1988) Geograpical variation in dormancy in a copepod: evidence from population crosses. *Hydrobiologia* 167/168: 367–374.

9. Taylor, B. E., G. Wyngaard, & D. L. Mahoney (1990) Hatching of *Diaptomus stagnalis* eggs from a temporary pond after a prolonged dry period. *Archiv für Hydrobiologie* 14–7: 21–278.

10. Wyngaard, G.A., I. A. McLaren, M. M. White, & J.-M. Sevigny (1995) Unusually high numbers of ribosomal RNA genes in Copepods (Arthropoda: Crustacea) and their relationship to genome size. *Genome* 38: 97–104.

11. McLaren, I. A., J.-M. Sevigny, & C. J. Corkett (1988) Body sizes, development rates, and genome sizes among *Calanus* species. *Hydrobiologia*, 167/168: 275–284.

12. Rasch, E. M., H. J. Burr, & R. W. Rasch (1971) The DNA content of sperm of *Drosophila melanogaster*. *Chromosoma* (Berlin) 33: 1–18.

13. Harding, J. P. (1963) The chromosomes of *Calanus finmarchicus* and *Calanus helgolandicus*. *Crustaceana* 6: 81–88.

14. Lin, Y. S., W.-Q. Cao, & J.-J. Yao (2000) Karyotype study of *Calanus sinicus* (Copepoda) inhabited in Xiamen Harbor. *Journal of Xiamen University (Natural Science)* 39(6): 826–830 [English abstract].

15. McLaren, I. A., J.-M. Sevigny, & B. W. Frost (1989) Evolutionary and ecological significance of genome sizes in the copepod genus *Pseudocalanus*. *Canadian Journal of Zoology* 67: 565–569.

16. Frost, B. W. (1989) A taxonomy of the marine calanoid copepod genus *Pseudocalanus*. *Canadian Journal of Zoology* 67: 525–551.

17. Bucklin, A., B. W. Frost, J. Bradford-Grieve, L. D. Allen, & N. J. Copley (2003) Molecular systematic and phylogenetic assessment of 34 calanoid copepod species of the Calanidae and Clausocalanidae. *Marine Biology* 142: 333–343.

18. Woods, S. (1969) Polyteny and size variation in the copepod *Pseudocalanus* from two semi-landlocked fjords on Baffin Island. *Journal Fisheries Research Board Canada* 26(3): 543–556.

19. Orgel, L. E., & F. H. C. Crick (1980) Selfish DNA: the ultimate parasite. *Nature* 284: 604–607.

21. Häcker, V. (1894) Über generative und embryonale Mitosen, sowie über pathologische Kernteilungsbilder. *Archiv für Mikroskopische Anatomie* 43: 759–787.

21. Boveri, T. (1888) Die Entwicklung von *Ascaris megalocephala* unter besonderer Berucksichtigung der Kernverhaltnisse. *Festschrift zum siebenzigsten Guburtstag von Carl von Kupffer*, 383–430. Jena, G. Fischer

22. Chinnappa, C. C., & R. Victor (1979a) Cytotaxonomic studies on some cyclopoid copepods (Copepoda, Crustacea) from Ontario, Canada. *Canadian Journal of Zoology* 57: 1597–1604.

23. Chinnappa, C. C., & R. Victor (1979b) Achiasmatic meiosis and complex heterozygosity in female cyclopoid copepods (Copepoda, Crustacea). *Chromosoma* 71: 227–236.

24. Grishanin, A. K., E. M. Rasch, S. I. Dodson, & G. A. Wyngaard (2005) Variability in genetic architecture of the cryptic species complex of *Acanthocyclops vernalis* (Copepoda). I. Evidence from karyotypes, genome size, and ribosomal DNA sequences. *Journal of Crustacean Biology* 25(3): 375–383.

25. Grishanin, A. K., E. M. Rasch, S. I. Dodson, & G. A. Wyngaard (2006) Genetic architecture of the cryptic species complex of *Acanthocyclops vernalis* (crustacea: copepoda). II. Crossbreeding experiments, cytogenetics, and a model of chromosomal evolution. *Evolution* 60(2): 247–256.

26. Beermann, S. (1966) A quantitative study of chromatin diminution in embryonic mitoses of *Cyclops furcifer*. *Genetics* 54: 567–576.

27. Beermann, S. (1977) The diminution of heterochromatic chromosomoal segments in *Cyclops* (Crustacea, Copepoda). *Chromosoma* 60: 297–344.

28. Clower, M. K., A. S. Holub, R. T. Smith, & G. A. Wyngaard (2016) Embryonic development and a quantitative model of programmed DNA elimination in *Mesocyclops edax* S. A. Forbes, 1891) (Copepoda: Cyclopoida). *Journal of Crustacean Biology* 36(5): 661–674.

29. Grishanin, A. (2014) Chromatin diminution in Copepoda (Crustacea): pattern, biological role and evolutionary aspects. *Comparative Cytogenetics* 8(1): 1–10.

30. Drouin, G. (2006) Chromatin diminution in the copepod Mesocyclops edax: diminution of tandemly repeated DNA families from somatic cells. *Genome* 49: 657–665.

31. McKinnon, C., & G. Drouin (2013) Chromatin diminution in the copepod *Mesocyclops edax*: elimination of both highly repetitive and nonhighly repetitive DNA. *Genome* 56(1): 1–8.

32. Sun, C., G. Wyngaard, D. B. Walton, H. A. Wichman, & R. L. Mueller (2014) Billions of base pairs of recently expanded, repetitive sequences are eliminated from the somatic genome during copepod development. *BMC Genomics* 15: 186 (14– pp.).

33. Zirbel, M. J., C. B. Miller, & H. P. Batchelder (2007) Staging egg development of marine copepods with DAPI and PicoGreen®. *Limnology and Oceanography Methods* 5(4): 106–140.

34. Tadros, W., & H. D. Lipschitz (2009) The maternal-to-zygotic transition: a play in two acts. *Development* 136: 3033–3042.

35. Robbins, L. G (1980) Maternal-zygotic lethal interactions in *Drosophila melanogaster*: The effects of deficiencies in the zeste-white region of the X chromosome. *Genetics* 96: 187–200.

36. Jue, N. K., P. G. Batta-Lona, S. Trusiak, C. Obergfell, & A. Bucklin (2016) Rapid evolutionary rates and unique genomic signatures discovered in the first reference genome for the Southern Ocean salp, *Salpa thompsoni* (Urochordata, Thaliacea). *Genome Biology and Evolution* 8(10): 3171–3186.

15

Copepodite Diapause

North Atlantic

Diapause defined

Diapause is a technical term for a life-cycle stage spent at rest, with reduced activity allowing for reduced metabolism. It is usually applied to invertebrates that develop through a series of anatomically and behaviorally distinct stages. The mechanisms of diapause have been best studied in insects, the eggs, pupae (often in cocoons), even adults of which can spend a long part of the year, even several years, doing nothing. Individuals move to secure locations: burrow into wood, hide on cave walls, cement to sticks, bury in sediment beneath a lake. Metabolism drops to a crawl or stops altogether; a few genes, maybe coding for internal antifreeze production, are up-regulated while most are down-regulated. Some sort of timekeeping or season-sensing begins to ensure eventual arousal. Hibernation (in winter) and aestivus (in summer) are related terms usually applied to vertebrates: bears sleeping through the winter in caves or burrows, toads burying themselves to wait out dry seasons. Different families of free-living copepods have several modes of diapause.

Many students of diapause restrict the term to cases in which the rest phase once established will not be broken until a "stratifying" period has passed. That term comes from agriculture or horticulture. Plant seeds are initially dormant, and those of many species will not germinate until subjected to at least a minimal duration and level of conditions during the plant's off season, say cold in winter. Northern farmers used to ensure this exposure by packing seeds between layers (strata) of leaves or wool and placing them where temperatures would be adequately cold for enough time to prepare them for germination, such as a cellar. In spring they were said to have been *stratified*. Insect rest very often has such "refractory" intervals; the diapause stage must experience some number of "degree days" at temperatures below (or above) some value. However, diapause of a population can be "leaky," with individuals emerging from the refractory phase after different stratifying periods or conditions. That allows for rapid evolution; natural selection can adjust the seasonal timing of diapause as seasons shift with latitude or with climate change. Global climate has always been changing, though human activity is now changing it with startling rapidity.

Oar Feet and Opal Teeth. Charles B. Miller, Oxford University Press. © Oxford University Press 2023.
DOI: 10.1093/oso/9780197637326.003.0015

Two different modes of diapause are found among pelagic copepods: (1) Many Heterarthrandrids of the "superfamily" Centropagoidea (*Acartia, Centropages, Diaptomus, Labidocera,* and so on) produce resting eggs that sink to the sediments and, thanks to the squirming about of worms and amphipods (bioturbation), get buried in. They do have leaky refractory periods for waiting through cold seasons in some species (e.g., *Acartia tonsa*), warm seasons in others (*Acartia hudsonica*). Most of the species studied produce both diapause eggs, with a special outer membrane as the active season ends, and eggs that hatch in a day or so termed "subitaneous" (meaning hatching after immediate embryo development). Resting eggs get their own Chapter 17.

And (2), some groups of the Calanoida, particularly the Calanidae (*Calanus, Neocalanus,* etc.) and many Cyclopoids (*Cyclops, Diacyclops,* etc.) diapause (used as a verb here) as late copepodites and sometimes as adults. Calanids prepare by accumulating a store of lipid in the upper ocean, then late copepodites descend to depths of 250 to 2,000 m, depending on location and species, adjust to nearly neutral buoyancy and stay there until some internal or external signal rouses them. Receiving that signal, they mature and spawn, either at depth or on returning to surface layers. In temperate climes, late copepodites of *Cyclops* and related genera bury themselves in lake sediments (sand, mud, or marl) at some time in fall and stay there until spring.

Cyclopoid Diapause, Burying into Sediment

Olga Dubovskaya, whom I met at a copepod conference in 2017, told me that in well studied lakes above the Arctic Circle in Siberia the buried diapause of *Cyclops* copepodites mostly occurs in summer. The active phases occur when the lakes are frozen over. In later correspondence she steered me to the work of the late Irina Rivier, who worked from Borok near Yaroslavl. Rivier summarized her 1996 description[1] of the life cycle of *Cyclops kolensis* as follows:

In spring, when the other planktonic crustaceans have not yet appeared, *C. kolensis* is an important food for fish. Its active reproduction occurs from April to the beginning of June. A rapid accumulation of stage IV copepodids in the plankton occurs in June and at 12–14°C they sink into the pelogene [sediment surface] in a diapause stage. During the summer stagnation period the diapausing copepodids are distributed evenly over the bottom at 0.7–0.8 million ind. m^{-2} (Rybinsk reservoir). During storms and in autumn active water mixing the copepodids together with detritus are disturbed and brought to the deepest, silt rich, part of the water bodies. After ice formation and at the beginning of bottom heating [due to sinking of the most dense 4°C water] the diapausing copepodids are transported by near bottom currents and are concentrated in depressions; their biomass reaches 60 g m^{-3}. After a thermo-oxy-cline forms they revive and begin to live actively. Copepodids feed, accumulate adipose matter and in February–March [under ice] they begin to moult [mature and reproduce].

Rivier and others have suggested that under-ice production of copepods and cladocerans is sustained by feeding on the abundant bacterial abundance that develops in layers where both oxygen and methane are available for "methanotrophy."

Dr. Dubovskaya has studied lakes in the Norilsk area of Siberia north of the Arctic Circle.[2] She observed winter activity of *Senecella siberica*, an unusual freshwater calanoid (family *Aetideidae*, not quite entirely marine). Her English abstract says, "During the winter of 2018, total zooplankton samples were taken under ice with a plankton net in Lake Sobachye, one of the large Norilsk lakes, in its western (April 13, May 6) and middle (May 6) parts at 70, 80 and 100 m depths. The previous assumption that *Senecella* in the large Norilsk lakes reproduces in winter and is a univoltine species with a long (one year) generation development time has been confirmed." That very likely does involve a summer diapause.

Dr. Dubovskaya has also pointed me to an essay about Lake Hövsgöl in Mongolia, just across the Russian border north of Lake Baikal. The diverse Hövsgöl zooplankton, including both diaptomid and cyclopid copepods studied there by her Russian colleagues,[3] all appear as the lake freezes over, reach population maxima, then disappear before ice out. Their stocks in the water column stay low all summer. *Arctodiaptomus* and its relatives presumably over-summer as resting eggs.

Calanid Diapause—Early Data

Sheina Marshall, OBE FRSE FRS (the British can pile it on. Figure 15.1), was a Scottish planktologist. She is most prominently known for work, with Andrew ("A. P.") P. Orr, on the biology and ecology of *Calanus* species in Scottish sea lochs. Marshall wrote an obituary for A. P. in 1963.[4] It covers much of her own scientific life more fully than the Wikipedia article about her, which gives her dates as 1896 to 1977. She obtained a zoology degree from Glasgow in 1919, spent some fellowship years there until 1922, then took a position at the Millport lab of the Scottish Marine Biological Association. She visited Scripps Institution of Oceanography in 1970–1971, and I actually met her there while joining a sampling cruise. I do not remember the details of our one conversation; something along the lines of "Hello, very nice to meet you," I suppose. She took an interest at Scripps in the plankton holography then being considered there.

The obituary does not say that Orr was born in 1898, but that can be found elsewhere. He was raised in Kilmarnock, Scotland, not far from the sea lochs

Figure 15.1 Sheina Marshall and A. P. Orr at the Millport Laboratory of the Scottish Marine Biological Association. Note the late Victorian-style dress. Permission, Scottish Marine Biological Association.

that later became his prime study subjects. He spent a year of study at Glasgow University, in about 1917, then enlisted in the Royal Fusiliers. He saw action in France late in World War I, was reported missing, and presumed dead. However, he later turned up healing from wounds in a German hospital. Returning to Glasgow to study, he graduated BSc and MA in chemistry and geology, then worked a year or so in a physiology lab. In 1923, after a vacation in Germany ("partly spent visiting the village where he had been a prisoner of war"), he took a position at the Millport lab.

The partnership with Marshall began immediately, with work done in a laboratory they constructed in the fish hold of the *Nautilus,* a converted herring seiner. After surveys of the entire Clyde Sea area, they reduced their study zone to a few stations in Loch Striven. At first the focus was phytoplankton, including the invention, attributed to Marshall, of the light-bottle, dark-bottle method for estimating water column photosynthesis with oxygen titrations. They also applied early methods for estimating the depletion of surface layer nutrients during the most active, spring growing season. Huge patience was involved, including twice-weekly cruises.

Then came a break. Both Marshall and Orr participated for a year from May of 1928 in the British Great Barrier Reef Expedition to Australia. In the obituary she writes, "To live for a year on a coral island, with such companions and under such circumstances was an enthralling and unforgettable experience." According to Marshall, A. P. met Rachel M. Stiles "on the way home from Australia" and married her. Another obituary reported that the Orrs had two children, a son and daughter. Back in Scotland, more than year was spent writing up the Great Barrier Reef results. But in the early 1930s they returned to study of Loch Striven. They added A. G. Nicolls to the team and set to work on the life and times of *Calanus finmarchicus*. Less than a decade of varied studies produced most of their working hypotheses and data. World War II produced a gap, and their piece of war work was to try to find a local and reliable seaweed source of agar, which was no longer available from the then traditional sources in Japan. It was very difficult pulling life and science back together in Europe after the war, and the obituary says little about that period.

In 1955 they published what copepodologists refer to as simply "Marshall and Orr," but which is titled *The Biology of a Marine Copepod*.[5] It reviewed their work, mostly in the Scottish sea lochs, and that of everyone else up to that point. It remains a foundational book about free-living copepods and is available as a Springer reprint. That "marine copepod" was a sort of amalgam of *Calanus finmarchicus* and *Calanus helgolandicus*, though they also distinguished them. A. P. Orr was widely recognized, granted a DSc by Glasgow University, and made a fellow of the Royal Society of Edinburgh. At the time of his death in 1962 he was deputy director and acting director of the Millport Laboratory.

A key feature revealed by their sampling of the *C. finmarchicus* life cycle is a diapause phase during the fifth (sometimes also fourth) copepodite stage. More decades of work by dozens of workers have gone into the study, but fundamental aspects remain to be understood. The sequence in Loch Striven is best illustrated by relative stage proportions, not actual numbers (Figure 15.2). The data are from 1933. From August to February most of the stock were fifth copepodites (C5). They had loaded their oil sacs and descended to near the bottom, where they stayed without feeding or much swimming until late winter. That is the calanid diapause: descent to depth after loading the oil sac, minimal swimming, and minimal metabolism. Being deep keeps them cold and probably minimizes loss of numbers by predation, though it also can concentrate them for predators. Starting in January, but peaking in February, they become "reactivated," mature as some males and more females, move to the surface, and then spawn to establish, in Loch Striven, the first of three generational cycles. The last of those return as C5s to rest. As shown, there can be a small minority of C4s and C3s in the resting population. I have even observed very small proportions of females in a newly established resting stock off New England; their oil sacs are reloaded

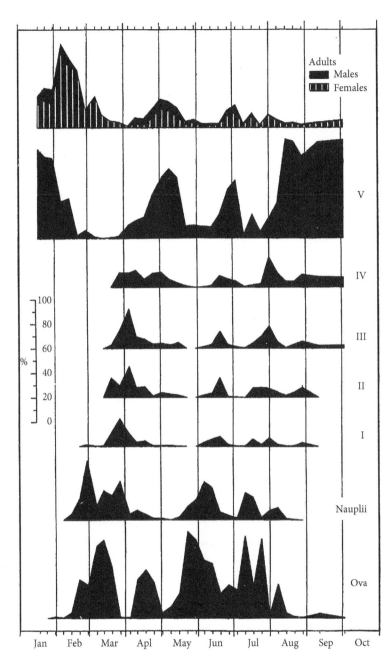

Figure 15.2 Proportions of *Calanus* life stages in Loch Striven, Scotland, part of the Clyde Sea. September numbers of C5 (V here) peaked at around 2,000 per net haul, which decreased to about 250 by January.[5] Stage-to-stage progressions are strongly evident. The March peak of eggs were adults by the end of April. Sampling issues affect all plankton data. For example, the April spawning seems to have disappeared, but maybe it was displaced by tidal exchange in the loch. August water temperatures were around 13°C, and many northern species of Calanidae descend to diapause when summer temperatures pass approximately 11°C. Permission, Springer Nature.

like those of C5s, and their oviduct walls bear imprints from ova previously matured there.

Calanus—Copepodite Diapause in the North Atlantic

Much interest, including my own, has been invested in trying to understand the regulation of this and related versions of copepodite diapause. The basic life cycle pattern, apart from timing, is well summarized in Figure 15.3 produced by two Woods Hole colleagues, Ann Tarrant and Mark Baumgartner,[6] a cartoon of the life cycle, with pictures of *Calanus finmarchicus* fifth copepodites advancing to immediate maturation (in b) and collected from diapause at depth (in c).

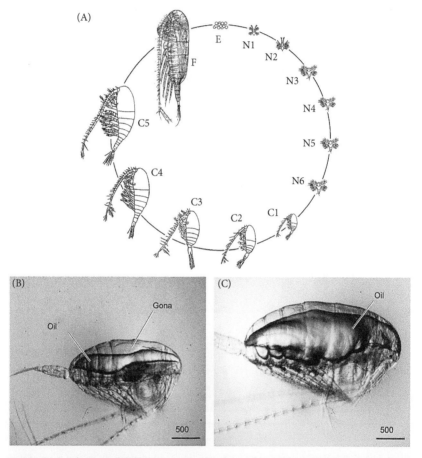

Figure 15.3 The basic life history cycle of *Calanus finmarchicus*. (A) basic cycle from eggs (E) through naupliar (N1 ...) and copepodite stages (C) to Adult female (F). There are also, indeed, males. (B) laboratory reared C5 approaching maturation. (C) C5 collected from the sea during diapause. From Baumgartner and Tarrant.[6] Permission, *Annual Reviews Marine Science*.

Richard Lee, then a graduate student at Scripps Institution of Oceanography, first showed in 1970 that the lipid in the oil sac is a wax,[7,8] not the triglyceride more common in animals. A festival of dissolving copepods in lipid solvents has extended the story since. A moment here of biological chemistry: "glyceride" refers to compounds of glycerol, a chain of three carbon atoms each bonded to a hydroxyl group (-OH) and one or two hydrogen atoms:

$$\text{glycerol}$$

By enzymatically removing the—OH groups as water (HOH), getting the additional hydrogens from three fatty acid molecules, those fatty acids (8- to 22-carbon chains) are bonded ("esterified") to the three carbon glycerol chain, becoming oils and fats. A wax, by contrast, is one long chain alcohol, CH_3-$(CH_2)_n$-CH_2OH, where n is perhaps 12 or more, and its terminal OH group can esterify to one fatty acid, making a two-chain, hairpin-shaped molecule. If those long chains each have several double bonds, they can almost coil, forcing loose packing. In bulk they remain liquid to low temperatures, even at high, deep-sea pressures. Thus, copepodites in diapause do not (well, probably would not) acquire a solid block running along their body center, something like a tube of congealed shoe polish. Flexing the prosome is important to escape swimming, which remains available to copepodites during diapause, although recovery for new jumps is very slow compared to stages actively feeding and growing.

David Pond has new ideas about the roles of liquid versus solid oil sac contents. He and Geoff Tarling identified liquid-to-solid transitions of the wax in *Calanoides natalis* (discussed in Chapter 16) occurring at depth, provided there are enough double carbon bonds in it. Pond tells the story well, and I leave the biochemically curious to chase down the articles[9, 10] (which in these online days takes about a minute). He suggests that makes it likely the wax and its bonding variants do have a part in achieving neutral buoyancy.

The extent to which *Calanus* species fill their prosomes with wax can be remarkable, exceeding half the available volume. The oil-sac filling level (Figure 15.4), however, varies with body size. The *C. finmarchicus* stock over Georges Bank and in the Gulf of Maine varies greatly in size, with prosome lengths of C5 ranging from 1.8 to 3.0 mm.[11] Jennifer Crain and I used video pictures and tracing software, along with several 2D-to-3D approximations, to estimate wax amounts in individual C5s. The smallest specimens can only fill to around 30% of their prosome volume, whereas the largest ones can pack in approximately 48% (see "diapausing C5" in Figure 15.3c). That does not seem so great a difference, but

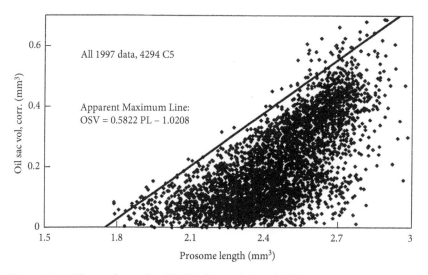

Figure 15.4 Oil sac volumes for C5 of *C. finmarchicus* of different prosome sizes collected from around Georges Bank in 1997. Collections were monthly from January to June. Most of the points are from specimens either filling their oil sacs or using the oil. The points along the apparent maximum line came primarily from specimens collected near the seafloor in May and June, and almost certainly, newly in diapause. From Miller, Crain, and Morgan.[11] Permission, Oxford University Press.

the biggest C5 carry around six times the oil of the smallest. If the oil is used at all during diapause to sustain metabolism, there must be substantial differences in how long it can last. One suspects that to some extent bigger C5 can sleep longer than small C5. That hasn't been proved.

In the North Atlantic the diapause depths of *C. finmarchicus* vary greatly. Working out of Woods Hole Oceanographic Institution, Tim Cowles and Peter Wiebe took a series of multiple opening-closing net (MOCNESS) hauls biweekly-to-monthly at night down to 800 and sometimes 1,000 meters in the North Atlantic Slope Water (NASW), a distinctive mass of cold water between Georges Bank, out to the east of Massachusetts, and the warmer Gulf Stream. But events intervened and they never thoroughly examined the samples. Later, when I had begun to think about the mechanics of diapause, they let me count the *Calanus* resting stages (Figure 15.5). The abundance profiles centered in the samples from 400 to 600 meters.[12]

Since some daylight reaches that level, we speculated that increasing day-length after the winter solstice could be the cue for maturation and rise to the upper water layers. Much spurious arithmetic about likely daily cycles of light levels at 450 m went into that speculation. It was spurious because in my enthusiasm I failed to check diapause depths for the rest of the Atlantic. Over much of

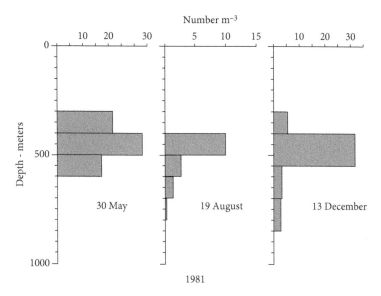

Number m^{-3}

Figure 15.5 Abundance estimates for *C. finmarchicus* C5 in a nets towed in sequential 100 m layers upward from 750 or 1,000 m just to the south of Georges Bank. Figure 4 from Miller et al.[12] Permission, Inter-Research. *Marine Ecology Progress Series.*

the range, those are much deeper and the depth-ranges of C5s can be much wider (Figure 15.6). There's a joke Groucho Marx often told: "Outside of dog, a book is man's best friend. . . . Inside of dog it's too dark to read." Beneath about 800 m at low winter sun angles there is no "daylength," only night darker than the inside of a cow. Michael Heath and his 17 European coauthors[13] made late autumn and early winter cruises to the Irminger Sea, towing an elaborate multi-depth sampler that took profiles including many depths to about 1,800 m. The net mouth was small, so the samples were small but adequate for animals as abundant as diapausing *Calanus*. The results show that the abundant stock to the southeast of Greenland takes its diapause rest from about 400 m on down to about 1,400 m in significant numbers.

It has been known at least since the work of Marshall and Orr in the 1930s that part of the diapause stock will be in C4, and in Heath's Irminger Sea data (Figure 15.6) their upper diapause depth range (about 1,000 meters) is deeper than that of C5, at about 400 m. August data of the same type showed some C5 still near the surface, with others scattering on down to about 800 m. It seems likely that once the C5 stop actively descending, they drift on down.

An Australian planktologist, Helen Grigg, who moved to Scotland and then stayed in the United Kingdom, showed that light does affect diapause in

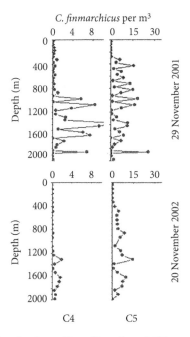

Figure 15.6 Iminger Sea depth profiles of fourth and fifth copepodites of *Calanus finmarchicus* during their autumn to winter diapause. Data from an ARIES sampler, each point integrating water filtered over 50–75 m. Sampling was near 64°29.6' N 32°36.5'W. From Figure 10 in Heath et al.[13] The paper has vast acres of detail. Permission, Elsevier.

C. finmarchicus. She kept resting C5, collected from Loch Etive, in beakers in a dimly and continuously lit cold room. They began to rouse from diapause immediately, and collections from August to November matured as C6 adults in about 10 days. The field stock stayed deep and resting. By mid-January many had started to rouse in the field, and maturations in her cold room began immediately, then continued over about 13 days.[14] I repeated that experiment[15] using C5 collected from a deep basin in the Gulf of Maine, only I used both continuous light and several photoperiods for animals in separate incubators. Helen Grigg came over to Maine for a brief visit to help me set up the experiments. We used very dim blue light. At least in autumn and winter only blue light remains below around 100 m. It turned out that continuous light roused all exposed specimens in about 10 days, just like Helen's serendipitous result. However, autumn collections exposed daily to *any* dark period did not arouse and mature. Animals from late January and February collections were the only ones that moved on to mature. When arousal was happening in the field, it also happened for some C5 at all photoperiods. Given their artifacts, the experiments have remained appropriately obscure.[15]

Because a vastly greater, oceanic part of the *C. finmarchicus* population spends its rest phase in complete darkness, which my enthusiasm induced me to ignore, the experiment does not need to be repeated. Helen's confirmed result seems to imply an adaptation for dealing with being upwelled into surface layers before the diapause season has fully passed. Once "you" (as a copepod) are in upper layers, you likely are over a shallow bottom. You might as well see what can be made of your remaining chances by maturing immediately, looking for a mate, copulating if not alone, and spawning if female. (A dose of anthropomorphic thinking and phrasing can save almost as many words as a picture can.) Except for the emergency of being upwelled, the cue for arousal likely comes from reaching some titer from an internal process. Many have supposed that comes when the wax supply is sufficiently drawn down. The problem with that idea is that the range of those supplies varies so much (Figure 15.4).

Attempts are being made to characterize the differences in gene activity among actively feeding and developing copepodites, late copepodites preparing for diapause, and diapausing copepodites. The very large genome of *Calanus* has not yet been fully sequenced. However, it is possible to extract the messenger RNA (mRNA) carrying the DNA codes for the amino-acid sequences of proteins from the nucleus to the ribosomes where enzymes are assembled. Once separated from tissues, mRNA can be copied as DNA (termed cDNA) and amplified (versions of polymerase chain reactions, PCR), sequenced and the sequences ("transcripts") identified with specific enzymes. Thus, it is possible to say which particular metabolic processes were active in a copepodite when it was captured and mixed into RNA extraction buffer. In fact, PCR can be modified to provide estimates of the amount of a given mRNA code, not just its identity, termed quantitative PCR (qPCR). Several groups have pursued this: Ann Tarrant and Mark Baumgartner at Woods Hole Oceanographic Institution (WHOI), Ann Bucklin (University of Connecticut), Petra Lenz (University of Hawaii), and Valerie Smith's group[16] in Scotland. I'll stick to the WHOI pair, because I have followed their work.

Mark Baumgartner: The Very Short, Whale-to-*Calanus* Food Chain

That following began when I was on Mark's dissertation committee at Oregon State University. He was and is an ecologist interested in whales, particularly the right whales (*Eubalaena glacialis*) of the North Atlantic.

Mark grew up in a small Upstate New York town, second son of an insurance agent and his golf-enthusiast wife. Mark was interested early in desktop computers and programming. His first job was developing a program for his mother to keep track of handicaps among her golfing set. He recalls receiving $300 for this task at age 12. He followed his admired older brother to Notre Dame University, also following his own interest in computer science. On graduating he got a more serious job programming actuarial tables and customer data for the Traveler's Insurance Company. He hated that and found more a more satisfying job at Stennis Space Flight Center in Mississippi, still based on computer skills but doing data management for a National Ocean and Atmospheric Administration (NOAA) study of marine mammals in the Gulf of Mexico. That left time for some graduate work at the University of Southern Mississippi, which had an office down the hall. From them he obtained both a masters degree in physical oceanography (oceanographers shorten that to PO) and a nascent interest in marine ecology. In 1994 he married a woman from his hometown, and she was soon offered a chance for graduate study in Boston. Needing a job in the vicinity, led Mark to a position at WHOI with prominent PO scientist Robert Weller, again managing data. Being around Woods Hole, however, enhanced his interest in ecology and marine mammals. Also, watching what the lead scientists around him did, plus publishing papers from his masters thesis, convinced him he could be and should be leading research.

Mark, noting that doctoral degrees are almost a required ticket for that, started applying to oceanography schools. An attractive offer came from whale biologist Bruce Mate at Oregon State University. At the cost of a two-year separation from his wife, who by then had a great job in Boston, Mark ran through the courses, including mine in biological oceanography. As a thesis project and to get back toward Boston, Mark organized a program tagging right whales in the Gulf of Maine with time-depth recorders. Such tags were a specialty of Bruce Mate. Mark did his tagging from NOAA fishery-science ships sailing from Woods Hole. In mid to late summer, the whales' dive profiles looked like Figure 15.7. The whales went straight down, turned to a horizontal path, and then cruised along at one sharply defined level. After 10 to 15 minutes (time at depth, Figure 15.7) they surfaced, took a few breaths, and then repeated the dive sequence. It had long been known that the gut contents of right whales, which were hunted from the seventeenth century on and almost to extinction, are *Calanus*. Mark used nets and profiling optical plankton counters (OPCs that count silhouettes passing an illuminated screen) to show that the layers whales visit were indeed where diapausing *C. finmarchicus* were most concentrated.[17]

For many whale experts, that would have been the end of the story. But Mark has an expansive curiosity and wanted to know every possible thing about *Calanus* diapause. He still does not know how a whale can sense when it is in the core of a layer of copepods. Visibility is limited, and even at the abundance peak the copepods are quite far apart. Do they stink? Do sleepy copepods tap

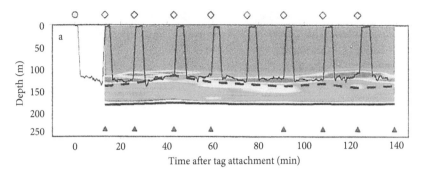

Figure 15.7 Time-depth record from a right whale during a feeding bout in the Bay of Fundy, northeast Gulf of Maine. The gray dot above left is at the time of recorder attachment by suction cup. White diamonds above are at end times of surfacing. Black triangles are times of optical plankton recorder (OPC) lowerings nearby. Red to pale blue colors represent more and less *C. finmarchicus* resting stages as estimated from the OPC. Feeding runs at ca. 130 m depth lasted 12 to 15 minutes. Based on the distance from descent to surfacing, the speeds in the copepod layer were about 0.4 to 1.9 km/hour, although they did not necessarily swim on straight tracks or move steadily. The dashed line just below the copepod maximum represents the limit above the bottom (solid black line) of strong mixing by the powerful, deep, tidal currents of the bay. From Baumgartner and Mate.[17] Permission, Inter-Research, *Marine Ecology Progress Series*.

discernibly against small hairs on the passing whale? If anyone can figure this out, it could well be Mark. Lately he has participated in a speculation that whales see copepods as the dimming of downwelling light by their aggregations.[18]

Ann Tarrant Adds Gene Activity Switching to *Calanus* Diapause Studies

Mark and Ann Tarrant, both on the WHOI staff, are mimicking in Figure 15.8 the Marshall and Orr pose above. Ann has advanced skills in molecular biology. She learned those just after graduate school, applying them to corals, work that she continues. She branched off recently into studies of the physiologic basis of copepod life cycles, initially working with Mark. Ann's family moved to the Chicago suburbs when she was three years old, and she grew up there. Her father was an air traffic controller. Along with all other US controllers, he went on strike in 1981, a strike that President Ronald Reagan stopped by firing all striking controllers and disqualifying them from any federal job ever. He replaced them with "scabs." That precipitated a family crisis. Her mother had to

Figure 15.8 Ann Tarrant and Mark Baumgartner modeling typical twenty-first-century laboratory styles. Picture by Jayne Doucette of Woods Hole Oceanographic Institution, with permission.

start work outside the home; her father entered nursing school. Ann, then six, and her younger brother became latchkey kids, home alone after school until their parents came home from work. There was never a great deal of money available at any point after that. Ann earned early pocket money by life-guarding; many marine biologists and oceanographers are "water babies." She and her father learned scuba diving together when she was in high school. Ann studied intently at her Catholic parochial school to qualify for college scholarships, and succeeded, obtaining one at the University of Miami. She majored in marine science, which required a double major. She chose biology and added a lot of chemistry and math.

Particularly inspiring to Ann was a graduate-level course at the University of Miami's Rosenstiel School of Marine Science led by coral ecologist Peter Glynn and coral physiologist Alina Szmant. That put her on track to become a coral scientist. Before graduation she spent a year studying the Great Barrier Reef in a program at James Cook University, later taking an internship at the Museum of Tropical Queensland for studies of coral taxonomy and spawning behavior.

All of these studies and jobs also extended her diving expertise and experience. After a few years in Australia her family was urging her to come home and she did, for a while. But graduate training called to her, and she was offered a visit to consider the oceanography program at the University of Hawaii. Visiting her way through the professors, she a got positive response from Marlin Atkinson, who was studying impacts of pollutants on corals. Steroids and steroid-mimicking compounds were a subject of great concern in the 1990s, and Ann's PhD thesis involved disruptions of coral growth and reproduction by estrogens and steroid-receptor blockers. Some of the work was with comparative endocrinologist (the hormones of everything) Shannon Atkinson. I will not review that work; it wasn't about copepods. But Ann has done so,[19] and has carried it far along with molecular work on corals and estuarine anemones.

Her PhD projects led to a visit at the lab of John McLaughlin at Tulane University in New Orleans, initiated at a seminar she attended. She worked awhile in his laboratory and on his field studies (he characterized the work as "bench to bayou"), where the methods were those of molecular genetics. Ann learned those, tutored by technician Tung-Chin Chiang, who understood the intense focus and constant checking required. Did the micropipette really load the intended solution? Did it fully drain to the reaction tube when the plunger was pushed? Endless potentials for errors had to be eliminated. Eliminating them led to identifying genes for steroid receptors in corals. One lesson from interviews, like those with Ann in September 2019, is that doctorates mostly lead to temporary postdoc positions. She met Mark Hahn of WHOI, who invited her to apply for one there. She took her steroid receptor interests along, working with Hahn on estrogen-related receptors (ERRs) in salt marsh fish[20] and with WHOI biochemist John Stegeman in lobsters.[21] Much of this work involved the early and developing methods for extracting and sequencing messenger RNA, then identifying the proteins they code for by comparison to vast collections of established gene codes from essentially every sort of organism. Surprisingly short stretches of gene similarity are thought to derive from common ancestry and to imply related (not necessarily identical) gene function.

In about 2003, while they were both still postdocs, Ann attended a seminar given by Mark Baumgartner. By way of explaining his data on whale diving and feeding patterns, he discussed the role of diapause in the life cycle of *Calanus finmarchicus*. Ann recognized that the profound changes in physiology between active growth of copepodites and the nearly motionless and nonfeeding torpor of deep-submergence diapause must involve radical shifts in the amounts of different enzymes needed, and thus in the production of mRNA to direct their production by ribosomes. Ann became excited by Mark's subject matter, which fit the methods to hand in her lab. The available background information on copepodite diapause is beautifully laid out by Ann and Mark's 2017 review.[6]

In preliminary studies, they compared active C5 with resting C5 from the Gulf of Maine. Quoting from their initial mRNA results:[22]

> Three genes associated with lipid synthesis, transport and storage . . . were upregulated [more highly expressed] in active copepods, particularly those with small oil sacs. Expression of ferritin was greater in diapausing copepods with large oil sacs, consistent with a [known] role of ferritin in chelating metals to protect cells from oxidative stress and/or delayed development. Ecdysteroid [molting hormone] receptor (EcR) expression was greater in diapausing copepods, highlighting the need for further investigation into endocrine regulation of [diapause].

In the austral winter of 2019 Ann worked off Antarctica, on a cruise led by Debbie Steinberg, on the molecular ecology of the large copepods there. She focused on *Calanus propinquus*, which does not diapause and does not store wax. She and coauthors have published interpretations for the reams of mRNA sequence from that and other Southern Ocean species, such as Berger et al.[23] on *Rhincalanus gigas*.

A 2019 interview with Ann in her laboratory revealed the diversity of her work. Studies on copepods were interspersed with trips to Saudia Arabia, where she worked on the biology of corals[24] at the Red Sea Research Center of King Abdullah University. After a taxi ride from the airport to a well-equipped lab, it was science like anywhere. She also sustained a program studying the molecular basis of daily activity rhythms in corals and other cnidarians, particularly the estuarine sea anemone *Nematosetella vectensis*.[25, 26] Most the interview focused on her career progress, but I did learn that life in Cape Cod, with rather lonely and often stormy winters, did not lead to meeting men with whom she wanted to rear children. She is quite satisfied with that. I thought perhaps her phenomenal scientific output filled all her time. For example, her publication list includes papers on impacts of ocean acidification.[27, 28]

However, after I shut off my recorder, we adjourned to lunch with Mark Baumgartner in one of Woods Hole's touristic cafes. Mark mentioned that Ann is progressing toward running a marathon in every one of the 50 United States of America. The COVID-19 pandemic held her to one in 2020, leaving 19 states still ahead, and she proposed then to work through them at three per year. Maybe the long stretches of running to stay continuously in marathon condition promote a stupendous level of scientific output.

Around 2012, Mark and Ann established a partnership with Bjørn Henrik Hansen's group at in Trondheim, Norway, who have developed a continuous culture system for *C. finmarchicus*. The key component likely was the use of 280-liter aquaria with slow but continuous water replacement. However, like all *Calanus* culture attempts so far, none of the copepodites in the Trondheim lab delay maturation for a resting stage. By sampling both the culture and the Trondheim field stock in May and June, Mark and Ann could compare individuals certain to mature (those in culture) with others almost certain to enter diapause. The work[6, 29] revealed many aspects of changing gene activity in fifth copepodites advancing toward immediate maturation. For example, using qPCR, Ann and Mark followed a dozen genes through C5 development toward immediate maturation. They showed that *ELOV*, a gene coding for fatty-acid elongation, was very active early in C5 and then tapers off, while *Torso-like*, a gene involved in preparing a new exoskeleton before molting, was producing enzyme late in the stage. Comparisons to the C5s in Trondheim fjord have yet to be carried very far, but eventually we will have an extended catalogue of the metabolic modifications allowing protracted diapause in *Calanus*. For now this is not in progress, since US funding agencies do not necessarily recognize the brilliant scientific opportunities presented to them. Maybe, if their plethora of other research subjects does not bury them, they will come back to it.

Next Things for Mark and Ann

Mark's career at WHOI has expanded in many directions, with a focus (at least in early 2019) on making submarine acoustic recordings and interpreting the whale calls captured. He has gathered a decade of hydrophone recordings (35,000 days' worth) from 19 labs around the northwest Atlantic and is analyzing spectrograms of specifically identified calls. With WHOI graduate student Genevieve Davis he published a study[30] (with 34 coauthors) showing how right whale distributions are changing as North Atlantic circulation rapidly shifts with climate change. They are now working up spectrograms from sei, fin, and humpback whales using programs Mark developed to automate the data searching. Working with the engineering department, he has developed systems for listening to patterns of marine mammal calls from autonomous Slocum gliders.

One glider active off Maine in winter of 2018–2019 was out for about four months, surfacing and sending in its recordings every two hours. To locations determined for right whale calls Mark directs a NOAA aircraft, and they obtain photos of the whales' backs and tails. With individual markings serving as tags, and using capture-recapture arithmetic, he can estimate the spring 2019 population as almost exactly 411 individuals, down unfortunately from reliable 2009 estimates of 481. Profoundly concerned about right whales and their potential extinction, Mark became chairman in 2017 of a North Atlantic Right Whale Consortium. That maintains databases about the population, and he chairs annual meetings with marine mammal regulators and representatives of fishing, shipping, and Navy interests. They discuss the problems the population faces and seek solutions. The main problems involving maritime industries are fishing-gear entanglements and ship strikes.

Ann, who wrote the resting-egg portion of their copepod-diapause review,[6] is looking into new ways to study populations of the *Acartia* species producing them in the coastal and estuarine habitats along the US East Coast. She may be able to provide high throughput identification from distinctive genes. For nauplii of the seasonally separated stocks of different species emerging from the sediments she could refine the timing and triggering of events with much less labor than rearing nauplii to more identifiable copepodite stages. Maybe she can work out the molecular basis of resting egg versus "subitaneous" (promptly developing) egg development and hatching (more on resting eggs in Chapter 17).

References

1. Rivier, I. C. (1996) Ecology of diapausing copepodids of *Cyclops kolensis* Lill: in rservoirs of the Upper Volga. *Hydrobiologia* 320: 235–241.
2. Dubovskaya, O. P. (2020). Life cycle and taxonomic status of *Senecella siberica* Vyshkvartzeva 1994 (Copepoda, Calanoida) in large Norilsk Lakes, Pyasina River Basin, Central Siberia. *Zoologichesky Zhurnal* 99(11): 1263–1267. [In Russian with English abstract.]
3. Pomazkova, G. I., & N. G. Sheveleva (2006) Zooplankton of Lake Hövsgöl. In *The Geology, Biodiversity and Ecology of Lake Hövsgöl (Mongolia)*, ed. C. E. Goulden, T. Sitnikova, J. Gelhaus, & B. Boldgiv, 179–200. Backhuys Publications, Leiden.
4. Marshall, S. M. (1963) Orr, M.A., D.Sc., F.R.S.E. (an obituary). Council of the Scottish Marine Biological Association, Annual Report for the year ended April 5, 1963.
5. Marshall, S. M., & A. P. Orr (1955) *The Biology of a Marine Copepod*, Calanus finmarchicus, *Gunnerus*. Oliver & Boyd, Edinburgh.
6. Baumgartner, M. F., & A. Tarrant (2017) The physiology and ecology of diapause in marine copepods. *Annual Reviews of Marine Science* 9: 387–411.
7. Lee, R. F., J. C. Nevenzel, & G.-A. Paffenhofer (1970) Wax esters in marine copepods. *Science* 167: 1510–1511.
8. Lee, R. F., J. Hirota, & A. M. Barnett (1971) Distribution and importance of wax esters in marine copepods and other zooplankton. *Deep-Sea Research* 18: 1147–1165.
9. Pond, D. (2012) The physical properties of lipids and their role in controlling the distribution of zooplankton in the oceans. *Journal of Plankton Research* 34(6): 443–453.
10. Pond, D. W., & G. A. Tarling (2011) Phase transitions of wax esters adjust buoyancy in diapausing *Calanoides acutus*. *Limnology & Oceanography* 56: 1310–1318.
11. Miller, C. B., J. A. Crain, & C. A. Morgan (2000) Oil storage variability in *Calanus finmarchicus*. *ICES Journal of Marine Science* 57: 1786–1799.
12. Miller, C. B., T. J. Cowles, P. H. Wiebe, N. J. Copley, & H. Grigg (1991) Phenology in *Calanus finmarchicus*; hypotheses about control mechanisms. *Marine Ecology Progress Series* 72: 715–791.

13. M. R. Heath, J. Rasmussen, Y. Ahmed, J. Allen, C. I. H. Anderson, A. S. Brierley, L. Brown, A. Bunker, et al. (2008) Spatial demography of *Calanus finmarchicus* in the Irminger Sea. *Progress in Oceanography* 76: 39–88.

14. Grigg, H., & S. J. Bardwell (1982) Seasonal observations on moulting and maturation in Stage V copepodites of *Calanus finmarchicus* from the Firth of Clyde. *Journal of the Marine Biological Association U.K.* 62: 315–327.

15. Miller, C. B., & H. Grigg (1991) An experimental study of the resting phase in *Calanus finmarchicus* (Gunnerus). *Bulletin of the Plankton Society of Japan*, Special volume: 479–493.

16. Clark, K. A. J., A. S. Brierley, D. W. Pond, & V. J. Smith (2013) Changes in seasonal expression patterns of ecdysone receptor, retinoid X receptor and an A-type allatostatin in the copepod, *Calanus finmarchicus*, in a sea loch environment: An investigation of possible mediators of diapause. *General and Comparative Endocrinology* 189: 66–73.

17. Baumgartner, M. F., & B. R. Mate (2003) Summertime foraging ecology of North Atlantic right whales. *Marine Ecology Progress Series* 64: 123–135 (2003)

18. Cronin, T. W., J. I. Fasick, L. E. Schweikert, S. Johnsen, L. J. Kezmoh, & M. F. Baumgartner. 2017 Coping with copepods: do right whales (*Eubalaena glacialis*) forage visually in dark waters? *Philosophical Transactions of the Royal Society B* 372: http://dx.doi.org/10.1098/rstb.2016.0067.

19. Tarrant, A. M. (2005) Endocrine-like signaling in cnidarians: Current understanding and implications for ecophysiology. *Integrative and Comparative Biology* 45(1): 201–214.

20. Tarrant, A. M., S. R. Greytak, G. V. Callard, & M. E. Hahn (2006) Estrogen receptor-related receptors in the killifish *Fundulus heteroclitus*, diversity, expression, and estrogen responsiveness. *Journal of Molecular Endocrinology* 37(1): 105–120.

21. Tarrant ,A. M., L. Behrendt, J. J. Stegeman, & T. Verslycke (2011) Ecdysteroid receptor from the American lobster, *Homarus americanus*: EcR/RXR isoform cloning and ligand-binding properties. *General and Comparative Endocrinology* 173(2): 346–355.

22. Tarrant, A. M., M. F. Baumgartner, T. Verslycke, & C. L. Johnson (2008) Differential gene expression in diapausing and active *Calanus finmarchicus* (Copepoda). *Marine Ecology Progress Series* 355: 193–207.

23. Berger, C. A., D. K. Steinberg, N. J. Copley, & A. M. Tarrant (2021) De novo transcriptome assembly of the Southern Ocean copepod *Rhinalanus gigas* sheds light on developmental changes in gene expression. *Marine Genomics* 58: doi.org/10.1016/j.margen2021.100835.

24. Cantin, N. E., A. L. Cohen, K. B. Karnauskas, A. M. Tarrant, & D. C. McCorkle (2010) Ocean warming slows coral growth in the Central Red Sea. *Science* 329: 322–325.

25. Reitzel, A. M., L. Behrendt, & A. M. Tarrant (2010). Light entrained rhythmic gene expression in the sea anemone *Nematostella vectensis*: the evolution of the animal circadian clock. *PLoS One* 5(1): 118–130.

26. Reitzel, A .M, A. M. Tarrant, & O. Levy (2013) Circadian clocks in the cnidaria: environmental entrainment, molecular regulation, and organismal outputs. *Integrative and Comparative Biology* 53(1): 118–130.

27. Thabet, A. A., A. E. Maas, G. L. Lawson, & A. M. Tarrant (2015) Life cycle and early development of the thecosomatous pteropod *Limacina retroversa* in the Gulf of Maine, including the effect of elevated CO_2 levels. *Marine Biology* 162(11): 2235–2249.

28. Breitburg, D. L., J. Salisbury, J. M. Bernard, W.-J. Cai, S. Dupont, S. C. Doney, K. J. Kroeker, A. Levin, et al. (2015) And on top of all that . . . coping with ocean acidification in the midst of many stressors. *Oceanography* 28(2): 48–61.

29. Tarrant, A. M., M. F. Baumgartner, B. H. Hansen, D. Altin, T. Nordtug, & A.J. Olsen (2014) Transcriptional profiling of reproductive development, lipid storage and molting throughout the last juvenile stage of the marine copepod *Calanus finmarchicus. Frontiers in Zoology* 11: doi:10.1186/s12983-014-0091-8.

30. Davis, G. E., M. F. Baumgartner, J. M. Bonnell, J. Bell, C. Berchok, J. B. Thornton, S. Brault, et al. (2017) Long-term passive acoustic recordings track the changing distribution of North Atlantic right whales (*Eubalaena glacialis*) from 2004 to 2014. *Scientific Reports* 7: doi:10.1038/s41598-017-13359-3.

16

Copepodite Diapause

In Other Oceans

North Pacific Versions

The subarctic Pacific has its own species of calanid copepods, each with a variant of the *Calanus finmarchicus* pattern of seasonal diapause. *Calanus pacificus* is distributed across the subarctic gyres, the Sea of Japan, and all along the east edge of the northern North Pacific. It exhibits *both* year round active maturation with mating then spawning *and* late-summer to late-winter diapause at depth. For example, its resting C5 can be found in a layer just above the upper boundary of anoxic water in the bottom of the Santa Barbara Basin off California, at the same time as females are producing eggs near the surface and larvae are developing through all of the life stages.[1] *Calanus marshallae*, living inshore from the Oregon-California border north into the Bering Sea, has an apparently more obligatory diapause at depth.[2]

Neocalanus plumchrus C5s (Figure 1.2) are found in diapause from June to January at depth in the Straits of Georgia (near Vancouver, Canada). Westward out to sea, *N. plumchrus*, the morphologically similar *Neocalanus flemingeri*, and the larger *Neocalanus cristatus* are distributed across the subarctic water masses to Asia, including the Bering Sea. All three have deep (mostly below 1,000 m) diapause stages, but with somewhat different seasonal timing, as revealed in 1980 from year-long, weekly sampling from weather ships located at Station P (50 °N, 145 °W).[3] Richard Conway, Martha Clemons, and Harold Batchelder alternated sequentially on eleven 49-day patrols of two Canadian Coast Guard weather ships sailing from Victoria, British Columbia. They took series of divided vertical net hauls to 2,000 m for more than a year. While one was out, the other two counted the samples. I went on sabbatical in San Diego.

At Station P, the C5 of *N. plumchrus* descend in May and June, then begin to mature in late August, the C5 stock slowly maturing to males and females all the way into March or April. Spawning occurs in layers below 500 m depth, most of it much deeper. Neither males nor females have teeth or useful feeding appendages, so gametes are made from stored nutrients. In fact, "spent" females are just empty exoskeletons. A few eggs may be stuck in the oviducts with no

Oar Feet and Opal Teeth. Charles B. Miller, Oxford University Press. © Oxford University Press 2023.
DOI: 10.1093/oso/9780197637326.003.0016

muscles left to push them out. Not only stored lipids, but also every organ is used up in making eggs.

I used to suppose that nauplii of *N. plumchrus*, hatching from eggs spawned by the early maturing adults, were doomed to winter starvation. However, Atsushi Tsuda (introduced in Chapter 10) and colleagues[4] devised a scheme for identifying *N. plumchrus* nauplii (N1 to N6) to species in large numbers by reading one of their mitochondrial gene sequences. Their results show that its N3 can be quiescent for long periods, accumulating from October until late winter between 500 and 2,000 m below the surface of the Oyashio. Based on a yet to be identified signal, probably an internal change, they move up to surface layers and proceed through N4 to C1 by May. Phytoplankton growth rates are sufficient then for them develop to C5, storing wax in preparation for a downward return migration into diapause. For now, this two-resting-stage annual life history is only known in *N. plumchrus*. Exactly what selective pressures led to it remains speculative.

Neocalanus flemingeri was confused with and called *N. plum*chrus through many studies by students of Pacific zooplankton (including me), until I examined a group of C5s alive in the middle of the Gulf of Alaska. Some had red markings, some had orange, and the pigmented patches were in different places. The orange ones agreed with the original description of *N. plumchrus*. The red ones on close examination had many distinctive characters for C5 and for both adult sexes. They needed a new name, and *N. flemingeri* was chosen. Martha Clemons, whom I had married by then, returned to the lab and recounted the entire year of weather ship samples. The species has a distinctive diapause pattern.[5] The C5 load with lipids in late May and June, then descend to around 500 m where they begin to mature, males first and then females (Figure 16.1). Mating occurs there in a relatively thick layer. Since in both May and June females are less abundant (about one in 10 cubic meters), the searching by males must require extended swimming. Once the females' seminal receptacles are full of sperm, they continue on down to below 1,000 m and rest, dispersed there from summer until late November. Then all of their ovaries begin to enlarge, and their oocytes start forming pretty much synchronously. After apparently swimming part way up, they spawn at the end of January. The eggs, like those of *N. cristatus* and *N. plumchrus* are loaded with lipid droplets[6] (Figure 16.2), so they are buoyant. They rise, hatching into nauplii somewhere along the way to the surface layer, where feeding and development proceed through the early spring.

Atsushi Tsuda and colleagues[7] have confirmed that *N. flemingeri* actually has *two* life-history versions in the western subarctic gyre, in the Oyashio Current and waters just to the east. In addition to the version with female diapause after one year, a "large form" (distinguished also by somewhat different mitochondrial genes) rests one year as fourth copepodites and another as females. You

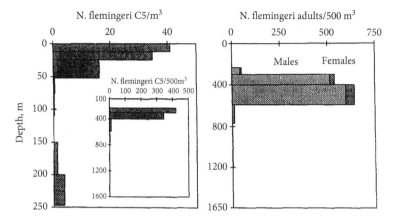

Figure 16.1 Vertical distributions of *N. flemingeri* C5 (left) and males and females (right) at Station P in the southern Gulf of Alaska during mid-June of 1987. Data are from separate tows of a multiple opening-closing net: 0–250 m left and 250–1,600 m right and inset-left. Note the different depth and abundance scales. By this date the abundances of adults were a little greater than those of remaining C5s. From Miller.[5] Permission, Elsevier.

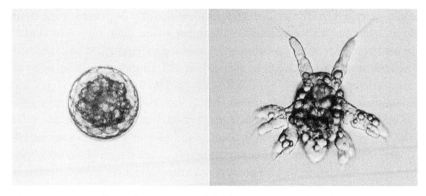

Figure 16.2 *Neocalanus flemingeri* egg, left (diameter 154 μm) and early nauplius, right (same magnification). The internal droplets are oil installed by the female before spawning. Eggs of *N. cristatus* (374 μm) and *N. plumchrus* (150 μm) are similarly loaded with lipid. From Saito and Tsuda.[6] Permission, Elsevier.

can run that story down if the amazing variations in copepod phenology (life cycle phases) are of interest. Recent studies of gene-activity in C5 *N. flemingeri* preparing for diapause are reported by Petra Lenz and Vittoria Roncalli, who examined spring samples from the northern Gulf of Alaska and Prince William Sound, helped by Russ Hopcroft, and others.[8]

Diapause in the Southern Ocean: Antarctic Copepods

Living in Antarctic waters there are at least five species of large copepods (4–6 mm) that might be expected to have a diapause. In particular, *Calanoides acutus* and *Calanus propinq*uus have close relatives with diapause in their life history, and *C. acutus* loads an oil sac with wax (up to half of body dry weight), descends, and goes quiet. It actually descends as C4, then molts at depth to C5 for overwintering.[9] In October–November it molts to adulthood with most mating occurring below 500 m (toothless males are only available for mating below 200 m). Females swim into the upper 200 m to feed and spawn starting in December or January (summer). Development takes place in those upper layers during the abundant summer phytoplankton blooms typical of the Southern Ocean.

Calanus propinquus, on the other hand, remains nearer the surface, under ice, some of its stock feeding and developing year round. That is likely allowed by the fact that the seasonal temperature range is very narrow and always cold, almost freezing. Once adapted to that cold, the seasonal differences are mostly in food stocks or predators. In spring, and presumably winter, most of the stock is found between 200 and 500 m, but some individuals are always in the upper 100 m. Presumably individuals move up and down through those upper layers. Grazing on ice algae and eating small copepods and krill larvae probably can carry *C. propinquus* through winter and allow it to skip diapause at great depth. In mid-summer, all stages reside in the upper 100 m. Both sexes feed; females spawn from November to February. *Calanus propinquus* C5s and adults do generate stores of lipid up to 50% of dry weight. In their case those lipids are triglycerides.

Calanus simillimus, *Rhincalanus gigas*, and *Metridia gerlachi* have other schemes for passing the dark, low production Southern Ocean winters.[10, 11] The first two are more prevalent north of the zone of seasonal ice cover. Like *C. propinquus*, *C. simillimus* stores triglyceride in late stages; *Rhincalanus gigas* stores wax. Apparently wax is associated with deep diapause and triglycerides are not, though the latter does not exclude a period at intermediate depths.

Tropical Copepodite Diapause: Sharon Smith and *Calanoides natalis*

Sharon Louise Smith is an emeritus professor of oceanography at the University of Miami (the one *in* Miami). She interviewed for a job at my Oregon State University (OSU) department in the early 1980s but was not selected for it. That was surely a mistake, because she has produced a lot of Bering Sea, Arctic and, of particular interest here, Arabian Sea ecology, much of it relating to copepods.

After growing up in Denver, next to mountains, she studied biology at and graduated from Colorado College. There are many US oceanographers who grew up far from the sea. I did not see an ocean (the Pacific) until I was twenty years old. Sharon, however, had retained a strong impression from a trip to South America at age ten, sailing with her family, porting in Brazil and elsewhere on a slow ocean freighter. The trip was her father's dream, and he took the family along. Sharon had a lot of time to soak up amazement about the ocean.

That fascination held, and after college with her biology degree centered on "whole picture" subjects like ecology, Sharon applied for marine science graduate study at the Duke University Marine Lab, a rather isolated station on the coast at Beaufort, North Carolina. Apparently she had been filling out every application posted on the Colorado College biology department bulletin board, because she got both a graduate fellowship from the National Science Foundation and a Fulbright for study in New Zealand. So not surprisingly, Duke invited her to come east. She did, but only for one semester with courses in invertebrate zoology (with Steve Wainwright) and animal behavior (with Peter Klopfer) that were eventually important to her. Then she moved to the University of Auckland in New Zealand, where she worked through 1968 on a masters degree based on a study of mollusc ecology in the rocky intertidal at the Leigh Marine Lab on the shore of Hauraki Gulf. Limpets and periwinkles have filled many marine station libraries with theses, including Sharon's at Leigh.

Back from New Zealand, Sharon needed a break from academia, so she found a job in commercial ecology to stay fed and housed. She worked for a consulting company in Connecticut, evaluating environmental impacts of nuclear power plants in New England. The power company actually did not allow women on the plant premises at the time, so Sharon examined larval settling plates, placed and retrieved by others qualifying as male, that had been "incubated" in the cooling water effluents. Preparation of environmental impact statements (EIS) has always been done by consulting firms, and the documents typically detail habitat destruction with a sort of vague honesty, but then say it will be the inevitable and only the necessary minimum. EIS volumes are often stuffed to thousands of pages to discourage anyone from reading them. Sharon, like many others, could feel the strain on integrity and got out after a couple of years. She does appreciate the lessons from having worked on that side of ecology.

Applying again in 1971 for more graduate work at Duke, she was accepted for study under Richard Barber. Many on the faculty were not accepting any women, and she recalls that Barber was one of very few helping women get their toe in the door to oceanography. Sharon took advantage of Barber's participation in Coastal Upwelling Ecosystem Analysis (CUEA), a nationwide program studying upwelling ecosystems. Her dissertation work was on the role of zooplankton in the nitrogen dynamics of the Canary Current off northwest Africa. She measured ammonia excretion by copepods and other small swimmers. She went on to post-doctoral study at Dalhousie's oceanography department in Nova Scotia, where she shared projects with several important copepod workers, particularly Carl Boyd and Gordon Riley. Dalhousie and Bedford Institute undertook a large oceano-graphic expedition to the Humboldt Current off Peru, and Sharon participated. She continued to investigate regeneration of nitrogen by zooplankton, including urea excretion. She also investigated copepod phototaxis and feeding satiation.

After that OSU interview in the late 1970s, Sharon took a job a Brookhaven National Lab, which at the time had an oceanography program. She worked as part of a team on the plankton ecology of the Bering Sea,[12] and characterized the gradients in copepod fauna from the very broad eastern shelf off Alaska to out past the edge of the deeper western basin. She says she liked working in a research-only system, and that the ability to fully concentrate on investigations was a greater advantage than she realized at the time. She had great colleagues, including the copepod-oriented marine ecologists Michael Dagg, David Judkins, and Julio Vidal. As the Brookhaven marine ecology group broke up, Sharon was hired at the University of Miami.

Immediately after it came out in 1982, I read a paper[13] from work in the Arabian Sea off the Horn of Africa that Sharon had started from her Brookhaven base. It was fascinating because she reported working in current velocities ex-ceeding 300 cm per second. Oceanographers often work and think in nautical miles (1.82 km), and thus speeds in knots: 1 knot = 1 nautical mile per hour. However, we often are required by journals to publish in metric units, so units for currents may appear as cm/sec. We also remember the conversion that 50 cm/sec is about 1 knot. Sharon was reporting current speeds of 6 and 7 knots, over three times faster than any current I had ever heard of. I wrote immediately to say she had to have added some decimal places. She was quick to reply that I must not have heard about the Somali Current pushed by the southwest monsoon along the Somali coast of the Arabian Sea. Not only does it typically run at 350 cm/sec in July, but the flow is coupled with nearshore upwelling of 17 °C water from below the offshore surface at around 23 °C.

Her paper does not include the word "wave," but Sharon vividly recalled in a 2018 interview that the 40 knot southwest monsoon winds raised 12-meter waves. Nets can sometimes be towed along waves that high from a large enough ship. The wave profiles can be smooth enough to ride along parallel to the crests and troughs, pulling a net behind. The view from the slowly rocking deck alternates from a vast field of ridges to walls of water on either side. Sharon managed to tow along five transects spaced at about 2° north-latitude intervals from the equator to the Horn of Africa at 11°N. Her story of the entire expedi-tion includes madcap adventures in the Suez Canal and foreign ports, problems with officials from Sudan to Somalia, breakdowns of ship's machinery, bizarre scientist-crew interactions (actually not so unusual in oceanography), and more.

The paper was also fascinating because it reported the presence close in along the Somali coast of adults of *Calanoides carinatus*, now termed *C. natalis*. This dovetailed with a story from the Atlantic Ocean I had studied with interest, because of the unusual diapause of this tropical zooplankter. In my personal logic, that comes here. I'll come back to Sharon and the Arabian Sea.

Calanoides natalis Brady, 1914, is a large (total length 2 to 4 mm) member of the family Calanidae, which lives in the equatorial and low temperate eastern Atlantic, most abundantly in the Gulf of Guinea and along the shores of southwest Africa: Angola, Namibia, and South Africa. The name *C. natalis* is old, but only recently[14] adopted in place of *Calanoides carinatus*, by none other than Janet Bradford-Grieve, who was introduced in Chapter 5 (with molecular data from Leocadio Blanco-Bercial and input from Irina Prusova). Copepods with the latter name are now known, based on morphological details and some gene sequences, to be restricted to the tropical *western* Atlantic. The range of *C. natalis* actually extends from Mauritania all around southern Africa and then to the north end of the Arabian Sea. Its life history was vividly described in a 1975 paper by Denis Binet and Eric Suisse de Sainte Claire,[15] who were working at a laboratory of the French colonial science organization ORSTOM. The lab was in Abidjan, the capital of the Ivory Coast. I do not know much about either author, except that Denis Binet (Figure 16.3) had a long career with ORSTOM, the organization of dedicated scientists serving in the

Figure 16.3 "Denis Binet, oceanographic biologist, when he was Director of Research at IRD (Institute de Recherches pour le Développement). He had worked with ORSTOM on the ecology of tropical zooplankton at Madagascar and in Africa (1965 -1975)." [Author's translation from French original] Permission, Maison de la Recherche en Sciences Humaines, Universite de Caen, Normandie.

widespread French colonial system. He carried out oceanographic and zoo-plankton studies off Madagascar, and he worked with Suisse de Sainte Claire from a base in Abidjan during 1969 to 1974. They sampled zooplankton in the bight (Golfe Ivoirien) between Cap des Palmes (8° W) and Cap des Trois Points to the east (2° W), carrying out 28 collecting cruises along and across the shelf from near the shore to deep water. They also collected weekly and sometimes twice weekly all year round, over 35 m depths just offshore from the barrier islands that protect the Ivory Coast.

The northern coast of the Gulf of Guinea runs east-west along about 5°N from Nigeria to Liberia and is subject to seasonally alternating winds. During the northern summer, usually from May to October, surface water flows away from the coast and the sea levels by upwelling of deep waters from the Gulf. Binet and Suisse de Sainte Claire showed[15] that the deep water coming up is cooler, around 18°C, replacing around 28°C (Figure 16.4), and it carries nutrients to support abundant phytoplankton growth. Also arriving are the late copepodite stages of *C. natalis* that have been in diapause at depth well offshore. Those late larvae, like the other species of calanidae in diapause discussed above, hold a store of nutriment in an oil sac along the central prosome axis above the gut. Feeding on the blooming diatoms, they molt to adults, mate, and copiously reproduce. Because

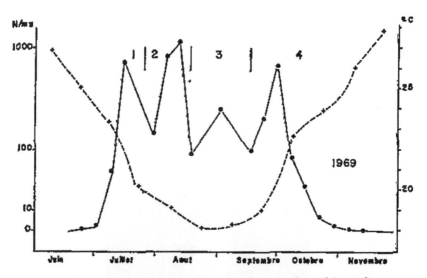

Figure 16.4 Temperature, °C, right scale, and total copepodites of *C. natalis* per cubic meter in weekly samples taken over the 35 m isobath off Abidjan in 1969, the $\log_{10}$ scale at left. Numbers above are generation cycles. Figure 2-1969 from Binet and Suisse de Sainte Claire.[15] Permission, Institut de Recherche pour le Développement (IRD), France.

temperatures are relatively high for a calanid, they run rapidly through development and complete a series of four to six life cycles in the alongshore bights (Figure 16.5). Some or most C5 individuals in the final cycles during September or early October load their oil sacs and return offshore and to depths below about 500 m. When the upwelling ends, the surface layers inshore again become warmer and oligotrophic.

Professor Smith Goes to Oman

Calanoides natalis exhibits similar cool-season outbursts in near-shore up-welling areas all the way around the African coast to Somalia and then east to Oman. Sharon Smith first reported them off Somalia from sampling during the southwest monsoon reported in her 1982 paper.[13] She's been studying the Indian Ocean stocks on and off since, whenever she can get near them. That paper reported her early feeding rate studies near the Horn of Africa, showing that *C. natalis* depends for rapid development on the abundance of large diatoms promoted by the monsoonal upwelling. Still thinking about the Somali area in 2003, even more difficult to visit now thanks to piracy, she participated in some physical-biological population modeling.[16] It was based on the notion that arousing *C. natalis* from diapause depends upon being carried by the monsoonal upwelling into the near-shore surface. That might be an adaptation similar to what Helen Grigg discovered for *C. finmarchicus* in diapause: when moved to a laboratory and the lights suddenly come on, they shift maturation into gear, molt to adults, and reproduce. Characterizing the effects on her mind of those cruises off Somalia, Sharon tells me that she became obsessed with the Arabian Sea and its zooplankton, particularly this large, oddly cycling copepod.

Maybe reflecting this obsession, she was the most determined organizer of a US and international Arabian Sea expedition involving many investigators and numerous agencies of several nations, several ships, moorings including deep sediment traps, satellite mapping of temperature and chlorophyll, and more during 1994–1996. The strength it takes to persist in such organizing, through endless rounds of proposals, rejections, revisions, meetings, funding limitations, and diplomatic arrangements, is Herculean (for Sharon, "Athenean" would be mythologically more accurate).

Several book-length journal issues presenting the results were edited by her.[17, 18] It also takes the patience of a saint to shepherd many authors to meet their deadlines. Much of the sampling was well offshore, where "*C. carinatus* [*natalis*] C5 is found at depths greater than 400 m from December to May, and in the upper 100 m abundantly in early June at the onset of upwelling"[17]

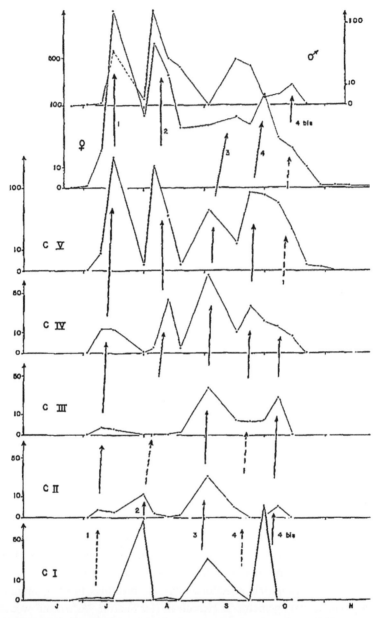

Figure 16.5 Copepodite stage-abundance progressions of the four most distinct and abundant generations of 1969, and a likely fifth is indicated. Abundances as individuals per cubic meter are on $\log_{10}$ scales at left. Abscissa labels, J J A S O N, are June to November. Figure 10 from Binet and Suisse de Sainte Claire.[15] Data from other years are in that paper. Permission, Institut de Recherche pour le Développement (IRD), France.

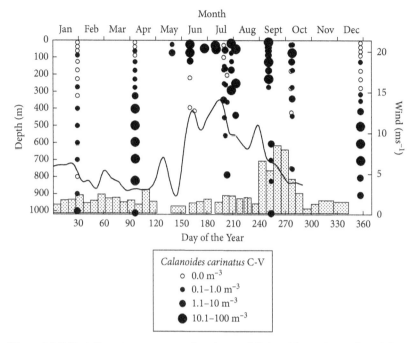

Figure 16.6 Dot diameters represent abundance of *Calanoides carinatus* (*natalis*) fifth copepodites (C-V), at various sites in the Arabian Sea. Abundant samples were mostly from inshore off Somalia and Oman. Black line represents wind speed averages at 15° N, half-way from Somalia to India, stronger from June to September during the southwest monsoon. Note: 10 m per second is 19 knots. The bars are sediment trap collection rates (scale not shown). From Smith.[17] Permission, Elsevier.

(Figure 16.6). Estimated abundances from depths of 400 to 800 m were 10 to 100 C5 per cubic meter both in April, at the start of strong inshore surface production during the southwest monsoon, and after that monsoon in December. Those are huge abundances for such great depths, and they may be sustained through the diapause season because it is spent in the mid-water layers of the Arabian Sea that are very low in oxygen availability. In diapause the copepodites do not need much oxygen. Its absence may protect them from predators.

Sharon's interests in the Arabian Sea and *Calanoides natalis* were not fully satisfied by the 1994–1996 expedition. However, her expertise led later to an invitation to Sultan Qaboos University (SQU) in Muscat to review their oceanography program. That produced an airline ticket and other expenses to get near the scalloped Omani coast where *C. natalis* thrives during the southwest monsoon. It helped her to become a long-distance and traveling member of a group at SQU doing some general biological oceanography. An initial project [19] was analysis of zooplankton collected in 2007–2008 on five seasonal fishery surveys aboard the 45 m *R/V Al Mustaqila-1*. The cruises, managed by a commercial team from New Zealand, sampled zooplankton with a small bongo net of 500 μm mesh size. Stations were all along the Omani shelf from the Ras al Hadd at the northeast to the border with Yemen, and over 20 to 250 m depths. For help working through the samples Sergey Piontkovski, then a staff scientist at SQU, established a partnership with Elena Popova, currently of the Russian Academy of Sciences and working in Sevastopol, Crimea. She produced seasonal lists of the large copepods retained by the relatively coarse net, and she continued to work with Sharon and Sergey for years.[20]

The southwest monsoon (August to September) induced strong upwelling in 2008, and thus very intense phytoplankton blooms in the coastal bights (Figure 16.7). Chlorophyll is also quite high in the bights during April to May, while it is very low farther offshore, but the temperature in the bights is warmer than during the peak upwelling period later. Samples from the August–September 2008 cruise had only a few large copepod species, with C4, C5, and adult female *Calanoides natalis* dominant. This suggests that their pattern is much as in the coastal bights of the Gulf of Guinea studied by Denis Binet; there must be a seasonal series of multiple and rapid generations.

Hoping to establish that population cycling, Sharon headed for the shore. Since the 1990s she had been considering Masirah Island (offshore at the arrow in Figure 16.7) as a possible base for routine sampling. Starting in 2007, she hired a Bedouin fisherman, Juma Mohammed al Farsi, to sample from there year round, weather permitting. She bought him a modest boat and outboard engines for it. Eventually she got National Science Foundation funding to sustain the time series. Like Binet, she now has samples from a decade. The sea is too wild for sampling in July at the peak of the southwest monsoon, but there is good coverage of August and September. Sharon visited Masirah Island annually between 2007 and 2016 to participate in sampling, and she has established friendships with al Farsi's family (Figure 16.8), learning extensively about Bedouin culture. Another partnership Sharon established among the remarkable copepod taxonomic expertise established in Russia. Key was Irina Prusova of the Russian Academy of Sciences working in Sevastopol. She, Sharon, and Elena Popova produced a lovely guide (published by SQU) titled *Calanoid Copepods of the Arabian Sea Region*.[21] Irina was engaged in initial identifying and counting the entire copepod assemblages of Sharon and al Farsi's Masirah shelf collections, with special emphasis on stage progressions of *C. natalis*. Sharon has now published some of the results, with additional counting by Argentine copepodologist Dora Pilz and by Sharon's students Maria Criales and Carolann Schack.[22]

A decade of sampling is enough to answer questions about changes in ocean conditions and plankton population processes.[22] Indeed, the situation close to Masirah Island does prove similar to that off Cote d'Ivoire. Copepods of tropical,

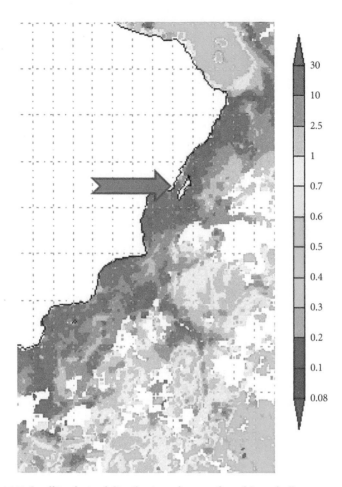

Figure 16.7 Satellite-derived distribution of sea-surface chlorophyll concentration in mg/cubic meter (false-color scale shown) during the SW monsoon, August to September 2008. White represents land or cloud masking. The highest phytoplankton concentrations are in Oman's coastal bights; more than 10 mg/m³, the ocean there looked distinctly green. The blue arrow indicates Masirah Island. From Piontkovski et al.[19] Permission, *International Aquatic Research:* http://creativecommons.org/licen ses/by/4.0/.

oceanic surface waters are dominant in winter and spring, particularly five spe-cies of the genus *Subeucalanus*. During the southwest monsoon, from July to autumn, dominance shifts to *Calanoides natalis*. I selected data (Figure 16.9) to represent the spring and summer data from two well sampled years. The full figure in their paper shows that my selection represents the general case. The spe-cies mixture shifts back in late autumn as upwelling lapses, and *C. natalis* enters diapause as fifth copepodites at depth beyond the shelf. Sharon sees a suggestion

Figure 16.8 Sharon Smith on a beach outing at the western tip of Masirah Island off Oman in the Arabian Sea with the wife of her Omani fisherman colleague, Juma Mohammed al Farsi, and a family friend (in red). Provided by Sharon Smith.

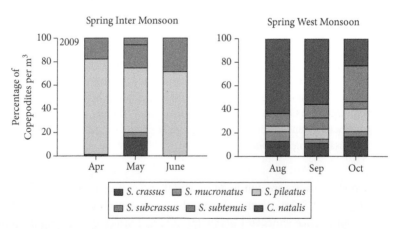

Figure 16.9 Proportions of total copepodite abundance among dominant large copepods during two seasons near Masirah Island, coastal Oman. Distinguishing species of *Subeucalanus* is quite difficult (considered in Chapter 19), which suggests the high level of taxonomic expertise applied to the counting by by Dora S. Pilz, an Argentine copepodologist working with Sharon. Ms Pilz counted samples from 2007 to 2012. Samples for 2013 to 2016 were counted by, Carolann Schack and Maria Criales. From Figure 7 in Smith et al.[22] Permission, Elsevier.

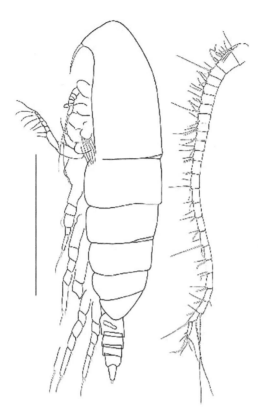

Figure 16.10 Habitus drawing of adult female *Calanoides natalis* Brady, 1914. From Figure 5 in Bradford-Grieve et al.[14] The scale bar at left is 1 mm. The antennule (at right) was removed to show the other limbs. Permission, Taylor & Francis.

that the southwest monsoon is lasting longer year by year, and it has become more intense. The *Calanoides* stock is responding by more prolonged inshore residence, adding two generations to its annual cycle. The sea-surface temperature difference between the seasons is remarkably small: about 28°C in spring, and 25°C in late summer. The paper contains a festival of statistical analysis.

The specimen of *Calanoides natalis* drawn (example in Figure 16.10) and measured by Janet Bradford-Grieve for her redescription of the species were from Sharon and al Farsi's station at (22.20° N, 58.75° E) on the seaward side of Masirah Island.

References

1. Johnson, C. L. (2004) Seasonal variation in the molt status of an oceanic copepod. *Progress in Oceanography* **62**: 15–32.
2. Osgood, K. E., & B. W. Frost (1994) Comparative life histories of three species of planktonic calanoid copepods in Dabob Bay, Washington. *Marine Biology* **118**(4): 627–636.

3. Miller, C. B., & C. J. Clemons (1988) Revised life history analysis for large grazing copepods in the subarctic Pacific Ocean. *Progress in Oceanography* 20: 293–314.

4. Fujioka, H. A., R. J. Machida, & A. Tsuda (2015) Early life history of *Neocalanus plumchrus* (Calanoida: Copepoda) in the western subarctic Pacific. *Progress in Oceanography* 147: 196–208.

5. Miller, C. B. (1993) Development of large copepods during spring in the Gulf of Alaska. *Progress in Oceanography* 32: 295–317.

6. Saito, H., & A. Tsuda (2000) Egg production and early development of the subarctic copepods *Neocalanus cristatus, N. plumchrus* and *N. flemingeri*. Deep-Sea Research I 47: 2141–2158

7. Tsuda, A., H. Saito, & H. Kasai (1999) Life histories of *Neocalanus flemingeri* and *Neocalanus plumchrus* (Calanoida: Copepoda) in the western subarctic Pacific. *Marine Biology* 145: 533–544.

8. Lenz, P. H., & V. Roncalli (2019) Diapause within the context of life-history strategies in calanid copepods (Calanoida: Crustacea). *The Biological Bulletin* 237: 170–179.

9. Schnack-Schiel, S. B., W. Hagen, & E. Mizdalski (1991) Seasonal comparison of *Calanoides acutus* and *Calanus propinquus* (Copepoda: Calanoida) in the southeastern Weddell Sea, Antarctica *Marine Ecology Progress Series* 70: 17–27.

10. Ward, P., R. Shreve, & G. C. Cripps (1996) *Rhincalanus gigas* and *Calanus similGlimus*: lipid storage patterns of two species of copepod in the seasonally ice-free zone of the Southern Ocean. *Journal of Plankton Research* 18(8): 1439–1454.

11. Ward, P., R. Shreve, & G. C. Cripps (1996) Mesoscale distribution and population dynamics of *Rhincalanus gigas* and *Calanus simillimus* in the Antarctic Polar Open Ocean and Polar Frontal Zone during summer. *Marine Ecology Progress Series* 140: 21–32.

12. Smith, S. L., & J. Vidal (1986) Biomass, growth, and development of populations of herbivorous zooplankton in the southeastern Bering Sea during spring. *Deep-Sea Research* 33(4A): 523–556.

13. Smith, S. L. (1982) The northwestern Indian Ocean during the monsoons of 1979: distribution, abundance and feeding of zooplankton. *Deep-Sea Research* 29(11A): 1331–53.

14. Bradford-Grieve, J. M., L. Blanco-Bercial, & I. Prusova (2017) *Calanoides natalis* Brady, 1914 (Copepoda: Calanoida: Calanidae): identity and distribution in relation to coastal oceanography of the eastern Atlantic and western Indian Oceans, *Journal of Natural History* 51(16–14): 807–836.

15. Binet, D., & E. Suisse de Sainte Claire (1975) Le copépode planctonique Calanoides carinatus répartition et cycle biologique au large de la Cote d'Ivoire. *Cahiers O.R.S.T.O.M., séries Océanographique* 14(1): 15–30.

16. Idrisi, N., M. J. Olascoaga, Z. Garraffo, D. B. Olson, & S. L. Smith (2004) Mechanisms for emergence from diapause of *Calanoides carinatus* in the Somali current. *Limnology & Oceanography* 49(4, part 2): 1262–1268.

17. Smith, S. L. (2001) Understanding the Arabian Sea: Reflections on the 1994–1996 Arabian Sea Expedition. *Deep-Sea Research II* 48: 1485–1402.

18. Smith, S. L., M. Roman, I. Prusova, K. Wishner, M. Gowing, L. A. Codispoti, R. Barber, J. Marra, & C. Flagg (1998) Seasonal response of zooplankton to monsoonal reversals in the Arabian Sea. *Deep-Sea Research II* 45: 2369–2403.

19. Piontkovski, S. A., A. Al-Mawali, A. Al-Kharusi, W. M. Al-Manthri, S. L. Smith, & E. A. Popova (2015) Mesozooplankton of the Omani shelf: taxonomy, seasonality, and spatial distribution. *International Aquatic Research* 7: 301–314.

20. Piontkovski, S. A., B. Y. Queste, K. A. Al-Hashmi, A. Al-Shaaibi, Y. V. Bryantseva, & E. A. Popova (2017) Subsurface algal blooms of the northwestern Arabian Sea. *Marine Ecology Progress Series* **566**: 67–78.

21. Prusova, I., S. L. Smith, & E. Popova (2011) *Calanoid Copepods of the Arabian Sea Region.* Sultan Qaboos University, Academic Publications Board, Muscat, Oman.

22. Smith, S. L., M. M. Criales, & C. Schack (2020) The large-bodied copepods off Masirah Island, Oman: An investigation of Southwest Monsoon onset and die-off. *Journal of Marine Systems.* https//:doi.org/10.1016/j.jmarsys.2020.103289.

17

Egg Diapause

Seasonal Diapause in Lakes and Temporary Ponds

There are odd disconnects among disciplines. Despite combining their societies (like ASLO, the Association for the Sciences of Limnology and Oceanography), for lake or ocean scientists to actually read in the sibling field must be rare. Thus, production by copepods of two kinds of eggs, those immediately developing and hatching and others prepared for extended rest phases, has long been known for species of *Diaptomus* (Figure 17.1) living in lakes. However, it was not recognized in marine genera until the late 1960s.

Valentin Hächer described two kinds of eggs produced by *Diaptomus laciniatus* and *D. denticornis* in 1902.[1] He also supplied German equivalents of two terms still applied to these types: *resting eggs* (Dauereier), those that sink to the bottom, delay development, then develop and hatch when conditions are again right for growth of the young; and *subitaneous eggs* (Subitaneier), those with immediate embryo development and hatching.

Diaptomus species are common in lakes generally and in temporary seasonal ponds, where their resting eggs can withstand some desiccation and long seasons of near total life suspension. Those species that inhabit seasonal ponds, like *Hemidiaptomus amblyodon* sampled in Ukraine,[2] only produce eggs that won't develop embryos and hatch *until* they have been buried, partially desiccated for a while, and then soaked. According to Larysa Samchyshyna and Barbara Santer, reports exist from the early 1900s until now of the simple, single-layer chitinous chorions (outer shells) of subitaneous eggs contrasting with resting egg chorions thicker and with as many as three added layers. The outermost layer of diaptomid diapause eggs is weakly rugose in some species, indistinguishable from subitaneous eggs in others. Many regional and local adaptations are evident. Nelson Hairston Jr. and coworkers[3] have estimated by isotopic dating downward in cores of lake sediments (deeper layers of mud or marl are older) that the few eggs still present up to 30 cm down-core can be more than 40 years, some over a century, from being spawned. Substantial fractions of those will hatch given the seasonally right conditions in a laboratory. They induced hatching of one egg collected from a layer dated at 333 years old. There is a vast literature awaiting the curious about the seasonality of life cycles in the Diaptomidae. The timing of active versus resting life-cycle phases is subject to very rapid evolution, and thus

Oar Feet and Opal Teeth. Charles B. Miller, Oxford University Press. © Oxford University Press 2023.
DOI: 10.1093/oso/9780197637326.003.0017

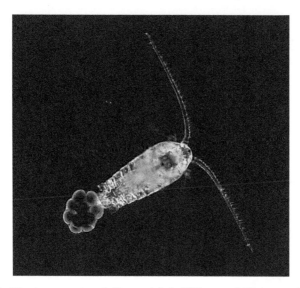

Figure 17.1 *Diaptomus castaneti.* Picture labeled "Tesoros de Esperanza"; the label says the specimen was from S.O.S. Lago de Sanabria in northeast Spain. Permission, Dr. Antonio Guillen, Proyecto Agua, Logoño, Spain.

can be fitted quite precisely to local conditions, not only to usual temperature or wet-dry cycles but to the seasonality of other fauna like predatory fish.[4]

The Solved Mystery of Seasonal Cycling in Estuarine and Nearshore *Acartia*

I was actually part of this story; thus, the temptation to tell it in more detail. It goes way back, but let's begin with observations from Narragansett Bay, in Rhode Island. The plankton there have been the subject of many sampling time series. A cleanly presented one (Figure 17.2) was by Elsie Hulsizer,[5] who was part of Ted Smayda's research group at University of Rhode Island (URI). Flow in and out of the north-south oriented Narragansett Bay is divided into two channels by Conanicut Island. Hulsizer's station was just to the northwest of the island's inner end. Plankton throughout the whole bay, however, are influenced by tidal exchange with the offshore coastal ocean. In the cold months from November to June, *Acartia hudsonica* Pinhey, 1926 (earlier recognized as *A. clausi*, now considered European) is dominant. In July those are gone, apparently altogether, replaced quickly though not in such dense numbers by *Acartia tonsa*. For a surprisingly long time, given knowledge of *Diaptomus* resting eggs, the seasonal alternation of these two copepods was difficult to explain.

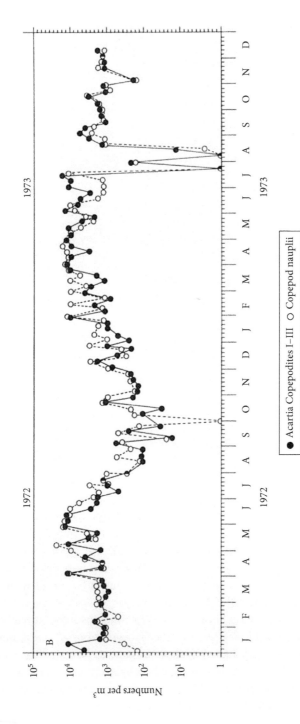

Figure 17.2 Redrawn from Hulsizer.[5] Abundance of the numerically dominant copepods at Station 2 in Narraganset Bay. The scale has logarithmic steps, but the scales between those, say 10^2 to 10^3 are linear, so the distance from 100 to 200 is the same as that from 900 to 1,000. The pattern survives that. Much of the jaggedness is typical of estuary time series. Gradients of abundance shift upstream and down with the tides. At a fixed station the abundance can shift many fold because it is measured at high spring-tide one week and at low neap-tide the next. Permission, Springer Nature.

Robert Conover,[6] from Yale University working in Long Island Sound, and URI instructor H. Perry Jefferies,[7] sampling both Narragansett Bay and Raritan Bay in New Jersey, found at all three sites that each species in its seasonal turn diminished to the point that feasible sample volumes were inadequate to find specimens. However, they thought it likely the species in low abundance were not simply gone, that a few persisted to resume production when conditions were again suitable. Perhaps the residual stock was offshore, available to wash into the bays again. The shifts were interpreted in terms of a difference in the "niche spaces" of the two species based on temperature and salinity. A sort of competition was invoked. A niche difference there certainly is, but it did not explain the very rapid appearance of each species, population bursts that were as evident in time series like Hulsizer's.[5]

In 1969, I was hired to replace the zooplankton professor at Oregon State University, the late Herbert Frolander, who had moved to several positions in the administration. Starting work at OSU in 1970, I learned that since about 1959 he had been conducting weekly sampling of zooplankton in Yaquina Bay (Figure 17.3), following a pattern from his thesis work[8] in Narragansett Bay. The sampling was continuing in 1970 with Sea Grant funding; one of Frolander's positions was Sea Grant director. The Clarke-Bumpus net samples were being taken by graduate students and were counted by technician Joan Flynn. She had learned to identify the abundant plankton species starting as an undergraduate. She had been at it through almost the whole series, so the counts were consistent. There were shelves filled with notebooks of counts from stations all up and down the bay, but nothing analyzed or published. The huge mass of data could, however, be attacked with computer processing that had only recently become feasible. Flynn, the students, and I converted the notebooks to cases and cases of IBM computer cards, so many that I still use them for bookmarks and shopping lists. We plotted up the data, station-by-station, for the dominant species and wrote a paper.[9]

The numerically dominant plankton were copepods, and *Acartia* species were most numerous, especially upstream from navigational marker 21. The lower bay had greater numbers of species washed in and out from the ocean. A float placed below marker 21 at high tide moved with the ebb into the ocean. One placed shortly upstream moved to about the bridge, then returned upstream on the flood. The names we used were on the order of *Acartia* (*Acartiura*) *clausi* and *Acartia* (*Acanthacartia*) *tonsa*. The first are like the *Acartia hudsonica* common on the US East Coast, without rostral filaments trailing from the copepodites' foreheads, and the latter are like *A. tonsa* with them. Lacking rostral filaments distinguishes subgenus *Acartiura* from the other subgenera.[10] More likely the copepods of Oregon estuaries and coastal waters related to *A. hudsonica* (and lately we use that name for them in Oregon) actually belong to several species. They have strongly different sizes in different seasons, different color patterns upstream and down. Much is left to learn. Eventually we learned that the *tonsa*-like

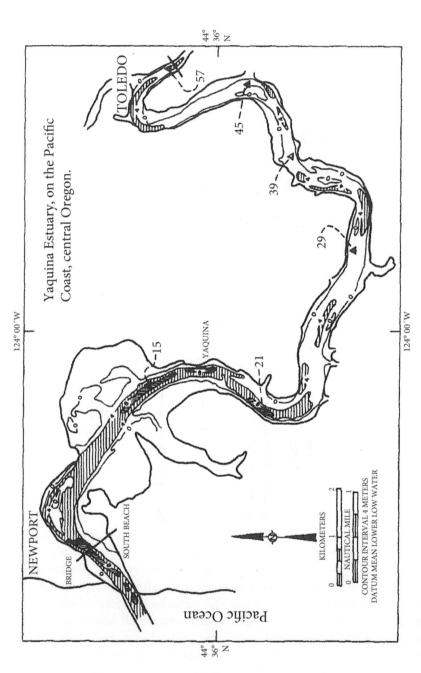

Figure 17.3 Bathymetric map of Yaquina estuary emptying to the Pacific on the central Oregon coast. Dr. Frolander's stations 1959–1972 were under the bridge and at navigation markers 15, 21, 29, and 39. Ken Johnson, during 1972–1974, sampled 21, 29, 39, 45, and 57, concentrating analysis on the summer stock of *Acartia californiensis* that is mostly restricted to that range. From Johnson.[17] US Government open access publication.

copepods with explosive population bursts upstream from marker 21 are, or are very close to, *Acartia californiensis* Trinast, 1976.[11] We call them that. It is found in upstream habitats in other Pacific estuaries. They are smaller (♀1.1 mm, ♂1.1 mm) than *A. tonsa*, including the species that moves north along the US Pacific Coast in winter (♀1.4 mm, ♂1.34 mm).

I'll show some of it below, but the Yaquina Bay data faced us with the same issue as the "*clausi* versus *tonsa*" shifts seen on the East Coast. How could substantial stocks dwindle to apparent extinction, then burst again into abundance after a long seasonal gap? The reappearances were not the months-long crawls expected given the development times of individuals and their modest fecundity. The answer came from a US government scientist, Edward Zillioux, in some meeting reports.[12, 13] The first was published in 1972 as part of a 500-page conversation among experts on culturing of marine organisms. Zillioux's presentation just pops up on page 73 as an offering to the 27 assembled wise men (yes, all of them men). In the course of developing means for culturing *Acartia tonsa*, he had discovered that it produces resting eggs at low temperatures and that they would not develop or hatch at temperatures below 4° C. The conversation then went something like this:

BATTGLIA (BRUNO): Is there a clear-cut difference in the eggs?"
ZILLIOUX: Yes. One type is obviously spiny while the other is smooth.

Zillioux also showed the assembly a micrograph (copied here from the book, Figure 17.4).

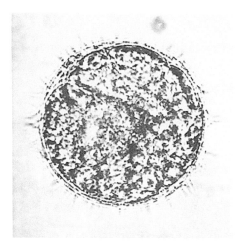

Figure 17.4 Resting egg of *Acartia tonsa* photographed by Edward Zillioux.[13] Size not given, but *A. tonsa* resting eggs are around 80 μm diameter, exclusive of the spines. Those spines cover the whole surface. Permission: Taylor & Francis, the only likely copyright holder, has no record of the book.

Dr. Zillioux holds a PhD in biological oceanography approved by a sort of combined committee from the universities of Rhode Island and Miami. He initiated his work on culturing *Acartia tonsa* while working at a US Naval Research Laboratory's biochemical ecology section. He had produced an earlier report of success at rearing *Acartia tonsa* through twelve generations. Successful culturing was a breakthrough for zooplankton research, and methods from a Zillioux and Wilson paper[14] and a later contribution[15] have been widely applied and modified, including by my students and me. Why would the US Navy care? Probably because they needed a "bioindicator" animal for testing the toxicity of ship's hull coatings. In the 1960s, and still for very large ships, coatings include some of the most toxic substances known, particularly tri-butyl tin. Bioindicators, more than copepods, were Zillioux's long-term interest.

He then moved to a position at the National Marine Water Quality Laboratory in Rhode Island, and it was from there that he attended the culturing symposium.[12] By the time of his second resting-egg report[13] he had moved to the Rosenstiel School at the University of Miami. Later work in many positions along the US East Coast concerned problems of chemical pollution, particularly by heavy metals like arsenic and mercury. He published often on those topics, at least until 2015 when he offered a thorough review[16] of mercury present in fish. He identified himself in that as the co-founder of an Environmental Indicator Foundation in Ft. Pierce, Florida (earlier names of his organizations specify *bio*indicators). He seems to be another exemplar that there can be life after copepodology. I have not met Dr. Zillioux, but we talked lately by telephone. He remains scientifically involved in his early 80s (Figure 17.5).

Ken Johnson and the Resting Eggs of *Acartia Californiensis*

The comings and goings of copepod stocks in Yaquina Bay suggested that some explanatory population dynamics might be accomplished there. Fortunately for achieving that goal, a professor of intertidal ecology had lost funding for a graduate student with obviously great talent, J. Kenneth Johnson, and I still had Sea Grant funding to keep the Frolander sampling program going in the bay a few more years. Ken took the stipend and went to work. Already in his late 20s, he was not quite the then-typical grad student. There were remarkable reasons for that, which he reviewed with me in 2019 at his pleasant home in Beaverton, Oregon.

Ken was born in 1943 in Green Bay, Wisconsin, only three years after his graduate advisor (me) was born in Minneapolis, and who in 1971 was still figuring out the faculty side of graduate education. It had taken Ken awhile to kick-start college and get to advanced studies. He was a member of the Church of Jesus

Figure 17.5 Edward Zillioux in 2017. This is a photo clipped from a YouTube video of him giving an invited talk at the International Society of Environmental Indicators. With his permission.

Christ of Latter Day Saints (LDS), a Mormon, and getting through Mormon youth can take a lot of time. His father, of Danish extraction, was raised a Lutheran on a family farm near New Denmark, Wisconsin. He had joined the LDS in his twenties when working in Arizona for a dairy farmer strongly of that faith. He married the farmer's daughter, which was connected to his LDS conversion, and in a few years the new family moved to New Denmark near the "ancestral home," as Ken puts it. Ken, is still active in the LDS faith, having served long as an Elder in all sorts of roles. In high school he excelled at and says he loved Latin; took pre-college classes in math, chemistry, and biology; and raced on the track and cross-country teams. His track coach recommended he join the National Guard to fulfill his military obligation, and he did. Basic training was at Fort Knox, Kentucky, and then he trained some months in Oklahoma, where he learned surveying (one of the military-engineering skills). He also got some experience driving "Deusenhalf" (2.5 tons) troop and supply trucks.

Completion of Guard training led to Ken's second educational delay, a two-and-one half year tour in Paris and western France as a church missionary. Off he went, was assigned to a more experienced missionary, and learned French in the streets. Returning to the United States in 1965, he needed a job to make money for college, but also to support a younger brother during his mission. His father ran a small business at that time, driving and repairing semi-trucks, which fed the family but without much left over. College tuition and room and board would not come from Ken's family. The work Ken found was 3,000 feet underground in a phosphorite salt mine in Moab, Utah, where he applied his surveying skills. He worked with drill-core data to figure out where shafts should go to maximize salt yields and minimize "spoils." Ken says the work was fascinating; indeed, it would have been meaningful, challenging work solving large-scale, three-dimensional puzzles. He mentions moving through the mine keeping track of "down" by lighted plumb-bobs hanging from the stope ceilings, anticipating collapsing shale layers and the constant potential for methane winning against the ventilators. "Even so, it was sciency. I loved it," he says.

College had to be at Brigham Young University, the LDS-founded school at Provo, Utah. BYU entrance was (and is) very competitive, but so was Ken, who got some scholarship help based on high school grades and extended life experience. He was exceptional at classroom scholarship there and later at OSU, scoring perfect A grades. He was much taken with an invertebrate-heavy oceanography course somehow taught out in the Utah desert by Prof. Lee Braithwaite. Another professor recommended that he a take summer course at the OSU Marine Science Center in Newport, Oregon, with Jeff Gonor, which he did. That clearly led to him eventually coming to study with us. He had worked as both a teaching assistant and lab tech on invertebrate studies while at BYU, and after his senior year he worked a summer with famous marine larval ecologist Rudi Scheltema at Woods Hole Oceanographic. He wrote a paper based on his study of feeding and growth in *Crepidula fornicata*, a snail that has fascinated many.[17] Ken arrived at OSU with wide experience.

That was the school part, but BYU is coed. Ken met his future wife Kathy at a fall student assembly and started a "root beer courtship." (Root beer does have a special place in BYU life.) Kathy had already graduated from the business program. They rented a small house, and together they got Ken through four undergraduate years and eight at OSU. A picture from their 2018 Christmas card (Figure 17.6) shows them still together. They have seven adult children (most born while Ken studied at OSU), 27 grandchildren, and (as of August 8, 2019, Ken's 76th birthday) one great-grandchild. A chat with them suggests that those progeny are their greatest sources of joy.

Three initial applications for graduate training in marine biology all failed, a bit odd for the department valedictorian at a major university. That led to a fifth year at BYU, much of it devoted to invertebrate paleontology. Utah is paradise for hunting fossils ranging from Cambrian trilobites to late Cretaceous

Figure 17.6 Kathy and Ken Johnson of Beaverton, Oregon, from their 2018 Christmas card. With their permission.

dinosaurs. Ken still chips a trilobite free occasionally. In 1972 he came to OSU, initially accepted for work with Jefferson Gonor. As mentioned, funding for that did not work out. So, after aceing all his basic oceanography courses, Ken began plankton sampling in Yaquina Bay, initially following but then doubling Frolander's weekly sampling frequency from June to November. He also concentrated on the upstream section (Figure 17.2, above marker 21) where the waters warm up every summer. Spring-summer is our upwelling season, which brings cold water, around 7 to 13°C, up from depth in the Pacific and exchanges it into the bay to above marker 15. It is also a season of very low rainfall, so that freshwater flow all but stops. Thus, higher salinity works its way upstream, reaching 30 grams salt/kg at marker 29. At the same time, downstream water and population losses are very limited. Sunny days warm the water at marker 29 to 15°C by June and 20°C in July–August; water at that site cools back to 12° in September and then to 10°C in October. With similar conditions from marker 21 to well inland above marker 45, a population of warm-water adapted *Acartia californiensis* (Figure 17.7) can develop. Ken sampled it with Clarke-Bumpus nets (Chapter 5) towed obliquely at each station with 112 μm mesh to catch first copepodites and older stages.

At least the females P5 of *A. californiensis* are very distinctive (Figure 17.7). Males with rostral filaments found with such females in upper Yaquina Bay are reliably *A. californiensis*. The population dynamics[18, 19] of this small copepod

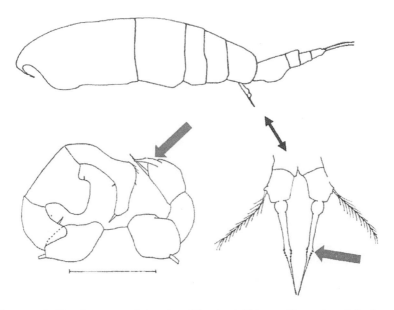

Figure 17.7 Key characters of *Acartia californiensis* Trinast. Above, female body, length around 1 mm. Below left: male P5 posterior view; red arrow indicates distinctive spines. Below right: female P5; red arrow indicates distinctive feature of terminal spines. P5 scale bar 0.05 mm From Trinast.[11] Permission, Brill, Leiden.

were just that, *dynamic* (Figure 17.8). Sampling in a tidal flow channel introduces odd cycling at any fixed station, as stated above regarding Narragansett Bay. Because high water and low water shift later in time by 50 minutes per day, a full cycle (from full moon to dark of the moon, or dark to full) is completed every two weeks. Thus, results of weekly sampling at one point, taken at the same and a convenient time of day (say 10 a.m.) will appear to cycle up then down, up then down, over and over through the year. A population with an along-stream distribution peak with tails upstream and down will have that peak at different stations each week, shifting back and forth. Ken dealt with that in two ways. One was to sample all stations twice per week, count the samples (no./m³) and then estimate the density at the abundance peak (Figure 17.8). The other was using then recent US Army Corps of Engineers sounding data, plus tidal height data at the times of sampling to convert the abundance estimates to an approximate number of copepods in the entire bay. Numbers at the seasonal peaks were around 100 billion copepodites and approximately 80 billion adults in 1972 and 1973. No reason was obvious for the lower peak stock numbers in 1974: around 90 million copepodites and 30 billion adults.

Patterns for both calculations look alike (Figure 17.8). There were pulses in 1972 of greater copepodite abundance about June 26, July 9 and 21, August 7,

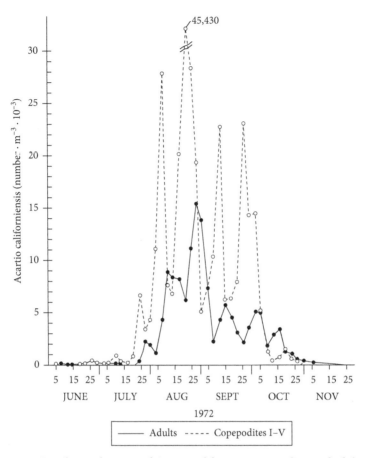

Figure 17.8 Population densities of *Acartia califoriensis* copepodites and adults in upper Yaquina Bay, Oregon in 1972. Scale is numbers per cubic meter averaged from water surface to bottom (mostly well mixed by tides). From Johnson.[18] US Government, open access publication.

and so on, roughly each two weeks until October. There was an extra adult peak around October 15, 1972, after which the population quickly disappeared. The intervals were close to the development times expected from laboratory rearing experiments at the temperatures in the upper bay at the time. Each of the eight copepodite peaks was followed, as expected, by an adult peak 4–7 days later. In fact, the cohorts in each of the larger five or six summer peaks, 1972, 1973, and 1974, could be progressively followed through all six copepodite stages. Vast detail is available in Ken's dissertation.[19]

Let's think a moment on that population sequence (Figure 17.8). One of history's most significant mathematicians, Leonard Euler (1707–1783), showed that given constant rates of mortality at each age in life and with constant rates of reproduction at each age of adulthood, the rate of exponential increase of a population of organisms can be calculated. Moreover, the age-frequency distribution (X percent babies, Y percent young adults, Z percent old folks) would remain the same from a modest initial cohort out to infinite numbers. For *A. californiensis* in Yaquina Bay, none of that is true; the age distribution snaps back and forth from mostly nauplii (though not counted; too much like the "*A. hudsonica*" nauplii also present), to mostly copepodites, to mostly adults. Euler's beautiful calculus is useless.

In fact, no population behaves so stably as Euler required, so the equations are nice thinking/teaching tools, sort of like the "perfect gas laws." It occurred to Ken and me that the explanation has to be that the adults of *A. californiensis* do not survive and reproduce for weeks as they can in the lab. Something takes them out almost as soon as they mature, so they only get a few days to pump out a batch of subitaneous eggs before crashing. Watching them grow in beakers, the females only became readily visible to us when their bodies were loaded with eggs. We came to believe that planktivorous fish, visual predators mostly abundant downstream, might get the news somehow that those very good meals were available and could migrate upstream to eat them. A sampling scheme would have been possible to demonstrate that, but we didn't have the time or resources. Something did remove them in each of three years, sustaining the strange and beautiful population cycling each summer into October.

Ken already knew, based on Zillioux's studies, that the autumn disappearance of copepodites must be caused by a shift to production of resting eggs. And he knew that return of the population to the water column must derive from nauplii hatching from those eggs in the sediment. Nevertheless, Ken checked by collecting the upper few centimeters of sediment at marker 39 in winter, sieving away the mud and sorting thousands of the spiny resting eggs of *Acartia* species. Holding those at various temperatures and salinities, then rearing the nauplii that hatched, he learned that both *A. californiensis* and *A. clausi* resting eggs were on the bay floor. He found no way of distinguishing the eggs of those two species. So back to the bay for careful work with sediment collected in February, when no *A. californiensis* and very few *A. clausi* were in the water. Eggs of both species hatched at 15°C and salinity around 20 g salt/kg. In those conditions, proportions of the nauplii reared to copepodites suggested about 55% were *A. californiensis*, and experiments at other temperatures showed that those should hatch readily from early June in the upper bay.

Next, Ken ran experiments to determine the conditions under which *A. californiensis* females produce resting eggs. He collected females near marker 39 in early October 1975, when temperature was 14.9°C, salinity 27 g salt/kg. After brief feeding and holding at 17 °C, groups of 50 ♀ and 25 ♂ were established in

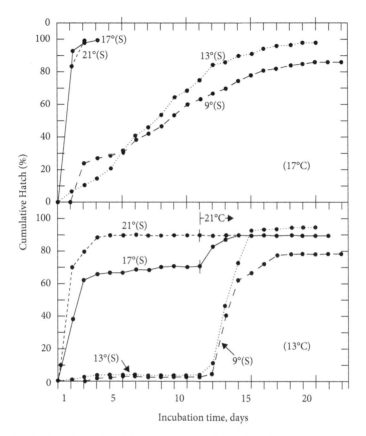

Figure 17.9 Cumulative hatching versus time of *A. californiensis* eggs spawned at four temperatures (e.g., 17º(S)), then held at 17°C (above) and 13°C (below). On day 11, the eggs held at 13°C were warmed to 21°C. From Johnson.[18] US Government open access publication.

containers and held at 21, 17, 13 and 9°C. After some days of acclimation to those temperatures with food, all their eggs were removed and discarded. Then new eggs were collected and distributed to different temperature and salinity conditions to determine hatching rates. A little sample of Ken's results (Figure 17.9) shows a shift in modes of egg production comes between resting eggs at 13°C and subitaneous eggs at 17°C. The comparisons (Figure 17.9) include all four of the holding or spawning (S) temperatures, in the figure: 21º(S), 17º(S), 13º(S), and 9º(S). All four were used to check times to hatching, but two are enough to show the effects of different spawning and hatching temperatures.

Only a few eggs spawned at 13°C and 9°C, and then held at 13°C, developed at all until warmed up on Day 11. Then large fractions developed as if primed to do so. That is a common feature of diapause states, though the priming was not particularly prolonged in this case (10 days only). The organisms do not resume activity (in this case development) until suitable conditions appear. However, experiencing unsuitable conditions for a time, primes them for rapid resumption when things do turn good. Eggs spawned at 17°C and 21°C, then held at 17°C, all hatched in a day or two, while those spawned by females in colder water initiated development to hatching at 17°C at widely varying intervals, some taking 20 days to hatch. Thus, cold-spawned eggs exhibited a sort of leaky diapause. That is, when the cold-spawned eggs were not primed by more cold (Figure 17.9 upper frame), onsets and durations of development were strongly variable. Ken's experiments with eggs collected from bay sediments suggested groups of "unprimed eggs" were present there, as well as cold primed eggs, and would gradually begin to develop with modest warming (e.g., 17°C).

Other studies have compared production of subitaneous and diapause eggs in similar experiments. The late Nancy Marcus,[20] working at Woods Hole with *Labidocera aestiva* (recall *Labidocera*'s large eye lenses from Chapter 3), used two-factor experimental designs examining the interactions of temperature and daylength on production of subitaneous versus resting eggs. For at least that copepod, photoperiod is controlling, weakly modified by temperature. It is active in summer, and it switches to production of benthic resting eggs in autumn when daylengths are short and the water cools. Marcus worked extensively on resting eggs for many years, mostly at Florida State University in Tallahassee. She produced many good reviews on copepod resting eggs.

In the early 1980s Barbara Sullivan (see Chapter 2, rediscoverer of opal teeth) and student Liana McManus,[21] working both on Narragansett Bay and with large (13 m³) experimental tanks, examined conditions of subitaneous and resting egg production of *Acartia hudsonica* dominant in winter to spring. Its strategy is to produce subitaneous eggs at cold season temperatures:

> During early summer *A. hudsonica* produces eggs that do not hatch at the temperatures at which they are produced (> 16°C), and they will hatch only after exposure to low temperature. These eggs were also present in the sediments of Narragansett Bay and are probably true diapause eggs. (from their abstract).

Narragansett is at the boundary between the distribution of the lower-latitude *A. tonsa* and that of *A. hudsonica* farther north. In their bay data for 1981 and 1982, *A. hudsonica* seems to erupt into the water column at <10°C in mid-winter, January and February, quickly reaching a few thousand per cubic meter.

In 1979, Ken Johnson found a job with the Pacific States Marine Fisheries Commission (PSMFC) in Portland, Oregon. He was clear that permanent work like that of most planktologists, who chase research grants, was not for him. At the PSMFC he served as the Regional Coordinator of coastwide coded wire tagging (CWT) of hatchery salmon and steelhead trout for every river draining to the Pacific from Alaska to California.[22] Tiny bits of wire with a laser etched code indicating hatchery origin, release group and date are injected into the nasal cartilage of salmon smolts just before they are released to the rivers. Magnetometers at the processing lines identify tagged salmon and steelhead captured in the fisheries or returning to the hatcheries. Their snouts are collected, the wires are removed and read, the data collated. As coordinator, Ken maintained the statistics and an on-line database. Thus, return rates can be accurately estimated. While on the job, Ken finished his dissertation, and (as Dr. Johnson) he stayed with that job for 29 years. Later he did the same work seven years for the Oregon Fish and Wildlife Department. Now he is retired, living outside Portland, enjoying family history research and some travels with Kathy.

Kasahara, Onbé, and Uye: Resting Eggs in the Inland Sea of Japan

A paper appearing in 1974[23] was the first of a series from Hiroshima University about copepod resting eggs in the sediments of the Inland Sea of Japan. It does not cite Zillioux, and it appears to be an independent discovery of resting eggs from marine copepods. Takashi Onbé started the work, which came from the Plankton Laboratory of the Department of Fisheries Science. Dr. Onbé was and is an expert on the taxonomy of marine cladocerans, and those have been long known to produce resting eggs that reside in sediments over seasons unsuitable for the pelagic life stages.

Curious what resting eggs might be found in sea bottom muds of the shallow (maximum depth ~11 m) Inland Sea of Japan, Onbé obtained some grab samples of that mud during winter when no cladocerans were in the water. Sieving the mud from the upper few centimeters, he examined what was left on his screen with a microscope. Tiny eggs were indeed there, so he put them into warmer water and many hatched. Some were cladocerans, but there were also copepod nauplii. He probably referred the matter to his associate Shogoro Kasahara and to Shin-ichi Uye, who was then a graduate student. Their first paper[23] represents an apparently independent discovery of *Acartia* resting eggs, and also those of *Centropages abdominalis*, *Tortanus discaudatus* and several other species. The methods section credits Dr. Onbé with the initial recognition. Those specific identifications (Figure 17.10) were partly from egg shapes and decorations described in the literature (e.g., *Tortanus*), partly from eggs spawned by females collected from the Inland Sea. Dr. Kasahara and Shin-ichi Uye went ahead with determining the life histories of all six of the species with identifiable resting eggs

Acartia erythaea   Acartia clausi   100 µm

Figure 17.10 Resting eggs from two *Acartia* species collected from seafloor mud of the Inland Sea of Japan. Microphotographs from Kasahara et al.[23] Permission, Springer Nature.

in the mud.[24] Dr. Onbé told me by email (2019) that he had been busy with the cladoceran aspects of the story.

Shin-ichi (Shin) Uye eventually became an internationally recognized student of zooplankton. I interviewed him in summer 2017. He is the son of a family that had a successful tofu-making business in Hiroshima. Spending a little time in Japan shows how central tofu is to the national diet, not so much so as rice but close. The family lived near the Inland Sea, and Shin says it was his boyhood playground for fishing and clamming. He sold clams for pocket money. He contemplated going to a maritime college, but with a high school teacher he reviewed the job market for ship's officers. At the time it wasn't good. Instead he took the entrance exam for fishery science at the University of Hiroshima, a department mostly active in mariculture. An initial project for an undergraduate thesis was to rear groupers from eggs. His then advisor, Takashi Onbé, provided some adults to maintain, but they never spawned. So Shin switched to culturing copepods as mariculture foods and readily succeeded with some *Pseudodiaptomus* (an estuarine genus that is not in the family Diaptomidae at all). That thesis work brought him into planktology, and Dr. Onbé suggested that Shin work on the newly discovered resting eggs.

With a small boat, the *Acartia*, Kasahara and Uye made collections in the central Inland Sea each two weeks for just over a year, May 1973 to May 1974. Those were a grab sample of mud and a vertical net tow for counting active stages. All of the species are distinctive as copepodites. Their paper [24] covers all the species with recognizable eggs. Each received a neat graph like those copied here (Figure 17.11) for *Acartia erythaea* (subgenus *Odontacartia*, not closely related to *A. tonsa*, but a late summer species in Asian waters) and *Acartia clausi* (winter, now called *A. omorii*). Numbers of *Acartia* copepodites of both species (including adults, thousands to tens-of-thousands per cubic meter) peak just before

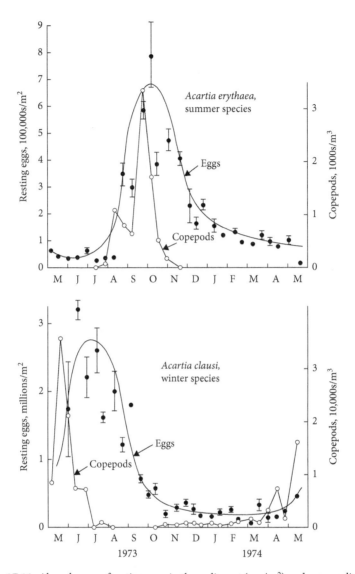

Figure 17.11 Abundances of resting eggs in the sediment (no./m²) and copepodites in the water column (no./m³) of the Inland Sea of Japan. *Acartia clausi* were present in the water from October to March, then increased in May–June to several tens of thousands per cubic meter. Relabeled from Kasahara et al.[24] Permission, Springer Nature.

the peaks of resting egg abundance in the sediment (hundreds of thousands to millions per square meter—based on sorting a few square centimeters). Then in each case the egg numbers decline over the months until the copepodites re-appear. That likely is because all marine sediments with oxygen dissolved in the interstitial water are crawling with invertebrate animals. Many worms in particular simply ingest mud as they move through, getting nutrition by digesting anything organic that will digest. That would certainly include resting eggs. Many worms include gizzard-like grinders in their foreguts to break open small meals like resting eggs. However, Nancy Marcus[25] has demonstrated that both subitaneous and resting eggs of *Labidocera* aestiva can pass completely through the guts of small polychaete worms (*Capitella* and *Streblospio*, for those who know worms), and a somewhat smaller proportion can still hatch. In any case significant ongoing hatching is not implied by the strong decline in abundance. Resting eggs contribute to the nutrition of the "benthos" (seafloor life).

By the time of their resting eggs II paper,[24] the Hiroshima planktologists were citing Sazhina,[26] Zillioux, and some others who had suggested there might be resting eggs without demonstrating them.* When discoveries are ready to happen, they sometimes come showering out from all over.

The research on calanoid copepod resting eggs in sea-bottom muds continued at Fukuyama (site of the agricultural and marine programs of Hiroshima University) through papers III, IV, and V, all in *Marine Biology*. Shin-ichi Uye did not participate in III, because he had taken the year 1975 out for study and work at the Scripps Institution of Oceanography (SIO) in La Jolla. He became close friends there with fellow plankton student and copepodologist David Checkley, who helped him perfect his English. Abraham Fleminger worked with him on environmental factors inhibiting and promoting hatching of eggs (some thought to be resting) of three species of *Acartia* that can be collected offshore from the SIO pier using small boats. They showed effects of temperature, salinity, oxygen availability, whether there was sediment present or not, and illumination on hatching.[27] Lack of oxygen can inhibit development of both resting and subitaneous *Acartia* eggs for a considerable time without killing them.

Shin says, about his time in La Jolla, that the lectures by the late Michael Mullin on biological oceanography opened his eyes to the research possibilities regarding the interactions of phytoplankton and zooplankton swirling together in waves and currents. The university at Hiroshima did not at that time offer a PhD in its Faculty of Fisheries. So next, Shin spent two years at Tohoku University in Sendai, advised by Prof. S. Nishizawa, working on the life history of two species of *Acartia* in Onagawa Bay.[28] He returned to Hiroshima to write his dissertation while starting new research work in the plankton laboratory. His early work as a young faculty member was extended study of the plankton of the Inland Sea: copepod egg production rates, much more on resting versus subitaneous eggs, feeding and seasonality. He was associate professor by age 34 and eventually became professor on Dr. Onbé's retirement.

---

* L. I. Sazhina was cited by Zillioux and Gonzalez.[13] She was actually first to report *marine* copepod resting eggs.

Shin is married to an artist talented in several media, Setsuko Uye, and during 1995 they came to Corvallis (home of Oregon State University) on sabbatical, enrolling their two sons in public school where they learned English under the intense pressure faced by immigrant children here for four centuries. Setsuko painted and kept them all healthy. Shin ran successful experiments on the viability of copepod eggs spawned by females fed diatoms versus those fed dinoflagellates.

There was a challenge before the world community of marine copepodologists set by Adriana Ianora and Serge Poulet.[29] They had noticed that something about diatoms of numerous species would interfere with cell division in copepod embryos. The challenge was to prove that diatoms were the source of the problem. After he got some diatom cultures going from local ocean waters, Shin and I set out to sea in a small coastal boat. He found that more viscerally challenging than the Inland Sea of Japan. However, we got good collections of *Calanus pacificus* females. An outstanding master of culture work, Shin held them in groups of 10 and fed some groups diatoms, others dinoflagellates. The result was that on day 5 or day 6 of eating *Chaetoceros difficilis* (a diatom) egg viability dropped, then went to zero. Unviable eggs then kept coming. However, Shin then switched the food of several groups fed the diatoms to *Prorocentrum minimum* (a dinoflagellate) on day 12, and, 一丁上がり (Japanese for *voila*), females in those groups produced mostly viable eggs after a day and then all viable eggs to day 18, when he ended the experiment.[30] One example is shown here (Figure 17.12). Ianora and Poulet were surely right.

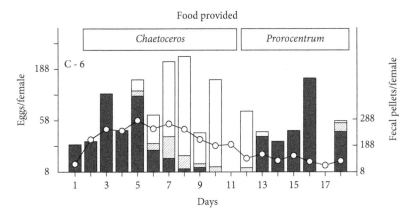

Figure 17.12 Daily egg production (bars) and feeding activity (connected circles) of *Calanus pacificus* females. Dark and shaded bars for eggs that hatched. Shaded bars for deformed nauplii. Open bars for no development. *Chaetoceros* is a diatom. *Prorocentrum* is a dinoflagellate. Connected circles were fecal pellet counts, a measure of feeding rates. From Uye.[30] Permission, Inter-Research, *Marine Ecology Progress Series*.

The females are not themselves poisoned by something in the diatoms, but when eating them they do supply something fatal to their eggs. Apart from gamete production, adult copepods do not have any cell division, and that is the sensitive function. Dr. Ianora and colleagues eventually showed that the toxin is the aldehyde 2-trans, 4-trans decadienal (dubbed DD). A great deal has been learned about this toxin.[31] It is not present in all diatom species.

Back in Japan, Shin's studies eventually turned to jellyfish, to moon jellies (*Aurelia*) and to the massive *Nemopilema*, which in outburst years makes serious problems for fisheries in the Sea of Japan and down the east coast of Honshu. Along with students and helpers, he sustained that work even through some years serving as vice president of Hiroshima University, helping it through some institutional issues. Shin is a member of the World Association of Copepodologists and has been its president. Now in retirement, he continues work on jellyfish and keeps track of copepodology around the world. At least occasionally, his path passes back through Corvallis as it did in October of 2009 after he attended a meeting in Portland (Figure 17.13).

Figure 17.13 Shin-ichi Uye at the author's home at Corvallis, Oregon, in October of 2009. Included with Dr. Uye's permission.

References

1. Hächer, Valentin (1902) Über die Fortpflanzung [reproduction] der limnetischen copepoden des Titisees. *Berichte der Naturforschschenden Gellschaft zu Freiburg* 12: 1–133.

2. Samchyshyna, L., & B. Santer (2010) Chorion structure of diapause and subitaneous eggs of four diaptomid copepods (calanoida, diaptomidae): SEM observations. *Vestnik Zoologii* 44(3): doi 10.2478/v10058-010-0017-9.

3. Hairston, N. B. Jr., R. A. Van Brunt, C. M. Kearns, & D. R. Engstrom (1995) Age and survivorship of diapausing eggs in a sediment egg bank. *Ecology* 76(6): 1706–1711.

4. Hairston, N. B. Jr., & E. J. Olds. (1986) Partial photoperiodic control of diapause in three populations of the freshwater copepod *Diaptomus sanguineus*. *Biological Bulletin* 171: 135–142.

5. Hulsizer, E. E. (1976) Zooplankton of lower Narragansett Bay, 1972–1973. *Chesapeake Science* 17: 260–270.

6. Conover, R. J. 1956. Oceanography of Long Island Sound, 1952–1954. VI. Biology of *Acartia clausi* and *A. tonsa. Bulletin Bingham Oceanographic Collection* 15: 156–233.

7. Jefferies, H. P. (1962) Succession of two *Acartia* species in estuaries. *Limnology and Oceanography* 7(3): 354–364.

8. Frolander, H. F. (1955) The Biology of the Zooplankton of the Narragansett Bay Area. Masters thesis, Brown University, Providence, RI.

9. Frolander, H. F., C. B. Miller, M. J. Flynn, S. C. Myers, & S. T. Zimmerman (1973) Seasonal cycles of abundance in zooplankton populations of Yaquina Bay, Oregon. *Marine Biology* 21: 277–288.

10. Bradford, J. M. (1976) Partial revision of the *Acartia* subgenus *Acartiura* (Copepoda: Calanoida: Acartiidae), *New Zealand Journal of Marine and Freshwater Research* 10(1): 159–202.

11. Trinast, E. M. (1976). A preliminary note on *Acartia californiensis*, a new calanoid co-pepod from Newport Bay, California. *Crustaceana* 31(1): 54–58.

12. Costlow, J. D. Jr. (1969) *Marine Biology, Vol. V.* Gordon & Breach, New York.

13. Zillioux, E. J., & J. G. Gonzalez (1972) Egg dormancy in a neritic calanoid copepod and its implications to overwintering in boreal waters. In *Fifth European Marine Biology Symposium*, ed. B. Battaglia, 217–230. Piccin Editore, Padova.

14. Zillioux, E. J., & D. F. Wilson (1966) Culture of a Planktonic Calanoid Copepod through Multiple Generations. *Science* 151(3713): 996–998.

15. Zillioux, E. J. (1969) A continuous recirculating culture system for planktonic copepods. *Marine Biology* 4: 215–218.

16. Zillioux, E. J. (2015) Mercury in fish: history, sources, pathways, effects, and indicator usage. In *Environmental Indicators*, ed. R. H. Armon & O. Ha'nninen, 743–766. Springer Science+Business Media, Dordrecht.

17. Johnson, J. K. (1972) Effect of turbidity on the rate of filtration and growth of the slipper limpet, *Crepidula fornicata* Lamarck, 1799. *The Veliger* 14(3): 317–320.

18. Johnson, J. K. (1980a) Effects of temperature and salinity on production and hatching of dormant eggs of *Acartia californiensis* (copepoda) in an Oregon estuary. *Fishery Bulletin* 77(3): 567–584.

19. Johnson, J. K. (1980b) Population dynamics and cohort persistence of *Acartia californiensis* (Copepoda: Calanoida) in Yaquina Bay, Oregon. PhD dissertation, Oregon State University.

20. Marcus, N. H. (1982) Photoperiodic and temperature regulation of diapause in *Labidocera aestiva* (Copepoda: Calanoida). *The Biological Bulletin* **162**(1): 45–52.

21. Sullivan, B. K., & L. T. McManus (1986) Factors controlling seasonal succession of the copepods *Acartia hudsonica* and *A. tonsa* in Narragansett Bay, Rhode Island: temperature and resting egg production. *Marine Ecology Progress Series* **28**: 121–128.

22. Johnson, J. K. (1990) Regional overview of coded wire tagging of anadromous salmon and steelhead in Northwest America. *American Fisheries Society Symposium* **7**: 782–816.

23. Kasahara, S., S. Uye, & T. Onbé. 1974. Calanoid copepod eggs in sea-bottom muds. *Marine Biology* **26**: 167–171.

24. Kasahara, S., S. Uye, & T. Onbé. 1975. Calanoid copepod eggs in sea-bottom muds. II. Seasonal cycles of abundance in the populations of several species of copepods and their eggs in the Inland Sea of Japan. *Marine Biology* **31**: 25–29.

25. Marcus, N. H. (1984) Recruitment of copepod nauplii into the plankton: importance of diapause eggs and benthic processes. *Marine Ecology Progress Series* **15**: 47–54.

26. Sazhina, L. I. (1968): On hibernating eggs of marine Calanoida. *Zoologische Zhurnal* **47**: 1554–1556.

27. Uye, S., & A. Fleminger (1976) Effect of various environmental factors on egg development of several species of *Acartia* in Southern California. *Marine Biology* **38**: 253–262.

28. Uye, S. (1980) Development of neritic copepods *Acartia clausi* and *A. steueri*. I. Some environmental factors affecting egg development and nature of resting eggs. *Bulletin of the Plankton Society of Japan* **27**(1): 1–9.

29. Poulet, S. A., A. Ianora, A. Miralto, & L. Meijer (1994) Do diatoms arrest embryonic development? *Marine Ecology Progress Series* **111**: 79–86.

30. Uye, S. (1996) Induction of reproductive failure in the planktonic copepod *Calanus pacificus* by diatoms. *Marine Ecology Progress Series* **133**: 89–97.

31. Ianora, A., A. Miralto, S. A. Poulet, Y. Carotenuto, I. Buttino, G. Romano, R. Casotti, et al. et al. (2004) Aldehyde suppression of copepod recruitment in blooms of a ubiquitous planktonic diatom. *Nature* **429**: 403–407.

18

Molecular Genetics Applied to Copepods

On the Insight Generating Power of DNA Sequence Data

Over the last five decades the study of evolution has acquired an array of tools from molecular genetics. They provide capacity to compare the DNA sequences of genes among sets of individuals and species known to be more versus less closely related based on anything, on "characters" that we observe about them. For copepods characters are most often from body form, but also from behavior, color patterns, habitat details, and geographic distribution. As discussion here moves to allozymes and DNA sequences of genes, keep in mind that for taxonomy and interpreting phylogeny variants of both are just character states, much like body length, setal details, and limb segmentation. However, genes also underlie expressed characters, and they feel particularly powerful.

In a sense, studies of genes go back to prehistoric trials at selective breeding of domesticated animals, to choosing more productive rice, wheat, and maize varieties. The name "genes" came much later, of course, as Mendel's rules of "particulate" inheritance (taken originally from crossing pea varieties) came into delayed prominence. According to Jim Endersby,[1] the term "genes" just gradually replaced "factors" and "Anlage," as Mendel called the character determinants that segregated independently in pea-seed formation. Hugo de Vries had termed them *pangenes* (after an old evolutionary notion, pangenesis). Endersby credits William Bateson (1861–1926) with coining the term "genetics."

Initially gene-sequencing tools were equivalent to hand saws, but over those five decades, figurative motors were added as means for actually reading the codes, for brewing multiple copies of genes, for automated sequencing, and now for rapid-fire, full-genome analyses. Also, practitioners can read and quantify messenger RNA (mRNA) transcribed from active genes, learning which genes are active, which not, under different circumstances. Equally important, we have the elaborate sequence-data analysis systems known as *bioinformatics*. Following the applications of it all to copepods may require you to understand the results of this progression fairly well. So some generalizations about molecular genetics are provided in Box 18.1. The scientific progression was followed in her professional career by Ann Bucklin. She applied the new tools to copepods soon after each appeared.

Oar Feet and Opal Teeth. Charles B. Miller, Oxford University Press. © Oxford University Press 2023.
DOI: 10.1093/oso/9780197637326.003.0018

Box 18.1 Basic Molecular Genetics

The key thing to know is termed the "central dogma of molecular biology" developed in and since the 1960s. If you know that, skip this. (1) Information for producing ribonucleic acid (RNA) polymers and protein amino-acid chains is stored as long sequences of four nucleotide bases in deoxyribonucleaic acid (DNA). Those genes are two chains of ribose sugars, each sugar ring bonded to one of four nucleotides: adenosine (A), Cytocine (C), Guaine (G) or Thymine (T). The two chains are held together by relatively weak bonds (hydrogen bonds) between the A and T bases and the C and G bases. The along-chain sequence literally forms a sentence in which each nucleotide can specify the bases to form an RNA molecule that an enzyme generates as a new sequence. That process is termed "transcription." The ribose in RNA has one more oxygen bonded to its ring, and the thymine of DNA is replaced by a fifth nucleotide, uracil (U). RNA can itself be a chemical catalyst, usually as parts of organelles called ribosomes. Or, a three nucleotide subset in a DNA sequence can specify one of the 20 amino acids that constitute almost all proteins. In eukaryotes (with cell nuclei), the DNA is wound rather complexly with proteins in chromosomes.

(2) Protein "readings" (by enzymes) of the DNA have an intermediate step, the formation of messenger RNA (mRNA), which transfers across the cell to complex molecular machines termed ribosomes. Those are macromolecular particles that take in the mRNA, collect amino acids from the cytoplasm (delivered by transfer RNA, tRNA), and bonds them one by one into the protein specified in the mRNA sequence. Yes, there are 64 permutations of four bases taken in triplets. So there are synonymous codes for some amino acids, and some codes mean start or stop in production of mRNA and, thus, proteins. The speed and precision of this ribosomal assembly, termed translation, are both astounding.

(3) DNA is not particularly stable; it is damaged often and at random by sundry radiation and most importantly by free-radicals (not all the chemistry is covered here) resulting from oxidative metabolism. DNA-repair enzymes move about in the nucleus, fixing some damaged chains and nucleotides, but not all.

(4) Cell division involves separation of the two chains of DNA and addition of new complimentary nucleotides to each by enzymes (DNA polymerase) running along the chains. Changes in the sequences resulting from unrepaired damage are the mutations upon which selective evolution depends. Changed ACGTTAGAGCT .... sequences result in changed mRNA and, thus, amino acid sequences.

(5) Other facts. The oxidative metabolic centers of cells, mitochondria, have their own small array of genes (mtDNA), transcription enzymes and ribosomes. In eukaryotes (animals with cell nuclei), those genes are usually only partial codes for the mitochondrion's functional enzymes, the other parts having moved to the nucleus. Also, only a few eukaryotes have any DNA repair enzymes in their mtDNA, so its evolution is much faster than that of nuclear DNA. Thus mtDNA can be excellent (with caveats) for tracing genetic variation between species, genera and beyond. Sequences of specific mito-chondrial genes, particularly cytochrome oxidase I (mtCOI), have become popular as DNA barcodes that usually identify animals at the species level. Nuclear genes that do not code for functional RNA or proteins, such as "inter-nally transcribed spacers" (DNA within functional genes that is edited out at translation), also evolve rapidly, and are popular for identifying changes be-tween species and even individuals. More stable genes, such as those coding for ribosomal RNA (rRNA) are excellent for tracking changes over longer spans of time.

(6) Current sequencing methods are based on DNA polymerases (extracted from bacteria) like those that cells use to copy, transcribe and translate DNA. A key technique is the polymerase chain reaction (PCR, Box 18.2). Sequencing depends on stopping DNA polymerizations (basically extended PCR) at different chain lengths by adding modified versions of the nucleotides (A & T, G & C) that both jam the polymerases producing replicates of as many lengths as the chain being amplified and also fluo-resce in four distinctive colors. The DNA produced is chromatographed, separating the chains by length. Then each link in the chain is identified from the four distinctive colors by reading along the chromatogram. Simple? No, but now automated.

Internet resources for all of this can be excellent and provide effective diagrams.

Ann Bucklin, Marine Molecular Biologist

Ann is a marine biologist and geneticist working currently at the University of Connecticut. She credits her interests in invertebrates and marine ecosystems to David Egloff, a biology professor at Oberlin College, where she had a schol-arship. It is a liberal arts school where science is considered one of the liberal arts. She started at Oberlin intending a math major, but, while taking some required biology, she studied invertebrates with Dr. Egloff. The forms were diverse, but Ohio is far from any ocean, and many of the marine organisms were preserved or represented by prepared slides. However, there were also fascinating things like cladocera and rotifers to study alive with microscopes. Moreover, she recalls Egloff as almost magically instilling enthusiasm, and she has learned that several generations of professional marine biologists emerged from his classes. Ann was captivated and not surprisingly did well with the course, so Egloff recommended her for a summer course on invertebrates at the Marine Biology Laboratory in Woods Hole.

She was impressed by the teachers in the MBL course, enjoyed their lessons on research and says, "it was MBL that got me to Berkeley." Simple research projects were required, and she tells of asking about the extent that sediment grain-size influences what "mud-eating" worms ingest. Many seafloor worms feed by digesting the organic coatings off sediment grains. Ann set up an experiment with different grain sizes on opposite sides of a petri dish and added a worm. One of the research stars on the faculty came by, removed the worm, mixed the sediments together, and then put the worm back. He explained that she could only learn whether the worm was selecting *if* it had to work at it. She could not make the choices for it. She remembers that introduction to research methods as inducing a sort of hunger to be asking answerable questions about marine animals. That hunger is still being fed (Figure 18.1).

So, after college, Ann tried out the very popular notion that it would be great to be a marine biologist in a formal, professional way. That idea pops up often for young biology students; it derailed my medical education. She was accepted for graduate study by the Zoology Department of the University of California, Berkeley (UCB), which operated (with the University of California, Davis) a marine station on Bodega Bay. She spent most of her time there, sustained by a UCB fellowship for four years. Her major professor, Cadet Hand, was a student of sea anemones, and he directed Ann to varied studies of *Metridium senile*, a white (without symbiotic algae to make it green) intertidal and subtidal species with a tall basal column and bushy tentacles.

Ann's Alloyzme Phase

Ann's dissertation is dated 1980, and was not based on papers already published. Her dissertation defended, she headed off for two postdoctoral fellowships: one from Woods Hole Oceanographic Institution (WHOI), and one from NATO for work in Europe. She spent some of the time on getting published. Her first paper,[2] based on the dissertation, reported a three-year

Figure 18.1 Ann Bucklin at sea, working in a lab van secured to the deck of a Norwegian research ship in the Barents Sea. Forceps in one hand, she is sorting copepods into microcentrifuge vials in the other. Photo by Peter Wiebe, with permission.

(1977–1979) evaluation of gonad activity in *M. senile* sampled monthly. They aren't "senile"; the Latin suggests they more or less sit still. Intertidal (in hip boots, Figure 18.2) collections from multiple sites in Bodega Bay and on the jetties just outside agreed on reproductive timing for females. Ann worked that out based on serial sections from the basal columns of her samples, the ovaries being inside. She measured ova diameters with a scale in a microscope ocular. The egg cells grew larger in summer and were spawned in August or September. Males are separate. Their development of mature sperm proved less regular, for no reason obvious from conditions in the bay. The techniques were classical tissue methods—rather nineteenth-century, but with better microscopes.

However, Ann's second paper[3] represents the state of genetic comparisons in the 1970s. Sequencing DNA was still to come; however, work had shown by

Figure 18.2 Ann Bucklin in graduate school days sitting next to Ralph Smith (1918–1993), prominent University of California, Berkeley intertidal ecologist and physiologist, who Ann says was an important mentor at Bodega. She does not remember the name of the lad at her right. Courtesy of Ann Bucklin.

then that genes determined differences in proteins, and methods had emerged to show when proteins differed even a little. Dennis Hedgecock, then a young professor at UC Davis, had been working with starch-gel electrophoresis to compare two or more versions of enzymatic proteins. Those are termed "allozymes," because they are thought to be derived from variant forms of genes for the same protein, different "alleles." Hedgecock introduced Ann to the method and let her work in his lab. They compared a suite of 13 enzymes between two species of *Metridium*: *senile* and *exilis*.

Allozyme analyses are rarely done now, so description here can be brief. More methodological detail was given by Francisco Ayala,[4] a student during that era of genetic variation across wide geographic provinces, such as among fruit flies from most of Latin America. Ann would place tissue-mash samples from, say, a dozen specimens of anemone in wells cut along one edge of a sheet, about 10 x 15 cm, of gelled potato or similar starch resting on a plastic platform with raised edges. The platform, which had electrodes at each end, was flooded with a saline buffer solution, and a direct current through the buffer was switched on. Because most enzymes carry a negative electric charge, they would migrate from the wells toward the positive electrode, making their way along the tortuous path through the starch fibers. The smaller a protein's size and the stronger its charge, the faster it moved. After a time determined to move an enzyme of interest along the sheet but not off the end, the buffer was drained but the sheet kept moist. Then Ann treated the sheet with a mixture of *substrate* for an enzyme of interest and an indicator dye for the *product* of the reaction catalyzed by that enzyme.

If the biochemistry was properly designed, along with suitable variations of starch treatments, conductive buffer, resultant pH, and temperature, then the enzyme would generate its product from the substrate and its migration band would be stained by the indicator dye. In studies such as Bucklin and Hedgecock's, different *allozymes* (coded by variants of the gene for the same enzyme) would migrate slightly different distances in given times: slightly but measurably, a millimeter or so in ten or more centimeters of gel). Papers from allozyme work in the 1970s, Ann's and those by Ayala and colleagues, typically do not show pictures of the marks on the gels. Thus, I do not have Ann's to show here. And by and large, the specific results have not been checked with later methods. Probably they would hold up.

When two animals had consistently different distances to their marks from numerous enzymes, those catalyzing the same reactions, it seemed very likely the animals were genetically distinct. The array of differences can be enough to justify that even rather similar looking organisms should have their own species names. That was the case for *M. senile* versus *M. exilis*.[3] The validity of their specific distinction had been questioned previously (the two anemones look only a little different, one white the other pink), but the allozyme distinctions supported it as likely valid. The names are still considered so.

Having mastered this tool, surprisingly difficult to handle in practice, Ann was set to join others in evaluating the allozyme status of virtually any organism. That was of interest to many active biologists during the next phases of her career. She started at WHOI, initiating a partnership with Dr. Nancy Marcus, who was working on the resting-egg ecology of the coastal pontellid copepod *Labidocera aestiva* (Figure 18.3).

Nancy's own graduate studies had been on allozyme variation in sea urchins. Her technique differed from Ann's mainly in using polyacrylamide for the electrophoresis matrix, not starch-gel. They applied that version to comparison of allozymes in three populations of *L. aestiva* scattered down the East Coast of the United States. However, in order not to lose the opportunity represented

Figure 18.3 *Labidocera aestiva*, female. The back dots in the head are eye pigment. She has mated and is towing an empty spermatophore sac and its coupler. Its total length (anterior tip to end of furcae) around 2.5 mm. Photo by Jonathan Cohen, University of Delaware, with permission.

by the NATO fellowship, Ann put the WHOI postdoc in park and headed to Britain. After a stop-off at the University of Reading, which turned out to be far from seashores and seawater, she transferred to the labs of the Marine Biological Association of the United Kingdom in Plymouth. She was able to do that because there was at the time a vacant "Reading Chair" at Plymouth. The MBA was welcoming, provided decent facilities, and from there she could run a new allozyme study of *Metridium* along British coasts.[5] She also gained the acquaintance of Roger and Marlene Harris (Chapter 13), recalling lovely visits at their home in Cornwall.

Next, Ann returned to Woods Hole to finish out her postdoc there, and she stayed on awhile as a guest investigator. Her main purpose was to hunt for a job in the states, but she resumed the partnership with Marcus on geographically separated populations of *L. aestiva*. Those were at Ft. Pierce Inlet in Florida, Beaufort Inlet in North Carolina, and waters adjacent to Woods Hole. They tested gels for 29 enzymes, finding six suitable for interpreting electrophoretic phenotypes. They followed a convention among "gel jockeys" at the time, terming the most common band distance (perhaps about 6 cm on acrylamide) 100, and bands differing by -1, -2, +1, +2, etc. millimeters as 98, 99, 101, 102. In the *L. aestiva* case, 100 was assigned to the most common distance at Woods Hole. To give you an idea, Table 18.1 shows the proportions of alleles thus identified for just two enzymes from the study:

Table 18.1 Proportions of identified alleles for two enzymes from *Labidocera aestiva* collected at three East Coast USA sites, Data from Bucklin and Marcus.[6]

(1) Succinyl aminopeptidase (that cleaves amino acids from peptide chains):

Allele	Beaufort	Ft. Pierce	Woods Hole
No. specimens:	48	32	104
100	0	0	0.50
101	0.96	0	0.01
102	0.04	0.45	0.49
103	0	0.41	0
104	0	0.14	0

(2) A malic enzyme (a metabolic enzyme extracting energy from malic acid):

No. specimens:	40	48	124
99	0.62	0.54	0.40
100	0.38	0.46	0.60

In both cases, the sample numbers are sufficient to say the proportions are significantly different. For the malic enzyme the overall probability of contingency of allele proportions on collection site was about 2% (under the null hypothesis of no difference). The stocks differed between sites. Bucklin and Marcus[6] interpreted the results (including all six enzymes) as indicating restricted gene flow (interbreeding) between the populations, allowing both random variations to accumulate (genetic drift) and natural selection for locally more suitable alleles. Which of those? Nothing about the method allows interpretation of that.

That was Ann's first published connection with copepods. She also worked at WHOI on a krill project with Peter Wiebe, a graduate-school comrade of Bruce Frost and me. As part of that she got her first shipboard experience, which she says she just loved. She must still love it, given the long list of cruises up to 2018 included in her present vita. The project documented substantial allozyme differences between two North Atlantic euphausiids.[7] For both species the proportions of allozymes in a small ocean region varied wildly and without any patterns from station to station, month to month, or year to year. As well as at geographic scales, plankton exhibit *patchiness* of all sorts locally, from genotypes (and so allozymes) to counts of development stages. Patchiness affects essentially all data from plankton sampling. Some of that is explained, but much of it is not. Ann found that then, at least, there was no explanation for allozyme patchiness in *Euphausia krohnii* or *Nematocelis megalops* in slope water, a discrete water type found between Georges Bank and the Gulf Stream.

At the same time Ann sent applications in response to every job advertisement on the WHOI bulletin boards and in the back pages of journals. She got an offer at the University of San Diego to serve a year substituting for a biology professor on sabbatical. During that year the Scripps Institution of Oceanography advertised for a faculty member to join its Marine Life Research Group (MLRG), and Ann was hired for that position. Immediately opportunities showed up to characterize allozyme variation. Physiologist George Somero had gotten himself aboard one of the second-generation submarine investigations of deep-sea hydrothermal vents, and he brought back a frozen collection of vestimentiferan worms, *Riftia*. Ann examined their allozyme variation.[8] Bathypelagic ecologist Ken Smith Jr. (he didn't live deep in the sea, he studied the biology there) had collected the giant amphipod *Eurythenes* from both a deep basin and higher up on a sea mount (a guyot). Ann compared their allozymes.[9] Skills with such analyses created a professional niche. Ann worked with those colleagues at Scripps for several years. The institution then had some great students of copepods: Michael Mullin, who worked on them alive; and Abe Fleminger, who worked with preserved specimens. Neither of them worked on allozyme distinctions.

Like all US oceanographic institutions at the time, Scripps received substantial program support from the Office of Naval Research (ONR). That connection netted Ann research support in the form of an ONR young investigator award, which came to involve a partnership with two physical oceanographers, Michelle Reinecker and Chris Mooers. They worked more directly for the Navy at the Institute for Naval Oceanography in Mississippi. The joint project was a study of onshore-to-offshore mixing by "coastal filaments" of flow reaching west across the California Current. The work was mostly classical hydrography done from a small ship, the *RV Point Sur*. However, Ann took nighttime zooplankton samples at 15 scattered stations with oblique net tows to 100 or 150 m depth. She caught and froze sufficient numbers of readily recognized female *Metridia pacifica* (Figure 18.4) for allozyme analyses.

Ashore she ran staining tests on electrophoresis gels for 30 enzymes, of which five suggested multiple alleles. For two tests, there were pairs of proteins that were strongly distinct versions, likely from different genes (loci) catalyzing the same reaction, raising the total to nine loci. Expanding analyses of those, she ran a total of 5,616 tests to obtain proportions (in sample sizes from 22 to 138 individuals) of two-to-four distinct allozymes in a dozen samples.[10] As I've emphasized for the projects of other copepodologists, vast amounts of work can be needed at the microscope and lab bench. The results suggested inshore and offshore subsets of *M. pacifica* (Figure 18.4) with distinctive allozyme proportions. The inshore group were being carried offshore in the filament.

The "two groups" result for *M. pacifica* could have been an early hint of a new species later discovered by Canadian planktologist Moira Galbraith

Figure 18.4 *Metridia pacifica*, female (right) and a smaller form, also a female *Metridia*, likely a new species. The grid is 1x1 mm. Photo by Moira Galbraith, Institute of Ocean Sciences, Sidney, British Columbia, with her permission.

of the Institute of Ocean Science (IOS) on Vancouver Island. She has found a smaller species of *Metridia*, shaped much like but not identically to *M. pacifica* (Figure 18.4). It appears in coastal waters near Victoria, B.C., when warm flow moves in from the south, as occurred during the 1997–1998 El Niño. Moira's draft paper[11] shows both some morphologic distinctions she has found and differences in specific genes from the two species (some of those sequences by Ann Bucklin). Would Ann have missed the shape differences? She has said to me that she cannot imagine how people like Bruce Frost and Janet Bradford-Grieve see the minute distinctions in form that they do.

By the time the *"Point Sur"* paper came out, Ann had moved to Washington, DC, where she worked at the ONR as a program manager under Eric Hartwig. She credits him with hiring her and teaching her the job. She spent time also at the Chesapeake Bay Institute of Johns Hopkins University, which long sustained close ties with Naval research. CBI provided lab space for scientists at ONR to maintain their own research. It was also a publishing address for distancing finished work from a scientist's military connection. Ann did not do that, listing both addresses on her next paper. Most oceanographers at the time just considered ONR support to be part of the federal funding of science, not part of preparation for war.

Ann's next paper[12] also concerned *M. pacifica* in the California Current, specimens of which were collected across a sampling grid between Cape

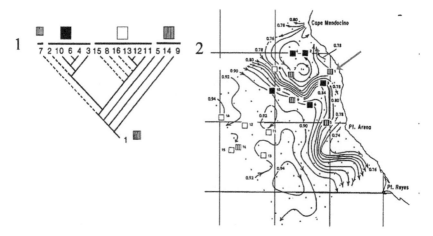

Figure 18.5 (1) Cladogram[12] of related sub-groups of *Metridia pacifica* females identified from the allozyme proportions in plankton samples. Solid black squares were sites with the inshore group. Open squares were sites with the offshore group. Vertically hatched squares were sites that did not group, especially sample no. 1 (arrow). (2) Current streamlines off central California in early May 1987. The swirls are dynamic height contours: lines close together represent fast flow; those wide apart slow. A meandering jet flows along the coast, enclosing an eddy below Cape Mendocino. Flow is slow and more random west of the jet. Numbered boxes are at the locations for nighttime *Metridia pacifica* collections. Box colors as in (1) Permission, American Geophysical Union.

Mendocino to the north and Point Reyes to the south (Figure 18.5, 2). The grid covered a pattern of California Current flow around an inshore eddy below Cape Mendocino, and it extended southwestward across the main current stream into relatively slow and random flow well offshore. Ann examined the allozyme variations in 16 samples of females from night tows on 16 dates in spring of 1987. Results were similar in character to those from the *Labidocera* study shown above, but involved more sites and 24 possibly allelic variants of six enzymes out of 30 tested for interpretable allozyme variation. Ann developed a table of allozyme proportions based on 3,358 individual specimens, each tested for just one enzyme in samples running from 13 to 97, averaging 36. She treated the table of allele proportions to a sort of cluster analysis and drew the result as a cladogram, a tree (Figure 18.5, 1). Six stations formed one group, mostly taken offshore, and five formed a second associated with the inshore eddy. The remaining five samples were each quite distinct from all the others. The sample numbered 1 in Figure 18.5, 2 (arrow) had very low similarity to any other sample and was plotted at the root of the cladogram (which really isn't "rooted" in any formal sense).

Moving to New England, Adopting PCR

Ann found the work at ONR satisfying. In the 1970s to early 1980s, ONR and the National Science Foundation (NSF) worked with oceanographic institutions under different models. NSF considered proposals from scientists and either funded them or did not, depending upon results of an elaborate community review process. By the 1970s it also funded some complex, multi-institutional projects including a coastal upwelling program on American and African west coasts. But those were basically organized and proposed by investigator groups. In contrast, ONR program managers would devise programs and propose them to investigators, singly or in groups. They also had to propose them to Navy and ONR leaders, including admirals and the Oceanographer of the Navy. The global strategic balance was reasonably static during those decades, and approvals went to projects of basic scientific interest, such as the Coastal Transition Zone Program that included Ann's work on *Metridia* genetics. Ann enjoyed the work of coordinating and promoting such ONR programs.

But then a series of incidents involving sales of large machine tools to the USSR by Toshiba corporation and others resulted in transfer of methods for making very quiet propellers for submarines. Starting around 1984, the Soviet nuclear submarine fleet progressively went silent. The tracking systems of the United States and its allies no longer provided data on where each of those subs, previously identifiable from sonar signatures, was located and moving. The focus of the navy's research support shifted inshore, like places submarines could hide in continental slope canyons. It moved away from effects of zooplankton on acoustics in seawater. Little along the new lines involved zooplankton genetics. Ann needed a new job, and left ONR.

NSF had provided Ann as she left ONR with a sort of "parachute" grant. The agencies cooperated in that way to help departing staff members re-establish their research careers. With that grant she secured a temporary position at the Marine Biological Laboratory, just across the Eel Pond from WHOI's Redfield Hall, where zooplankton expert Peter Wiebe had his laboratory. That led, in the course of human events, to Ann marrying Peter (who appears lower left in Figure 18.14). The timing was the same as the wide introduction of the polymerase chain reaction (PCR) to molecular genetics after 1985, and Ann ran a first trial while at MBL.

Throughout the 1970s and 1980s several methods had been developed for sequencing short bits of DNA, a few hundred bases. They depended on obtaining large quantities of just the DNA for a specific gene or part of one. The method initially available for getting those is termed *cloning*, multiplying the pieces of interest by inserting them into the DNA of bacteria and culturing masses of those. The copies of the inserts are then extracted, say a few nanograms. Cloning is complex and slow, but much genetics progress was made using it. At the same time, thousands of biologists were working worldwide on the molecular mechanics of DNA replication and of protein production from its codes (Box 18.1). Activities of many of the enzymes involved in the copying of DNA were characterized, including the DNA polymerases.

The polymerases were extracted and characterized from all sorts of things, including bacteria that lived at near-boiling temperatures in hot springs. Those enabled their use by the late 1980s in PCR (Box 18.2) to make huge copy numbers of target DNA sequences. The method was simpler and cheaper (especially in scientist time) than cloning. Initially, sequencing those sets of copies remained tedious, and the sequences had to be short. But at this point Ann grabbed the brass-colored PCR ring beside the molecular genetics carrousel. Her first such project[13] was very simple. She and a colleague amplified two mitochondrial genes (termed mtDNA) from three species of *Calanus* from widely different sites. They used primers based on cytochrome *b* from vertebrates and primers she found for Cytochrome Oxidase-I in a manuscript she cited as "B. Kessing et al. 1989. The Simple Fool's Guide to PCR (*unpublished*)." Later versions of that guide indicate its origin was the lab of Steve Palumbi, another early grabber of the PCR brass ring.

The amplifications worked, and to characterize them Ann used five restriction endonucleases (Hinc II, Mbo I, Ssp I, Taq I, and Alu I). Those are bacterial enzymes that chop DNA molecules apart at the locations of specific bits of code, presumably bits occurring in bacteriophage (viruses) likely to attack the bacterium if not neutralized. Hinc II, for example, chops (▼) everywhere any DNA sequence includes

$$5' \ldots GTY \blacktriangledown RAC \ldots 3'.$$

The Y means C or T, and R means A or G, which produces four combinations. By 1990 there was a substantial catalogue of such cutters, and they were widely used and commercially available. Ann tested many enzymes and chose those that cut her sequences in only one place. She generated data from the different samples by running gel electrophoresis measures on the restriction products, the shorter chunks moving faster through the gels than the longer ones. Differences in lengths obtained from partial gene copies with restriction sites in different places are termed restriction fragment length polymorphisms (RFLPs).

Box 18.2 Polymerase Chain Reaction (PCR)

The late Kary Mullis (he died at age 74 in 2019) worked in the early 1980s for Cetus, a biotech company. In 1983 he had the idea for PCR. He said he was driving at the time and pulled off to write the idea down. The notion was to use the known sequence for some gene, selecting two well-separated strings of bases, perhaps 15 or more each, from the code (those commonly used for the mtDNA of invertebrate Cytochrome *c* Oxidase subunit I are 25 and 26 bases[43]). Mullis termed those strings *primers*. He could, he thought, put sampled DNA from, well, any organism in a test tube, add some single-strand DNA for two primers known to be well separated in the gene of interest, add some DNA polymerase from hydrothermal bacteria and a mix of phosphate- and ribose-bonded bases, for example, deoxy*adenosine* triphosphate:

On heating the mixture, the hydrogen bonds of A to T and of G to C would release, the DNA becoming single stranded. Then Mullis would cool the tube some. The primers would find their matches on the organism's DNA strands and anneal. With a little rewarming, the heat-stable (the key thing) polymerase would use the ribose-bonded bases to fill in the space between the primers, perhaps generating a whole gene, perhaps only part of one. After much development research by Mullis and others, a temperature for opening the sample DNA is typically 94°C for 1 minute, that for annealing the primers 45°C to 55°C for 2 minutes and that for extending polymerization 74°C for 3 minutes.

By repeating that sequence, the copies would multiply by 2, then 4, 8, and so on, until some serious nanograms of just the DNA between the primers would be in the tube. The chain reaction would stop when the raw materials were used up, but Mullis realized he should stop the heating before that, to end with double-stranded copies The steps are fast, so a typical 40-cycle run would be 240 minutes, or 4 hours. All of this typically runs in capped 2.5 cm long tubes sized for micro-centrifuges and for the holes in the metal blocks of thermocycling machines.

That is the polymerase chain reaction. Once announced,[44] researchers and biotech companies landed on PCR like a swarm of flies on a dead possum (PCR, possum cadaver reaction), and *they also* multiplied geometrically. Details were ironed out, programmable thermocycling machines were invented, manufactured and sold, and soon everybody was copying genes, even me. Electrophoresis techniques were developed to sequence the copies, and the copies sequenced got longer and longer. Eventually, those systems (all involving PCR one way or another) were developed to sequence whole genomes, like the human one for which a final draft was announced in 2003.

Here are her results in base pair counts from just Hinc II for a 360-base "amplicon" of Cytochrome *b*:

Table 18.2 RFLP lengths (in base-pair counts) from Hinc II digestion of Cytochrome Oxidase I (mitochondrial DNA) from four species of *Calanus* collected at different sites.

	C. finmarchicus Gulf of Maine	C. pacificus Puget Sound	C. pacificus San Diego	C. marshallae Puget Sound
Hinc II RFLPs:	280, 80	190, 170	210, 150	180, 180

Those, and similar RFLP differences for the other restriction endonucleases, implied that these species were different in respect to the genetic coding for common genes,[13] as well as to the morphologic characters by which they were long recognized as distinct species. It was a first step toward more sophisticated molecular studies. So far as I know, it was a first application of PCR to copepods, though not to genes of vertebrates, *Drosophila*, or sundry other organisms listed in the "The Simple Fool's Guide to PCR."

WHOI had a strict rule against nepotism, which kept it from hiring numerous promising dual-career couples, particularly a problem because it is at a distance from other sources of scientific jobs. And now Ann was married to one of its prominent senior scientists. The rule led Ann to hunt for a position somewhere in New England close to Peter Wiebe. It turned out that Thomas Kocher at the University of New Hampshire (UNH) had been doing genetic work on fish and other fauna using PCR.[14] He was willing to promote the hiring of a new colleague doing similar work on plankton. So Ann took a position at UNH in 1990. She and Peter bought a house in Durham, and they began a long siege of commuter marriage. Peter lived in North Falmouth, just outside Woods Hole. For almost three decades they traded married weekends at their respective homes and carried on very productive work lives 131 miles apart (eventually Ann moved to Mystic, Connecticut, which is almost as far). They agree that was difficult. However, I suspect that the separated workweeks reduced distractions from family life and contributed to their remarkable productivity through that long time.

Ann's job was connected to Kocher's zoology department and also to a UNH oceanographic program titled the Ocean Process Analysis Laboratory (OPAL). Her background as an ONR-program-manager probably suggested to somebody at UNH that she could be director of its new Sea Grant Program. Sea Grant is nationally administered through the National Ocean and Atmospheric Administration (NOAA) and supports marine research in so-called Sea Grant universities. The name was meant to reflect some similarity to the Land Grant colleges established under Abraham Lincoln in 1862. Public colleges back then were given federal lands to sell in order to establish themselves, particularly in respect to agricultural research and advice to farmers. Sea Grant policy was also to support *applied* marine science and extension work. Now Ann had a whole basket of hats to don in different offices and in Falmouth.

Genes of *Calanus*, the Classic Model Copepod

Tom Kocher and Ann worked with Bruce Frost over some years on the genetics of *Calanus*. Bruce (introduced in Chapter 4) checked the identifications of the specimens, and Ann and Tom worked out DNA gene sequence similarities and differences among them. Initially they examined codes for a short region of *16S* mtDNA, part of the DNA coding the RNA in their mitochondrial ribosomes.[15] By 1990 early versions of automated sequencers were available, and Kocher's UNH laboratory operated one. First, they ran a PCR on the readily amplified *16S* genes, for which conserved invertebrate primers were by then well tested. But this time a small fraction of each DNA base precursor (each dideoxynucleotide) in the PCR mix was labeled with a molecule that would fluoresce at a base-specific color in ultraviolet light. If, in a given PCR cycle, the polymerase added one of those, rather than a normal base, the PCR process would jam. That created mixtures of DNA molecules of different, progressively longer lengths, each labeled at its end with a fluorescent tag. The sequencer would then run an acrylamide gel electrophoresis, and the gel would be automatically passed under a UV light with color-sensitive photocells recording the end fluorescence of the chains from longer lengths to shorter (from slower in the gel to faster). That generated a record of the sequence, such as AAGCTTACGTAC.

Fluorescent-labeling and extension-blocking during PCR are a version of "Sanger sequencing" (invented by Fredrick Sanger). It was the basis for many

```
          151                                                         200
C.pac-N   CCCATATTGC GAAATTTTAT TCTGAGTGAA AATACTCAGC AGATATATTT
C.pac-S   T......... .......... .......... .......... ..........
C.fin     ..TCAT.... ..........A ...A...... ......T... ..T.G..C.A
C.mar     T.T.AT.... ..........A .......... ..........A ..........
Metridia  GGG.AT.ACT C.....AA.A .T.TGA.... ...TTCA.AA TCTA.A.CC.
Drosoph   ATTTA.AATA ...T...AT. .T.T...C.. ..AG..A.AA TT.AT.TAAA


          201                                                         250
C.pac-N   AGACGAGAAG ACCCTATGAA GCT------- ---GCTAGGC CAAAAAGATA
C.pac-S   .......... .......... ...------- ---....... ..........
C.fin     G......... .......... ...------- ---.GC.AA. T.TT..T-AC
C.mar     .......... .......... ...------- ---.G...A. TTCC..TG..
Metridia  ......T... .......... ...------- -----.ACT A....G.CC.
Drosoph   .......... .......A.. T..TTATATT TTATT..TTT T..TT.T.A.
```

Figure 18.6 One hundred aligned DNA base sequences coding 16S-ribosomal RNA from four *Calanus* populations, *Metridia lucens* (the Atlantic species) and *Drosophila yakuba*. Dots indicate a match to *C. pacificus*-North. Data from Bucklin et al. (1992).[15] Permission, Springer Verlag, *Molecular Marine Biology & Biotechnology*

later DNA sequence studies, including the human genome, until the advent of next-generation sequencing (Chapter 21). The automation is essential to making the system work for more than a few hundred bases. Bucklin and Kocher were applying it right when it became available; so far as I can find, they were first to study DNA sequences in a copepod gene.

So, what did they learn? A good way to answer is to look at a stretch of the data, the thing itself, which they actually published (Figure 18.6). From the simpler 1992[15] study I chose bases 151 to 250 of the 468 bases that could be reasonably aligned in most of the sequences. "Aligned" means that an optimizing scheme was applied, supplying the greatest number of matches to one of the populations arbitrarily chosen as a standard, in this case *Calanus pacificus*-North (from Puget Sound). They included *16S* sequences from their data for *Metridia lucens* and from the literature for the fly *Drosophila yakuba*. As might be expected, for those 100 bases, the sample for *C. pacificus*-South (from ~90 nautical miles south-west of San Diego) differed by only one Thymine (T) instead of Cytosine (C) at position 151 in the set shown, and three base differences in the whole set. For *Calanus finmarchicus* from the Gulf of Maine, there are 23 of those differences (count them). *Calanus marshallae* (also from Puget Sound) was different at 14 sites. Between *C. finmarchicus* and *C. marshallae* the count is 18 differences. You can count even more *Drosophila* differences, and that fly has a stretch at sites 24 to 33 that isn't present in the *Calanus* species at all.

For the whole set of sequences, the *similarities* can be presented as a matrix of proportions. Out of orneriness, I counted them using slightly different rules. Bases in the published alignments ranged from 408 to 437 pairs. As percentages of identical nucleotides, the similarity matrix is shown in Table 18.3:

Table 18.3 Percentages of 408 to 437 base pairs of aligned mtDNA coding 16S ribosomal RNA common to sampling-site pairs and species pairs of *Calanus* and *Metridia*.

	C. pacificus-S	C. finmarchicus	C. marshallae	Metridia lucens
C. pacificus-N	98.6	82.9	85.8	64.5
C. pacificus-S		81.2	86.4	64.4
C. finmarchicus			87.6	65.2
C. marshallae				66

The two populations of *C. pacificus*, long classified as one species, differed by < 2%. The three named species of *Calanus* differed by 11 to 17%. The two genera (in fact in different calanoid families) differed by ~35%. So the levels of difference generally agree with the taxonomy based on morphology.

In their next *Calanus* study,[16] they started with more bases and more species. The color-terminated autoanalyzer sequences were translated for similar sets of hundreds of DNA bases for the 16S mtRNA genes from five, independently sequenced individuals from each of eleven samples identified to seven species of *Calanus*, plus a sample of *Nannocalanus minor*. The last provided data for a species more distantly related to serve as an "outgroup." They presented the new data in terms of percentages of *differences* in base sequences. Why the switch? I have no idea. Their data for about 387 aligned bases showed 0.3 to 2.6% differences within named species and 7.3 to 25.2% differences between them. The among-species similarities and distinctions were (and are) sufficient to draw a tree of relationships. There are a handful of ways to draw these progressions of relationships. The easy part is to put pairs with small differences at the tips of the branches. You see that in the figure for the pairs sharing species names. More difficult is how to quantify relationships among those pairs and any remaining singletons in order to add more connections to the trees. We will go into a now extensive literature below, but most simply imagine comparing an average for groups already linked to values for species not yet combined. Say for three *C. pacificus* sequences, the new, single value would be (0.8% + 0.3% + 1%)/3 = 0.7%. Using that strategy, one keeps going. Figure 18.7 shows what the data suggested to the team via a formal algorithm.[17]

This left more to be done, since branching of *C. pacificus* with the *finmarchicus* group did not agree with its long placement in a group with *helgolandicus*. Some of that is related to closely related combinations (at right in Figure 18.7) being more reliable than "higher order" branchings (at left). Ann did not give up on

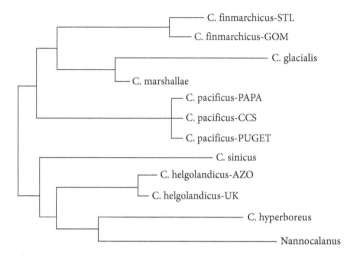

Figure 18.7 Molecular phylogenies of seven *Calanus* species plus *Nannocalanus minor* (as outgroup) constructed by neighbor joining.[17] Branch lengths are roughly proportional to differences in similarity (e.g., *glacialis* is more different from *marshallae* than the two samples of *finmarchicus* are from each other). From Bucklin, Frost, and Kocher.[16] Permission, Springer Nature.

getting morphological and genetic notions of the phylogeny of copepod genera to match, but she was a while before coming back to it. So far as I can find, no one has gathered a global collection of suitably preserved *Calanus* species to add more and the same genes for all its species at once.

For now, at least, *Calanus* phylogeny in the far north has acquired a twist. Based on repeating Ann's *16S* mtRNA sequences plus some nuclear "microsatellite" sequences, Geneviève Parent and colleagues[18, 19] suggested that *C. glacialis* and *C. finmarchicus* were hybridizing. Their molecular data even suggested the hybrids could be interfertile. However, work with nuclear sequence insertions and deletions by Norwegian geneticists[20] found no evidence of hybridization near Greenland or elsewhere in the far northern Atlantic and its fjords. Torkel Nielsen, Irina Smolina, and Marvin Choquet suggest the microsatellite sequences are subject to much higher rates of mutation and possibly convergent evolution ("homoplasy"), while insertions and deletions are much less so. They found no evidence of hybrids west of Greenland, in the Norwegian Sea or in several fjords. Dr. Parent and colleagues have published a spirited rejoinder in 2021.[22] I leave further evaluation of the issue to people more qualified. It is harder to imagine mating between *C. hyperboreus* and the other two Arctic species, given its relatively enormous size (Figure 18.8).

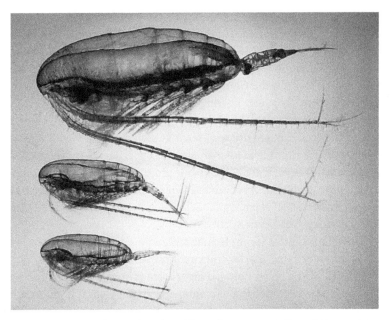

Figure 18.8 Arctic *Calanus*: females of *hyperboreus, glacialis* and *finmarchicus*, all with full oil sacs and no mature oocytes showing. Adult females of *Calanus* species do participate in diapause in some areas, sometimes after previously spawning. Total lengths, though quite variable, would typically be around 6, 4, and 3 mm. Photo: Ida Beathe Øverjordet and Dag Altin with their permission. Front cover of *Journal of Plankton Research*, March 2012. Permission, Oxford University Press.

Ann's First Multi-Investigator, Multi-ship, Multi-Institutional Project

In the mid- to late-1990s Ann began participating in a series of multi-investigator programs; the first was the Georges Bank project of the US Global Ocean Ecosystem (GLOBEC) program, funded jointly by US agencies NSF and NOAA. GLOBEC (eventually international), was an effort to promote studies of distinctive ecological relationships in regional ecosystems. Georges Bank is a submerged, but shallow, glacial deposit at the southern edge of the Gulf of Maine, separating it from the deeper Atlantic to the south. The cod and had-dock fishery there had suffered a near total collapse in the 1980s, probably due to overfishing. It was decided to examine aspects of the Bank ecosystem from physics to fish. Some copepods were designated as "target species," including *Calanus finmarchicus* (Figure 18.8) and species of *Pseudocalanus*.

An implementation team was created in 1992, including Peter Wiebe and Ann;[23] funding was promised, proposals written, and investigators selected on the basis of proposed projects. In addition, a collaboration was created called TASC (TransAtlantic Studies of *Calanus*), with plankton experts from the United States, Canada, Iceland, the United Kingdom, and Scandinavia. Ann was part of that from the early going, offering an initial paper on genetic variation of *C. finmarchicus* from work already in progress with Norwegian colleagues.[24] TASC was the organizing basis for a "Year of *Calanus* in 1997," with ships at sea sampling and labs running measurements and experiments from Cape Cod to Norway's North Cape. The Georges Bank project, with "process" cruises and monthly winter-spring surveys over and around the Bank, lasted five years from 1994. Ann wrote a sheaf of papers about her work with many others on Georges Bank and all across the North Atlantic.

Of particular interest for molecular techniques applied to copepods were methods for identifying species from gene sequences. Georges Bank is at the southern end of the ranges of all *Pseudocalanus* species. Those commonly found over and around it are *P. moultoni* and *P. newmani* (Figure 18.9, left), both species originally recognized by Bruce Frost.[25] Males make up tiny fractions of their stocks. The females have nearly indistinguishable shapes,

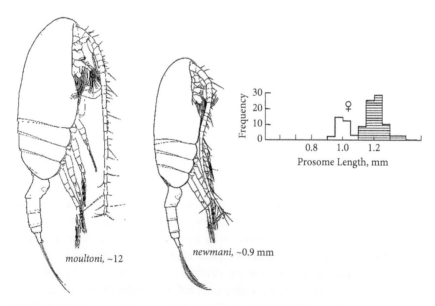

moultoni, ~12

newmani, ~0.9 mm

Figure 18.9 Size comparison of females between *Pseudocalanus moultoni* (hatched) and *P. newmani* (open) over Georges Bank in April 1981. Mean size difference is typical, but in other samples overlap occurs frequently. From Frost.[25] Permission, *Canadian Journal of Zoology.*

though they differ in mean size and possession (*newmani*) or not (*moultoni*) of minute sensory hairs dorsally on some urosome segments. However, sizes of both vary (Figure 18.9, right) with season, and with sampling happenstance, of course. What's more, no simple routine allows ordinary mortals to see those hairs. Frost did not find the two species to be close kin; he saw *P. newmani* as sharing more characters with *P. acuspes*, while *P. moultoni* shares more with *P. elongatus*. That comes up again in the next section. In order to examine distributional differences, Ann developed a DNA-based identification scheme.[26]

Specifically, Ann's team used the by then widely used PCR primers ("Folmer" primers) for mitochondrial Cytochrome Oxidase I (mtCOI) to amplify and sequence a section of that gene region for both *moultoni* and *newmani*. The standardizing specimens were from Puget Sound and identified by Bruce Frost. The sequences were different, because mitochondrial genes in most metazoans have no DNA repair enzymes and, therefore, they gather mutations quickly. To pair with the Folmer primer for the 5'-end of mtCOI DNA, they selected a new primer sequence for the 3'-end of each species, one closer the to 5'-end for *moultoni,* one farther from it for *newmani* (Figure 18.10, b). Then switching to Georges Bank specimens, her team smashed up samples of specimens (Figure 18.10, a), each in its own PCR tube. They added PCR reagents, ran the thermal cycles and cleaned up the DNA. The resulting mtCOI DNA was then evaluated with electrophoresis (Figure 18.10, c). The original specimens had, indeed, been identified by the lengths of their amplified mtCOI segments.

Next, they applied the test to samples of 18–30 female *Pseudocalanus* specimens from each of a dozen May 1996 stations over and around Georges Bank (a GLOBEC Broadscale Survey, Figure 18.10, d).[26] That gave rough proportions of the species at the stations, which could be multiplied by abundance estimates from the volumes the nets filtered and counts provided by Ted Durbin's University of Rhode Island group of identifier-counters, "the Planktoneers." In a later paper,[27] Ann, working with modeler Dennis McGillicuddy, compared samples from different depths and all across the Bank from February to June 1997. Monthly mappings like Figure 18.10, d for 20 specimens from each station in the Broadscale layout showed the two species were carried in the persistent clockwise flow from roughly the April 1996 pattern progressively farther around the bank edges. I suppose that proves that *Pseudocalanus* are, indeed, planktonic.

Ann and her teams also put considerable effort during and after the TASC and GLOBEC programs into quantifying genetic variability in *Calanus finmarchicus*, work well beyond that reported at the TASC planning workshop in 1996.[24] Ann's studies of variation across the whole species range from

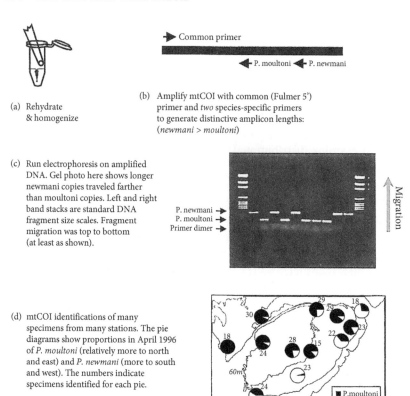

(a) Rehydrate
& homogenize

(b) Amplify mtCOI with common (Fulmer 5')
primer and *two* species-specific primers
to generate distinctive amplicon lengths:
(*newmani > moultoni*)

(c) Run electrophoresis on amplified
DNA. Gel photo here shows longer
newmani copies traveled farther
than moultoni copies. Left and right
band stacks are standard DNA
fragment size scales. Fragment
migration was top to bottom
(at least as shown).

(d) mtCOI identifications of many
specimens from many stations. The pie
diagrams show proportions in April 1996
of *P. moultoni* (relatively more to north
and east) and *P. newmani* (more to south
and west). The numbers indicate
specimens identified for each pie.

Figure 18.10 Explanation of species-specific mtCOI amplification and
identification. Method steps are explained by cartoons (parts a and b) and by a photo
(part c). Part d shows a relative proportions result for a station-by-station survey
of Georges Bank, April 1996. Parts a, b & c are from Bucklin et al.,[26] permission,
Springer Nature; part d is from Bucklin et al.,[27] permission, Elsevier.

Georges Bank to Norway's North Cape ended later with a study[28] of single
nucleotide polymorphic sites (SNPs) done with graduate student Ebru Unal
(Figure 18.11). Recall that the DNA codes for adding specific amino acids to
proteins are three base pairs long. There are more three-letter codes, 64, than
amino acids, 20. Thus, there are many synonyms. SNPs in protein-coding genes
are mostly changes from a code for a given amino acid to one of its synonyms.
Even in nuclear genes, they can change quickly among synonyms; for phenylalanine, say, from GAT to GAC. With automated sequencing systems Ann and
Ebru could determine the frequency of SNPs at specific sites in very large numbers of specimens. They did that for 24 SNPs in three nuclear genes at 16 widely

Figure 18.11 Ebru Unal, doing bioinformatics. At present she works at the Mystic Aquarium, and remains affiliated with the University of Connecticut. Photo by Laura Thompson. With permission from Ebru Unal.

spaced stations. After much sophisticated statistical analysis, their simplest statement of their conclusions was the following:

> Overall, this study confirms [three] earlier studies indicating high levels of gene flow among *C. finmarchicus* populations and also provides clear evidence of small but significant population genetic differentiation at both sub-regional (areas) and large scales (gyre) across the N. Atlantic Ocean.[28]

Next, Phylogenetics: Many Species, a Standard Gene, Mid-Length Sequences

In the early 2000s Ann returned to discerning evolutionary relationships using gene sequences, closely examining the families Calanidae (*Calanus, Neocalanus* and more) and Clausocalanidae (*Clausocalanus, Pseudocalanus*).[29] She was the gene sequencer for three high-level morphological taxonomists: Frost, Bradford-Grieve, and Nancy Copley of WHOI. They examined 34 species in those two families collected by many helpers from wherever those species live.

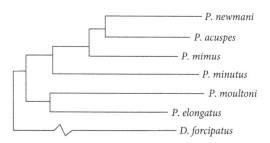

Figure 18.12 Neighbor-joining relationship[17] tree for six of the seven species of the copepod genus *Pseudocalanus* based on 639 aligned base pairs from their mtCOI genes. From Figure 3 in Bucklin et al.[29] Permission, Springer Nature.

Ann and a student, Lisa Allen, sequenced a 639 base-pair sequence from the mtCOI of all the fourteen described species in *Clausocalanus* and six of the seven species of *Pseudocalanus* (plus *Drepanopus* as an outgoup). To get that many base pairs at the time required extending beyond the by-then classical primers, designing new primers to reach well beyond their eventual suite of 639 bases. That suite emerged after sequences for all 20 species were aligned and trimmed to match the smallest fully covered one. They applied the neighbor-joining algorithm to get evolutionary trees. That for *Pseudocalanus* can serve here as an example (Figure 18.12). This tree exactly matches Frost's evaluation[25] of the likely relationships based on details of morphology. The missing species, *P. major*, is restricted to shelf waters in the Arctic Ocean off Siberia and Northern Alaska and could not at the time be obtained for the study. In Frost's scheme it was most closely related to *P. acuspes*, showing very small differences in segment shapes.

A relationship tree for all 14 species of *Clausocalanus* matched some of the original morphology-based grouping by Frost and Fleminger (1968; check back to Chapter 4), but far from all of it. If you feel driven to think about that, check a later paper (2009) by Ann and Bruce.[30] Ann sequenced more genes. They ran several different cladogram algorithms on both morphological characters and gene sequences. No simple evolutionary sequence emerged. A partial explanation was that perhaps several of the species simply did not pair well, particularly in extensive molecular data, with any of the others. (Surprise: more research is suggested.)

Mitochondrial sequence did reprise morphological relationships well for *Pseudocalanus*. However, recall that those genes do not have DNA repair genes linked with them in the mitochondria, so they are without the generational DNA repair provided to nuclear genes by meiosis. Mitochondrial genes gather

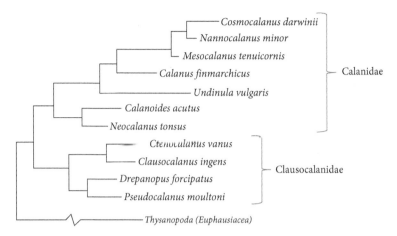

Figure 18.13 Cladogram calculated from similarities among 660 bp sequences of seven genera of the Calanidae and four genera of the Clausocalanidae, which are families of Calanoid copepods. From Figure 4 in Bucklin et al.[29] Permission, Springer Nature.

mutations much faster than nuclear ones, which have both. Over generations mtCOI gets "saturated" with changes that do not destroy function (changes after changes in many locations). Once that happens, as perhaps with *Clausocalanus*, they lose their virtue for tracing evolutionary history.

For a longer-term view,[29] Ann used protocols from the "tree of life" project[31] to develop a long sequence, 1,749 base pairs, from the *Calanus finmarchicus* nuclear gene for the *18S* RNA component of their cytoplasmic ribosomes. From that she selected primers that gave good PCR amplifications of 660 base pairs for the equivalent gene in 11 genera and one euphausiid (to supply an outgroup). From the twelve sequences she could generate a tree (Figure 18.13) reaching farther back in time, at least probably so.

For that study, Janet Bradford-Grieve had checked two things: (1) that the genera are all well-founded morphologically, implying the species in each are closely related; and (2) that identifications of the specimens for sequencing were correct. The genera shown in the cladogram (Figure 18.13), each based on one or two representative species, were well separated, and the two families separated at the first branching of the tree. To an extent that validates the generic and family levels of classification. Long ago people like John Lubbock, G. O. Sars, and Andrew Scott got it right that *Cosmocalanus* Lubbock, 1860; *Nannocalanus* G. O. Sars, 1925; and *Undinula* Scott, A., 1909 are all distinctive and yet relatively closely related.

Multiple-Everything Projects

At the beginning of the twenty-first century, Ann took on organizing more projects involving many collaborators. ZooGene was one, funded by NSF from 2000 to 2004. She calls it her "first international leadership opportunity," though demonstrably it was not. She was project coordinator, with co-investigators Peter Wiebe, Bruce Frost, and fisheries ecologist Michael Fogarty. The idea was to combine precision morphological identifications of planktonic copepods and euphausiids with species-specific DNA sequences in a comprehensive database. ZooGene sought extensive "systematic range," meaning many species, rather than large numbers of genes. The sequences were all to be for mtCOI using the Folmer primers. Seven international partners, all recognized taxonomists, were recruited from New Zealand, Australia, Mexico, and Japan. The "Folmer sequence" had come to be considered nearly universal, and sequences between them came to be called a *barcode* for animal species identity. The term was formally suggested by an insect taxonomist, Paul Hebert, and his colleagues in 2003.[32] By then it was already in use by many, including Ann, and led to several Cold Spring Harbor symposia. A picture of the assembly at one of them shows Ann, two other women, and eighteen men, almost all old white guys. That still happens to women scientists. There are now barcode catalogues for many groups, from nematodes and butterflies to, indeed, copepods and euphausiids. ZooGene had a substantial role in initiating barcoding of marine zooplankton. I'll show a barcode below.

ZooGene gave Ann a basis for convincing program managers of the Sloan Foundation that zooplankton needed further work on the actual levels of species diversity in all its groups. The Sloan Foundation is based on the wealth of an automobile company chairman, who established it to promote advances in science and technology. In the early 2000s it was interested in expanding knowledge of global biodiversity, with a focus on marine life, particularly "charismatic megafauna" (think whales and giant squid). The program was called the Census of Marine Life (CoML). Ann stayed in the ear of a key Sloan program manager until he agreed to extend CoML to zooplankton, creating the Census of Marine Zooplankton (CMZ) as one of its field projects. The foundation primarily provided start-up money and insisted it be used to develop a large, globally distributed steering committee with actual in-person meetings. So Ann recruited a distinguished group and held a meeting in 2004 at the University of New Hampshire, in Portsmouth (Figure 18.14). The list of her 22 colleagues on the CMarZ steering committee includes some names mentioned in this book: Russ Hopcroft, Janet Bradford-Grieve, Shuhei Nishida, and Erica Goetze (Chapter 19). All of the others are recognized experts on other groups, like ostracod taxonomist Martin Angel and chaetognath and coelenterate expert Erik Thuesen. A key member, becoming a close friend to Ann, was the late copepod

Figure 18.14 CMarZ Steering Committee at their March 2004 meeting in Portsmouth, New Hampshire. Photo provided by Ann Bucklin, with her permission.

ecologist Sigrid Schnack-Schiel of the Alfred Wegener Institute in Germany. Field work, mostly funded by many national science agencies, included over 80 sampling cruises in every part of the oceans. Like ZooGene, a major focus of the effort was gathering mtCOI barcodes for large numbers of species, and to do it using fresh specimens with automated sequencing aboard ship. Sequences for other genes were generated as well.

Also in 2004, Ann moved from her several duties at the University of New Hampshire, where she says things were going well and she was happy, to the Department of Marine Sciences of the University of Connecticut (UConn). The offer was too tempting to resist. The department has a field station in Groton, at Avery Point, with lovely buildings, views of Long Island Sound, and a small ship and facilities for it. It had many capable staff, including marine planktologist Hans Dam. They invited Ann to locate at Avery Point and become UConn's Director of Marine Sciences. The offer included a handsome salary increase and generous lab start-up money. That enabled her to move her lab, including its staff and students. She remains at UConn in 2022. She is no longer director, but serves as an active professor. The start-up funds were enough to buy several automated sequencing machines, which enabled her to take the barcoding process out to sea during CMarZ, from collection to sequencing.

Figure 18.15 Leo Blanco-Bercial of the Bermuda Institute of Ocean Sciences. Photo by Tiffany Wardman, from the institute website (with its and his permission.

Indeed, CMarZ discovered several new species. For example, eleven new members of the Megacalanidae[33] were found, a calanoid copepod family largely restricted to waters below 1,000 m. Sampling with nets at depths down to 4,000 m is very time consuming, hours to wind cable out and back for long tows of large nets to catch sparse animals. But the folks of CMarZ recognized that new species were most likely to be found down deep, so that is where they sent their fine-mesh trawls. Through much of this CMarZ interval, a young investigator, Leocadio Blanco-Bercial (Figure 18.15), was a postdoc then fellow investigator in Ann's lab. He has become an intellectually gifted and hard-working star of plankton molecular biology. His doctorate was from Universidad de Oviedo, Spain, based on studies in a lab run by some of Ann's by then global connections among "molecular systematists." Eventually Leo moved to the Bermuda Institute of Ocean Sciences (BIOS), and at this writing is still there.

Leo generated barcodes for many of the Megacalanidae collected by CMarZ. For the four new *Megacalanus* species, amplifying the mtCOI gene only worked for specimens collected above 1,000 m. However, a nuclear *28S RNA* gene sequence did amplify consistently well for four specimens of *M. princeps*, five of *M. frosti*, and seven of *M. ericae*, the last two among the newly named forms. There were 786 base pairs for most of 16 specimens and minimum of 644 base pairs. Below is a table of its within- and between-base-pair differences out of 644 for a total of 16 specimens in three species of *Megacalanus*. Differences are 0 or 1 for specimens that Janet Bradford-Grieve assigned the same names, more for those in different species. The genetic differences are distinctive characters, just as is the crest of *Megacalanus frosti* (Figure 5.6 and Figure 18.16). I choose that one from the amazing Bradford-Grieve, Blanco-Bercial, and Boxshall

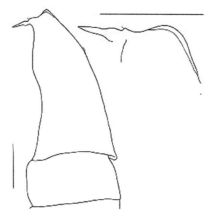

Figure 18.16 *Megacalanus frosti*, male: cephalosome carapace (left), rostrum and crested forehead (right). Scale bars are 1 mm and 0.1 mm. Drawings by J. Bradford-Grieve. From Bradford-Grieve et al.[33] Permission, Magnolia Press, *Zootaxa*.

monograph[33] on the Megacalanidae, because I originally discovered it.[34] That makes it fun, at least for me. I was afraid to assign a new name based on just one obvious character (the forehead crest is unique in this species). Janet and Leo and Geoff Boxshall (brief biography in Chapter 20) found more characters, especially its nuclear *28S RNA* sequence.

I said I would show you an mtCOI barcode. It was successfully amplified and sequenced during CMarZ from three specimens of *M. frosti* (mentioned in Chapter 5). Below is the Fulmer mtCOI code for specimen number Co439.2.2, which was collected at 34.13° N 120.97° W (off California) by Mark Ohman (of Scripps) in 2008 and identified by Janet Bradford-Grieve:

```
ORIGIN - 5' Not including the primers. Starts and ends in the middle
of the Folmer sequence of 710 base pairs of mtCOI typical in many
invertebrates.
  1 ttctctttac ctattggcag gaatgtggtc gagtatagtt ggaacgggac tgagaatact
 61 tattcgacta gagctaggac aagcaggacc attaatcgga aatgatcaaa tttataacgt
121 tattgtaaca gctcatgctt ttattatgat ttttttatg gtgatgcctg ttcttattgg
181 agggtttggt aattgattaa ttccattaat gattggtgca tctgacatag catttcctcg
241 tttaaataat ataagatttt gatttttaat acccgctctt attatattat tatctagtgc
301 attagtagaa agaggtgcag ggacaggttg aacagtatat cctcctcttg ccagtaatat
361 tgcgcacgct gggaggtctg tagactttgc tattttttca ttgcacctgg caggaattag
421 atctatttta ggggctgtta attttattag gaccttgggc aacttgcgag tatttggtat
481 acttatagat cgcataccat tatttggttg agcgacatta attacggctg tattactact
541 cctatcttta cctgtattag caggtgcgat taccatgtta ttgacggatc gaaatttaaa
601 tactacattt tacgatgtcg gaggggggtgg agatcctatt ttatatcaac
```

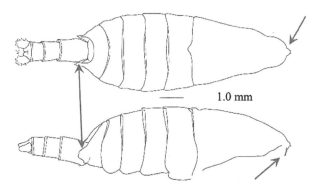

Figure 18.17 *Bathycalanus bucklinae*, Bradford-Grieve et al., 2017,[33] dorsal and lateral habitus drawings. Distinctive characters are the lappets alongside the genital segment (double arrow) and the knob-like protrusion bearing the rostrum (arrow). Other characters are on the limbs. Permission, Magnolia Press, *Zootaxa*.

Not very exciting in itself, I must say. Anyone can get these barcodes by query to the National Center for Biotechnology Information (NCBI) GenBank. The paper told me this one was accession number KU053559. I requested that from the GenBank website, and a lot of background data showed up followed by the sequence. If you get a bunch of those you can readily count the differences, and computer routines will do it for you faster. This sequence was submitted to GenBank from the Bermuda Institute of Ocean Science by Leo Blanco-Bercial.

One of the new species described by Bradford-Grieve et al. was *Bathycalanus bucklinae* (Figure 18.17). It is daring to name a species based on one specimen, but this one seemed unique and likely to have very similar relatives. If anyone is qualified to say so, it would be Janet Bradford-Grieve and Geoff Boxshall.

The description includes this etymology: "This species is named for Professor Ann Bucklin, University of Connecticut, who conceived and led the CMarZ programme." That honorific for Ann is much deserved, and she can (and is) very proud of CMarZ. She is in a way a hidden coauthor of the monograph and of many other CMarZ studies. Unfortunately, no male of her namesake species was found, and, with only the holotype specimen, no genes could be amplified. The sample came from the South Atlantic, 25.08°S, 9.58°W (centered between Brazil and Namibia), and from depths between 2,062 and 2,990 m. Near that is where to sample for more specimens and males.

CmarZ Wasn't Everything, and a Side Trip to Visit Petra Lenz

CMarZ led to a cascade of papers, 20 of them included Ann as first author or coauthor in 2010–2012. However, not all of her activity before 2010 was on CMarZ. She participated in other studies. Her international network had connections everywhere. There was a study with Italian colleagues[35] of genetic variation in the euphausiid *Meganyctiphanes norvegica*, which does not live only in Norwegian waters. Its distribution extends to the Mediterranean. Based on variation in its mtND1 (NADH dehydrogenase) gene, there is genetic divergence (at least in rapidly evolving mitochondrial genes) between the Scandinavian and more southern stocks. Another example was a genetic comparison of the *Acartia (Odontacartia) ohtsukai* n.sp. population isolated in Ariake Bay, Japan, with its relative *A. pacifica* in more open coastal waters. That was published[36] together with Hiroshi Ueda's "flattened fauna" drawings of both species (Figure 18.18).

Ann participated in another interactive project[37] led by Petra Lenz of the University of Hawaii. Petra had collaborated over several summers with, among others, David Towle of Mount Desert Island Biological Laboratory. It involved a technique that was maturing in the early 2000s: evaluation of the activity of specific genes by quantifying the messenger RNA transcribed from them. The method was sketched in Chapter 15 in respect to gene activity before and during diapause in *Calanus finmarchicus*. Lenz, Bucklin, and others also applied it to

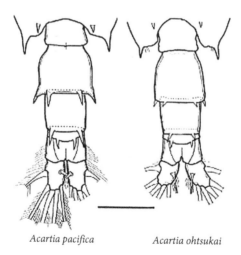

Acartia pacifica   *Acartia ohtsukai*

Figure 18.18 Urosomes from males of two *Acartia* species from Japanese waters, showing distinctive characters. Scale = 0.1 mm. Ueda and Bucklin.[36] Permission, Springer Nature.

C. finmarchicus (Figure 18.8). As a first step,[38] they used "quantitative PCR" to compare the activity of the gene for the heat-shock protein hsp70 in fifth copepodites (C5s) maintained at 8°C to its activity in other C5s moved for 30 minutes to 20°C, then returned to 8°C. Heat shock proteins protect metabolic and other proteins from denaturing (getting cooked) at temperatures well above those of an animal's usual habitat. The result was not surprising: hsp70 mRNA was four times higher in the groups subjected to 20°C than in those not. Some but not all individuals survived accidental exposure to 18°C *for two days*, and those surviving had the same four-fold elevation of hsp70 mRNA as the 30-minute groups.

Even before that study was complete, molecular technique had advanced to the point that nearly all of the messenger RNA in an animal can be copied as cDNA, and at least fragments of many genes can be sequenced together in a single commercialized process. Lenz and colleagues did that for *Calanus finmarchicus*, creating a cDNA "library" (or a *transcriptome*) of functionally identifiable sequences for 995 genes, publishing a first notice in 2009 in a Mount Desert bulletin and a detailed description in 2014.[39]

Lenz and colleagues have repeated mRNA processing in *C. finmarchicus*, making separate extractions and cDNA libraries from multiple life stages. As was expected, testing of different stages showed they have different gene activity profiles. More recently, Lenz and her student Vittoria Roncalli, have created a "reference transcriptome" for diapausing female *Neocalanus flemingeri* collected from a deep basin in Prince William Sound, Alaska. They are working with Russ Hopcroft (who helps with collections and photography, Figure 18.19) to compare gene activity between late larvae and diapausing females.[40] The art form of gene-activity comparisons remains to be widely applied in copepods, except for studies of diapause (see Chapter 15).

Bucklin Takes Barcodes to Another Level: Metabarcoding

Ann's most recent and continuing projects are an extension of the barcode concept. The notion is termed *metabarcoding*. Applying it to ocean zooplankton, Ann takes a net tow and breaks up the whole catch with a blender so that genes (and every other molecule) are suspended in a soup. A DNA-extraction procedure is applied, and the mixture is then sequenced by a high-throughput DNA sequencer to provide sequences for a barcode gene from more-or-less every organism that was in the sample. If that seems fascinating, I leave you two references for chasing down how it works and what the results look like.[41, 42] Copepods are involved, since they are so persistently dominant in plankton samples, but metabarcoding is not so explicitly about them as this book aims to be.

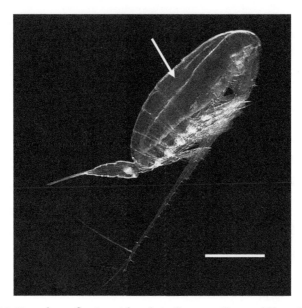

Figure 18.19 *Neocalanus flemingeri* female in diapause. She would have been collected from a very dark deep layer. Her oil sac is quite full and her quiescent ovary (arrow) is discernible. Scale bar 1 mm. Photo by Russ Hopcroft, with permission.

If you happen to be enthusiastic about molecular genetics applied to species identification and relationships, you would enjoy the 2019 review by Dr. Bucklin and others of applications to zooplankton: Ann Bucklin, Kate DiVito, Irina Smolina, Marvin Choquet, Jennifer M. Questel, Galice Hoarau, and Rachel J. O'Neill, "Population Genomics of Marine Zooplankton," in Marjorie Oleksiak and Om P. Rajora, eds., *Population Genomics: Marine Organisms*, Springer Verlag, 2019, pp. 61–102. The review extends to the phyla and crustacean classes other than copepods contributing to ocean plankton.

References

1. Endersby, J. (2007) *A Guinea Pig's History of Biology.* Harvard University Press, Cambridge, MA.
2. Bucklin, A. (1982) The annual cycle of sexual reproduction in the sea anemone *Metridium senile. Canadian Journal of Zoology* **60**: 3241–48.
3. Bucklin, A., and D. Hedgecock (1982) Biochemical genetic evidence for a third species of *Metridium* (Coelenterata, Actiniaria). *Marine Biology* **84**: 175–182.
4. Ayala, F. J., J. R. Powell, M. L. Tracey, C. A. Mourão, & S. Pérez-Salas (1972) Enzyme variability in the *Drosophila willistoni* group. IV. genic variation in natural populations of *Drosophila willistoni. Genetics* **70**: 113–139.

5. Bucklin, A. (1985) Biochemical genetic variation, growth and regeneration of the sea anemone, *Metridium*, of British shores. *Journal of the Marine Biological Association of the United Kingdom* 65: 141–157.

6. Bucklin, A., & N. H. Marcus (1985) Genetic differentiation of populations of the planktonic copepod *Labidocera aestiva*. *Marine Biology* 84: 219–224.

7. Bucklin, A., & P. H. Wiebe (1986) Genetic heterogeneity in euphausiid populations: *Euphausia krohnii* and *Nematoscelis megalops* in North Atlantic Slope Water. *Limnology and Oceanography* 31: 1346–1352.

8. Bucklin, A. (1988) Allozymic variation of *Riftia pachyptila* populations from the Galapagos Rift and 21°N hydrothermal vents. *Deep-Sea Research* 35: 1759–1768.

9. Bucklin, A., R. R. Wilson Jr., & K. L. Smith Jr. (1987) Genetic differentiation of basin and seamount populations of the deep-sea amphipod *Erythenes gryllus*. *Deep-Sea Research* 34: 1795–1810.

10. Bucklin, A., M. M. Reinecker, & C. N. K. Mooers (1989) Genetic tracers of zooplankton transport in coastal filaments off Northern California. *Journal of Geophysical Research* 94(C6): 8277–8288.

11. Galbraith, M. (in litt.) *Metridia* species from the West Coast of British Columbia.

12. Bucklin, A. (1991) Population genetic responses of the copepod *Metridia pacifica* to a coastal eddy in the California Current. *Journal of Geophysical Research* 96(C8): 14,799–14,808.

13. Bucklin, A., & L. Kann (1991) Mitochondrial DNA variation of copepods: markers of species identity and population differentiation in *Calanus*. *Biological Bulletin* 181: 357.

14. Kocher, T. D., W. K. Thomas, A. Meyer, S. V. Edwards, S. Pääbo, F. X. Villablanca, & A. C. Wilson (1989) Dynamics of mitochondrial DNA evolution in animals: amplification and sequencing with conserved primers. *Proceedings National Academy Sciences* 86: 6,196–6,200.

15. Bucklin, A., B. W. Frost, & T. D. Kocher (1992) Mitochondrial 16S rRNA sequence variation of *Calanus* (Copepoda: Calanoida): intra- and interspecific variation. *Molecular Marine Biology & Biotechnology* 1: 397–407.

16. Bucklin, A., B. W. Frost, & T. D. Kocher (1995) Molecular systematics of six *Calanus* and three *Metridia* species (Copepoda: Calanoida). *Marine Biology* 121: 655–664.

17. Saitou, N., & M. Nei (1987) The neighbor-joining method: a new method for reconstructing phylogenetic trees. *Molecular Biology and Evolution* 4(4): 406–425.

18. Parent, G., S. Plourde, & J. Turgeon (2012) Natural hybridization between *Calanus fimmarchicus* and *C. glacialis* in the Arctic and Northwest Atlantic. *Limnology and Oceanography* 57(4): 1057–1066.

19. Parent, G. J., S. Plourde, P. Joly, & J. Turgeon (2015) Phenology and fitness of *Calanus glacialis*, *C. finmarchicus* (Copepoda), and their hybrids in the St. Lawrence Estuary. *Marine Ecology Progress Series* 524: 1–9.

20. Nielsen, T. G., S. Kjellerup, I. Smolina, G. Hoarau, & P. Lindeque (2014) Live discrimination of *Calanus glacialis* and *C. finmarchicus* females: can we trust phenological differences? *Marine Biology* 161(6): 1299–1306.

21. Choquet, M., G. Burckard, S. Skreslet, G. Hoarau, & J. E. Søreide (2020) No evidence for hybridization between *Calanus finmarchicus* and *Calanus glacialis* in a subarctic area of sympatry. *Limnology and Oceanography* (preprint), 12 pp.

22. Parent, G. J., S. Plourde, H. I. Browman, J. Turgeon, & A. Petrusek (2021) Defining what constitutes a reliable data set to test for hybridization and introgression in

marine zooplankton: Comment on Choquet et al. 2020 "No evidence for hybridization between *Calanus finmarchicus* and *C. glacialis* in a subarctic area of sympatry. *Limnology and Oceanography* 66: 3597–3602.

23. Wiebe, P. H., D. Mountain, R. Beardsley, & A. Bucklin (1996) Global Ocean Ecosystem Dynamics: initial program in Northwest Atlantic. *Sea Technology* 37: 67–76.

24. Bucklin, A., R. Sundt, & G. Dahle (1996) Population genetics of *Calanus finmarchicus* in the North Atlantic. *Ophelia* 44: 29–45.

25. Frost, B. W. (1989) A taxonomy of the marine calanoid copepod genus *Pseudocalanus*. *Canadian Journal of Zoology* 67(3): 525–551.

26. Bucklin, A., A. M. Bentley, & S. P. Franzen (1998) Distribution and relative abundance of the copepods *Pseudocalanus moultoni* and *P. newmani* on Georges Bank based on molecular identification of sibling species. *Marine Biology* 132: 97–106.

27. Bucklin, A., M. Guarnieri, D. J. McGillicuddy, & R. S. Hill (2001) Spring evolution of *Pseudocalanus* spp. Abundance on Georges Bank based on molecular discrimination of *P. moultoni* and *P. newmani*. *Deep-Sea Research II* 48: 589–608.

28. Unal, E., & A. Bucklin (2010) Basin-scale population genetic structure of the planktonic copepod *Calanus finmarchicus* in the North Atlantic Ocean. *Progress in Oceanography* 87: 175–185.

29. Bucklin, A., B. W. Frost, J. Bradford-Grieve, L. D. Allen, & N. J. Copley (2003) Molecular systematic and phylogenetic assessment of 34 calanoid copepod species of the Calanidae and Clausocalanidae. *Marine Biology* 142: 333–343.

30. Bucklin, A., & B. W. Frost (2009) Morphological and molecular phylogenetic analysis of evolutionary lineages within *Clausocalanus* (copepoda: calanoida). *Journal of Crustacean Biology* 29(1): 111–120.

31. Maddison, D. R., K.-S. Schulz, & W. P. Maddison (2007) The tree of life web project. In *Linnaeus Tercentenary: Progress in Invertebrate Taxonomy*, ed. Z.-Q. Zhang & W. A. Shear, 19–40. *Zootaxa* 1668: 1–766.

32. Hebert, Paul D. N., A. Cywinska, S. L. Ball, & J. R. de Waard (2003) Biological identifications through DNA barcodes. *Proceedings Royal Society of London B.* 270: 313–321.

33. Bradford-Grieve, J., L. Blano-Bercial, & G. A. Boxshall (2017) Revision of family Megacalanidae (Copepoda: Calanoida). *Zootaxa* 4229(1): 1–183.

34. Miller, C. B. (2002) A variant form of *Megacalanus longicornis* (Copepoda: Megacalanidae) from deep waters off Southern California. *Hydrobiologia* 480: 129–143.

35. Papetti, C., L. Zane, E. Bortolotto, A. Bucklin, & T. Patarnello (2005) Genetic differentiation and local temporal stability of population structure in the euphausiid *Meganyctiphanes norvegica*. *Marine Ecology Progress Series* 289: 225–235.

36. Ueda, H., & A. Bucklin (2006) *Acartia (Odontacartia) ohtsukai*, a new brackish-water calanoid copepod from Ariake Bay, Japan, with a redescription of the closely related *A. pacifica*. *Hydrobiologica* 500: 77–91.

37. Lenz, P. H., E. Unal, R. P. Hassett, C. M. Smith, A. Bucklin, A. E. Christie, & D.W. Towle (2012) Functional genomics resources for the North Atlantic copepod, *Calanus finmarchicus*: EST database and physiological microarray. *Comparative Biochemistry & Physiology, Part D Genomics Proteomics* 7(2): 110–123.

38. Voznesensky, M., P. H. Lenz, C. Spanings-Pierrot, & D. W. Towle (2004) Genomic approaches to detecting thermal stress in *Calanus finmarchicus* (Copepoda: Calanoida). *Journal of Experimental Marine Biology and Ecology* 311: 2037–2046.

39. Lenz, P. H., V. Roncalli, R. P. Hassett, L.-S. Wu, M. C. Cieslak, D. K. Hartline, & A. E. Christie (2014) De Novo Assembly of a Transcriptome for *Calanus finmarchicus* (Crustacea, Copepoda): The Dominant Zooplankter of the North Atlantic Ocean. *PLoS One* **9**(2): doi.org/10.1371/journal.pone.0088589.

40. Roncalli, V., M. C. Cieslak, S. A. Sommer, R. R. Hopcroft, & P. H. Lenz (2018) *De novo* transcriptome assembly of the calanoid copepod *Neocalanus flemingeri*: A new resource for emergence from diapause. *Marine Genomics* **37**: 114–119.

41. Bucklin, A., P. K. Lindeque, N. Rodriguez-Ezpeleta, A. Albaina, & M. Lehtiniemi (2016) Metabarcoding of marine zooplankton: Progress, prospects and pitfalls. *Journal of Plankton Research* **38**: 393–400.

42. Bucklin, A., H. D. Yeh, J. M. Questel, D. E. Richardson, B. Reese, N. J. Copley, & P. H. Wiebe (2019) Time-series metabarcoding analysis of zooplankton diversity of the NW Atlantic continental shelf. *ICES Journal of Marine Science* **76**: 1162–1176.

43. Folmer, O., M. Black, W. Hoeh, R. Lutz, & R. Vrijenhoek (1994) DNA primers for amplification of mitochondrial cytochrome *c* oxidase subunit I from diverse metazoan invertebrates. *Molecular Marine Biology and Biotechnology* **3**(5): 294–299.

44. Mullis, K., F. Faloona, R. Saiki, G. Horn, & H. Erlich (1986) Specific enzymatic amplification of DNA in vitro: the polymerase chain reaction. *Cold Spring Harbor Symposia on Quantitative Biology*. **51**(Pt 1): 263–273.

19

Beta Taxonomy I: How Speciation Works

Copepod Sprigs on the Tree of Life

If alpha taxonomy (Chapters 4 and 5) is the practice of discerning and describing distinctive animal and plant species, then *beta taxonomy* would be the next step: determining how discernibly related groups of individuals (species) relate to other groups as genera, families of genera, and so on. Some general notions for evaluating levels of relationship were introduced in Chapter 18, emphasizing there the value of genetic tools for taxonomy and tracing of evolutionary history. By general consensus the most informative standard for degree of relatedness is position (close or distant) in the sequence of evolutionary progression, the familiar tree of life.

Darwin's notion of selection for greater relative fitness does imply such a tree. Out on the extant branches of the trees, there should be progress toward better adaptation for survival in the conditions experienced by current generations. Because those conditions do not remain constant, evolution can wander, selection shifting characteristics back and forth over time. But different new patterns (color, body form, tooth shapes, limb segmentation, ecological specialization, disease resistance, and behavior) can become fixed over long periods for populations in different circumstances. That happens particularly, but not only, for interbreeding groups in different *locations*. Thus, geographic separation is frequently the basis by which one species becomes two or more distinguishable populations. When those diverge until their genetic makeup is too incompatible to allow interbreeding, there is less potential for "backing up," thus leading to new biological species. Then, new and substantively distinctive species can diverge again. The biological definition of species as at least potentially interbreeding groups is generally in biologists' minds, though naming them is more often typological (introduced in Chapter 4).

Abraham Fleminger and Allopatric Speciation in Marine Planktonic Copepods

That key branching process leading to a family tree (a *cladistic pattern* in evolutionary jargon) is *speciation*, one species becoming two or more. Abraham

Oar Feet and Opal Teeth. Charles B. Miller, Oxford University Press. © Oxford University Press 2023.
DOI: 10.1093/oso/9780197637326.003.0019

Fleminger, some of whose taxonomic work was mentioned in Chapter 4, advocated the importance of a sequence of geographic events in the speciation of oceanic copepods, versions of an *allopatric* ("different country") process. For copepods, step-by-step and in the ocean, that starts when some change of circulation, climate, or tectonics divides the geographic range of a species into two or more areas. Each of the isolated populations experiences both different genetic drift (random and eventually widespread mutations in their individuals) and different selective pressures. With time the genetics of the stocks become incompatible, and mating between them, if they occur, would be sterile.

Next, new climate, current shifts, or tectonic events could allow conditions in the ranges to again merge and overlap, with the several populations spreading into each other. Then those sterile matings would indeed happen. Individuals of previously isolated stocks would not necessarily have developed any way of rejecting matings likely to produce infertile offspring. So there would be a strong selection for something that would allow such recognition, anything that would mitigate against the sterile matings. In many arthropod groups it is common for the selected characters to be divergent shapes and armatures of the structures involved in copulation. That is why entomologists are often accused of looking too long and too fondly at wasp and beetle genitalia. It is why copepod identifiers and systematists often make their most careful examinations of male grasping antennules, of male and female fifth legs, and of the shapes and likely sensory structures on female genital segments. The refined and consistent differences between congeneric species are often located there.

Abe, working at Scripps Institution of Oceanography, developed many detailed examples. He worked extensively with genera and species groups of the family Pontellidae (*Labidocera,*[1] *Pontella, Pontellina*) living mostly in coastal zones. He and fellow copepod systematist Kuni Hulseman also developed an oceanic example[2] from the species of the circumglobal, warm-water genus *Pontellina*. They used specimens from about 2,000 samples collected above 200 m in the Atlantic, Indian and Pacific Oceans from 40°N to 45°S. They had some help with sorting all those samples.

As adults, *Pontellina* specimens are around 1.5 mm total length, large enough to sort readily, and the body shape of the prosome is distinctive (Figure 19.2). So spotting them in a mass of plankton is not challenging, especially given the shiny round eye lenses of the males. However, distinguishing the species takes sophisticated attention to specimen details (as I have emphasized). That had not been done before Abe and Kuni gathered their all-oceans sample array. In their acknowledgements, they credit Gillian Maggert for drawing various specimens so accurately that the distinctive characters became evident.

They found characters in both sexes, though more for males, to distinguish four species, so they named three of them: *P. morii* (seen and drawn before by

Figure 19.1. Abe Fleminger explaining a sampling location offshore from San Diego Harbor (visible on his chart) to students on a 1964 demonstration cruise. Born in 1925, Abe died in early 1988. Picture by the author (who was one of the students).

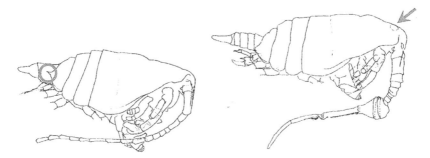

Figure 19.2 Female (left) and male of the circumglobal, upper-ocean copepod *Pontellina plumata* (Dana) 1849. The circle (arrow) on the male's head represents the lens of its right dorsal eye below the clear exoskeleton. Average total lengths of the male and female specimens were 1.54 and 1.42 mm, respectively. The female's posterior spine (circled in blue) is discussed in the text. Thoracic legs I-IV not shown. From Fleminger and Hulsemann.[2] US Government open access publication.

Takamochi Mori, whose 1937 identification was *P. plumata*), *P. sobrina* (Latin for "cousin," since it is closest in characters to *P. morii*), and *P. platychela*. To sort through the wealth of details, you have to look at their papers.[2, 3] I have selected one character for each sex. Since they named *P. platychela* for the distinctive shape of the claw on the male's right fifth leg (used to hold a female's urosome while attaching a spermatophore), consider some drawings (Figure 19.3) of that leg from males of the four species.

Characters separating females are fewer, which is true of many copepod (and bird) species pairs. Part of the presumption for pairing the sexes of both *morii* and *sobrina* came from stations with the males of only one by assigning females caught together with them to their species. However, the spine on the fused 4th

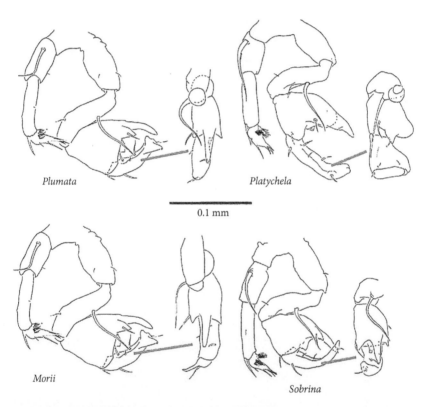

Figure 19.3 Male fifth legs from four species of *Pontellina*, posterior views (right leg on the right). Blue lines connect to an "into the chela view" of the same leg. You can see the reason, upper right, for the name *platy*(broad*)chela*(claw). Note also the differing relative proportions of the segments. Possibly the mating chela shape of a mismatched male simply would not allow clasping a female effectively. From Hulsemann and Fleminger.[3] Permission, *Bulletin of Marine Science*

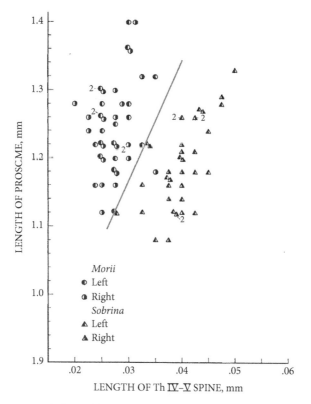

Figure 19.4 A morphological feature, size-comparison graph (see Chapter 4) for females of *Pontella morii* and *P. sobrina*. Lengths (x-axis) of their thoracic spines (circled in Figure 19.1) are plotted versus their prosome lengths (both in mm). The blue line separates all save one specimen thought (by Abe and Kuni) to be one, but possibly the other.[2] US Government open access publication.

and 5th thoracic segments (circled in Figure 19.2) of *sobrina* is distinctly longer (relative to body size) than in *morii* (Figure 19.4).

From the levels of similarity, Abe and Kuni deduced that the order of speciation almost certainly was as shown here:

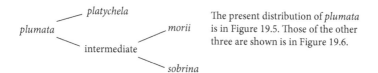

The present distribution of *plumata* is in Figure 19.5. Those of the other three are shown is in Figure 19.6.

Looking at those distributions (Figures 19.5, 19.6), Abe and Kuni then proposed that changing conditions, changes they thought likely through the ice

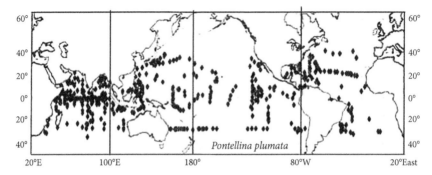

Figure 19.5 Circumglobal warm-water distribution of *Pontellina plumata*. Each diamond is a sampling station at which the species was found in a near-surface (epipelagic) net tow. There were also many tows without it in the same regions. None were caught outside 40°N to 40°S. From Fleminger and Hulsemann.[2] Permission, US government open access publication.

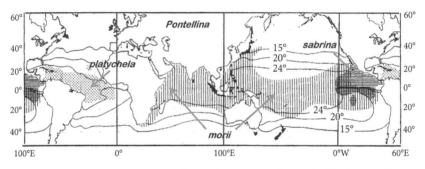

Figure 19.6 Distribution patterns of *Pontellina* species descended from *Pontellina plumata*. Station-dot maps for each are in Fleminger and Hulsemann's paper.[2] Permission, US government open access publication.

ages, would have separated the stock into three: Atlantic (diverging to *platychela*), Indian (perhaps remaining *plumata*), and Pacific (becoming the intermediate). Water withdrawn from the oceans into glaciers created a land or shoal-water barrier from Australia across to Malaysia. After a span of time long enough for the divergences, rewarming could have spread *plumata* to all three oceans again and allowed the Pacific intermediate to spread into the Indian basin, taking the present *morii* pattern. Possibly next, further warming established conditions somewhat more extreme than now: very warm surface waters in the western Pacific and eastern Indian Ocean excluded the intermediate. Its residual stocks would then have been in a pattern known at present among other plankton, populations divided between the cooler upwelling, and thus more productive, areas of the eastern tropical Pacific and the Arabian Sea. For example, the krill species *Euphausia distinguenda* in the eastern tropical Pacific has a very close cousin, *Euphausia sibogae*, in the Arabian Sea. Moreover, *Euphausia diomedeae* has a pattern now exactly like that of *P. morii*. Thus, with gene flow (interbreeding) stopped between the *morii* and eventual *sobrina* stocks, they diverged.

Finally (at least for now), some cooling again of the west Pacific "warm pool" would allow *morii* to spread all the way east into sympatry with *sobrina*. There, in the speciation process that Fleminger supposed, selection would have produced mating barriers. Exactly so, he and Kuni found most of the specific characters defining the *morii-sobrina* species pair in male grasping antennules, male and female fifth legs, spermatophore couplers, and so on. Questions arise. For instance, could thoracic spine length (Figures 19.2 and 19.4) be involved in identification of *sobrina* females by *sobrina* males and their rejection by *morii* males? A test would be very difficult. There are other possible range expansion-contraction scenarios that you can reasonably imagine. However, almost certainly, repeatedly shifting geography with barriers arising and breaking, was a major feature of the process.

I should say here that nearly exclusive focus on secondary sexual characters seems misplaced to me. Yes, those characters often exhibit refined differences between closely related species. But so do mouthpart limbs, behavior, even vertical distributions of congeners living in the same lakes or swaths of ocean. Consider *Neocalanus plumchrus* and *Neocalanus flemingeri*, which I have studied. Those two similar and closely related copepods, endemic in the subarctic Pacific, were identified as one species for many decades. However, when a difference was finally recognized (because they have different colors when alive), both mating parts and mouthparts provided distinctive characters. So did the setal fans on their caudal rami. Moreover, copepodite diapause of *N. plumchrus* occurs in C5, while that of *N. flemingeri* occurs in C6-females. There are dozens of other examples of divergence in detail among close congeners other than in reproductive structures. Once stocks with sufficient infertility on crossing to develop barriers to mating are living mixed together (*sympatric*), there is probably also sufficient competition for resources to force selection for character displacement generally.

Abe Fleminger was a more private person than most of the other zooplankton experts at Scripps. He was married, and we knew his wife was named Joyce. They had a son about whom Abe frequently expressed pleasure, but I do not recall ever seeing either the boy or his mother. They lived in Pacific City, a suburb on the way toward San Diego proper, and Abe commuted by city bus. There was also a dog in the family named Job. It is immortalized (sort of) in the name of *Clausocalanus jobei* Frost and Fleminger. During my first winter at Scripps Abe was one of the faculty leading a student cruise. He demonstrated net-towing techniques, pointing out groups of plankton animals in samples. A good teacher with whom we could feel at ease, one-on-one sessions moving back and forth with him between microscopes were very productive.

Abe really knew the marine Calanoida, instantly naming the genus of any specimen set before him on a slide. It was not the same with most harpacticoids; he was truly a specialist. However, he did keep a completely sealed glass cylinder about 20 cm tall atop a filing cabinet in his lab. It held air above, some sediment below, some seawater with algae and the tide-pool harpacticoid copepod *Tigriopus*. It got just enough sunlight from the window for the algae to grow, supporting an alga-herbivore food chain. There must also have been some recycling nematodes, bacteria or something converting the copepods back to nutrients for the algae. It did not "balance"; rather, the populations cycled. That was obvious across successive visits for help. People who miss Abe personally are fewer and fewer, of course. Maybe we should get together and give Scripps a likeness bust with the inscription:

```
      Abraham
      Fleminger
   Copepodologist
     1925-1988
```

Erica Goetze: Extending and Modifying Fleminger's *Eucalanus* Study

Erica Goetze, now Professor of Oceanography at the University of Hawaii (Figure 19.7), became a graduate student at Scripps in 1998. So she never met Abe Fleminger. However, with the new genetic tools by then available, she took up an unfinished problem that Abe had spent years studying. That was the species diversity and biogeography of the copepod genus *Eucalanus*. In the early 1960s he had directed a dissertation by Bui Thi Lang[4] about *Eucalanus* and *Rhincalanus* species and their distributions in the Pacific. The two genera are quite obviously related, though now classified as families *Eucalanidae* and *Rhincalanidae*. Lang had come from Vietnam to study at Scripps. Her thesis is dated 1965, and she probably defended it earlier. With her doctoral degree she returned to Vietnam, where she seemed for a long time to have disappeared in the fog of war. However, shortly after getting there, she published a paper (cited by Fleminger) about part of her *Eucalanus* research.[5] I have not seen that. I learned decades later that she had eventually become a teacher of human anatomy at a medical school in Saigon. She did survive the 1965–1975 war, and

Figure 19.7 Erica Goetze (right) in the plankton lab aboard *RRS James Cook* with Katja Peijnenburg. Their collections of upper-ocean plankton from over the Mid-Atlantic Ridge are stacked before them. Clearly the ship was not rolling much at the moment. Their collection of "mystery plankter of the day" photos are on the bulkhead behind them. With permission from Erica and Katja.

we corresponded briefly about her hope of returning to studies of copepods. So far as I know, that did not materialize. Erica says Lang attended the 100th anniversary celebration at Scripps in 2005, where they met and talked. Unfortunately, Erica then lost track of her.

Abe Fleminger spent considerable energy trying to revise Lang's dissertation for publication, but various issues kept him from finishing, particularly that the data for the genera were incomplete seen only from a Pacific perspective. Recall the global analysis that he and Bruce Frost developed for *Clausocalanus*. Abe was consistent that all oceans should be included in studies of genera with species in all of them. So he gathered Atlantic and Indian Ocean specimens from sample collections everywhere and expanded Lang's study. Looking at the usual leg segmentation and setation patterns, and particularly secondary sexual features, he did not find enough characters to fully separate the species that he could intuitively discern as different. Those intuitions were driven most effectively by the shapes of female genital segments, but that did not seem enough.

To fill the gap, Abe went to work on the patterns of gland openings and other pores over the surfaces of the body.[6] He wrote that the idea came from R. B. Seymour Sewell, a medical doctor in the British army and amateur (although FRS and director of the Zoological Survey of India from 1925 to 1935) copepodologist. Sewell[7] had used pore patterns to distinguish *E. attenuatus* (Dana, 1849) from *E. pseudattenuatus* Sewell, 1947 (which Abe decided should be a "new synonomy" with *attenuatus*; sorry Seymour). Some pores that Abe checked were just thinned spots to allow sensory hairs to extrude; others were gland openings. So far as I can find, little is known about what might be emitted from such glands (in other families there are pores for bioluminescent secretions, but not in the Eucalanidae).

What Abe did was soak a specimen in hot lye, dissolving everything except the exoskeleton. Then he snipped open its ventral side, stuck it flat to a slide like a tiny bear-skin rug, stained it intensely with a drop of chlorozol black, rinsed and then examined it with a compound microscope at 600X. He drew maps of points where light came through (Figure 19.8). Indeed, there were distinctive patterns, and they could be associated with the female genital segment shape variations.[6] Abe had the added characters he needed, and he identified and named three previously unrecognized species (including *E. langae*).

Also, he recognized related subgroups of the seventeen total species based on both the pattern variations, and some of the species distinctions long recognized by others, including Lang. All three of the new species were close relatives of one of the first *Eucalanus* species described, *E. attenuatus* (Dana, 1849).[†] Except for the attenatus group,[6] Abe never did publish maps of the global distributions, but he provided brief statements telling where each species lives, for example:

> *E. hyalinus* (Claus, 1866). Tropical-subtropical, circumglobal especially in eutrophic, oxygenated waters adjacent to boundary currents; deep epiplankton to upper mesoplankton.

Ocean habitats are *epipelagic* (with epiplankton) from the surface down to about 200 meters, *mesopelagic* from there to about 1,000 or 1,200 meters. Deeper waters are *bathypelagic*. *Eutrophic* means nutrient-rich with abundant phytoplankton in the epipelagic, and its opposite is *oligotrophic* meaning low nutrients, low biological production rates.

Frequently when a taxonomist recognizes distinctive groups of species within a genus, someone who agrees will raise one or more groups to named-genus

---

[†] J.D. Dana named it *Calanus attenuatus* in 1849, placing it later in *Eucalanus*, which he established in 1852.

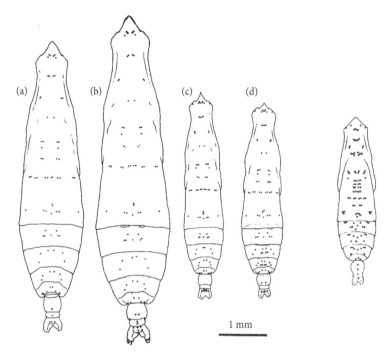

Figure 19.8 Dorsal pore patterns of the attenuatus group of *Eucalanus*, from Fleminger.[6] Left to right they are (**a**) *E. parki* Fleminger, 1973; (**b**) *E. langae* Fleminger, 1973; (**c**) *E. sewelli* Fleminger, 1973; and (**d**) *E. attenuatus* Dana, 1849. Those pore pattern distinctions are subtle and most obvious on the urosome. *Eucalanus subtenuis* (subtenuis group) is placed at right to show the greater between-group contrast. The arrowhead shape of the head is characteristic of most species in the genus. The bodies are very thinly muscled and even more transparent than most other upper ocean copepods. Permission, US government open access publication.

status, also extending their list of papers. Genus designations are, as widely recognized, more arbitrary than those of species, with divisions partly based on convenience. Some convenience comes from having modest numbers of species in most genera. In that spirit Y. V. Geletin[8] created two new genera *Pareucalanus* (Abe's attenuatus group) and *Subeucalanus* (Abe's subtenuis and pileatus groups), adding some developmental characters he found for the urosomes of *Eucalanus*. So it was not purely to extend Geletin's bibliography.

When Abe died, his office/laboratory was packed with filing cabinets of notes and preliminary manuscripts, his microscopes, and rows of boxes holding slides horizontally in trays with preserved copepods waiting attention in drops of glycerin/water. Is the cycling *Tigriopus* culture still there? I doubt it would be. Mark Ohman, Scripps faculty member and collection curator, has been in charge of access, and some serious attempts have been made to work further on Abe's suspended projects. Scripps graduate student Erica Goetze got Mark's help in that lab, tackling issues with *Eucalanus* that Fleminger had left unaddressed. She told me (in November 2020) that looking through Abe's files and books, she found scraps of paper with notes on all sorts of insights about copepods that he never got to pursue. She also held up her dog-eared copy of Lang's dissertation, saying that also is full of preliminary observations never pursued or published. Before showing some of her results, lets meet Erica (Figure 19.7).

Erica told me that her father, a geophysicist at MIT, died when she was a child. Her mother, widowed at age 36, taught biology at Phillips Andover Academy, a prominent, mostly residential, coeducational prep school in a small Massachusetts town. They were housed on campus, so she grew up in an intensely academic community. Erica's connections to biology started then: the frog eggs to tadpoles to frogs progression that recurs in these bios, bird watching with her aunt. Her maternal grandfather was a boat builder in Acadia (Down East Maine), and much time in summer was spent there messing around in boats, cruising the New England coasts, and clamming. She became a biology major at Wesleyan University, in another small New England town. In the summers she took several jobs as a biological research assistant. One was at the Farallon Islands Wildlife Refuge off the California coast, studying seabirds with William Sydeman. Studies of their gut contents were her initial introduction to zooplankton.

Wesleyan apparently was not quite rigorous enough, so Erica added an overloaded semester of science and math courses at Cornell University. They included an ornithology course taught by David Winkler. She did not miss any of the field trips around Cornell. After finishing that semester in December 1994, she continued on working in "Wink's" lab as a research assistant on tree swallows (migration, population dynamics). She loved being part of Wink's lab:

> He was inspiring, encouraging, asked good questions—got us very invested and engaged. He introduced me to the idea that math was essential to ecological research and sent me off to learn about Bayesian methods.

She worked with Dr. Winkler from about March to August 1994, and then went to work in Antarctica from September 1994–March 1995 (the Southern Ocean field season). She did graduate from Wesleyan. The Antarctic work[9] was with ornithologist Wayne Trivelpiece. She took a research assistant position that she applied for, not knowing him in advance.

Prof. Trivelpiece and his wife Susan had been working since 1976 on the biology of a penguin colony at Point Thomas, King George Island in the South Shetlands. The colony is home to three distinct species of those mostly krill-eating birds: Adélie, chinstrap, and gentoo penguins. The work included a lot of stomach pumping to look at penguin diets, so Erica got to know partly digested krill pretty well. She says the interesting thing was that stomachs were often full of just one species, just one age group, or even mostly just one sex. Papers by Wayne Trivelpiece and others state that food in penguin stomachs is layered, each layer almost entirely of just one food item. That agrees with things known about schooling in *Euphausia superba* and other regional krill species. The US Copacabana Station is close to Poland's Arctowski Field Station. During exchange visits among their scientists, Erica got to know a Polish colleague, Grzegorz Dziadurski. That led to some extended travel to Poland, and after Erica moved on to graduate school they were married.

In 1997–1998 Erica applied to some oceanography graduate programs, and Scripps, seeing her potential, flew her from Warsaw (where she had taken the Graduate Record Exam) for an interview. Both Mike Mullin and Mark Ohman urged her to come. So she did, starting there in 1998. Realizing that no dissertation project was going to be handed to her, she undertook an intense reading program on zooplankton, along with the oceanography core courses. From study of Fleminger's papers and Bui Thi Lang's Scripps dissertation, she decided she would add some genetic sequence data to extend the work on diversification in *Eucalanus*. To do that she needed hands-on training in DNA techniques, which by 2000 included PCR and automated sequencing of reasonably long gene segments. That training was generously provided by Professor Ronald Burton and his students and colleagues in their laboratory. Then she wrote a dissertation proposal, and, using the ideas, she and Mark Ohman obtained an grant from the National Science Foundation to support the work. Mark Ohman became her principal advisor, and she set out on a substantial sea-going adventure.

For her project[10, 11] Erica needed specimens not preserved with formaldehyde, the most typical plankton-sample preservative. It mostly destroys the structure of DNA making it almost unavailable for gene sequencing methods. And sequences were what Erica proposed to use. Her goal was to address the following questions:

(1) Do morphological species correspond to genetic clades? To what extent are there genetic subdivisions within species?
(2) Do continents function as barriers to gene flow in circumglobal species?
(3) Do evolutionary relationships among species of Eucalanidae match expectations based on morphological similarity?
(4) Are the Eucalanidae monophyletic?

She went on numerous research cruises, often on research ships deadheading between regions for completely different studies by other scientists. She volunteered to work on cruises led by others for their purposes, sandwiching her sampling into any gaps in their programs. She was able to collect *Eucalanus* species where they are common. She also received specimens from colleagues, extending her collections to all three oceans and from 60°N to below 60°S, all preserved in ethanol or frozen. In this initial work she managed to collect 16 of the 17 *Ecalanus* species that Fleminger recognized (not *E. dentatus*), as well as all of the recognized species of *Rhincalanus*, a related genus. To identify species morphologically she used standard characters for those easily separated and used Fleminger's pore patterns for seven others.

Then she extracted DNA by standard techniques, often from just one end to preserve the specimens' pore patterns. She obtained gene sequences from all of them. Those genes were mtCOI, mt*16S* rRNA, nuclear *18S* rRNA, and nuclear internally transcribed spacer 2 (ITS2). Not every species produced sequences for all four genes, but the data base was big, and Erica made the most of it using bioinformatics algorithms to analyze the similarities and distinctions among the sequences. She aligned the sequences, trimmed them some to assure alignment, and counted DNA base differences (A-T vs. G-C). Finally she applied algorithms quantify levels of sequence similarity, moving from closest resemblances to more distant ones. She chose two modes of comparison, using both sequence alikeness indices (e.g., neighbor-joining, Chapter 18) and indices of numbers of changes needed to get from one sequence to a different but related one. The latter methods are said to be based on "parsimony," since fewer changes likely represent closer relationships. (A little more on bioinformatics in Chapter 20.)

Erica had data[10, 11, 12] demonstrating strong differences among all of the morphologically recognized species. That is represented here by a redrawn version of her diagram of the genetic clades (Figure 19.9), from which I left out Erica's analysis of the related eucalanid genus *Rhincalanus*.

More important than the cladogram to our understanding of copepod evolution was that four of the *Eucalanus* species Fleminger recognized (from detailed examination of body parts), had substantially different gene sequences in different ocean areas, or in some cases the same areas:

- ➢ North Atlantic versus North Pacific versions of *Pareucalanus sewelli*
- ➢ An unnamed species related to *Parecalanus langae* and *P. parki*
- ➢ Two surprisingly sympatric (same geographic patterns) versions of *Eucalanus hyalinus*
- ➢ North Atlantic versus North Pacific versions of *Subeucalanus pileatus*
- ➢ North Atlantic versus Southwest Pacific versions of *Subeucalanus crassus*

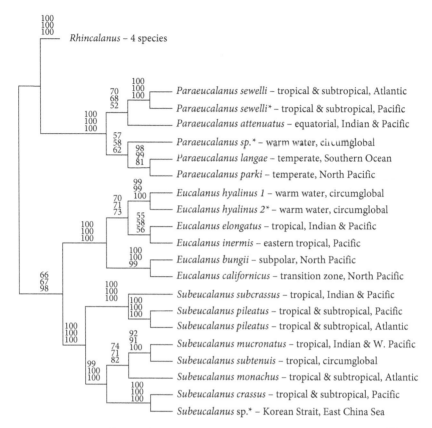

Figure 19.9 A *Eucalanus* relationship tree reduced from Figure 2a in Goetze.[10] Numbers at the branch points are estimates of likelihood based on three modes of phylogenetic analysis applied to similarities in their sequence data. The asterisks designate four lineages distinct from the species most similar. Given their levels of genetic distinction, it is possible they are cryptic species, separately interbreeding populations not morphologically distinct, at least for current levels of resolution. Permission, Royal Society London.

It was important to obtain sequences from single individuals, which assured that no confusing mixture was sequenced, which could result in a sort of contest winner in the PCR gene amplification before sequencing. Thus, "cryptic species" that look as closely as possible to identical, but are genetically at least somewhat distinct, could be recognized in collections, even cryptic pairs caught in the same net tow.

Just how different were the DNA sequences between specimens of a presumably recognized species and its cryptic congener? Let's look at just one gene, 326 base pairs of mt16S rRNA, and just two species pairs. I use their *p*-distances, the percentages of base pairs differing between the distinct stocks:[10]

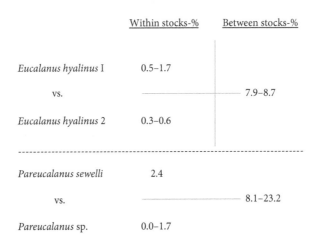

	Within stocks-%	Between stocks-%
Eucalanus hyalinus I	0.5–1.7	
vs.		7.9–8.7
Eucalanus hyalinus 2	0.3–0.6	
Pareucalanus sewelli	2.4	
vs.		8.1–23.2
Pareucalanus sp.	0.0–1.7	

The first pair share similar ranges, at least now, both living in the temperate zones of all three major oceans. The between-stocks genetic differences are large, but since they both occur in the genomes of what must be *functioning* mitochondria, why would they represent a barrier to mating success? The answer has been suggested by Scripps geneticists Christopher Ellison and Ronald Burton from work with the tide-pool copepod *Tigriopus*.[13] Mitochondrial genes do not provide the full data for constructing the completed protein structures (in this case mitochondrial ribosomes); the rest of the code has transferred to nuclear genes, which evolve much more slowly. Thus, barriers to mating success (not mating, per se) can arise when the mRNA from the two sources no longer generates functional mitochondrial ribosomes or enzymes when meeting in the mitochondria. Similarly, in mtCOI, population differences in mtDNA as large as 13.5%, which Erica found by comparing *E. bungii* to *E. californicus* with overlapping North Pacific ranges, could often lead to such mismatches and, thus, dysfunctional mitochondria. On the other hand, it is not necessary to worry much about the level of differences in the barcodes; differences in Goetze's data for nuclear 18S-rRNA and nuclear ITS2 were also large.

From her genetic data suggesting the existence of at least four cryptic species hiding inside names assigned to species (now sister species) of *Eucalanus*, Erica chose the pair *E. hyalinus* 1 and *E. hyalinus* 2 for refined morphological examination. She teamed up with Janet Bradford-Grieve, and they examined every body dimension, limb segment, and skin pore on specimens from all three oceans.[14] At that new level of precision they found distinctive morphologic characters, for some of which they could plot Fleminger graphs (Chapter 4 and Figure 19.4)

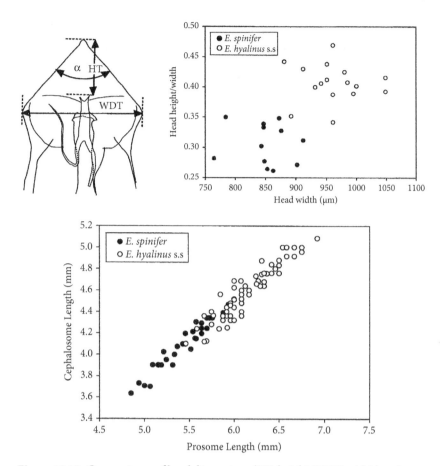

Figure 19.10 Comparisons of head dimensions (HT, height, WDT, width) and cephalosome versus prosome lengths between *E. spinifer* and *E. hyalinus*. From Goetze and Grieve.[14] Permission, Elsevier.

showing the difference. One example is shown in Figure 19.10. The Eucalanidae are characterized by their anterior arrowhead shapes, and the height to width ratio for the heads of gene-group 1 and gene-group 2 are strongly distinct.

Working back through all of the descriptions of *Eucalanus* in the literature, they discovered that *E. hyalinus* 2 agreed, particularly in body size (Figure 19.10), with one described from the Gulf of Guinea by Thomas Scott in 1894, which he had named *Eucalanus spinifer*.[15] That name had been declared a synonym of *E. hyalinus* by Giesbrecht and Schmeil (1898), but now it could be resurrected, which they did. In fact, there were distinctive characters beside head shape, the easiest to use being prosome length: 4.8 to 6 mm for *E. spinifer*, 5.5 to 6.8 mm for *E. hyalinus*.

Scott said the male fifth legs were distinctive. The drawing of them by Scott's son, Andrew Scott, detailed but without a size scale,[15] agreed well enough (Figure 19.11) with much more detailed examinations of new specimens. Perhaps the general similarity and 0.5 mm overlap in lengths had served to distract Giesbrecht, Fleminger, and others from noticing the additional distinctions. Goetze and Bradford-Grieve went to the added trouble of checking for their redefined *E. spinifer* in net tows (Danish Atlantide Expedition, 1945–1946) from the Gulf of Guinea, Scott's type locality. Indeed, they found them. Scott also stated that, along with body size, *E. spinifer* was characterized by the sharp projections at the back of the prosome (Figure 19.11), clearly the basis of

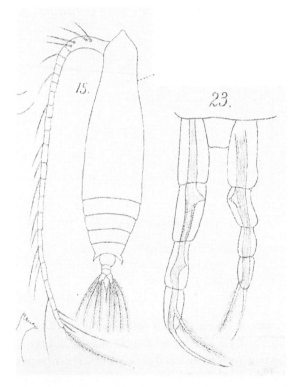

Figure 19.11 From Scott, Thomas.[15] Enlargements of pictures drawn by Scott's son Andrew of *Eucalanus spinifer*, at that time a "new species." Taken from the 39 tiny, randomly scattered, but numbered etchings in Scott's Plate I, ten of them of *E. spinifer* or its parts. Left (15), the only habitus figure (prosome must have been around 4.7–5.7 mm, see Figure 19.10). Right (23), the male P5. No scales or magnifications were given. The drawings on the plate are turned at various angles and closely tucked together, so the "business end" of a mandible pokes in at lower left. Permission, Wiley & Sons, *Transactions of the Linnean Society, London, Series 2*.

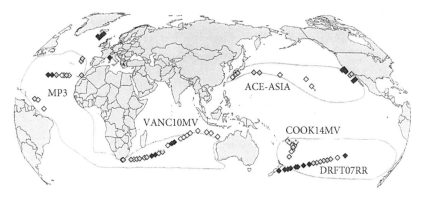

Figure 19.12 Station sites at which *E. hyalinus* s.l. (s.l. stands for *sensu lato*, meaning as earlier defined) were captured on five cruises (MP3, etc.) in net tows from 400 or 1,000 m to the surface. Symbols: ◇ - *E. spinifer*, ◆ - *E. hyalinus*, ◈ - both together. The swirling outlines are distribution patterns for *E. hyalinus* s.l. described by Fleminger and Hulsmann.[17] From Goetze.[16] Permission, The Society for the Study of Evolution.

his name choice. There are similar projections in the new drawings, though a bit blunter. The three other genetically distinctive eucalanid species (two from *Paraeucalanus*, one from *Subeucalanus*, Figure 19.9) are ready for similar work.

Just later, Erica published[16] more detail on the global variation of genetic patterns within *spinifer* and *hyalinus* populations. The distributional patterns of the two species were not completely overlapping (Figure 19.12), although it is difficult to say what, if anything, distinguishes the waters with one of the species, from those with the other one or with both.

In addition to learning, primarily from her nuclear sequence data, that those two *Eucalanus* species were certainly two so-called cryptic species, Erica learned from her mtCOI data that its *haplotype proportions* varied among the parts of its range. Those were, of course, sampled for *E. spinifer* and *E. hyalinus* from the three major oceans and the North Pacific and South Pacific subtropical gyres. A lot was said in Chapter 18 about mtCOI, but nothing about its haplotypes (or those of other mitochondrial genes). Chromosomes in mitochondria are a *single* sequence of double-stranded DNA, a sequence actually bonded end-to-end in a circle. They are haploid, distinct from the two-copy or diploid nuclear genes wound in two, paired eukaryote chromosomes. The mitochondria of zygotes are nearly all supplied from the mother in the egg. Thus, if a mitochondrial gene, say mtCOI, has different sequences by even a few base pairs, those are termed distinct haplotypes. Let's look just at one of the two species, *E. hyalinus*, since the general conclusions for each are the same. Among 429 specimens, Erica found 239 distinct haplotypes. Careful comparisons of their degrees of difference allowed grouping

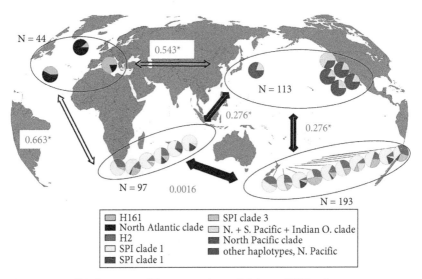

Figure 19.13 Pie diagrams of the proportions of eight mtCOI haplotype groups (or clades) of the temperate zone copepod *Eucalanus hyalinus sensu strictu* at 25 circumglobal stations. H161 was only in the Atlantic, North Pacific clade only in the North Pacific. Double arrows are dark for more similar regions (circled with N = total specimen numbers) less dark for more distantly related station groups. Numbers with the arrows (Φ_{ST} = 0.543*, 0.002, etc.) are basically ratios of within-group (the ovals) haplotype variability to between-group variability.* denotes statistical significance, at 5% level. Color version supplied by Erica Goetze of Figure 19.13 In Goetze.[16] Permission, The Society for the Study of Evolution.

those into eight groups (with some leftovers). Relative haplotype proportions are commonly displayed as pie diagrams. Erica has given me a color version of the Goetz[16] scatter-map of her pie diagrams (Figure 19.13). Notice, again, the level of work involved in obtaining 429 sequences from individual copepods.

The implications are vastly reduced gene flow between ocean regions, except for the southern Indian and South Pacific transects, where the mixtures of haplotypes were substantially closer. Oceanic mixing times among these regions are long, many decades (South Pacific with northeast Pacific) to centuries (North Pacific with North Atlantic, at least in their temperate zones). Erica has written extensively since about the implications of these observations, but much of that has been based also on new studies of regional variations in genetic divergence for other copepods, for pteropods, chaetognaths and more.

Erica defended her Scripps dissertation in 2004. She and her husband, Grzegorz, decided they needed some time in Europe. So Erica organized two postdoctoral terms: six months at the British Natural History Museum (BNHM) in London and three years at the Danish Institute for Fisheries Research in

Copenhagen (DIFR). At the BNHM she worked with Adrian Glover, helping with genetic comparisons of benthic annelid worms. That included studies of polychaetes living in the bacterial mats that grow on the amphipod-cleaned bones of whales lying on the seafloor. Erica and Adrian had both participated in collection of the worms on a cruise off California with Craig Smith of the University of Hawaii (UH), with whom Dr. Glover had been a postdoctoral fellow. Glover was Erica's postdoctoral adviser at BNHM, and they worked with a group fascinated by these odd benthic communities.[18] Then it was on to Denmark for a position at DIFR with planktologist Thomas Kiørboe (introduced in Chapter 12).

Erica's Postdoc in Denmark

At the time Erica visited it, the DIFR was literally housed in a palace, Charlottenlund Slot (mentioned in Chapter 12). It was constructed originally for the restful and luxurious summering of Danish royalty. There were many rooms for offices and meetings, stables usable for laboratories of several sorts. Thinkers and observers important for copepod studies included Thomas Kiørboe, Sigrún Jónasdóttir, and André Visser. If I have it right, Sigrún makes great observations and André mostly thinks. Thomas Kiørboe does some of each, but also cultivates partnerships with people of Erica's high caliber. Erica had obtained a Marie Sklodowska-Curie European Fellowship to support her for a three-year term working at the Slot. She says it was good to spend a couple of years in such talented company, and that the crew there sustained lively social interactions, particularly group lunches ingested with side dishes of intellectual nutrition. The Danes, she found, worked very hard, but also took more time off than US scientists, particularly for her in contrast to Scripps. Among other things, the open time allowed Erica and Grzegorz to get acquainted with a daughter born in Denmark; she had her fifteenth birthday in 2021.

Erica came to DIFR thinking about one of Abe Fleminger's notions: that species become distinct (1) when gene flow (successful mating) stops between geographic subpopulations of its parent population; (2) when those diverge genetically enough to have reduced hybrid fitness; and then (3), when they are again pushed back into contact, adaptations arise allowing them to avoid mating. That is because sterile or less successful matings have a cost. That cost selects for any recognizable and hybridization-stopping difference. Erica must have been reading Jeannette Yen's papers (with others) on mate-finding in coastal copepods (Chapter 10, where the video methods are explained). By the time Erica worked in Charlottenlund, Kiørboe and Espen Bagoien had been repeating Yen's observations and theorizing from the results.[19] Dr. Bagoien had recently moved on, but there was video equipment for study of copepod mate-finding in species tolerant of small spaces. Erica wondered whether avoiding interbreeding could depend upon mate-attracting pheromone scents becoming distinctive for a species. So she applied their system of X-looking and Y-looking, high-speed video recordings of swimming copepods in 1-liter, cubic aquaria.[20]

Temora longicornis were collected from the North Sea and *T. stylifera* from the Mediterranean off Barcelona. Those congeners have regions of population overlap west of Normandy and south of Georges Bank where males and females of these distinct species could possibly attempt to mate. Video sessions were devoted to both conspecific and "heterospecific" (their term) mate-finding observations. For heterospecific pairings the attempted crosses were *stylifera* males with *longicornis* females and then with the sexes reversed. The tracks were remarkable, producing some lovely illustrations (Figure 19.14, showing only one).

As you see in that example, a *T. longicornis* male tracked and found a *T. stylifera* female, made contact and captured her with his geniculate antennules. There were replicate videos with the same result. However, the opposite was not true. When *T. stylifera* males were placed with *T. longicornis* females, "little mate-finding activity [was] observed." According to the text, conspecific trials for both species were similar to the successful heterospecific ones. So once a *longicornis* male clasped a *stylifera* female, what happened? Spermatophore placement only occurred once (in 30 observed captures), but then it was also rare in conspecific contacts of both species.

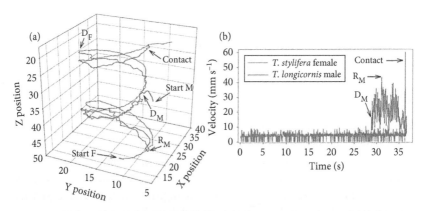

Figure 19.14 (A) Track of male (M, blue) *T. longicornis* following a pheromone track from a *T. stylifera* female (F, red). The male found the track at D_M, followed it in the wrong direction to R_M, reversed and backtracked past D_M, touching the female at "contact." Axis units are mm within a 1 liter cubic aquarium. (B) Velocities of both sexes during the sequence. From Goetze and Kiøboe.[20] Permission, Inter-Research, Marine Ecology Progress Series.

That was also the case for observations of *Calanus marshallae* by Atsushi Tsuda and me. Usually after contact, the males followed the females over a few short jumps, but then usually the female just took off, leaving the male bewildered (anthropomorphically is the only way to think about some animal behaviors). The conclusion about avoiding interspecific mating by *Temora*, because effective pheromones are different, is mixed. Possibly the pheromones from females of the two species are different in a way that *stylifera* males can detect, whereas they seem the same to *longicornis* males.

Before the *Temora* experiments, Erica had run another pair of heterospecific mating trials with *Centropages typicus* and *C. hamatus*.[21] Here's what she found:

> Isolation between *C. typicus* and *C. hamatus* is incomplete, since males of each species were observed to successfully detect and follow pheromone trails, make contact, capture, and place spermatophores on the urosomes of females of the other species. . . . *Centropages* heterospecific mating attempts closely resembled conspecific mating in these species, as the same characteristic behaviors were observed.

Erica even looked for and reported tracking and capture of *C. typicus* females by *T. longicornis* males. It is not safe to conclude from just these cases that pheromones are never strongly distinct between species, not even between copepod genera. But at least in these small coastal copepods, mate-attracting pheromones must be closely similar. There is no doubt, however, that the second stage grasping gear on the fifth legs of the two kinds of males are different (Figure 19.15). Recall (Chapter 10) that capture is accomplished with the toothed folding joint in the right antennule. Erica tells me that the *Centropages* males she observed mostly grabbed the females' caudal setae. The two bodies then swing so the male can grasp the female's urosome with a claw on his fifth leg (P5). As Fleminger thought likely about structures involved in mating, Erica thinks that just the differences among the males' fifth-leg grasping clamps would usually stop attempts at hybridization.

In 2007, shortly after leaving Copenhagen, Erica participated in a Nordic Marine Academy graduate course at Stockholm titled "The Challenge of Pelagic Feeding: From Prey Detection to Secondary Production in Contrasting Pelagic Food Webs." She and fellow participants eventually wrote a paper on a genetic method for identification of *Pseudocalanus acuspes*, apparently the only species of that genus in the Baltic Sea. Papers with dates shortly after Erica's time at Charlottenlund resulted from work there with Guillaume Drillet and others using the laboratory's large-volume *Acartia tonsa* cultures. There was also a

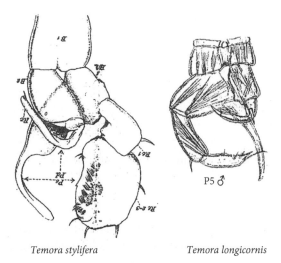

Temora stylifera   Temora longicornis

Figure 19.15 Male fifth legs from two species of *Temora. T. stylifera* (anterior) from Giesbrecht;[23] *T. longicornis* (posterior) from Sars.[22] Both about 0.1 mm across the basis.

paper on picoeukaryotes (single cell organisms, <10 μm with nuclei) based on collections from an Indian Ocean cruise (VANC10MV) important to her thesis.

A Real Job in Hawaii and Return to High-Seas Copepod Genetics

To some extent connected with her earlier work with UH Professor Craig Smith, Erica was invited to join its Department of Oceanography. She did, leaving Copenhagen a year early. At UH, she very shortly got back to long sea cruises collecting specimens for new studies of genetic variation in copepods. Through the whole time from graduate school to moving to Honolulu, she was developing connections with copepodologists around the world, particularly those applying genetic techniques. Important colleagues not mentioned elsewhere in the book were Katja Peijnenburg (Figure 19.7), Ryuji Machida, and Carol Lee. Erica participates in a global exchange among virtually all the molecular sorts in the World Association of Copepodologists and others beside. Her publications since joining Oceanography at UH have been coauthored with many of them.

One of the early papers from Hawaii[24] was based largely on a return to the samples used in her *Eucalanus* studies at Scripps. She had also sorted *Pleuromamma xiphias* from her samples, the largest species in this temperate-tropical, oceanic

Figure 19.16 *Pleuromamma xiphias*, female, lateral. Note the black knob indicated by the blue arrow, the red pigment near the mouth and around the foregut (see Chapter 9), and the fifth leg swinging away in the manner of an escape leap. Photo by Erica Goetze, with her permission.

genus of consistent diel vertical migrators (Chapter 8). Recall that *Pleuromamma* is readily recognized, because all the species have a dark black, spherical knob on just one side (Figure 19.16). Most *P. xiphias* carry it on the right side, a few on the left. Erica sequenced the mtCOI barcodes for 651 specimens from most of the major distribution areas. Recall that mtCOI is haploid. She found 140 haplotypes in 651 specimens (smallest sample 22 specimens, most about 25).

After grouping haplotypes with minor differences, she produced a map of pie diagrams demonstrating the spatial variation of the groups (Figure 19.17). Members of two groups, colored black (Clade 3) and dark blue in the "pies," occurred everywhere. The others were almost specific to distinct ocean areas: yellow slices in the North Pacific; tan slices in the south Indian; red (Clade 4), rose, and gray slices in the eastern equatorial Indian; white slices most prominent in the south Pacific; and slices of three unique colors in the tropical Atlantic. There were some khaki slices in the south Pacific for haplotypes grouping neither with each other nor with any major cluster. On average Clade 4 specimens, both male and female, were a little smaller than the others. Clade 3 females had relatively more of their photophores on the left (174 of 456) than Clade 4 females (13

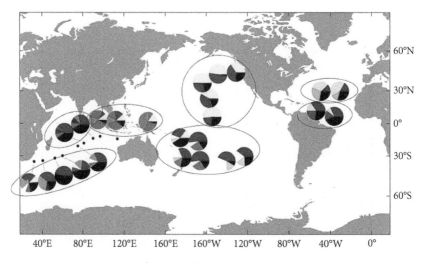

Figure 19.17 Proportions of groups of specimens with substantially similar mtCOI haplotypes in females of *Pleuromamma xiphias*. There were 140 haplotypes among 651 specimens. Slicing of these pies begins at 90° from vertical and proceeds *clockwise* (opposite to Figure 19.12). From Goetze.[24] Permission, Oxford University Press.

of 63). That is a significant contingency ($p < 0.01$), but the two distinctions could have a common cause or just happen to occur in the same spatial pattern.

What does all that variation in mtCOI or the correlated differences in morphology among these major groupings mean? The morphology differences could have been ecophenotypic (nuture, not nature), or both morphologic genes and mtCOI respond similarly to changing rates of gene flow at increasing distances. That occurs even along stretches of continuous current flow (e.g., south Indian samples), but more so where there are strong circulation barriers, North versus South Pacific, North versus South Atlantic, North versus South Indian, and, of course, between major ocean basins.

Basically, Fleminger must have been right, gene flow is restricted by those two kinds of barriers: continents and nearly confined current circuits. It is reasonable to suppose there could be cryptic (not morphologically obvious) species among the populations of *Pleuromamma*. It is not possible to say for sure that there are from mitochondrial genes. Erica and colleagues pursued variations in *Pleuromamma* species through several more studies, including a long section of samples taken above the mid-Atlantic ridge and crossing the equator. There were strong mtCOI haplotype differences in *P. abdominalis*, *P. piseki*, and *P. gracilis* between North and South Atlantic temperate gyres. Colder equatorial flow partially separates their populations, members of all of them are present in it, but sparse.

Temperature difference is not necessarily the cause of that effect. Similar patterns of haplotype distinction occurred on equator crossings from different years.

At about this point, while still getting fully established at UH, Erica and Grzegorz welcomed a second daughter to their household. We did not discuss how they manage the added complexity. That daughter turned ten in 2020, the year the COVID-19 pandemic arrived in the United States. Erica's list of duties during the epidemic's several phases included parenting, some long cruises with a sea-bottom pump sampler she has invented and covering her UH professorial duties from home. Computers have kept people at work in a way that sometimes seems merciless. But of course, the virus has also been merciless.

A conclusion similar to that for *P. xiphias* came from a project conducted largely by graduate student Emily Norton.[25] It was partly based on the mid-Atlantic section, partly on mining again (and why not?) the same Pacific and Indian Ocean samples as for the *P. xiphias* sequences. It was titled "Equatorial dispersal barriers and limited connectivity among oceans in a planktonic co-pepod." That copepod was *Haloptilus longicornis*, an upper mesopelagic (200 to 1,000 m) species found in subtropical gyres of all three oceans. The study was based on mitochondrial Cytochrome Oxidase II (not I), for which they developed PCR primers. The results, not shown here, were regional distinctions much like those for *P. xiphias* (Figure 19.17).

Then, with new colleagues, they studied *H. longicornis* again.[26] They added Atlantic samples and a suite of nuclear sequences, specifically *microsatellites* (mentioned in Chapter 18). These are not genes per se, but important in the miniscule machinery of the chromosomes. They are repeating sequences of 1 to 5 base pairs, such as CGAAT:CGAAT:CGAAT.... They can run out to dozens or hundreds of tandem repeats. And there are dozens of them with different base pair lists in almost any eukaryote genome. They do seem to be functional. For example, they make up large parts of chromosomal centromeres, the attachment hooks along chromosomes that attach to mitosis spindle tubules during cell division. Their sequence lengths vary so widely that a suite of 18 or 20 of them can be amplified with PCR, and the combination of lengths of the products can be measured to serve as DNA fingerprints nearly specific to individuals. They are the basis of most forensic identifications by DNA.

So, Goetze and company[26] sequenced a very large part of the *Haloptilus* genome (using next-generation 454 sequencing) and found 16,742 microsatellite regions. Selecting PCR primer sets for some and testing their reliability (consistency of amplification), they selected eight sets producing "amplicons," each of variable lengths that could be measured. That allows a substantial mass of combinations, and (after a tour de force of bioinformatics) it appeared there are two strongly distinctive groups. Moreover, they mapped neatly against a mtCOII clustering shown here (Figure 19.18). There are two cytochrome *c* oxidase genes

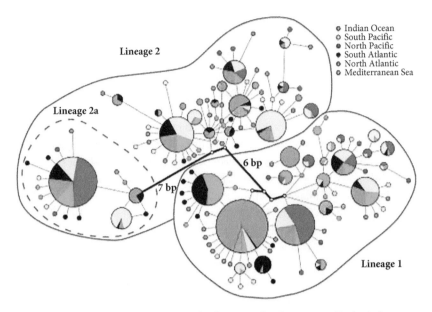

Lineage 2

Lineage 2a

○ Indian Ocean
◐ South Pacific
● North Pacific
● South Atlantic
◑ North Atlantic
○ Mediterranean Sea

6 bp

7 bp

Lineage 1

Figure 19.18 "Median-joining network of mtCOII haplotypes. . . . Each circle represents a different haplotype; the sizes of the circles represent the number of specimens with that haplotype; colors represent the proportions from different sampling locations; and line lengths connecting the circles represent the number of mutational differences between haplotypes." The thin solid lines surround lineages (1 and 2) that were divergent at microsatellite loci. The thick black line joining Lineages 1 and 2 indicates a 6 base-pair difference in haplotypes. That between Lineages 2a and 2 indicates a 7 base-pair difference in haplotypes. The figure and caption quote are from Andrews et al.[26] Permission, Wiley & Sons.

(I and II) in eukaryote mtDNA. After some further testing on 63 new specimens, the identical lineage clusterings merged. Andrews et al.[26] are confident there are two cryptic species of *Haloptilus longicornis*, temporarily designated 1 and 2, and essentially fully sympatric (both present wherever they live). Apparently, the microsatellites and mtCOII do not evolve fully independently. That is the implication of the 9 to 12 base-pair shifts in mtCOII from Lineage 1 to Lineage 2. Lineage 2a underwent an additional 7 base-pair shift, either someplace, and then spread everywhere, or somehow and eventually everyplace at once.

There are more papers on this theme for H. *longicornis* from Goetze's group. One[27] shows that North Atlantic versus South Atlantic differences in its mtCOII haplotype proportions were stable, within the expected variation of multiply sampled proportions, between an initial section above the mid-Atlantic Ridge in 2010 and a follow-up section in 2012. I haven't found checking by anyone for re-fined morphological distinctions between the groups distinct according to their microsatellite genetics. The two species, if that's the right interpretation, have not yet been given separate names.

After 2015 Erica has kept at her demonstrations of substantial genetic variability in widely distributed, named species of high-seas zooplankton, both within regions and even more so between regions. However, she and her associates largely moved on from copepods: to pteropods, heteropods, hyperiid amphipods, microzooplankton, bacteria, and even a snail fish (Liparidae). Since this book is about copepods, I only mention that these studies are out there awaiting your attention. The mechanisms of this genetic divergence within sympatric populations remain a subject of research, much of it Erica's or shared with her. On the other hand, her studies do demonstrate that a large part of the genetic divergence between populations of oceanic plankton species is indeed allopatric, much as Fleminger thought from morphologic data alone.

What Is the Significance of This Demonstrated Genetic Variability?

Readers not already steeped in current evolutionary doctrines may wonder exactly that: What does it matter? Why should I care? And, "doctrines"? Yes, they converge on the intensity of religious beliefs. The variation of essential genes across wide ocean reaches shows that the basic mechanics of evolution and speciation remain fully active in wild populations. The changes demonstrated in, say, mtCOII (Figure 19.18) are not proved to have changed the function and viability of the proteins translated from that DNA. In fact, the sequence data came from specimens that had developed and survived to adulthood, and, thus viable. Partly that results from the very large number of synonymous triplets in the genetic codes for particular amino acids. Nevertheless, the observed shifts show that evolution is proceeding, with genetic mutations providing new combinations and, when viable, passing them along the chains of interbreeding (of gene flow) sustained by along-current mixing. It also shows that more isolation between current gyres and between oceans allows greater divergence to emerge. The raw materials of evolution, of adaptation to changing conditions, are forming and changing again. Those raw materials are simply variation itself, as Wallace and Darwin recognized some 170 years ago.

You probably do not need to be reminded that the unprecedented current rates of climate change have already warmed most ocean sectors about one centigrade degree, some wide areas more. Even small warming increases all physiologic rates in animals like copepods, which do not thermally regulate. Most impressive are the impacts on developmental rates of eggs and larvae, the tendency for faster development to carry copepods from stage to stage after a little less

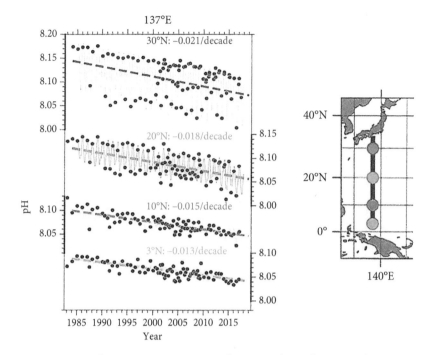

Figure 19.19 Hydrogen ion concentration along a north-south transect (map at right) in the western Pacific. Data from Japan Meteorological Agency. pH is the *negative* logarithm of the hydrogen ion molarity (concentration), so larger numbers represent *lower* concentrations. Acidity is rising at the rates shown, faster to the north where concentrations are lower. From Figure 2-11 in Christian and Ono.[28] Permissions: PICES, Japan Meteorological Agency, Drs. Jim Christian, Hisashi Ono, and Masao Ishii (who drew the graph).

growth. That means somewhat smaller specimens and lower egg-clutch numbers, but also faster stock turnover and, perhaps, more generations annually. You also won't need reminding that rising atmospheric levels of carbon dioxide push its dissolution into the oceans as carbonic-acid, lowering their pH levels on the order of -0.02 units per decade from 1983 to 2018 (Figure 19.19),[28] though with regional variability of initial pH levels and change rates. In experiments, marine copepods are reasonably resistant over those and wider ranges. But a good deal of adaptive evolution will be needed to bring them through in the long run.

Copepods have indeed survived through many and certainly worse ocean acidity changes over geologic time, but always over million-year intervals. Present changes have become measurable over a few decades. The data from Erica Goetze and others show that genetic variability is available for adaptive selection to work upon, which is perhaps reassuring—if you take a very long-term perspective. The

greater the stresses, the faster selection can work, but the durations required to re-
build diversity stretch far beyond familiar human time scales.

References

1. Fleminger, A. (1975) Geographical distribution and morphological divergence in
   American coastal-zone planktonic copepods of the genus *Labidocera* *Estuarine
   Research* 1: 392–419.
2. Fleminger, A., & K. Hulseman (1974) Structures and distributions of the four sibling species composing the genus *Pontellina* Dana (Copepoda, Calanoida). *Fishery
   Bulletin* 72: 63–120.
3. Hulsemann, K., & A. Fleminger (1990) Some aspects of copepodid development in
   the genus *Pontellina* Dana (Copepoda: Calanoida). *Bulletin of Marine Science* 25(2):
   174–185.
4. Lang, Bui Thi (1965) Taxonomic Review and Geographical Survey of the Copepod
   Genera *Eucalanus* and *Rhincalanus* in the Pacific Ocean. PhD dissertation, University
   of California, San Diego.
5. Lang, Bui Thi (1967). The taxonomic problem of *Eucalanus elongatus* Dana. *Annals,
   Faculty of Science, University of Saigon* 1967: 93–102.
6. Fleminger, A. (1973) Pattern, number, variability and taxonomic significance of in-
   tegumental organs (sensilla and glandular pores) in the genus *Eucalanus* (Copepoda,
   Calanoida). *Fishery Bulletin* 71(4): 965–1010.
7. Sewell, R. B. S. (1947) The free-swimming planktonic Copepoda. *The John Murray
   Expedition 1933–34, Scientific Reports*, 8, 303 p.
8. Geletin, Y. V. 1976 The ontogenetic abdomen formation in copepods of genera
   Eucalanus and *Rhincalanus* (Calanoida, Eucalanidae) and [a] new system of these
   copepods. *Issledovaniya Fauny Morei* 18: 75–93. [In Russian; English title from
   original.]
9. Mader, T. R., P. Ciaputa, E. Goetze, N. J. Karnovsky, W. Z. Trivelpiece, & S. G.
   Trivelpiece (1996) Antarctic seabird ecology and demography in Admiralty Bay, King
   George Island, Antarctica. *Antarctic Journal of the United States* 31: 110–111.
10. Goetze, E. (2003) Cryptic speciation on the high seas; global phylogenetics of
    the copepod family Eucalanidae. *Proceeding of the Royal Society of London, B* 270:
    2321–2331.
11. Goetze, E. (2004) Speciation in the Open Ocean: The Phylogeography of the Oceanic
    Copepod Family Eucalanidae. PhD dissertation, University of California, San Diego.
12. Goetze, E., & M. D. Ohman (2010) Integrated molecular and morphological bi-
    ogeography of the calanoid copepod family Eucalanidae. *Deep-Sea Research II* 57:
    2110–2129.
13. Ellison, C. K., & R. S. Burton (2006) Disruption of mitochondrial function in
    interpopulation hybrids of *Tigriopus californicus*. Evolution 60(7): 1382–1391.
14. Goetze, E., & J. Bradford-Grieve (2005) Genetic and morphological description of
    Eucalanus spinifer T. Scott, 1894 (Calanoida: Eucalanidae), a circumglobal sister spe-
    cies of the copepod *E. hyalinus* s.s. (Claus, 1866). *Progress in Oceanography* 65: 55–87.
15. Scott, T. (1894). Report on the Entomostraca from the Gulf of Guinea. *Transactions of
    the Linnean Society, London, Series 2,* VI(Part I): 1–161.

16. Goetze, E. (2005) Global population genetic structure and biogeography of the oceanic copepods *Eucalanus hyalinus* and *E.* spinifer. *Evolution* 59(11): 2378–2398.
17. Fleminger, A., & K. Hulsemann (1973) Relationship of Indian Ocean epiplanktonic calanoids to the world's oceans. In *The Biology of the Indian Ocean*, ed. B. Zeitzschel, 339–347. Springer, New York.
18. Glover, A. G., E. Goetze, T. Dahlgren, & C. R. Smith (2005) Morphology, reproductive biology and genetic structure of the whale-fall and hydrothermal vent specialist *Bathykurila guaymasensis* Pettibone 1989 (Annelida: Polyoidae). *Marine Ecology* 26(3–4): 223–234.
19. Kiørboe, T., & E. Bagoien (2005) Blind dating: Mate finding in planktonic copepods. I. Tracking the pheromone trail of *Centropages typicus. Marine Ecology Progress Series* 300: 105–115.
20. Goetze, E., & T. Kiørboe (2008) Heterospecific mating and species recognition in the planktonic marine copepods *Temora stylifera* and *T.* longicornis. *Marine Ecology Progress Series* 370: 185–198.
21. Goetze, E. (2008) Heterospecific mating and partial prezygotic reproductive isolation in the planktonic marine copepods *Centropages typicus* and *Centropages hamatus.* Limnology & Oceanography 53(2): 433–445.
22. Sars, G. O. (1903) *An account of the Crustacea of Norway, Vol. IV. Copepoda Calanoida.* Bergen Museum, Bergen.
23. Giesbrecht, W. (1892) *Systematik und Faunistik der Pelagische Copepoden des Golfes von Neapel,* Tafel 17, Figure 19. In *Fauna und Flora des Golfes von Neapel, XIX,* W. Giesbrecht. Verlag von E. Friedlander & Sohn, Berlin. 831 pages.
24. Goetze, E. (2011) Population Differentiation in the open sea: insights from the pelagic copepod *Pleuromamma xiphias. Integrative and Comparative Biology* 51(4): 580–597.
25. Norton, E., & E. Goetze (2013) Equatorial dispersal barriers and limited population connectivity among oceans in a planktonic copepod. *Limnology and Oceanography* 58(5): 1581–1596.
26. Andrews, K. R., E. L. Norton, I. Fernandez-Silva, E. Portner, & E. Goetze (2014) Multilocus evidence for globally distributed cryptic species and distinct populations across ocean gyres in a mesopelagic copepod. *Molecular Ecology* 23: 5462–5479.
27. Goetze, E., K. R. Andrews, K. T. C. A. Peijnenburg, E. Portner, & E. L. Norton (2015) Temporal stability of genetic structure in a mesopelagic copepod, PLOS ONE, doi:10.1371/journal.pone.0136087.
28. Christian, J. R., T. Ono, eds. (2019) Ocean acidification and deoxygenation in the North Pacific. *PICES Special Publications* 5: 116 pp.

20

Beta Taxonomy II: Copepods in the Stream of Time

Species, Genera, Families, Orders, . . .

We have been dealing with *phylogeny* steadily above. Those with some biology lodged in their forebrains are familiar with the term, if only that *ontogeny* recapitulates it. And that is the only way most of us ever use "recapitulates." Ontogeny (development of embryos and larvae) does have stages shared among wide stretches of the animal realm, including at least all multicellular forms. That was known in the nineteenth century. Lately the relationship of evolutionary sequences and developmental ones has reemerged as "evodevo." We aren't going there. While evolutionary trees are shaped at their branch tips by division of species populations into distinctly different sorts, that is "new" species, conceptual phylogenetic trees, reach back through periods and eras as bushy shrubs, or "cladograms," of relationships. That term shares its root with "cladistics," the study of phylogeny, also termed phylogenetics. In this chapter I will expand the view from that above, which mostly has been marine calanoids, *Tigriopus* among harpacticoids and some Oithonids. Here we will take on all the orders, then examine the tree of order Calanoida.

Allow me a note here about terminology for higher-level taxonomic groups. The classical systematic sequence from more to less inclusive groups in zoology was kingdom (all animals), phylum, class, order, family, genus (species appearing closely related simply because of alikeness), species, and often subspecies. Those are still used, though the complexity of many groups has required more terms. For genus *Calanus* among the Copepoda, which is a *subclass* of class Crustacea, we have orders (Calano**ida**), super families (Calano**idea**), families (Calan**idae**), genera (*Calanus* and *Undinula*), and often, though less formally, species groups (e.g., the helgolandicus group within *Calanus*), and species. The bold-type is to show the spellings that distinguish them.

I propose to skim the cream from recent molecular and morphology-based cladistics results in regard to copepods, and I will leave some trail blazes into the literature for anyone who wants more. Let's begin, however, by backing out to the levels of phylum, classes, and orders. A *Nature* paper published in 2010[1] by collaborating groups at the University of Maryland and Duke University

Oar Feet and Opal Teeth. Charles B. Miller, Oxford University Press. © Oxford University Press 2023.
DOI: 10.1093/oso/9780197637326.003.0020

had collated masses of DNA sequence data for selections of species from many subsets of the phylum Arthropoda (animals with stiff, chitinous exoskeletons). They were led by Jerome Regier (of UM) and Clifford Cunningham (of Duke). The work depended on many published studies, but took things further and added some art. It was based on a phylogenetic analysis of DNA sequences for 62 protein-coding genes from 75 species of arthropods, plus five species long thought somewhat related to arthropods: tardigrades and onychophorans (e.g., *Peripatus*). Those five were "outgroups," which allows a sort of rooting to trees of gene similarity because they are considerably more distinctive. The 62 sequences for each species were all concatenated as one long and aligned series, a total of 41,000 base pairs. Again, aligned means the nucleotides, and the amino acids they specify, were in equivalent places for genes of all the species. Determining that is feasible without overwhelming error rates.

Consider right off that the commonality of gene sequences (and functional protein structure) has to be, and conveniently is, close enough to say that each gene is doing the same thing across the entire array of 80 species. The consensus of three modes of cladistic analysis (which agreed reasonably well) is shown in the paper as an elaborate cladogram for the entire species set. For the unwashed (me, maybe you), the authors also provided a very nice diagram (Figure 20.1). The three genera of copepods in the list were *Mesocyclops*, *Acanthocyclops* (both Cyclopo**ida**), and *Eurytemora* (Calano**ida**). The analyses separated all of those, and the subclass distinctions (copepods vs. others) were stronger than the ones within subclasses. For you, now copepodologists all, that lends credibility to the whole diagram.

A thing fun for fans of team carcinology is that the Hexapoda (the wildly diverse and mostly terrestrial insects) prove to be a side branch, one not only within Pancrustacea but of the rather advanced group that Regier et al. termed the Altocrustacea. Copepods (at least those three copepods) take an interesting spot between the clade for Malacostraca (crabs, shrimps, and so on) and Branchiopoda (a range from fairy shrimp, represented by *Artemia*, to Cladocera, represented by *Daphnia*). Tardigrades and *Peripatus* were way to one side of the scheme, so the tree was rooted, but simplification for Figure 20.1 left them out.

The ink was barely dry on that 2010 *Nature* paper when critical comments erupted. That is common in this particular enterprise. Seeking also to obtain an "all arthropods" phylogeny was a group working in Germany and Austria. Karen Meusemann and others, led by Bernard Misof, opened their paper[2] by pointing out that there were relatively few amino acids (13,000) in the Regier et al.[1] data set. By 2010 vast data were available on arthropod genes, and Muesmann et al. could examine relationships based on sequences derived from messenger RNA (expressed sequence tags, ESTs). They started with data for (paraphrasing) 775 putatively related protein-coding genes representing 350,356 amino acid

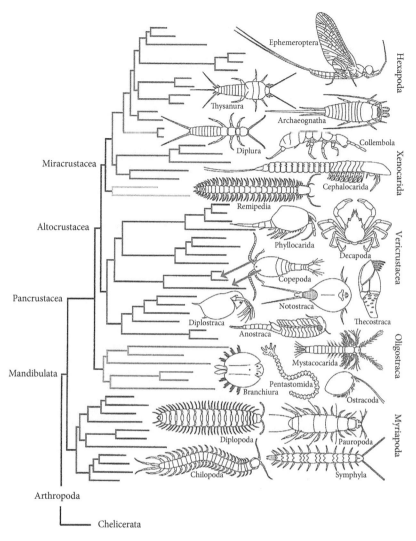

Figure 20.1 A proposed phylogeny based on similarities (and differences) among 62 aligned, nuclear genes for proteins identifiable in all arthropods. Only the Pancrustacea and its sister group, the Myriapodia (millipedes and centipedes) are included here. A large clade (Chelicerata), including horseshoe crabs, spiders, scorpions and pycnogonids, was deleted from the figure, because this book is about copepods (their clade indicated by red arrows) and relatives closer than chelicerates. Extracted from Figure 2 in Regier et al.[1] Coauthor Jeffrey Shultz created the figure. Permission, Nature

positions for a set of 233 taxa (214 true arthropod taxa plus 3 onychophorans, 2 tardigrades, and 14 even more distantly related taxa). So yes, they had more data, but after eliminating species and genes for low information content, they reduced it to 117 taxa, 129 genes (32 coding for ribosomal proteins and 97 for other proteins) and 37,476 amino acid positions.

All in all, the Meusemann tree looked much like that from Regier et al. The copepods in the German study were *Calanus finmarchicus* (an oceanic calanoid), *Tigriopus californicus* (an upper intertidal harpacticoid), and *Lepeophtheirus salmonis* (the parasitic salmon louse, a siphonostomatioid). They grouped closely together, at least relative to all other crustacea, and seemed to be a sister group to barnacles. Those two branches were, as in Regier et al., intermediate between that of the more complex malacostraca and the strongly different branchiopods. The US group had found Pancrustacea (including Hexapoda) to be a sister group with Myriapods (millipedes and centipedes); that is, they initially branched together with them. The German group found those clades to be more distantly related and emphasized that as being a substantial difference. The body plans are strongly different, and there never has been doubt their ancestors diverged long, long ago. One of the problems was that the terrestrial millipedes and centipedes have tracheas (hollow tubes with an opening in the exoskeleton) to move air internally for oxygen exchange, though not tracheas located like those of insects. Nevertheless, that suggested a common ancestor. However, phylogeneticists working with genes want things nailed down. The same German group, publishing in 2014 as Rehm et al.,[3] generated new sequence data (so-called 454 transcriptomes) for six more myriapods, thus a total of eight, and they reran the analysis. Several cladogram programs branched them as a sister group with Pancrustacea. Rehm et al., a little coyly, avoided saying that was the claim of Regier et al. Nevertheless, it's probably settled.

A greater interest for copepodology in the later Rehm et al. paper[3] is an attempt at a generalized rate of change analysis based on the same large sets of sequence data from many forms. Eleven time points were established from geologic ("stratigraphic") records for reasonably well classified fossils of close relatives of the sequenced species. Also, some specific times were assigned for early branchings based on hunches. Complex models with "relaxed clocks" were then used to set numerous divergence times based on the relative amounts of genetic change between earlier and later branchings. The result was a similar cladogram, but along a time scale (Figure 20.2). Evidently, massed genetic-sequence change data, coupled with a few solid fossil dates, suggest, as do also a few fossils, that copepods date back to the Cambrian, maybe more specifically to the relatively late Cambrian. An even more elaborate cladogram with divergence timelines was produced a few years later by a Eurozone team, Omar Rota-Stabelli et al.[4] showing that essentially all still extant animals with surface

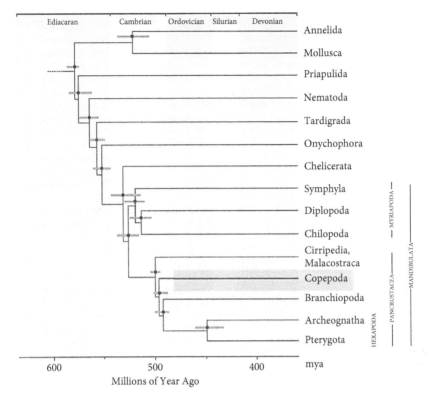

Figure 20.2 Approximate times of phylogenetic branches to some major phyla and nine arthropod groups, as reconstructed by Rehm et al.[3] Permission, Elsevier.

cuticles ("ecdysozoa") diverged before or during the Cambrian. They had added sequences for nematodes, kinorhinchs, and others. Again, all the major groups of pancrustacea were judged to have diverged by the end of the Cambrian.

All of the groups in Figure 20.2, including copepods, emerge from their initial branching as long straight lines, which must extend all the way to the present, since the gene sequences were collected from *living* descendants. Also in the 2014 study, copepods were represented by only three species in two families. Yet the actual history of copepods can look nothing like that. There have been five major extinction and diversification cycles since the Cambrian. There have been snowball-Earth episodes, massive lava and gas eruptions producing global warming and later cooling, explosive meteor landings, and cycling ice ages. The agreement of two studies, each including only three copepod genera, suggests that aspects of the copepod molecular program and thus their body form, modes of reproduction and life history (varied as those are), have persisted since their earliest ancestor that was distinctly a copepod. There is a strong "just-so story"

Figure 20.3 A few players in the arthropod molecular phylogeny games. From left above: Clifford Cunningham, Jeffrey Schultz, Karen Meusemann. Left below: Bernhard Mishof, Omar Rota-Stabelli, Björn von Reumont. From online sources, except Meusemann and von Reumont, who provided photos. Karen's photo by Robin Hood Casalla Daza. Jerome Regier chose not to appear, and photos for Peter Rehm have not been found. With permissions from all six.

quality to that insight, one Rudyard Kipling would have appreciated. However, this one is likely true, and it applies to all living animal groups, including genus *Homo*.

Copepod Group Distinctions at the Orders Level: Rony Huys and Geoff Boxshall Address Higher-Level Classifications

Geoffrey Boxshall (Figure 20.4), now retired, was a Merit Researcher in life sciences at the British Museum of Natural History (BMNH) in London, specializing in studies of copepods. Sharing that specialty there is the Belgian copepodologist Rony Huys, who also produces remarkable drawings of the flattened-fauna sort standard in copepod taxonomy. Working both together and not, they have done more than anyone else, save Janet Bradford-Grieve, to stabilize our view of the higher order systematics of copepods. Geoff grew up in Hampshire County in the United Kingdom, where he attended Churchers College, at the time a private boys school from about 11 to 18 years of age. His father was a bank manager and

Figure 20.4 Photograph, provided by Geoffery Boxshall. Given his cascade of papers (more than 270 with a distinguished array of coauthors) and books, mostly on copepods, the intense focus radiating from this 2004 photo is certainly real and effective. His face, according to later photos, remains little changed.

his mother a purchasing agent for the military. As a senior, Geoff was captain of the Churchers field hockey team, and he played some rugby for the Hampshire County team in those school days. Geoff is listed by Churchers as one of its notable alumni. He must have had decent A-levels and exams, because he moved on to the University of Leeds.

In his last undergraduate year, Geoff took a field course on parasites with Professor R. Wynne Owen. Even after the class, he kept bringing fish carrying copepod parasites to Owen, which they studied together. Parasites were not his only interest; another led to marriage in 1972 with Roberta Smith, a singer-songwriter. The connection with Owen remained strong, and Geoff received a scholarship for PhD work under him. A first project on parasite effects on the cell biology of blood in fishes ran into problems, and with encouragement from Owen, Geoff switched to a survey of the copepod parasites of North Sea flatfish, completing the degree in 1974 (at age 24). His thesis was titled "Studies on the Copepod Parasites on North Sea Marine Fishes, with Special Reference to *Lepeophtheirus pectoralis* (Müller, 1776)." He immediately published seven papers on that work, starting with a parasite of the thorny skate (*Raja*).[5] He also looked for jobs and found none in fish pathology, fish parasites or anything else connected to fish.

Soon enough, a position did open up at the BMNH for a copepodologist willing to work on free-living planktonic copepods, calanoids at first. Geoff was hired and never left. Pretty soon, starting in 1975, he and Roberta added four children to the family. Geoff identifies several key experiences over the next years. One was a cruise with Howard S. J. Roe of the National Institute of Oceanography (NIO), who was using large nets to take vertically stratified hauls in the eastern Atlantic. During his first stint of five or six years at BMNH, Geoff worked on pelagic copepods using the samples from NIO. There was a calanoid phase and long specialization on copepods other than calanoids: oncaeids,[6] deep-living misophrioids,[7] *Sapphirina*, and others. Some of the early work was on details of copepod muscular anatomy, showing, with remarkable precision for animals so small, exactly how muscle fibers move the limbs and segments within the limbs.[8]

Geoff's work up to the late 1980s on all of the parasitic and free-living groups of copepods gave him an extensive overview of their distinctions and similarities. So he teamed up with Rony Huys, then a Belgian colleague (Figure 20.5), who had already drafted a book on the subject. They revised it and published in 1991 a volume titled *Copepod Evolution*.[9] Dr. Huys trained at and was then working at the Rijksuniversiteit Gent. He had a similar range of scientific experience, though a decade younger. His studies had been more focused on both harpacticoids (including meiofaunal species) and parasitic forms. The goals of the book were to establish a terminology for limbs, limb parts and segments consistent with

Figure 20.5 Dr. Rony Huys, a photo cropped from one taken near the Korean Sea after the 12th International Conference on Copepods in July 2014. Photo and permission provided by Rony Huys.

names by then standard for all crustacea (not "pancrustacea," just shrimp, crabs, amphipods, and so on), to define the higher-level divisions of the copepods in a stable way, and finally to suggest a general proposal for the branching evolutionary paths connecting the copepods.

The book was successful in the first two of those respects, partly because of the excellent body and limb drawings (Figure 20.6) that characterize all of Huys's work. In respect to terminology, new descriptions no longer refer to first and second antennae, but to antennules and antennae, respectively. Similarly, first and second maxillae are now quite consistently referred to as maxillules and maxillae. In fact, the only terminology currently in use is that in *Copepod Evolution*. A major step in examining overall copepod evolution was provision of defining morphological distinctions for ten orders of copepods. Some rearrangements have come along in the relationships Huys and Boxshall suggested for the orders. With that book as evidence that he and Boxshall were a strong team, the BMNH hired Dr. Huys, who moved to London in 1992. They have worked together on few projects since, but each has produced many papers and insights about copepods working with fellow copepod experts from everywhere on Earth.

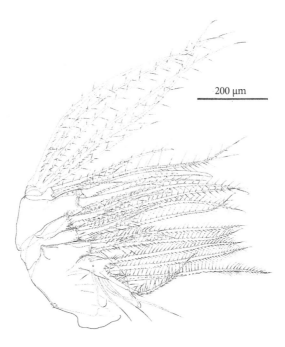

200 μm

Figure 20.6 The maxilliped of *Pleuromamma xiphias*, a calanoid copepod from the family Metridindidae, drawn by Rony Huys for *Copepod Evolution*.[9] Permission, The Ray Society.

Copepod Orders Defined: Simplified from Huys and Boxshall

Copepodologists, like entomologists and other specialists, have long traditions of taxonomy with progressively more refined observation of everything about every classified individual. With much careful observation, more divisions than the phylum-to-subspecies sequence have been added to our system for copepods. At present we have stacked names for almost every group such that exact equivalences to the older class-order-family scheme aren't possible. We do have, however, better classifications and more realistically phylogenetic ones. At present the overall diversity of copepods divides quite neatly and convincingly into ten or eleven groups of families, reasonably termed "orders." Table 20.1 briefly characterizes ten named orders that Huys and Boxshall accepted in their book.

In Table 20.1, *gymnoplea* are copepods with narrowing to the urosome starting after five thoracic segments (after Th5), and *podoplea* are those in which the narrowing comes between Th4 and Th5, quite often retaining small legs at the sides of their first urosomal segment (Th5). After reading this chapter for me, Dr. Boxshall pointed out that "strictly" there are six thoracic segments, "because the maxilliped-bearing segment is thoracic." There is a solid evolutionary argument for that. Maxillipeds are clearly modified thoracic legs among shrimp, for example. Goeff suggests using "leg-bearing segments," for which the long-standing P1 to P5 abbreviation for legs posterior to the cephalosome (the pereopods) would be correct.

Most of those orders have long been recognized, though with species and families moved about some until more defining characters held in common within groups could be determined. However, Gelyell**oida**, copepods that live in subterranean waters, were only discovered in 1976 by Raymound Rouch and F. Lescher-Moutoué.[10] The genus name of the first-found species, *Gelyella droguei*, refers to the type locality, a 60 m well outside the village of Saint-Gély-du-Fesc, close to Montpellier, France. The species was named for the authors' colleague C. Drogue, who helped in finding it. Sampling was by pump (80 hours at 100 m³/hr produced 9 ♀, 11♂and 2 copepodites) from a layer of Eocene limestone laced with subterranean lakes well below the land surface (karst topography). The animal does, indeed, look like an harpacticoid, but some major features differ, particularly the point at which urosome segments are reduced in size. Rouch and Lescher-Moutoué described the species accurately (Figure 20.7), helping Rony Huys to recognize how fully distinct the new genus is. He established the new order.[11]

The extreme diversity of copepod form is quite well compartmentalized among those orders, so little change has come about since *Copepod Evolution*.[9] However, Goeff Boxshall and Sheila Halsey[12] (more on their book below) moved

Table 20.1 Orders of the Subclass Copepoda of the Arthropod Class Crustacea

The first two orders are gymnoplea:

Platycopioida Bulbous bodies with narrow, 5-segmented urosomes in both sexes. Gymnoplea, with the female genital segment not fused with the second urosomal segment. Thoracic legs 2 and 3 with two lateral spines on the first exopod segment (Ex1). Such seeming details have been long retained within these orders. Currently about a dozen species (in four genera, including *Platycopia* Sars, 1911) from seafloors and some tropical anchialine (flooded with seawater) caves.

Calanoida Gymnoplea, with genital and second urosomal segment fused in females. Legs 2 and 3 with one spine on Ex1. Mostly free-living forms of diverse shapes in oceans and lakes. Forty-two families and many genera.

The following eight orders are all podoplea:

Misophrioida Deep-sea, especially near-bottom, but also near-bottom species in coastal waters and tropical seawater caves. Thorax segments may have "expansion joints" to allow ingestion of whole, relatively large prey (termed "gorging"). Relatively few known species.

Harpacticoida Primarily bottom-dwelling, many small enough to move among sediment grains. Some planktonic forms can be found crawling among clustered algal cells, e.g., *Macrosetella gracilis* in *cyanobacterial mats (Trichodesmium)*. Many symbiotic and parasitic forms. Urosomes in males 6-segmented, in females mostly 5-segmented (genital segment U2 fused with U1). Antennules short: 9 segments in ♀, at most 14 in ♂. Very great diversity at genus and species levels.

Monstrilloida Parasitic as nauplii and copepodites in mollusks and polychaetes. Last copepodites molt to free-living, pelagic adults, but nonfeeding with no mouthparts, antennules of only a few segments, and five stubby legs. Females have long spines extending from near the genital opening to which eggs become attached.

Mormonilloida A few species of very distinctive copepods living below 400 m in the oceans. Apparently particle feeders. Males with greatly modified antennules: both geniculate with big aesthetascs.[9] Female antennules with very long, setae.[10]

Gelyelloida Just three thoracic legs with their first segments (coxae) fused beneath the body. Narrowing of prosome to urosome not distinct. Less than 1.0 mm, living in subterranean ground water (Figure 20.7). Only a few species found to date, likely because sampling is difficult in suitable habitats. The generally similar males and females are both known.

Cyclopoida Males with both antennules geniculate *and* the antennae lacking a definite exopod. Some mouthparts have fused segments not seen in other groups. Hugely diverse in both the ocean (e.g., Oithonidae) and freshwater (e.g., Cyclopidae). A dozen families in all, including many divergent parasitic forms.

Siphonostomatoida Parasites with ventrally extended "oral cones" formed by the labrum and fusion of the labial palps (or paragnaths). The mandibular gnathobase extends into the cone from the sides, piercing the surface of a host to which the copepod holds itself by one or more piercing limbs. Many whole individuals are reduced to simple tubes. More than 1000 named fish parasites and hundreds of parasites of marine invertebrates.

Poecilostomatoida Free-living families (*Oncaea, Lubbockia, Corycaeus* are common) with one or another anterior limb modified for attaching to surfaces (jellyfish, appendicularian houses, chaetognaths). Many families of diversely modified parasites.

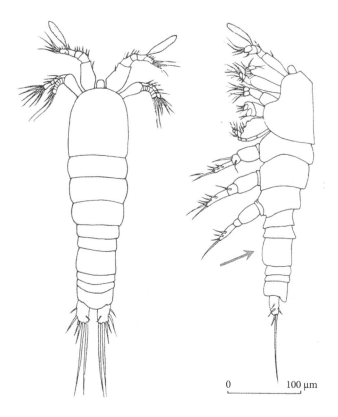

0   100 μm

Figure 20.7 Habitus drawings of a female *Gelyella droguei* from the original 1977 description by Raymond Rouch and F. Lescher-Montoué.[10] Arrow points to the genital segment (with relatively simple ventral openings) believed to be a fused double segment. Length of four measured females ranged from 0.308 to 0.327 mm, very small. Permission, Cambridge University Press.

all of Poecilostomatoida into Cyclopoida, a demotion (and combination) with which not all equally serious copepodolgists agreed.[13] Some previous groupings combined them with the Siphonostom**atoida** (see Figure 20.8).

Recently a group working in Germany and led by Pedro Martinez Arbizu produced a DNA-based, phylogenetic cladogram for the orders of copepods that they presented at a copepodologists' conference in 2017. A paper about the study (Kohdami et al.[14]) has since been withdrawn (in October 2020), following criticisms of the level of statistical confidence that should be placed in its points of cladistic division, and other problems. Sahar Kohdami was the lead presenter at the meeting and first author of the paper. Her drawings of adult specimens from the groups they identified as Orders are in Figure 20.8. She placed them

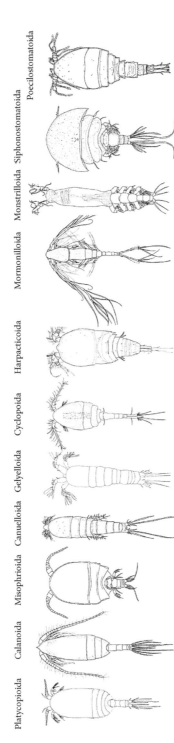

Platycopioida Calanoida Misophrioida Canuelloida Gelyelloida Cyclopoida Harpacticoida Mormonilloida Monstrilloida Siphonostomatoida Poecilostomatoida

Figure 20.8 Ten possible orders of Copepoda (and one family, Poecilostomatidae) illustrated by their free-living forms (or forms most like those in other groups that are free-living). The drawings are based on typical female adults. They are similar to those in a Huys and Boxshall[9] cladogram (their Figure 4.1.6), also in about the same order left to right. That relationship tree was one of the several suggested in 1990 by Ju-shey Ho,[15] a prominent student of parasitic forms. He suggested the Poecilostomatidae are closely related to the Siphonostomatoida, combining them under the latter name. The DNA sequence data of Kohdami et al.[14] agree with Boxshall and Halsey[12] that they be considered part of the Cyclopoida. It also suggests that some groups adjacent in this row are more related to each other that to the others: Platycopioida with Calanoida; Canuelloida, Gelyelloida, Cyclopoida and Harpaticoida are cousins; so are Mrmonilloida, Monstrilloida and Siphonostomatidae. The drawings are by Sahar Kohdami, who graciously provided them and permission to publish them.

over a tree diagram like one in Huys and Boxshall[9] of the orders according to basic appearance. The drawings are useful, even though the cladogram might not hold up. That is because the pictures give a clear idea about aspects of the morphology that copepods classified in each order have in common. Not exactly, however: the form shown is that for adult females of each order's families and genera that are least diverged from the general shape and limb details likely primitive in the order. That general copepod shape is that shown throughout this book; it is even shared by some of the less modified parasitic copepods. Not shown is that all the orders share development as nauplii, which in some parasitic groups is the strongest clue to copepod identity.

A likely valid result from Khodami et al.[14] is their suggestion to promote Canuellidae, previously thought to be a harpacticoid family, to ordinal level, Canuelloida. That was supported by the suggestion of strong divergence in the sequences from two specimens each of two Canuelloida species (*Canuella perplexa* and *Longpedia* sp. from the North Sea) from sequences found among the Harpacticoida. While a couple of genes from two species may seem slim evidence, the many distinctions of Canuelloida among the harpacticoids have been long recognized. They were named as a distinct group in 1948 by Karl Lang, a serious student of harpacticoids, and Francis Dov Por[R16] actually drew an evolutionary tree separating the Canuelloida from the Harpacticoida in 1984. Dr. Kohdami has pointed out to me that Hans Uwe Dahms[17] also declared them separate (using different terminology) on the basis of beautifully drawn distinctions in naupliar morphology. *Canuella* and related genera mostly live in muddy coastal sediments. The Kohdami et al. study needs repeating, which could be as difficult as doing it the first time. That first time included a long and sometimes arduous adventure to get the needed sequences from species in multiple representatives of each order and from wildly different habitats. The tale is well told in the withdrawn paper, still available online from Springer Nature.

Royal Society Recognition Spurred Geoff Boxshall to Expanded Efforts

Very few copepodologists have received the sorts of recognition that seem common among particle physicists and molecular geneticists: a Nobel Prize, Fellowship in the Royal Society of London (FRS), election to the National Academy of Sciences in the United States, or similar acclaim. However, Geoff Boxshall reports being surprised and taken aback at the relatively young age of 44 to become Geoff Boxshall, FRS. Somebody recognized his excellence and potential and did the work of nominating him. Geoff tells me he felt it required extended effort on his part to justify the honor. Indeed, excellent work on every aspect of copepod existence followed this appointment, as well as preceding it.

Among the work most useful to ecologists and others faced with abundant and diverse collections of copepods from all marine and freshwater habitats, from water columns to interstitial spaces in sediments, from the North Pole to the shoreline of Antarctica, is a 2004 book, Boxshall and Halsey's *Introduction to Copepod Diversity*.[12] At 966 pages, it is so large that it had to be bound in two volumes. There is a chapter for every copepod family named at that time. It opens with keys to the copepod orders, then proceeds to sections on each with chapters on each recognized family. Those chapters have keys to the genera, new keys and others revised from sources in what had become an immense literature and illustrated with hundreds of drawings of key characters. There are notes guiding readers to resources for extending identifications to species level. There are notes on the biology and ecology of every family group.

Geoff completed *An Introduction to Copepod Diversity* in partnership with Sheila H. Halsey. She assembled and checked the bibliography (105 pages, over 3,000 references). Equally important was her work on the drawings. Geoff felt that getting permissions for including every literature figure they needed would likely be impossible. So Ms. Halsey redrew them all: 1,819 of the by now familiar "flattened fauna" figures by which we communicate copepod form. Even Geoff Boxshall needs dedicated help. Together they created not simply an introduction, but a now indispensable compendium of the knowledge of copepods to that date.

An Early Look at Classifying the Families of Calanoida Using Genes

Dr. Diego Figueroa was my last graduate student, an advising role I shared with a fellow Oregon State University oceanographer, Harold Batchelder. I will introduce Diego in the next chapter and explain there the role Boxshall and Halsey's book played in his early career. Here, I want to show his early use of gene sequences to evaluate the phylogeny of the Calanoida. Diego carried my lab into the molecular biology era, so he is my personal connection to current reports on the molecular phylogenetics of copepods.

Why is that family structure of Calanoida of particular interest? At least for biological oceanographers, the majority of copepod species we catch in our samplers belong to its families. In general, its specimens are quite readily recognized under our stereomicroscopes in regard to their family identity. There are a few freshwater families of Calanoida. The Diaptomidae and some of the Temoridae are examples, and they are also quite readily recognized. Moreover, the Calanoida are almost the only gymnoplea (see Table 20.1) caught in ecological studies (the Platycopioida being very deep-living and seldom abundant), so we come to know them. Interest in relationships and evolution among its families arise immediately. Finally, studies of both morphological details and genetic distinctions suggest reasonable groupings of the families into super families, and that in turn probably reflects the order's steps of repeated population divisions (*vicariance events*) and species divergence.

Working in the Galapagos, Diego discovered that several, maybe numerous, species of pseudocyclopid copepods (e.g., genera *Ridgewayia* and *Pseudocyclops*) can be found in anchialine waters all around the Galapagos Archipelago. Anchialine habitats are sea caves and various cracks near ocean shores that hold pools of seawater exchanging with the nearby ocean. He has published descriptions of those copepods, and during his dissertation work he generated gene sequence data from them to confirm their species distinctions and relationships from a perspective separated from their morphology. He published those results, too.

Then, during his postdoctoral work in Florida, Diego ran a GenBank search for gene sequences from a significant number of families in the Calanoida. It was an attempt to see where the subject genera of his study might fit in the family phylogeny. He found 32 species from 16 of those families for which the *18S* RNA gene could be aligned over a usefully long sequence. From Ridgewayiidae that included *Exumella mediterranea* and from Pseudocyclopidae a then unnamed species of *Pseudocyclops* identified by Rony Huys.[18] The latter's *18S* RNA had been sequenced to provide an outgroup for a study of parasitic copepods. Diego ran analyses using four cladogram-generating algorithms. He[19] selected one of the resulting and mostly agreeing cladograms (Figure 20.9) using, the Akaike information criterion. A *little* more is said about these analyses just below. The interesting aspect of that result, and it remains interesting, was the close relation of Ridgewayiidae and Pseudocyclopidae, recalling suggestions by Taisoo Park[20] and Vladimir Andronov[21] that they are close enough in detailed structure to be treated as one family.

Note that this is not necessarily a phylogeny; rather, it is a cladogram from work in progress toward one. The three numbers on each branch represent, according to four caladistic algorithms, the statistical strength of the vicariance points, an approximate estimate of the confidence that separation occurred when (and usually where) one ancestor species population divided into two with distinctive subsequent evolution. Good is 100; "only maybe" is 50. Early vicariance events can be suggested by later splits (all the branching more to the left).

Let me try your patience here to explain roughly the logic of the algorithms. I started that in Chapter 18. Diego Figueroa used four algorithms: neighbor-joining (NJ), maximum-parsimony (MP), maximum-likelihood (ML), and

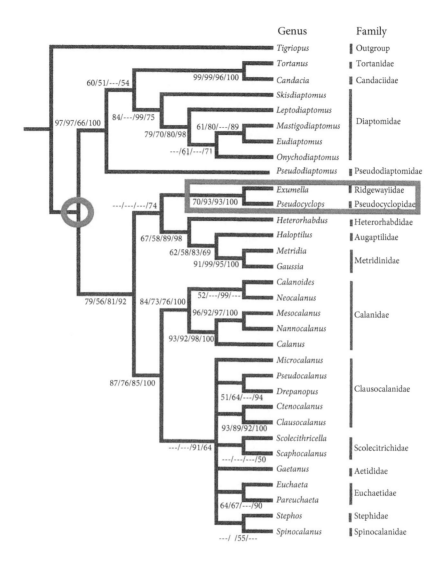

Figure 20.9 Diego Figueroa's cladogram[19] showing (in the red box) the close relation of one species from Ridgewayiidae species (*Exumella mediterranea*) to one from Peudocyclopidae and of those together to an incomplete, necessarily random selection of genera from 14 families of Calanoida. The *Exumella–Pseudocyclops* pair is a sister clade to the combine of *Heterorhabdus*, *Haloptilus*, and two metridinids. And that clade is sister to the remaining genera from *Calanoides* to *Spinocalanus*. The two main clusters separated at the red circle are also sister clades. Permission, Oxford University Press.

Baysian (e.g., the "Mr. Bayes" program) methods. There are others. They all start with a table of, say, species (or individuals) listed by rows and their gene sequences (nucleotide-by-nucleotide; sometimes concatenated segments from several genes), protein amino-acid sequences or morphological character-state codes in the columns. Some analyses begin at the right of the final tree, comparing sequences (codes) of each species with all of the others one-by-one, assigning similarity or difference scores. In NJ, a simple distance based on numbers of differences is calculated (Chapter 18). The results are stored and those pairs with the greatest similarity are linked, often with some left over (e.g., *Pseudodiaptomus* and *Calanus* in Figure 20.13). In MP the pairs are those with the smallest number of sequence (character) changes (hence, most *parsimonious*) needed to make them the same (changing the sequence of one species scores the same as changing the other). In both NJ and MP, some crafty logic is used to decide what sequence should represent the pairs moving left. Convincing logics have been found and applied. I know, that's cheating, and, if you actually care, you can chase those aspects down.

Methods of ML are computationally more complex. They are based on statistical distribution models for the rates of change of the characters over arbitrary time intervals in the past, with less change implying less time, less divergence. Most of the possible trees, given the number of rows, are set up in a computer. Then measures of the probabilities in each of their many vicariance events are calculated: low with many changes, high with few. Finally, an over all probability is calculated by multiplying all the probabilities of their individual vicariance events (actually adding logarithms of those probabilities). Consensus trees are assembled from those trees with high overall likelihood.

Mr. Bayes is supposedly based on a sixteenth-century probability theorem by one Reverend Bayes (as an inveterate frequentist, for those who know what that means, I dislike going near the Rev. Bayes). Two trees are chosen at random from all possible ones and given a "posterior probability" score based on the data, posterior meaning here that the probability comes from the data after the tree is set up. The tree in the pair with the highest scores is kept, and then another generated and compared to it. If that scores higher, then it is the one kept. Then on and on through the night, basically like King of the hill, until new trees stop being better.

A typical strategy is to apply more than one method, sometimes using different decision criteria, then take a poll among all the results to find a most phylogeny-like cladogram. In a book about copepods, I can only touch on these algorithms, apart from the two examples below. The whole field has been very popular for about three decades, and talented mathematicians have devoted long, long thoughts to it. Those mathematicians have determined ways to estimate values characterizing the certainty for each vicariance event in a tree as to whether it is the best possible step left in the cladogram, given the data combinations at that point. Often those are set down near those branchings; again, 100 if the best possible, 50 or less when there is serious doubt. There are good book-length treatments.[22, 23]

Why was Diego's gene-based cladogram exciting? We had worked together for a few years, so of course it was exciting to see him out in the science scramble, seeking new facts and approaches. But scientific importance? First, it demonstrated that even one partial, nuclear gene sequence could be a basis for determining likely phylogenetic structure among the calanoid families, and that it would be rather like one based on morphology. Thus, the available character set for phylogenetic study was greatly expanded. Second, it suggested a very close relationship between *Exumella mediterranea* (classified in Ridgewayiidae) and *Pseudocyclops*, and thus between Ridgewayiidae and Pseudocyclopidae. If you follow that, you are getting a feel for this jargon of genera gathered into families. That close relationship recurred in later studies by others.

Third, the cladogram showed a likely sister-group relationship between a diverse clade below the red box in Figure 20.9 and the clade above it, all of its member families and genera are in super family Centropagoidea. Specifically, that branching identified greater similarity of *18S* RNA within those entire groups than between them. Diego said nothing about either group being "basal," more closely related to primitive ancestors. Partly that was because so many Calanoid families were not included. However, it is widely noted[24] that all extant genera are just that, *extant*: they all have living representatives. So all of their family trees reach as far back (are just as basal) as every other. You can see that idea in Figure 20.2. Nevertheless, and fourth, the substantial number of likely vicariance events (the cladogram's bifurcations) from the genus level to those largest sister groups suggest long sequences of independent evolution and subdividing speciation in both groups.

So, how did Diego's 2011 paper go over? It had been noticed during the review cycles before it was published. A paper published the same month as Diego's[25] pointed out, gently enough, that all Diego's Clausocalanoida species (lower cluster, Figure 20.9) would be better not thought of as a sister clade of Centropagiidae. Fair enough. If enough more families were brought in, the cladogram would look different, and those authors (Blanco-Bercial, Bradford-Grieve and Bucklin) said so. They used concatenated sequence segments from four genes (*18S* and *28S* rRNA, mtCOI and mtCOb) for reasonably representative genera in 29 calanoid families and 7 superfamilies, generating a cladogram that was later expanded by Silke Laakmann and colleagues[26] by adding sequences for a suite of Bradfordian genera (introduced in Chapter 5).

Family-Phylogeny in the *Calanoida* Using Morphological Characters

Both Janet Bradford-Grieve and Geoff Boxshall have taken strong interest in the overall phylogeny of the Calanoida over their whole careers. In the twenty-first century they have worked together on it, cooperating with others. They wrote a paper with Shane Ahyong and Susumu Ohtsuka in 2010,[27] considering the likely phylogenetic relationships among the order's then 40 families. It was based on the defining morphologic features of the families, a total of 116 features. More study proved that analysis preliminary, and they came back with a new one in 2014. In a new paper[28] Janet described a new copepod species, *Pinkertonius ambiguus* (Figure 20.10), the species name "referring to the ambiguous nature of the [detailed$_{CM}$] morphology of this species as not exactly fitting into any of the previously described families."

Then they included *P. ambiguus* in developing a morphology-based clado-gram based on much the same characters as in 2010 (Figure 20.11), but focusing on a sister-group set from that earlier study and adding *Pinkertonius*:

Epacteriscidae (18 genera, including *Caiconectes antiquus*)
Ridgewayiidae (11 genera)
Pseudocyclopidae (2 genera)
Boholinidae (1 *Boholina parapurgata*)
Pinkertonius ambiguus

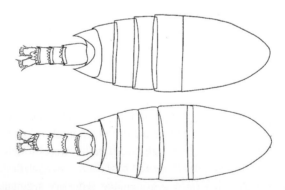

Figure 20.10 Female (above) and male of *Pinkertonius ambiguus*, dorsal views. This is here just to show that the female and male do look like calanoid copepods (at least to me and a perhaps a thousand others). From Bradford-Grieve et al.[28] Permission, Oxford University Press.

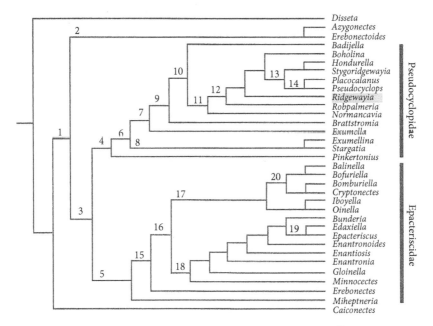

Figure 20.11 Cladogram from Figure 11 of Bradford-Grieve et al.:[28] "Strict consensus [among many runs of different programs] tree . . . after one round of successive character weighting." *Ridgewayia* is highlighted in red to emphasize its location. Permission, Oxford University Press.

They chose *Disseta palumbi* from Heterorhabdidae as the outgroup.

It is fascinating that *Caiconectes* and *Disseta* diverged essentially as far from each other (implied by long lines extending right from the origin), as each was from the other sister clades. The authors noticed it, but did not do any family naming. *Caiconectes* is usually grouped with the Epacteriscidae. The overall result convinced them to revise the family structure as labeled in Figure 20.11. All genera alphabetically from *Badjiella* to *Pinkertonius* (branching no. 4 in the tree) were declared to be cousins in the family Pseudocyclopidae. Diego's and my favorite family, the Ridgewayiidae (*Ridgewayia*, red box in Figure 20.10), lost its high status:

> Clade 4 contains the oldest described family, the Pseudocyclopidae Giesbrecht, 1893, but also includes genera from the Boholinidae and Ridgewayiidae, which are therefore considered to be junior synonyms of Pseudocyclopidae.[28]

Experts on a group's systematics at the level of Bradford-Grieve and Boxshall can make such a declaration, and it can be widely accepted until new data (new species or new characters) come along. Diego can be proud of anticipating this demotion (Figure 20.13), although in correspondence he states that he isn't so sure the statistical strength of all the vicariance events in the later molecular cladograms[26, 29] makes them fully convincing as phylogenetic. The term "junior synonym" feels a bit like Mr. and Mrs. Theophilus Bainbridge Snodgrass naming their son Theophilus B. Snodgrass Jr. It isn't quite the same. Note in Figure 20.11 that two unnumbered vicariance points separate *Ridgewayia* within clade 12 from sub-clades 13 and 14. Maybe that reflects the authors' confidence that no "family-grade" characters separate *Ridgewayia* from *Pseudocyclops*.

In addition to revising their morphological analysis, Janet and Geoff recruited Leocadio Blanco-Bernal to get some *Pinkertonius* sequence and run a somewhat parallel genetic cladistic analysis. He obtained its sequences of 18S and 28S rRNA, mtCOI, and mtCytochrome *b*. The results were informative and convincing enough that Silke Laakmann and colleagues also used their sequences for phylogenetics, as shown below. First, however, let's look at a yet later trial of morphology-based analysis of just the phylogenetics within superfamily Clauocalanoidea by Janet and Geoeff published in 2019.[30]

Bradford-Grieve and Boxshall Run Time Backward for Clausocalanoidea

How does such an analysis proceed? Janet and Geoff first selected 52 genera from three of the families they had earlier determined were likely basal, in the sense that they became distinct early in calanoid diversification, possibly retaining more of the order's primitive characters. Formally, these characters are now termed "plesiomorphic," recognized usually because widely shared within a diverse group, and so likely present in its common ancestors. The basis for inclusion of a genus was that anatomy for a sure representative of its species was thoroughly described, usually as drawn in an exhaustive set of figures of body and limbs, right down to every last seta and bump. Often, when the status of a feature was yet uncertain, they got out type specimens and determined that status. Elena Markasheva of the Institute of Zoology in St. Petersburg and others helped. Eighty-nine features were included. Just two examples are shown in Figure 20.13.

Almost all were chosen so they could be coded in a species versus character table as 1 = present or 2 = absent. There were a few characters with three states (coded, 1, 2 or 3) and a few missing data (coded with a ?). Those numbers can readily (by computers) be indexed for similarity levels across the whole suite of 52 genera (none were represented by two species) generating levels for 1,326

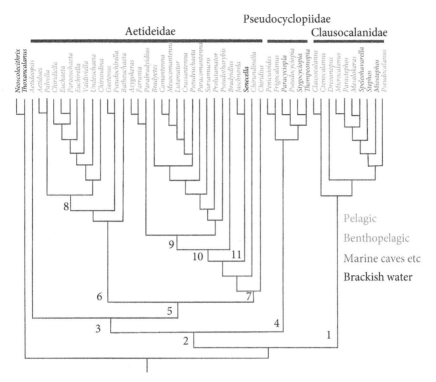

Figure 20.12 A "50% majority rule . . . maximum parsimony tree after one round of successive reweighting" (some characters allowed to be more influential than others). Numbers to the left of vicariance points are assigned clade numbers. The family groups are as labeled at the top. Note that *Euchaeta* and *Paraeuchaeta* used to be the Euchaetidae (Giesbrecht, 1892). Taisoo Park also felt Euchaetidae appropriately held family status. It may now be headed for "junior" synonymy status, since it was given family status on a later page of Giesbrecht (1892) than was Aetideidae. From Bradford-Grieve and Boxshall.[30] Permission, Oxford University Press.

pairings. The similarities allowed what is sometimes called a graphic analysis to generate cladograms. Just one example is shown here (Figure 20.12), which I chose because Janet and Geoff color coded the genera according their usual and different habitats.

The color coding in Figure 20.12 connects that cladogram to my own experience. All of the genera listed in green are marine, and they live their whole lives well up in the water column, never contacting surfaces, except perhaps touching jellyfish or salps. As plankton they swim at small scales, sometimes a few hundreds of meters, and drift with currents at geographic scales. All of these green genera are familiar to me, a student of the ecology of ocean and estuarine plankton. Unless our samplers come very close to the ocean bottom, mostly in quite deep water, the Benthopelagic genera (those in blue) are not seen. Sampling systems for doing that have mostly been deployed in recent decades. Working to considerable depth but not near the bottom, I have never seen any of the genera coded in blue. I have seen pseudocyclopiids from marine caves (red genera here), thanks to Diego Figueroa. Most estuarine Calanoida (and some more oceanic ones) with which I am also familiar belong to Centropagoidea (e.g, *Centropages*, *Acartia*), so they were not included in this Clausocalanoidea analysis at all.

The Clausocalanoidea paper[30] includes a few examples of a different sort of cladogram: character-state distributions (Figure 20.13). Again, they show the proposed phylogeny, but connecting lines are colored to represent the status of each genus for just one character. To illustrate here, I chose the results (a) for genera having or not having two setae on the female's antennule segment 20, and (b) for having 4 versus only 3 setae on the first praecoxal endite (side lobe) of the maxilla. What they demonstrate is that each of the two characters is consistently the same through the entire list of genera for proposed families. At minimal risk of error, they are considered likely to be homologous, or "synapomorphic," in recent phylogenetic jargon. This is true so far at least as the selected genera represent them. Some other characters also were likely synapomorphic, but not all. The state of almost any such character can change. In the crowds of species and genera that these families are, some characters are very likely to switch to the opposite status. Not only can one nucleotide in a species gene sequence mutate to another, later it can, even often, switch back. The idea of those changes saturating in mitochondrial sequences was introduced above. Similarly, seta presence versus absence or relative segment lengths can also change over time. It seems likely that modest changes in morphology can saturate over enough generations.

A Later, Molecular Analysis of *Calanoida* Phylogeny

Let's take one more look, using gene-sequence characters, at the likely evolutionary history of families in the order Calanoida, including most of its long list of families from A for Acartiidae to T for Tortanidae. In papers[25, 28] from 2011 and 2014, Blanco-Bercial, working with Bradford-Grieve and Bucklin, had developed a relational cladogram for the Calanoida based on gene sequences for *18S* and *28S* rRNA, mCOI, and mtCOb from 29 families in 7 superfamilies. I did not show that cladogram here, because in 2019 a German and Russian team, Silke Laakmann, Elena Markhaseva, and Jasmine Renz,[26] repeated the analysis

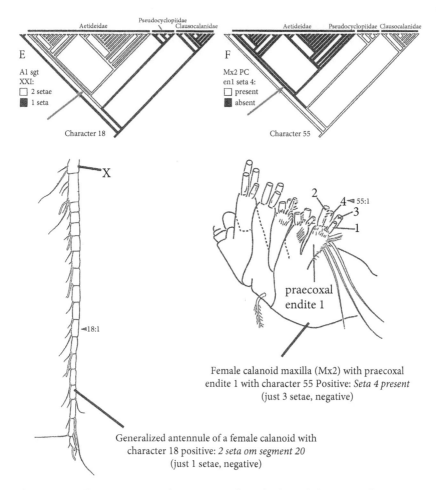

E

A1 sgt
XXI:
☐ 2 setae
■ 1 seta

Character 18

F

Mx2 PC
en1 seta 4:
☐ present
■ absent

Character 55

55:1

praecoxal
endite 1

Female calanoid maxilla (Mx2) with praecoxal
endite 1 with character 55 Positive: *Seta 4 present*
(just 3 setae, negative)

Generalized antennule of a female calanoid with
character 18 positive: *2 seta om segment 20*
(just 1 setae, negative)

Figure 20.13 Demonstration that some single and selected characters do separate families in the Calanoida. Part of Figure 5 from Bradford-Grieve and Boxshall.[30] The characters even work for the two outgroup species, the branch from the 'root' at the far left of both figures. Those outgroups are not closely related, the branch at the tip is short because there were only two of them. Both cladograms make *Aetideopsis* (red arrows) look like a separate family as well as genus. Permission, Oxford University Press.

using the same concatenated sequences (not all of them complete for every included species). However, they added their own data for more of the bradfordian genera and families in the superfamily Clausocalan**oidea**. They actually used many of the sequences used by Blanco-Bercial et al., getting them from GenBank (or similar databases). Thus, much of their cladogram (Figure 20.14) is also the same. The exceptions, of course, are (1) in the newly added bradfordian families,

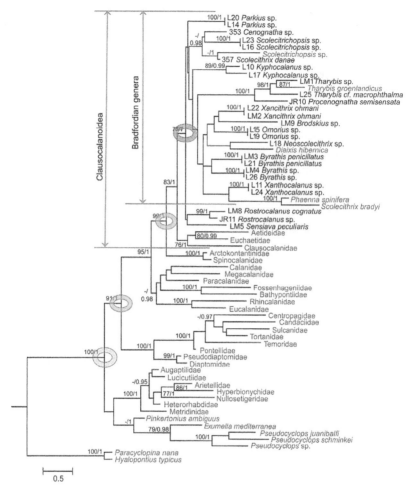

Figure 20.14 Figure 3, from Laakmann et al.,[26] cladogram for 28 calanoid families with multiple genera represented from the bradfordian families among the Clausocalanoidea superfamily (all its families fall above the lower blue line). Based on concatenated partial sequences from *18S* and *28S* rRNA, mtCOI, and mtCytb. Sequences for the highlighted genera (dark type for now and in the paper) were amplified by Laakmann, et al. All the others were the same as sequences amplified and evaluated for family relationships by Bradford-Grieve et al.[28] The blue and pink circles are explained in the text. Permission, Elsevier.

and (2) the early sister-clade relation of the Pseudocyclop**idae** and the families in Augaptilo**idea**.

One answer is No to the question posed in the title chosen by Laakmann et al., "Do molecular phylogenies unravel the relationships among the evolutionary young "Bradfordian" families [of] Copepoda: Calanoida)?" Those species formed a single clade (at the top of their cladogram diagram, Figure 20.18). Rather, the genetics suggest they diverged as a group from all other Calanoida at most recent major vicariance event, circled in pink. So they appear to be "evolutionary young." Too many of the identified vicariance events *within* the group were too weakly supported to be thought reliably phylogenetic.

The first vicariance event in the diagram (Figure 20.14) is between the two outgroup genera, *Paracyclopina* and *Hyalopontius*, respectively a cyclopoid and a siphonostomatoid, and all the rest. Then next (circled in pale purple) was the vicariance event between the sister clades of Augaptiloidea with Pseudocyclopidae (plus *Pinkertonius*) and all the rest. The statistical support is reasonably good. Indeed, the Pseudocyclopidae and Augaptiloidea appear to be, though not necessarily the most "primitive" (they are still living, still evolving), among the earliest diverged groups, while still distinctively subsets of the Calanoida. The usual pseudocyclopid ways of living, near surfaces, many in sea caves or associated with shell debris on continental shelves (like oyster cultch) may have been the starting places for emergence to more fully pelagic habitats like those of most of the families in the cladogram from Calanidae (*Calanus* and its relatives) to Metridinidae (*Metridia* and its). The Augaptiloidea (the cluster from Augaptilidae to Metridinidae) if their relation to Pseudocyclopidae is indeed phylogenetic, have already moved to fully pelagic. A suggestion that a seafloor-to-pelagic progression may have been general for free-living copepods, followed sometimes by reverse transfers, had been offered in 1991 by Huys and Boxshall in *Copepod Evolution*,[9] and Janet Bradford-Grieve[31, 32] has reviewed other indications that epibenthic to fully pelagic habitat transfers to have been common in copepod evolution, specifically among Calanoida.

The Value of Knowing Copepod Phylogeny

At least for out-to-sea ecologists, like me, and for limnologists, these are profound contributions to our understanding of how the copepod communities we study came to be the animals at key food-web exchange points in oceans and estuaries. I repeat, the majority of herbivorous and predatory copepod species (and very often individuals) in those habitats are from the order Calanoida. They are also abundant in freshwater, though in that respect calanoids share key roles with cyclopoids. So it is good that so much focus has been on them. If I may offer it, our gratitude goes to everyone involved in these recent analyses from Janet Bradford-Grieve, Rony Huys and Geoff Boxshall to Leo Blanco-Bercial,

Ann Bucklin, Diego Figueroa, Silke Laakmann, Elena Markhaseva, and Jasmine Renz. The lists should be much longer, including people from well back in time like Wilhelm Giesbrecht and Georg O. Sars, and on through more recent contributors like Vladimir Andronov and Taisoo Park. A second helping of thanks goes to Chad Walter and Geoff Boxshall, who maintain (with fourteen associate editors) an on-line listing of higher-level systematic connections among copepods, from species to orders at the World of Copepods database: http://www.marinespecies.org/copepoda

References

1. Regier, J. C., J. W. Shultz, A. Z. Zwick, A. Hussey, B. Ball, R. Wetzer, J. W. Martin, et al. (2010) Arthropod relationships revealed by phylogenomic analysis of nuclear protein-coding sequences. *Nature* 463: 1079–1084.

2. Meusemann, K., B. M. von Reumont, S. Simon, F. Roeding, S. Strauss, P. Kuck, I. Ebersberger, et al. (2010) A Phylogenomic approach to resolve the arthropod tree of life. *Molecular Biology and Evolution* 27(11):2451–2464.

3. Rehm, P., K. Muesmann, J. Borner, B. Misof, & T. Burmester (2014) Phylogenetic position of Myriapoda revealed by 454 transcriptome sequencing. *Molecular Phylogenetics and Evolution* 77: 25–33.

4. Rota-Stabelli, O., A. C. Daley, & Davide Pisani (2013) Molecular timetrees reveal a Cambrian colonization of land and a new scenario for Ecysozoan evolution. *Current Biology* 23: 392–398.

5. Boxshall, G. A. (1974) The validity of *Acanthochondrites inflatus* (Bainbridge, 1909) a parasitic copepod occurring on *Raja radiata* Donovan, in the North Sea. *Journal of Natural History* 8: 11–17.

6. Boxshall, G. (1977) The planktonic copepods of the northeastern Atlantic Ocean: Some taxonomic observations on the Oncaeidae (Cyclopoida). *Bulletin of the British Museum Natural History, (Zoology)* 31: 103–155.

7. Boxshall, G., & H. S. J. Roe (1980) The life history and ecology of the aberrant bathy-pelagic genus *Benthomisophria* Sars, 1909 (Copepoda: Misophrioida). *Bulletin of the British Museum Natural History, (Zoology)* 38: 9–41.

8. Boxshall, G. (1986) The comparative anatomy of the feeding apparatus of representatives of four orders of copepods. *Syllogeus* 58: 158–169.

9. Huys, Rony, & G. A. Boxshall (1991) *Copepod Evolution*. The Ray Society, London.

10. Rouch, R., & F. Lescher-Moutoué (1977) *Gellyella droguei* n. g., n. sp., Neuvelle curieux harpacticide des eaux souterranines continentals de la nouvelle famille des Gelyellidae. *Annals de Limnologie* 13(1): 1–14.

11. Huys, R. (1988) Gelyelloida, a new order of stygobiont copepods from European karstic systems. In *Biology of Copepods*, ed. G. A. Boxshall & H. K. Schminke, issued as *Hydrobiologia* 167/168: 485–495.

12. Boxshall, G. A., & S. H. Halsey (2004) *An Introduction to Copepod Diversity*. The Ray Society, London.

13. Ferrari, F. D., & H.-U. Dahms (2007) Post-embryonic development of the copepoda. *Crustaceana Monographs*. **8**: 1–226.

14. Kohdami, S., J. Vaun McArthur, L. Blanco-Bercial, & P. M. Martinez Arbizu (2017) Molecular phylogeny and revision of copepod orders (crustacea: copepoda). *Scientific Reports (Nature)*. [Withdrawn, October 14, 2020; available at *Scientific Reports* 7:9164 doi:10.1038/s41598-017-06656-4.]

15. Ho, J.-S. (1990) Phylogenetic analysis of copepod orders. *Journal of Crustacean Biology* **10**(3): 526–536.

16. Por, F. D. (1984) Canuellidae Lang (Harpacticoida, Polyarthra) and the ancestry of the copepoda. *Crustaceana*. *Supplement No. 7, Studies on Copepoda II* (Proceedings of the First International Conference on Copepoda, Amsterdam, the Netherlands): 1–24.

17. Dahms, H. U. (2004) Exclusion of the Polyarthra from Harpacticoids and its reallocation as an underived branch of the Copepoda (Arthropoda, Crustacea). *Invertebrate Zoology* 1(1): 29–51.

18. Huys, R., J. Llewellyn-Hughes, P. D. Olson, & K. Nagasawa (2006) Small subunit rDNA and Bayesian inference reveal *Pectenophilus ornatus* (Copepoda incertae sedis) as highly transformed Mytilicolidae, and support assignment of Chondracanthidae and Xarifiidae to Lichomolgoidea (Cyclopoida). *Biological Journal of the Linnean Society of London* **87**(3): 403–425.

19. Figueroa, D. F. (2011) Phylogenetic analysis of *Ridgewayia* (Copepoda: Calanoida) from the Galapagos and of a new species from the Florida Keys with a reevaluation of the phylogeny of Calanoida. *Journal of Crustacean Biology* 31(1): 153–165.

20. Park, T. S. (1986) Phylogeny of calanoid copepods. *Syllogeus* **58**: 191–196.

21. Andronov, V. N. (2007) New genus and species of copepods (Crustacea, Calanoida) from the central-eastern Atlantic and problems of classification of the Superfamilies Pseudocyclopoidea and Epacteriscoidea. *Zoologicheskii Zhurnal* **86**: 671–683. [In Russian.]

22. Nei, M., & S. Kumar (2000) *Molecular Evolution and Phylogenetics*. Oxford University Press, New York.

23. Felsenstein, J. (2004) *Inferring Phylogenies*. Sinauer Associates, Sunderland, MA.

24. Krell, F. T., & P. S. Cranston (2004) Editorial: Which side of the tree is more basal? *Systematic Entomology* **29**: 279–281.

25. Blanco-Becial, L., J. Bradford-Grieve, & A. Bucklin (2011) Molecular phylogeny of the Calanoida (Crustacea: Copepoda). *Molecular Phylogenetics and Evolution* **59**: 103–113.

26. Laakmann, S., E. L. Markhaseva, & J. Renz (2019) Do molecular phylogenies unravel the relationships among the evolutionary young "Brafordian" families (Copepoda; Calanoida)? *Molecular Phylogenetics and Evolution* **130**: 330–345.

27. Bradford-Grieve, J. M., G. A. Boxshall, S. T. Ahyong, and Susumu Ohtsuka (2010) Cladistic analysis of the calanoid Copepoda. *Invertebrate Systematics* **24**: 291–321.

28. Bradford-Grieve, J. M., G. A. Boxshall, & L. Blanco-Bercial (2014) Revision of basal calanoid copepod families with a description of a new species and genus of Pseudocyclopidae. *Journal of the Linnean Society* **171**: 507–533.

29. Figueroa, D. F. (2011) Phylogenetic analysis of *Ridgewayia* (Copepoda: Calanoida) from the Galapagos and of a new species from the Florida Keys with a reevaluation of the phylogeny of Calanoida. *Journal of Crustacean Biology* 31(1): 153–165.

30. Bradford-Grieve, J., & G. A. Boxshall (2019) Partial re-assessment of the family struc-ture of the Clausocalanoidea (Copepoda: Calanoida) using morphological data. *Zoological Journal of the Linnean Society* **185**: 958–983.
31. Bradford-Grieve, J. M. (2002) Colonization of the pelagic realm by calanoid copepods. *Hydrobiologia* **485**: 223–244.
32. Bradford-Grieve, J. M. (2004) Deep-sea benthopelgic Calanoid copepods and their colonization of the near-bottom environment. *Zoological Studies* **43**: 276–291.

21

Modern Taxonomy and Copepod Phylogeny

Learning Copepod Identities from Boxshall and Halsey

The importance of Geoff Boxshall and Sheila Halsey's *An Introduction to Copepod Diversity* was brought home to me by the dissertation project of Diego Figueroa, now a faculty member at University of Texas, Rio Grande Valley (UTRGV) in Brownsville.

After introducing Diego, I'll come back to that project and why it fits here. Diego (Figure 21.1) was born in Ecuador and initially raised in Quito. At an early age, twelve in Diego's case, students in Ecuador (at least fortunate ones), are taken to the Galapagos on a natural history field trip. That lived on in Diego's mind, and he has returned to that archipelago as an adult for several projects. In his early teens his family moved near to Washington, DC, settling for a long time in Virginia. Diego's father, Eduardo, an engineer, had become an officer of the Inter-American Development Bank (IDB) and had taken a position in the bank's Washington, DC office.

Diego attended junior and senior high schools in Virginia learning more English as he studied. He credits prize-winning success in science fair contests with his sustained interest in science. Graduating from high school, he started toward engineering training, a tradition from his grandfathers, father and at least some of his aunts and uncles. However, he found biology a better fit. He completed an undergraduate major in marine biology in 2000 at the University of Alaska, Juneau, which is pretty far from Ecuador on the equator or Virginia. Along the way he became a citizen of the United States, but he retained his Ecuadorian citizenship as well.

Diego's application to Oregon State's oceanography graduate program was not accepted, so he came anyway. He did very well in postbaccalaureate course work, and Dr. Hal Batchelder and I brought him on board, adding him to a project we had in hand. He started by writing a statistically sophisticated master's thesis on the reef-associated fish and invertebrate community survey he had done in 2002 as a visiting scientist at the Charles Darwin Research Station on Santa Cruz Island in the Galapagos.

Oar Feet and Opal Teeth. Charles B. Miller, Oxford University Press. © Oxford University Press 2023.
DOI: 10.1093/oso/9780197637326.003.0021

Figure 21.1 Diego Figueroa, marine science faculty member at University of Texas, Rio Grande Valley. Photo from his doctoral study of plankton in the Galapagos Archipelago. He is climbing the stairs out of Tunel del Estero carrying his belly-board, swimmer-pushed sampling system. Photo by K. Hoefel, permission D. Figueroa.

Diego's fascination with the archipelago grew along with that analysis, and he proposed a dissertation study to fit the zooplankton studies usual in my lab. His goals were season-specific faunal analyses of the plankton surrounding the equatorial Galapagos archipelago. The planktonic species present were known to vary through the year, because currents impinging on the islands flow from the northeast in the northern winter and from the southeast in northern summer. There are also contributions from the Pacific Equatorial Undercurrent as it rises on hitting the Galapagos Rise. Diego proposed sampling to document the species-level shifts that were not well documented. It was not particularly certain that he could get suitable samples, but Hal and I agreed to let him try. So off he went. In 2004–2005 he was working with his biologist friend, Kellie Hoefel, in a one-room laboratory building near the Galapagos National Park Service (GNPS) Headquarters in the town of Puerto Ayora on Santa Cruz Island. They lived nearby in an apartment, where I stayed during two seasonally separated visits. Diego's native-level Spanish helped him negotiate with the Park Service for access to that lab a short walk away. The park service even loaned him decent microscopes.

The lab was an almost mystical place: a roofed, stucco box with windows and some counters, built on stilts among mangroves, Sally Lightfoot crabs running among the roots, the equatorial Pacific splashing gently below the floor. Diego also negotiated to take plankton samples from a fishery patrol boat cruising all around and through the islands. Mostly he sampled by hand-hauled, oblique tows. With the keys and character drawings in Boxshall and Halsey, he could work his way to genus names for nearly every specimen, learning about copepods and improving his dissection skills as he went. Then, using their directions to species descriptions, I could email those descriptions to him from Oregon. Diego's dissertation would have been much more difficult without *An Introduction to Copepod Diversity* as a resource. Many recent studies of copepods, both ecological studies and others, similarly depend upon it. The results from his fishery patrol trips only lately emerged from Diego's 2009 dissertation.[1,2] After graduating, his attention was drawn away from copepods, first by postdoctoral work on the molecular genetics of deep-sea octocorals (formally Octocorallia, a cnidarian subclass with three orders) at Florida State University with Amy Baco, and later by establishing himself at UTRGV.

The Galapagos work had led Diego, via a chance rerouting, to expertise in DNA-sequence studies. While living in Puerto Ayora, he lost seagoing access to plankton offshore for several months, because of a strike staged by commercial fishermen. They were outraged by catch limits, closed areas and required reporting, as fishermen can be everywhere. Their picket lines blocked access to Diego's lab, to all the GNPS offices and labs, even the fishery patrol boat. There was no way to make progress of any kind, so Diego went swimming. That was at a site locally called Las Grietas. He had recalled swimming there during that early school trip. The name translates as "the cracks." In earlier times it was called La Grieta Delfin, after Hotel Delfin, then nearby but long gone by 2004. Las Grietas is a long, narrow, steep-walled fault in the volcanic rock, with two pools, each about 10 m long, 5 m wide, and filled with seawater to sea level from a depth of around 10 m. That water exchanges with the tides, the lower pool through a fully submerged tunnel of about 300 m at the shoreward end, the upper pool through some jumbled boulders. The pools are a sort of open-top, anchialine habitat (where seawater intrudes underground).

Having seen Las Grietas again, Diego went back with Kellie and a small plankton net and a coil of line about the length of the pools. She swam the net to the far end, then Diego pulled it through the water to a rock platform at the inland end. The plankton he caught looked nothing like that in the nearby ocean. He preserved them. Once he again had access to the lab, he ran the copepods through the key system in Boxshall and Halsey. There were species of both *Ridgewayia* and *Pseudocyclops*. Diego showed them to me when I was visiting after their initial discovery. We repeated the sampling, collecting live

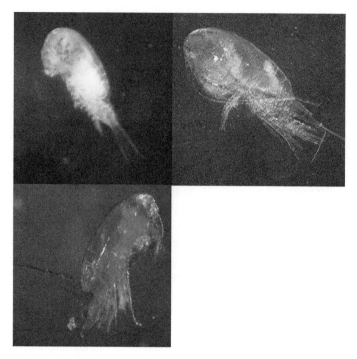

Figure 21.2 Upper left: *Pseudocyclops juanibali*, Lower left, *Ridgewayia delfine*. Right: *Ridgewayia tunela*. Note the intensely pigmented and frontally located eyes. Each about 0.7 mm, prosome length. Photos by D. F. Figueroa, with permission.

specimens. Though looking uninjured, neither species survived the walk back to the lab. It was likely we could not keep the water temperature low enough on the long walk, but these genera did not seem likely to be good subjects for culture studies. The *Pseudocyclops* were pale blue with orange eyes that glowed brightly when illuminated by a microscope lamp (Figure 21.2, upper left). The species of *Ridgewayia* from Las Grietas is solidly red (Figure 21.2, lower left). A species collected later from Tunel del Estero on Isla Isabela (Figure 21.2, right) is more transparent. Alive or recently dead, they too had orange eyes behind their clear, anterior "viewing domes." The strikingly distinctive thing about *Ridgewayia* species is the segmentation of the female fifth legs (Figure 21.3); the terminal segment of the exopod attaches in the middle of its second segment. Boxshall and Halsey used that as the key character to what for a time was a family (save for *Stargatia*, a one-species genus).

Thinking about the Galapagos finches studied by Darwin and his successors, we decided to devote some effort to sampling more cracks and caves, at first with me along, then later by just Diego and Kellie. They managed to sample

Figure 21.3 *Ridgewayia delfine* Figueroa & Hoebel, 2008,[3] female. Anterior aspect of left fifth leg. Unusual relation of Ex3 to Ex2 indicated by the arrow. Length ~ 0.25 mm. Permission, Inter-Research, *Marine Ecology Progress Series.*

20 anchialine cracks and nearly vertical crevices on three islands. Some of them, including Las Grietas, had been visited before by sea-cave diver and explorer Thomas Iliffe.[4] He had sampled plankton and epibenthic fauna in many sea caves around the tropics, and he had formed a partnership with Norwegian copepodologist Audun Fosshagen to describe his copepod finds. In Iliffe's collections, Fosshagen found many species that he placed in the family Ridgewayiidae, which Mildred Stratton Wilson had defined in 1958.[5] Fosshagen frequently assigned his new species generic rank. By 2009 the family included ten genera worldwide (currently there are 16 or more species of just *Ridgewayia*).

Diego captured hundreds of specimens of his first species in La Grieta Delfin. From the site on Isabela Island, he found an apparently different and also abundant *Ridgewayia* species. That one lives in the pool at the dark, seaward end of a "lava tunnel" (Tunel del Estero). It is a tube of basalt with a three-meter lumen extending down into the sea. Its opening (Figure 21.1) is approached by a stone stairway provided by the GNPS for visiting tourists. Diego's eventual and thorough investigation of the literature produced no species fitting the *Ridgewayia* specimens from either site, so he described and named them: *R. delfine* and *R. tunela*.[3]

To supply a comparative species from much farther away, Diego later visited the Dry Tortugas in the Florida Keys, the type locality for *Ridgewayia gracilis* and *R. schoemakeri* described by Wilson in 1958.[5] Wilson had found those copepods in samples from 1933 collected by Clarence Shoemaker, who was after amphipods. Shoemaker had rinsed rocks and coral rubble atop a fine mesh screen. That was how Scottish oceanographer William A. Herdman had collected the first *Ridgewayia* described, *R. typica*, by rinsing oyster cultch from shallows at Sri Lanka (then Ceylon). The species was described and named by Thompson Scott and Andrew Scott from those samples.[6] They could not know it was typical when it was the only one they had seen; the species name was meant to designate it the "type species" for their new genus. Diego did not find Wilson's species, but he found a third new one by using Herdman's sampling technique. Life has both ups and downs. Diego and Kellie parted ways on return to the United States, but Diego and I gratefully recall how helpful she was in the Galapagos.

Diego wanted to try for gene sequence data to strengthen the case that his three new *Ridgewayia* species and two new *Pseudocyclops* species, *P. juanibali* and *P. saenzi* from Las Grietas, are indeed distinctive. So he took over my PCR equipment and initially obtained mtCOI barcodes for *R. delfine* and *R. tunela* using the Folmer primers (Chapter 18). Despite their distinct, if refined, differences in morphology, their barcodes were exactly the same. A result like that screams "contamination!" So he cleaned the entire lab and its machines, used new reagents, and repeated the sequencing—with the same result. We still do not understand that, except that (as again noted below) mitochondrial genes of *Pseudocyclops* just do not amplify readily. So probably the identity problem was contamination from *R. delfine*. Instead of repeating the repetition, Diego figured out primers for the other end of the mtCOI gene and sequenced 656 bases there. He also described the new Dry Tortuga species as *R. tortuga* and eventually published that with some genetic data for all three of the new *Ridewayia* species.[7] The Las Grietas and tunnel species differed by 1–2% among individuals within the species and 2–5% between the species from the two sites.

In the midst of this maelstrom of study, drawing of copepod limbs, molecular genetics and writing, Diego met Nicole Scott, at the time an undergraduate biology student at Oregon State University (OSU), whom he eventually married. They lived in a house in the close-by town of Philomath that had been financed by Diego's parents, Eduardo and Manuelita Figueroa. It was very large for a young graduate student and his wife, but it had to be large partly because his parents wanted their own room for visiting. His parents also convinced their naturalized

American daughter, Viviana, to move to Oregon to finish a degree for which she was studying. After Diego and Nicole left Oregon, his brother David moved in while he studied at the newly established medical school in Lebanon across the Willamette Valley.

Diego, meanwhile, set out to get molecular data for the new *Pseudocyclops* species. However, at the time he could not get their mitochondrial genes to amplify, and later Leo Blanco-Bercial[8] had the same problem using specimens given him by Diego. Diego tells me in 2021 that he still cannot get mitochondrial sequences, especially the Folmer-primer bar codes, from any *Pseudocyclops* species. In 2008 he simply switched molecular gears. From both his *Ridgewayia* and his *Pseudocyclops* species, he amplified and sequenced an "internally transcribed spacer" (nITS1), a nuclear DNA sequence, that was coming into prominence for phylogenetic analysis. It sits between 18S rDNA and 28S rDNA in the nuclear code (rDNA) for ribosomal RNA (rRNA). nITS1 is transcribed with the functional rRNA genes, but it is enzymatically edited out before ribosome assembly. Because it does not function in the ribosomes, its mutations more readily survive than those in the functional segments of rDNA. However, because it is nuclear, much of its DNA damage is repaired during meiosis (Chapter 13), so it is not so labile over generations as mitochondrial DNA. Its variability increases at some intermediate rate. Diego also obtained, for an outgroup, nITS1 sequence from a quite distantly related copepod that he had also collected from Las Grietas, *Enantiosis galapagensis* Fosshagen, Boxshall and Iliffe, 2001. It is classified in the family Epacteriscidae.

With the *nITS1* sequence data for the *Ridgewayia*, 569 aligned bases from 10 specimens, Diego generated a mini-cladogram (Figure 21.4, left).[7] Using the same 569-base sequences, he generated a similar cladogram for two species of *Pseudocyclops*, which he described and named *P. junibali* and *P. saenzi*, after two Ecuadorian engineers and university professors, his grandfathers.[9] In two separate anchialine pools on Isla Santa Cruz, Cactus Forest and Deep Grieta (named by Illife in 1990), he had found populations of *Pseudocyclops* morphologically similar to *P. junibali*. Sets of specimens from all four stocks were included in the ITS1 cladogram (Figure 21.4, right).

The ITS1 cladograms agree with the apparent morphological species distinctions found by Diego; they similarly represent the levels of distinction among sets of individuals. The anatomical differences are spelled out in the two papers, the body shapes and limb details represented with classical, flattened-fauna drawings. At the level of alpha taxonomy, the value from proving such genetic distinctions is simply to add confirming characters to those from morphology. As seen from the work of Erica Goetze and others (Chapter 19), it can also suggest population-genetic divergence *not* also expressed as morphologic differences. That appears to be the case for *P. juanibali* from different pools; each

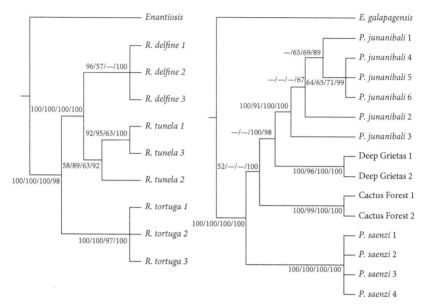

Figure 21.4 Species level cladograms for collections of two pseudocyclopid genera from anchialine habitats on the Galapagos and *R. tortuga* from coral rubble at Dry Tortugas Key, Florida.[7] The data were 569 base-pair *ITS1* sequences from individuals: 1, 2, 3, etc. Numbers on the vicariance events are confidence levels from four cladistic algorithms. Permission from OUP.

stock is genetically distinct, though ITS1 seems not likely to affect anything morphological. Adding such genetic characters to the descriptions of species has yet to become required. However, the impact of genetic characters on copepod taxonomy (and all taxonomy) is obvious from the previous two chapters as well.

Diego might have stopped there, but an issue about the family level distinction between the Ridgewayiidae and Pseudocyclopidae had remained unsettled for several decades, at least since remarks by Taisoo Park in 1986.[10] So Diego did some data mining in GenBank, and generated the cladogram (Figure 20.9)[7] for some of the families of Calanoida as introduced in Chapter 20.

A Transcontinental Move: Oregon to Florida

While still a graduate student, Diego began to make connections with other copepodologists, and he got particular help from Frank Ferrari at the Smithsonian Institution and Ryuji Machida of the University of Tokyo. Ferrari had done much previous work on *Ridgewayia* species,[11,12] and helped Diego with formatting species descriptions. Machida had developed the PCR primers for nuclear ITS1[13] and had given the sequences to Diego. None of those connections developed into a job or a postdoctoral position. What did come up was a chance to help with molecular genetic studies of deep-sea corals conducted by Amy Baco (later she added to her name: Amy Baco-Taylor) at Florida State University (FSU) in Tallahassee, Florida. That had many advantages. Professor Baco was funded to apply advanced sequencing techniques to her study animals, so Diego could learn about those and advance in bioinformatics.

Dr. Baco's project included sampling of deep-sea corals from research submarines in the Pacific to the north and west of Hawaii. Diego could participate in that, and he could learn from an expert how to identify species in a whole new phylum of animals. Once the collected specimens had yielded extensive sets of partial gene DNA sequences, Diego could master the bioinformatics for extracting meaningful genomics from those. And not so trivial, he and Nikki got to move to a new place far from Oregon and the storm and stress of graduate school. That led them to expand their family, adding a daughter and then a son. Those kids (Figure 21.5) were grade-school age as of 2020.

Figure 21.5 Nikki and Diego Figueroa with their offspring. The small ones are not my academic grandchildren; they are actual descendants of Eduardo and Manuelita Figueroa of Quito, Ecuador. Photo by D. F. Figueroa, with permission.

Figure 21.6 Left, Diego peering from a port on PICES-IV ready to launch. Right, Diego with some of the Baco and Figueroa hauls of Octocorallia (and likely other attached epibenthic fauna). Photos from D. F. Figueroa, with permission.

Diego plunged into the work at the FSU Baco lab. When he arrived, nothing was working by way of PCR amplifications; that happens readily in molecular genetics. Diego applied his OSU experience with odd or failed PCR results and cleaned everything from floor to ceiling. He eliminated suspect reagents and pipette systems, got some new enzymes, and started over. Pretty soon amplifications were rolling out as planned. Then he got to go on several sea voyages, including some dives in a submersible named PICES-IV (Figure 21.6, left) and participated in collections by remotely operated vehicles. Amy and Diego collected a sizeable suite of new octocoral specimens from ridges and seamounts deep in the North Pacific (Figure 21.6, right). They subjected those to total sequencing of their mitochondrial genomes with the goal of contributing to octocoral phylogenetics.

Total Mitochondrial Sequences: Another Key to Phylogeny?

Mitochondria are the powerhouse organelles of eukaryotic cells. They generate and export adenosine triphosphate (ATP), which (by dephosphorylation to ADP) provides the energy for almost all metabolism and muscle contraction. The endosymbiotic theory of the origin of mitochondria, as published by Lynn Margulis in 1967, has a lot of circumstantial evidence supporting it. Mitochondria are the result of a long-back capture and functional incorporation of a bacterial cell (one using oxidative phosphorylation) by a still primitive but more complex cell (possibly an Archean derivative) using phagocytosis. That is, a fold of cell membrane wraps around a particle or droplet, then pulls it inside the cell.

Part of the proof of this origin mechanism is that mitochondria retain a chromosome with partial gene sequences for 13 or 14 protein coding genes: oxidative cytochromes, ATP synthases, nicotinamide dehydrogenases (NADH), and in octocorals MutS (a DNA-damage repair enzyme). There are also two genes coding parts of the structural RNA for the distinctive ribosomes in the mitochondrion itself, and (finally) a complement of 22 transfer-RNA (tRNA) genes. In mtDNA, those genes have some characteristics of bacterial genes. Save for the tRNA genes, all of them are fragments, insufficient alone to construct a functional mitochondrial protein. Parts of each mitochondrial gene have moved to the nucleus over long evolutionary time. In fact, a functioning mitochondrion has perhaps 1,500 proteins, most of them translated from DNA in the nucleus; the others with some of their amino acid complements encoded in its mtGenome. Several theories explain why that might be advantageous. Another feature harking back to the captive ancestral prokaryote is that mtDNA gene sequences are bonded end-to-end in circles with around 15,000 to over 19,000 base pairs in different species. How all that became known involved interaction of electron microscopy, discoveries of "non-Mendelian evolution" of characters involving mitochondria (such as diseases). There are books, even more technical that this one, dealing with that.

By the early 2000s, complete mtDNA sequences were becoming available for species in many groups, including Cnidaria, the phylum to which the Octocorallia belong. There are more than 3,000 described species of that subgroup, the polyps of which have eight-fold symmetry (eight tentacles around the edge). There are still more to be found on deep, rocky seafloors around the world. Amy Baco was working to contribute more distinctive DNA sequences in that group. Some next-generation sequencing of whole genomes produced more code than their software could handle, but in some cases the entire mitochondrial sequence (the *mtGenome*) could be extracted. That produced likely primers to get those from more species. So she and Diego sequenced the entire mtDNA of seven species from their new collections. They did that by basic PCR sequencing, working step by step around their whole mtGenomes using primers for overlapping sections of them, starting with mtCOI. The entire "trip around" required 42 pairs of primers to cover the full ranges from 18,715 to 19,124 base pairs in their 14 specimens. The literature (with sequences available in GenBank) provided 54 more examples of complete mtGenomes from both Hexacorallia and Octocorallia to help with primer design. Only 5 primers in the 84-primer set had to be newly selected.

After the 42 PCR runs for each of their specimens (mostly not new species, though two *Anthomastus* specimens could be), they sent the whole array of DNA amplicons to a sequencing center at the University of Washington and waited for the data files to come back. When those arrived, they identified

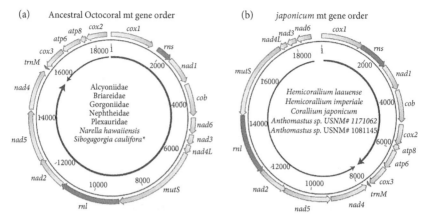

(a)   Ancestral Octocoral mt gene order

(b)   *japonicum* mt gene order

Figure 21.7 The two most distinctly different gene orders in the mtDNA of deep-sea Octocorallia. The third general version (in *Corallium konojoi* and others) is a less dramatic change from that considered (and labeled here) to be more ancestral. All the species listed in italics were sequenced by Figueroa and Baco.[14] Transcriptions of the genes move in the directions indicated by the arrowheads, from the 5' ends of the molecules to the 3' ends. The red arrows represent the small (*rns*) and large (*rnl*) ribosomal sequences. Enzyme genes are parts of ATP synthetase (*atp*), of cytochrome oxidases (*cox* and *cob*), of nicotinamide dehydrogenase (*nad*) and of a DNA repair enzyme (*MutS*). All of the very small transfer-RNA genes are a cluster at *trnM*. Note that most genes, except the *cox1* to *cob* genes, have had their 5' to 3' coding direction reversed in the *japonicum* order. The genes beside the thick, black, curved arrow have more GC code, so are literally heavier. The genes beside the thin black arrow have less GC and more AT, so are lighter. From Figueroa and Baco.[14] Permission, Elsevier.

the overlaps among all the genes and matched those around the loops. They found three basic patterns of gene arrangement,[14, 15] two of which are shown in Figure 22.7. They are decidedly not the same. To students of mtDNA in vertebrates, that would be a surprise, because all of phylum Chordata, "from amphioxus to the meanest human cuss" (from a song once popular with marine biology students), have very similar mtDNA gene orders. That is not true of the invertebrate phyla.

In a final step, Figueroa and Baco ran a phylogenetic analysis. That was not done based on the gene order. Rather they aligned the sequences for each of the genes in their 14 specimens and also aligned sequences of 69 other species available from the literature, including many Hexacorallia, some other Cnidaria (*Hydra*, moon jellies, etc.) and five sponge species. Each of those aligned sequences was then concatenated into one long, aligned chain (around 18,000 base pairs), providing a huge number of character states. Then, using a maximum-likelihood algorithm, they derived a phylogenetic cladogram. The consensus cladogram is large, but its main sister groups for Octocorallia were the three mtDNA gene-sequence groups: the Ancestral, the *japonicum* and the *konojoi*. The conclusion, oversimplified for our purposes, is that not only gene sequences but the orders of genes in eukaryote mitochondrial chromosomes carry phylogenetic information. However, Figueroa and Baco[15] had used the even more powerful possession of *all* the mtDNA sequences, showing that gene orders changed early in the history of a very old lineage, with vicariance events affecting both the gene nucleotide sequences and the mitochondrial gene ordering.

Dr. Figueroa Moves to the University of Texas, Rio Grande Valley

After about three years at FSU, Diego was offered a faculty position in marine science and biology at the University of Texas branch in Brownsville (UTRGV), a town at the very southern tip of Texas looking across the Rio Grande (the international border) to Matamoros in the Mexican state of Tamaulipas. The river drains into the Gulf of Mexico about 15 miles east of the city limits. The Gulf is the site for most research activity of the UTRGV Marine Science group. Diego worked hard to establish a teaching and research program there. Many of the university's student body have Latin American connections and hail from Spanish-speaking households. Diego has told me that right away his bilingual talents were in demand, clarifying ideas expressed in one language by knowing the other. He currently teaches courses in oceanography, pelagic ecosystem dynamics, marine biology field methods, and evolution. Working with students, he established some stations for recurring observations on offshore reefs, providing them with research projects.

Diego's own initial UTRGV research projects were more work on Hexocorrallia and Octocorrallia collected from banks and oil-drilling platforms on the south Texas continental shelf. He and fellow faculty member David Hicks, working with Nikki Figueroa and graduate student Amelia McClure, identified a soft coral living at a number of Gulf sites that had normally been found only in the Pacific, probably traveling as larvae with shipping.[16] Diego, Dr. Hicks, and Nikki collected a black coral (Hexacorallia) from Aransas Bank off Corpus Christi and ran out its *mtGenome* using Illumina (Box 21.1) next-generation (NGS) sequencing.[17]

That got Diego going at UTRGV with NGS as a basis for extended phylogenetic work with copepods based on *mtGenomes*. A major part of the work in such research is raising the funds to pay for both the lab work and the commercial (or commercialized academic lab) sequencing expenses. Texas Sea Grant, local

Box 21.1 Event Series in Next-Generation DNA Sequencing (NGS)

The first next-generation genome sequencing (NGS) techniques became available about 2005. Geneticists seeking long and nearly complete DNA sequences from a cactus or copepod prepare the DNA for a machine that runs DNA sequencing reactions on very large numbers of short lengths of single strands. The new process has many steps:

(1) All DNA is extracted from an organism or tissue sample.
(2) That is fragmented into relatively short lengths, around 200–500 base pairs, chemically or with ultrasound.
(3) Short molecular adapters are added to the strands ("tags") such that each length can stick to a suitably coated, flat surface. Steps 2 and 3 can be combined as "tagmentation." The strands are then rendered single stranded with hydroxide.
(4) A solution of tags is placed on a flow cell surface; they attach, extending from it.
(5) The surface is flooded with DNA synthase and a mixture of "terminator" deoxynucleoside triphosphates (dNPTs of the two purines and two pyrimidine bases) with attached fluorophores (molecules that will fluoresce in different colors for each base) in warm buffer. Terminator means the bases are also modified to block additional nucleotides from being added until the fluorophores are read (a version of a camera). Then the terminator moieties are washed off.
(6) The cycle repeats, the dNTPs and DNA synthase are again added, new pictures are taken . . . Much variation and detail are involved in the back and forth among adding dNTPs, getting the fluorescent signals, then repeating.
(7) The recorded sequences of color events at all spots on the surface are then decoded as ATCCAGGTTGA, and so on then saved for the many strands (to many millions) as digital records or files. Those are returned to the geneticist.
(8) Using computer algorithms, identical code stretches of usable length can be identified, allowing assembly by overlapping of long stretches of the original DNA before step (1).
(9) Those can be compared to the now vast array of known sequences to identify likely functional identities of the genes, spacers, control regions, even junk DNA.

That is a semblance of the Illumina NGS version. Different NGS versions use different modes of signaling which nucleotide was added at each step of laying down the DNA complement of the attached single strands. In some versions the nucleotides are identified as they are *removed* one by one from the short strands. There are also different modes of holding and complementing bases on the tagmented short single strands.

UTRGV funds, and the National Academy of Sciences Gulf Research Program were important initial sources for him. Diego has now directed several masters degree projects using Illumina NGS, including one on Ridley's sea turtle[18] and another on octocorals from the Gulf.[19]

Published recently, but reported in 2017 at the 13th International Conference on Copepoda, was a masters thesis project directed by David Hicks and conducted by Diego's wife Nikki (Nicole).[20] She used 689 base pairs of mtCOI to examine the geographic variability of *Acartia tonsa* in the Gulf off southwest Texas, but also took account of virtually all reported bar code data for that species and other *Acartia* species around the world. It will not surprise anyone who has worked with these small copepods, but the gene sequences suggest that quite a few misidentified *Acartia* have been crushed in PCR tubes, leading to odd discrepancies in their mtCOI sequences. Where that is not the case, mtCOI from *A. tonsa* indicates four clades with quite distinctive haplotype groups, each of which is widespread along North American coasts. Within those clades there was mostly modest along-coast haplotype variation, and the shifting was smooth from Texas around the Gulf to the north and east, then around Florida and on north to New England. Several cycles of differentiation are suggested, with vicariance events in the past to separate the major clades geographically, and then slow, continuous population mixing or gene flow within those clades.

Complete mtGenomes from Copepods: Diego's Ambition for More

Complete mtDNA data are published for only a few copepods. In 2002, Ryuji Machida and coworkers[21] published the complete mtGenome for *Tigriopus japonicus,* one of the small harpacticoids, mentioned frequently in these chapters. It lives in splash-zone pools high in the intertidal (an image of a close relative is shown in Figure 13.4). The technique was similar to that used by Figueroa and Baco for octocorals: PCR with "primer walking" along the chromosome. It seems more efficient, because they managed to amplify very long sequences, covering

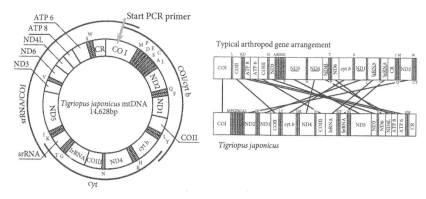

Figure 21.8 Mitochondrial gene sequence of *Tigriopus japonicus* in two formats. Left: genes in the order coded on the circular chromosome with arc-length proportional to base-pair numbers. Codes as in Figure 21.6; CR is the noncoding control region, location of the starting points for replication and transcription. Dot-filled genes code for t-RNAs identified by 1-letter amino-acid codes. Stripes around the circumference represent the three long-range PCR amplifications producing the whole genome, progressing clockwise and from 5'- toward 3'-DNA. Right, the same codes compared between *T. japonicus* and the very different sequence typical of most arthropods, here from a fruit fly (*Drosophila*) and a cladoceran (*Daphnia*). The tangle of connecting lines indicates the many relative position shifts. From Machida et al.,[21] figures 1 and 3. Permission, Springer Nature.

the whole chromosome of 14,628 base-pairs in just three sections (three primer pairs). Sequencing those required primer-walking ("step-out" sequencing), developing the necessary primers progressively. They graphed their results in two ways (Figure 21.8). The map of the chromosome at the left, is a less colorful version of the maps for octocorals shown above. I added an arrow for the starting point of the initial PCR in the mtCOI gene. The positions of all 22 transfer-RNA genes are shown explicitly. In *Tigriopus*, all of the genes are transcribed in the same direction (clockwise in Figure 21.7, left) from just one strand of the chromosome. As the figure shows, the gene order differed greatly from that more typical in arthropods.

In 2007 Ronald Burton and two colleagues[22] obtained a complete mtGenome for *Tigriopus californicus* (yes, from California), finding its gene order closely comparable to that of *T. japonicus*, with only a few shifts of the short sequences of tRNA. In a 2004 study, Dr. Machida and his coauthors[23] showed that species from two families of Calanoida, *Eucalanus bungii* and *Neocalanus cristatus*, also had mtDNA sequences substantially different from the arthropod "norm" and, in fact, quite different from each other. A 2005 study of *Lepeophtheirus salmonis*, a copepod parasite of fish[24] showed it to have yet another gene order. In 2009 Jang-Seu

Ki and colleagues[25] from Korea published an mtGenome for the high-salinity tolerant cyclopoid copepod *Paracyclopina nana*. Its gene order was sharply distinct from that of *Tigriopus* and, again, from the canonical arthopod mtGenome.

Evidently, unlike mitochondria in some classes of arthropods, including other crustacea, copepod orders and families show strong rearrangements of their mitochondrial gene orderings. Diego thinks that enough complete mitochondrial genomes could provide distinctive insights about copepod phylogeny. He is now applying NGS toward getting them.

As stated several times already, many copepods have huge nuclear genomes. For example, about 12.7 billion base pairs in *Calanus finmarchicus*.[26] Thus, there is much less mtDNA than nuclear DNA, and the mitochondrial fraction can be overwhelmed in next-generation sequencing. So it is proving difficult to get complete *mtGenomes*. However, Diego is making progress, focusing on families in the Calanoida, but adding others as they become available. He told me in 2021 that he had them for *Undinula, Centropages, Mecynocera, Acartia, Pseudodiaptomus, Lucicutia, Subeucalanus, Ridgewayia* (3 species), and *Enantiosis* among calanoids, and two from *Corycaeus* and *Sapphirina*. He added, "I have not been able to do *Pseudocyclops*, at least not for several species. There is something about that genus making it extremely difficult to get mitochondrial sequences." Of course, as discussed above, Diego has special interest in the species of *Pseudocyclops*. He also said the three *Ridgewayia* species, have extensive rearrangements, but he isn't ready to publish them.

Diego, with help from others banking mtGenomes from copepods, hopes to sequence and acquire enough of them for a new look at copepod phylogeny, particularly that of calanoids. If Illumina doesn't work, primer walking certainly will for most genera, though requiring much trial and error in primer searches. The mitochondrial gene sequences themselves will have phylogenetic implications, but perhaps more important will be the potential of those 15,000 to 19,000 base-pair character states for quantifying relationships within genera and families. At that level, base-by-base differences may not be saturated by changing and changing back over really long-term evolution. That could lead to better understanding of how copepod evolution progressed in terms of vicariance events far from the recent tips of the copepod tree branches.

Other workers are also digging through next-generation genome data for complete mtGenomes. A nice example involving Atlantic *Calanus* species comes from three teams scattered across the globe. It was summarized by a Polish team led by Agata Weydmann and Arthur Burzyński.[27] They obtained mtGenomes for *C. finmarchicus* and *C. glacialis*. Earlier work by a group in China headed by planktologist Sun Song[28] produced one for *C. sinicus*, and a Korean group headed by Han-Gu Choi [29] one for *C. hyperboreus*. The methods differed, and readers with deep interest can compare those in their papers.

Dr. Burzyński is the Department Head of Genetics at the of Institute of Oceanology of the Polish Academy of Sciences, located at Sopot. Dr. Weydmann was formerly a postdoctoral scholar with Ann Bucklin, and in 2021 is associate professor of plankton research at Gdansk University. She is also associated with the Institute of Oceanology. Their paper reports several failed starts toward a step-out sequence. Those failures point out the special difficulties species with very large mtGenomes present. Those are especially large, 27,342 and 29,462 base pairs in *C. gla* and *C. fin*, respectively, a third to almost half larger than in most animals. That is largely because, as the eventual results showed, they have multiple strings of NCR ("not protein coding") DNA, and unusually long ones. Some NCRs are "control regions" (the matching initials are unfortunate) and some may direct transcription or duplication.

Weydmann et al. tried a method used earlier by Sun Song's group to get around the quantitative dominance of nuclear DNA and get mtGenome NGS data for *C. fin*. from the Norwegian Sea and *C. gla* from Hornsund fjord at Svalbard Island. They macerated groups of specimens and centrifuged the tissue slurries to isolate mitochondria. But in these two species that did not separate mtDNA from nuclear DNA in separate layers. So they collected and purified all the DNA from the mixed band and processed it for NGS. Sending the products off to San Diego, California, for Illumina HiSeq-Ten sequencing, they waited in Sopot (or Gdansk) for the results. Time out for a moment to grasp the scale of just this one company busy with DNA sequencing and other biotechnology: Illumina's 2020 revenue was $3,236,000,000 US dollars.

When Weydmann's data arrived, it included around 3×10^8, 150 base pair reads, which occasioned a festival of bioinformatics. If you know the code names for the procedures, all that they used are listed in the paper.

Weydmann et al. examined the results in all the commonly deployed ways, including gene-order comparisons with *C. hyperboreus* and *C. sinicus*. To get an idea of how much of the orderings are in common, how much unique to each species, they used a parsimony analysis (the smallest number of changes to find identity) termed CREx from Matthias Bernt et al.[30] According to Weydmann et al., CREx "confirmed greater rearrangement distance between *C. hyperboreus* and the pair of *C. sinicus* and *C. glacialis*, the latter two consistently presenting a closer relationship." Suspending any cynicism temporarily, let's assume that the resulting common ancestor sequences are reasonable. Figure 21.9 shows the different mtGenome sequences of the four species and common ancestor.

To get a realistic sense of the phylogenetic implications of the gene reorderings, or for that matter the detailed sequences of mtDNA, it will be necessary to obtain mtGenomes for all the known *Calanus* species. None of the papers to 2021 citing Weydmann et al. included another. There has been time, but doing this work requires abundant resources, and possibly it seems that the cream has been skimmed. It remains unclear why the ordering of invertebrate mitochondrial

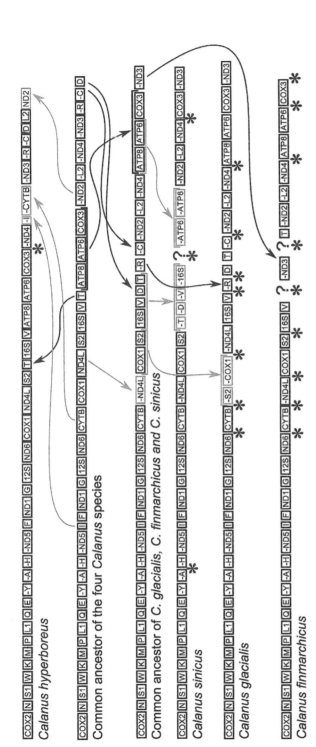

Figure 21.9 Mitochondial sequences among *Calanus* species, from Figure 3 in Weydmann et al.[27] Each little box has the genetic code name for the gene in its position. For examples, COX2 (cytochrome oxidase II) has been placed at the left of all the sequences, and they are all the same from COX2 to ND5 (nicotinamide adenine dinucleotide dehydrogenase, component 5). Presumably that stretch has greater order stability that the rest. The black box genes are transcribed from left to right on the so-called H-loop of the DNA; the gray boxed genes are transcribed from right to left on the L-loop. The arrows show the transfers implied by the data from the CREx model of the common ancestral sequence to the actual ones. Compared to vertebrate *mtGenome* variations within genera and more inclusive groups, these are very large rearrangements. The asterisks are locations of NCR sequences of various lengths. Many more added NCR blocks were found for *C. fin* and *C. gla*, mostly accounting for the greater overall size of their *mtGenomes*. The data had several gaps, also likely to be NCRs, indicated by ? marks. Permission, Springer Nature.

gene sequences is so labile. Copepod mitochondria, showing strong variations within genera, seem particularly susceptible to shuffling of their genes. Perhaps there is some selective advantage involved. Unsolved puzzles are, of course, the fuel we burn to make new discoveries. Diego Figueroa, Agata Weydmann, Sun Song, Han-Gu Choi, and many others are surely thinking about it and testing their hypotheses.

What Does Copepod Phylogeny Add to the Human Outlook?

How did humankind come to be as we find ourselves? And how did the world come to be of interest and seriously discussed once self-consciousness and language developed in *Homo*? Those questions, metaphysical ones, have been at the foundation of our religions, of our existential understanding of who we are. They are also a starting point for answering "Why?" When the answers are sketchy, myths develop and are shared. Creation myths have always come to hand as needed. Earliest written accounts of the creation of the world came from the Sumerians, who scratched their story of being summoned into existence by their gods onto clay tablets, then baked them. Many among us continue to believe the creation story in *Genesis*.

Lately, we have the Big Bang as a sort of beginning, which, myth or fact, drips mythological character, including its name. There is metaphysical satisfaction from understanding almost any detail of the story of the universe, our galaxy, star, solar system, the Earth, the origin of life, the patterns of evolution and diversification along the way to now. Just as important as knowing human life in suitable detail is understanding the widest possible range of living things. Since copepods are key creatures in the ecological dynamics of all wet habitats, since they are beautiful in form and life ways, it is wonderful and gratifying that evolutionary understanding of them is expanding. Of course, it is expanding also for cherry trees and potatoes, kangaroos and people, even for the bacteria and viruses that have plagued us since before memory. I think such metaphysical satisfaction is the most important value in these understandings, which are improving for virtually all groups of organisms.

References

1. Figueroa, D. (2009) Zooplankton of the Galapagos Islands. PhD dissertation, Oregon State University.
2. Figueroa, D. F. (2021) Environmental forcing on zooplankton distribution in the coastal waters of the Galapagos Islands: spatial and seasonal patterns in the copepod community structure. *Marine Ecology Progress Series* **661**: 49–69.

3. Figueroa, D. F., & K. L. Hoefel (2008) Description of two new species of *Ridgewayia* (Copepoda: Calanoida) from anchialine caves in the Galapagos Archipelago. *Journal of Crustacean Biology* 28: 137–147.

4. Iliffe, T. M. (1991). Anchialine cave fauna of the Galapagos Islands. In *Galapagos Marine Invertebrates*, ed. M. J. James, 209–231. New York, Plenum Press.

5. Wilson, M. S. (1958) A review of the copepod genus *Ridgewayia* (Calanoida) with descriptions of new species from the Dry Tortugas, Florida. *Proceedings of the United States National Museum* 108: 137–179.

6. Scott, T. I. C., & A. Scott (1903) Report of Copepoda collected by Professor Herdman at Ceylon in 1902. *Ceylon Pearl Oyster Fisheries, Suppl. Rep.*, No. 7. Report to Colonial Government, part 1: 227–307.

7. Figueroa, D. (2011) Phylogenetic analysis of *Ridgewayia* (Copepoda: Calanoida) from the Galapagos and a new species from the Florida Keys with a re-evaluation of the phylogeny of the Calanoida. *Journal of Crustacean Biology* 31(1): 153–165.

8. Bradford-Grieve, J., G. Boxshall, & L. Blanco-Bercial (2014) Revision of basal calanoid copepod families, with a description of a new species and genus of *Pseudocyclopidae*. *Zoological Journal of the Linnean Society of London* 171: 507–533.

9. Figueroa, D. F. (2011) Two new calanoid copepods from the Galapagos Islands: *Pseudocyclops juanibali* n. sp. and *Pseudocyclops saenzi* n. sp. *Journal of Crustacean Biology* 31(4): 725–741.

10. Park, T. S. (1986) Phylogeny of calanoid copepods. *Syllogeus* 58: 191–196.

11. Ferrari, F. D. (1995) Six copepodid stages of *Ridgewayia klausruetzleri*, a new species of calanoid copepod (Ridgewayiidae) from the barrier reef in Belize, with comments on appendage development. *Proceedings of the Biological Society of Washington* 108: 180–200.

12. Ferrari, F. D., & A. Benforado (1998) Setation and setal groups on antenna 1 of *Ridgewayia klausruetzleri, Pleuromamma xiphias,* and *Pseudocalanus elongatus* (Crustacea: Copepoda: Calanoida) during the copepodid phase of their development. *Proceedings of the Biological Society of Washington* 111(1): 209–221.

13. Miyamoto, H., R. J. Machida, & S. Nishida (2010) Genetic diversity and cryptic speciation of the deep sea chaetognath *Caecosagitta macrocephala* (Fowler, 1904). *Deep-Sea Research II* 57: 2211–2219.

14. Figueroa, D. F., & A. R. Baco (2014) Complete mitochondrial genomes elucidate phylogenetic relationships of the deep-sea octocoral families Coralliidae and Paragorgiidae. *Deep-Sea Research Part II* 99: 83–91.

15. Figueroa, D. F., & A. R. Baco (2014) Octocoral mitochondrial genomes provide insights into the phylogenetic history of gene order rearrangements, order reversals, and cnidarian phylogenetics. *Genome Biology and Evolution* 7(1): 391–409.

16. Figueroa, D. F., A. McClure, N. J. Figueroa, & D. W. Hicks (2019) Hiding in plain sight: invasive coral *Tubastraea tagusensis* (Scleractinia: Hexacorallia) in the Gulf of Mexico. *Coral Reefs* 98: 395–403.

17. Figueroa, D. F., D. Hicks, & N. J. Figueroa (2019) The complete mitochondrial genome of *Tanacetipathes thamnea* Warner, 1981 (Antipatharia: Myriopathidae). *Mitochondrial DNA Part B* 4(2): 4109–4110.

18. Fransden, H. R., D. F. Figueroa, & J. A. George (2019) Mitochondrial genomes and genetic structure of the Kemp's Ridley sea turtle (Lepidochelys kempii). *Ecology and Evolution* 10: 1–14.

19. Silvestri, S., D. F. Figueroa, D. Hicks, & N. J. Figueroa (2019) Mitogenomic phylogenetic analyses of *Leptogorgia vigulata* and *Leptogorgia hebes* (Anthozoa: Octocorallia)

from the Gulf of Mexico provides insight on Gorgoniidae divergence between Pacific and Alantic lineages. *Ecology and Evolution* 9(24): doi: 10.1002/ece3.5848.

20. Figueroa, N. J., D. F. Figueroa, & D. Hicks (2020) Phylogeography of *Acartia tonsa* Dana, 1849 (Calanoida: Copepoda) and phylogenetic reconstruction of the genus *Acartia* Dana, 1846. *Marine Biodiversity* 50: doi.org/10.1007/s12526-020-01043-1.

21. Machida, R. J., M. U. Miya, M. Nishida, & S. Nishida (2002) Complete mitochondrial DNA sequence of *Tigriopus japonicus* (Crustacea: Copepoda). *Marine Biotechnology* 4: 406–417.

22. Burton, R. S., R. J. Byrne, & P. D. Rawson (2007) Three divergent mitochondrial genomes from California populations of the copepod *Tigriopus californicus*. *Gene* 403: 53–59.

23. Machida, R. J., M. U. Miya, M. Nishida, & S. Nishida (2004) Large-scale gene rearrangements in the mitochondrial genomes of two calanoid copepods *Eucalanus bungii* and *Neocalanus cristatus* (Crustacea), with notes on new versatile primers for the srRNA and COI genes. *Gene* 332: 71–78.

24. Tjensvoll, K., K. Hodneland, F. Nilsen, & A. Nylund (2005) Genetic characterization of the mitochondrial DNA from *Lepeophtheirus salmonis* (Crustacea; Copepoda). A new gene organization revealed. *Gene* 353: 147–240.

25. Ki, J.-S., H. G. Park, & J.-S. Lee (2009) The complete mitochondrial genome of the cyclopoid copepod *Paracyclopina nana*: A highly divergent genome with novel gene order and atypical gene numbers. *Gene* 435: 13–22.

26. Choquet, M., I. Smolina, A. K. S. Dhanasiri, L. Blanco-Bercial, M. Kopp, A. Jueterbock, A. Y. M. Sundaram, & G. Hoarau (2019) Towards population genomics in non-model species with large genomes: a case study of the marine zooplankton *Calanus finmarchicus*. *Royal Society Open Science* 6: http://dx.doi.org/10.1098/rsos.180608.

27. Weydmann, A., A. Przyłucka, M. Lubośny, K. S. Walczyńska, E. A. Serrão, G. A. Pearson, & A. Burzyński (2017) Mitochondrial genomes of the key zooplankton copepods Arctic *Calanus glacialis* and North Atlantic *Calanus finmarchicus* with the longest crustacean non-coding regions. *Scientific Reports* 7: doi.org/10.1038/s41598-017-13807-0.

28. Minxiao, W., S. Song, L. Chaolun, & S. Xin (2011) Distinctive mitochondrial genome of Calanoid copepod *Calanus sinicus* with multiple large non-coding regions and reshuffled gene order: useful molecular markers for phylogenetic and population studies. *BMC Genomics* 12: doi.org/10.1186/1471-2164-12-73.

29. Kim, S., B. J. Lim, G. S. Min, & H. G. Choi (2013) The complete mitochondrial genome of Arctic *Calanus hyperboreus* (Copepoda, Calanoida) reveals characteristic patterns in calanoid mitochondrial genome. *Gene* 520: 64–72. doi.org/10.1016/j.gene.2012.09.059.

30. Bernt, M., D. Merkle, K. Ramsch, G. Fritzsch, M. Perseke, D. Bernhard, M. Schlegel, et al. (2007) CREx: inferring genomic rearrangements based on common intervals. *Bioinformatics* 23(21): 2957–2958.

Index

For the benefit of digital users, indexed terms that span two pages (e.g., 52–53) may, on occasion, appear on only one of those pages.

Tables, figures, and boxes are indicated by *t*, *f*, and *b* following the page number